Maintenance Planning and Scheduling Handbook

Maintenance Planning and Scheduling Handbook

Doc Palmer

Second Edition

McGraw-Hill

New York   Chicago   San Francisco   Lisbon   London   Madrid
Mexico City   Milan   New Delhi   San Juan   Seoul
Singapore   Sydney   Toronto

The **McGraw·Hill** Companies

CIP Data is on file with the Library of Congress

2 3 4 5 6 7 8 9 0 DOC/DOC 0 1 0 9 8 7

ISBN 0-07-145766-6

The sponsoring editor for this book was Kenneth P. McCombs and the production supervisor was Richard C. Ruzycka. It was set in Century Schoolbook by International Typesetting and Composition. The art director for the cover was Anthony Landi.

Printed and bound by RR Donnelley.

 This book was printed on recycled, acid-free paper containing a minimum of 50% recycled, de-inked fiber.

McGraw-Hill books are available at special quantity discounts to use as premiums and sales promotions, or for use in corporate training programs. For more information, please write to the Director of Special Sales, McGraw-Hill Professional, Two Penn Plaza, New York, NY 10121-2298. Or contact your local bookstore.

To Jesus and Nancy

Contents

Foreword

We are witnessing a major change in maintenance. It is moving from an equipment repair service to a business process for increasing equipment reliability and ensuring plant capacity. Its practitioners are trading their reactive cost center mentality for a proactive equipment asset management philosophy.

As editor of a technical business magazine covering the maintenance and reliability field, I have had an opportunity to track maintenance during its move from craft to profession. I have had the pleasure of writing about its leaders, the people, and organizations who are continually extending the benchmark for maintenance excellence. Many are well on their way to establishing themselves at a level where maintenance performance is measured not by simple efficiency, but by contributions to plant productivity and profitability.

One of my favorite jobs as an editor is the reporting of best practices to the maintenance community. I first met Doc Palmer during such an assignment—a magazine cover story on a plant maintenance improvement program. Since then, I have published some of his articles and heard his conference presentations, and found that he has a superb understanding of the practices leading to maintenance excellence.

One belief that the leading organizations hold in common is that maintenance is a business process and that formal planning and scheduling is key to its success. Yet, there is a dearth of practical references on the subject. Most articles and conference papers on planning and scheduling stress its strategic importance, but they do not delve into the practical details because of limitations imposed by article length or conference programming. Doc has leaped over this hurdle with his *Maintenance Planning and Scheduling Handbook*. There is now a ready reference to take the action oriented maintenance practitioner to the level of understanding needed to install a planning and scheduling function and make it work.

The book positions planning in maintenance operations and then proceeds logically to introduce the principles of planning and scheduling and explain how to make planning work. Additional sections cover the nuances of planning preventive maintenance, predictive maintenance, and project work. The book concludes with helpful information on how to get started.

Maintenance Planning and Scheduling Handbook is a welcome addition to the body of knowledge of maintenance excellence and how to achieve it.

ROBERT C. BALDWIN
Former Editor
Maintenance Technology Magazine
Barrington, Illinois
and Current Senior Editor
EDTRON.com
Technical Business Communications
Palatine, Illinois

Preface

The welcome demand for the *Maintenance Planning and Scheduling Handbook* around the world and repeated printings have encouraged this second edition. The author is profoundly grateful that maintenance practitioners across a wide spectrum of industries have found the handbook and its principles universally applicable.

The second edition has revised most pages throughout the entire handbook to clarify and amplify discussions based on the author's experience of the last seven years and practitioner feedback since the first edition. Perhaps the most fundamental of these revisions is moving the planning strategy from simply relying on craft skills to more of a procedures-based organization in terms of job plan detail. The second edition also adds a definitive aids and barriers analysis to virtually every key aspect of planning. Furthermore, it adds cause maps to investigate low schedule compliance and priority system problems. Two new appendices add much discussion on the soft side of maintenance planning (dealing with people) and the controversial issue of subcontracting maintenance. In addition, the second edition delves much deeper into implementing and using a CMMS. Finally, readers should welcome improvements in overall text format (larger font) and chapter organization that make the handbook easier to use. All of these changes make the handbook even more comprehensive and helpful. These changes should enhance the handbook's usefulness and unique contribution to this key segment of maintenance.

DOC PALMER
Neptune Beach, Florida
palmerplanning@bellsouth.net

Preface to First Edition

The *Maintenance Planning and Scheduling Handbook* shows how to improve dramatically the productivity of maintenance. For example, a group of 25 maintenance technicians should be performing the work of 39 persons when aided by a single maintenance planner. This book clearly and simply sets forth the vision, principles, and techniques of maintenance planning to allow achievement of this type of improvement in any maintenance program.

When I began writing articles and publishing papers describing the success we had achieved in maintenance through maintenance planning, I was not surprised by the requests for information I received. We had revamped our existing planning organization and the result was a total clearance of a large backlog of work that had some work orders in it as old as 2 years. The clearing took less than 3 months thus freeing up in-house labor and allowing a scheduled major overhaul to commence without costly contractor assistance. We had been through a learning journey in the course of our success. Before we got planning "working" we had to unlearn about as many false notions about planning as we had to learn principles to support what it really was. Most of the requests for information I received primarily centered on a need just to get a handle on exactly what maintenance planning was. Eventually McGraw-Hill asked that I write this book.

I believe that maintenance planning has remained an undeveloped area of tremendous leverage for maintenance productivity for several reasons. The planning function is positioned down in the maintenance group and does not command the plant manager's attention, so it is "beneath the plant manager." The techniques require an increased degree of organization, coordination, and accountability as well as a loss of some control (which some maintenance supervisors might not find appealing), so it is "unnecessary to the maintenance manager." Finally, the principles of planning are not technical in nature, so it is "uninteresting to the plant engineer." Nevertheless, a company seeking to be more competitive would do well to exploit such an area of leverage. A common saying states that for any endeavor, 1 hour of planning will save 3 hours of work. Maintenance planning saves more. *After a work order system, planning is the biggest improvement one can make to a maintenance program.*

This book considers "planning" as the preparatory work given to individual maintenance work orders before assigning them to specific craftpersons for work execution. This preparatory work, when properly done, greatly increases maintenance productivity. There exist few actual books in print for maintenance planning and most do not actually address planning the way the *Maintenance Planning and Scheduling Handbook* does. Each of these other books is excellent, but they portray maintenance planning as overall maintenance strategy or preventive maintenance instead of as preparatory work before work order execution. For example, one book focuses on planning maintenance management rather than planning work orders. That book emphasizes having detailed work plans for routine preventive maintenance, but the actual planning described in detail primarily shows how to schedule outage time for working on the equipment. Another book defines and presents planning as preventive maintenance or other work decided upon well in advance of execution. In other words, there exist two types of maintenance, planned versus reactive, so by definition there is no planning of reactive work. In contrast, my book presents planned versus unplanned and reactive versus proactive as two separate considerations. Planning of reactive work is essential. A third book also addresses overall maintenance management with little distinction on what type of work is being planned. It compares the accuracy of different types of planner job time estimates, but without much comment on how using them affects crew productivity. So the few books available define "planning" in varying manners. Overall maintenance management and preventive maintenance are not the maintenance planning to which the *Maintenance Planning and Scheduling Handbook* speaks. Even though these areas are important and my book touches on them in several ways, they are not "work order planning." Maintenance management is using the right tools and using them correctly. Preventive maintenance is a tool involving some of the right jobs to do. Work order planning is a tool to get the right jobs "ready to go." The *Maintenance Planning and Scheduling Handbook*, authored by an actual practitioner, fills the gap in the literature for work order planning.

This book is also important because even when considering work order planning, industry has a significant problem with the concept. *Most maintenance organizations do not have a planning function and many that do are frustrated and not getting anywhere near the improvement they should.* Just like learning the computer, planning has been made needlessly over-complicated. Thomas Sowell (1993) said, "If driving an automobile were taught the way using a computer is taught, driving lessons would begin with an elaborate study of the internal combustion engine, then move on to the physics of the transmission system and the chemistry of rubber tires before finally getting around to explaining how to put the key in the ignition and start the car." People have not seen a clear vision of the work order planning function. Until now, most of the literature that targets the work order planning concept has presented merely the responsibilities of planning without actually defining planning in a practical, bottom-line manner. Because that literature does not know the vision, it fails to translate the responsibilities into exactly what a planner does. The few previous attempts illustrate this lack of clear vision: their ideas about everything a planner should do lead to a job description impossibly full, even for Superman. The great truth is that the Pareto Principle is alive and well. 20% of what planners should be doing contributes to 80% of the impact of planning. In actual practice, organizations have crowded this most important 20% out altogether with another group of activities imposed upon the planners. Some companies even mistakenly think of planning as simply a software project. Imagine the computer industry reinforcing this problem where there is an explosion of software to manage maintenance without a clear understanding of what maintenance planning is. As a result, the many companies who have not implemented a planning function miss a great opportunity and the companies frustrated with planning need to back up and relearn what planning should be. *Because of insufficient understanding, maintenance planning remains an undeveloped area of intense leverage for maintenance productivity in industry.*

The *Maintenance Planning and Scheduling Handbook* clearly sets forth both the vision and the how-to specifics of maintenance planning. The handbook carefully explains what maintenance planning is all about and then nails down what is expected from a planner and why. It also shows how to measure the success. The handbook includes specific directions for planners. Readers can grasp the fundamentals of planning and make an impact in their organizations.

Typically, maintenance managers and plant engineers have called me with questions after I make a presentation or write an article. One maintenance manager called me saying he had 38 craftpersons and was going to hire 13 more for a major mill expansion. My opinion was that he should hire no one new (savings of over $500,000 per year) and transform two of his existing craftpersons into planners. His resulting productivity would be as if he had hired 20 more persons. I have also received a significant number of queries from computer system personnel and planners themselves. My favorite was from the planner who called from across the nation just to say I was "right on target and to keep up the good work." The book is primarily intended for maintenance managers (including plant engineers responsible for maintenance) and for plant managers. It is equally valuable for maintenance planners themselves, as well as for management information personnel working with a computerized maintenance management system (CMMS). Contractors and consultants helping others will benefit as well as risk management professionals interested in the care of physical assets. Corporate executives, of course, would benefit in learning what a difference planning can make as they play a role in getting the interest of these other persons.

The maintenance manager really determines whether planning will be successful, although to support planning, the plant manager must know what improvement can be expected, what impact will be made on production personnel, and what the maintenance manager must do. The maintenance manager will use this book to understand and believe the vision and to apply the principles to obtain the dramatic improvement. This maintenance manager is usually one of two persons: one is the degreed person, typically with an engineering background, who has been placed in charge. The other is the craftperson who worked up into this latest promotion. This latter manager is many times the only nondegreed manager on site. (A third possibility, especially internationally where the distinction between engineering and maintenance is blurred, is the craftperson who earns a technical degree while working. The resulting maintenance manager or plant engineer has much hands-on experience.) All of these persons need the concepts of maintenance planning very clearly expressed so that they can easily grasp them, apply them, and communicate them to others.

Maintenance managers and plant managers are not the only ones who will appreciate this book. The actual planners themselves will benefit from better understanding their mission and duties. Planners are typically maintenance personnel directly from the crafts or craft supervision, but sometimes they are engineers or construction technologists. In addition, computer system administrators and management information persons will better understand planning and be more able to help maintenance with a computerized maintenance management system. Companies that do maintenance for others and consultants that help improve programs will also use this book to establish superior maintenance performance. Finally, there is the risk management professional. Those persons in industrial insurance companies are taking a more active role spreading good ideas and asking pertinent questions. They have a vested interest in clients adequately protecting their assets; that is what maintenance is all about.

The *Maintenance Planning and Scheduling Handbook* is valuable to any person who wants to pick up a few good maintenance ideas. Nevertheless, the book is a handbook in its completeness of coverage and all readers will be able to use it to make their maintenance program dramatically more effective and productive. The readers will finally understand the simple truths about maintenance planning. Managers will be able to implement a new planning group or decisively redirect an existing one. Planners and supervisors will use this book as a training and reference tool. Because formalized planning can help any organization with more than ten maintenance persons, the resulting maintenance program can be a competitive edge for the company for utilization of labor and equipment assets.

Just after this Preface and the Acknowledgments, the Prologue narrates several typical scenarios of maintenance, some with and some without planning. These scenarios all have significant problems which many readers will recognize in their own organizations. See if you recognize your own organization. After the book develops the planning function, the Epilogue describes these accounts again, but as the events should have transpired, flourishing with a properly executed maintenance planning effort, a tool leading to superior maintenance.

DOC PALMER
Neptune Beach, Florida
palmerplanning@bellsouth.net

Acknowledgments

I gratefully wish to acknowledge several persons who made this book possible. My loving wife, Nancy, who generously gave me time, encouragement, and support to write apart from my full time maintenance job. Dr. and Mrs. Richard C. Palmer (Dad and Mom), who raised me with a work ethic including a sense of responsibility that things should work. The late Ralph McCallum (Atlanta lawyer) and Dr. Gary Poehlein (Georgia Institute of Technology) for patiently explaining the concept of "bringing something to the table." Richard Johns, whom I consider the utility father of planning. Bill Jenkins (utility vice-president, retired) for empowering me to make planning work. Les Villeneuve and David Clemons, the planning supervisor and one of the planners who supported the changes and did the working. David was the model planner. Hicks and Associates, who helped us format and conduct the first work sampling studies. Bob Anderson, who gave me valuable advice for outage and overhaul scheduling. Bob Kertis of Fluor Daniel who taught me the weekly scheduling routine using a worksheet such as in Figure 6.5. Pastor Tom Drury for his prayers and encouragement. David Stephens and my parents-in-law, Mr. and Mrs. Paul F. Peek Jr., who allowed me to spend several weeks of my vacation time in their vacation cabins not vacationing, but writing. Finally, one never learns maintenance in a vacuum. I wish to thank all the other persons who taught me along the way including those members of the Society for Maintenance and Reliability Professionals (SMRP). I am especially appreciative of SMRP members, Bob Baldwin and Keith Mobley, for recommending that such a book as mine be published to help others.

For the second edition, I would also like to acknowledge six additional men. Jack Nicholas of Maintenance Quality Systems for his excellent method of identifying and displaying barriers and aids and for his encouragement about becoming a "procedures based organization." Mark Galley of ThinkReliability.com for his excellent method of cause mapping. Todd Seitz of Shell who encouraged me to address the people side of maintenance. And finally, Edward Yourdon (author of *Death March*), Kevin Tyler, and Fred Kerber who made surviving a CMMS project bearable.

Prologue:
A Day in the Life—
May 10, 2010

This section has four short narratives typical of maintenance with inadequate or no planning help. These accounts—unfortunate and frustrating—are sure to be recognized by many hands-on managers who know what can go on in a workforce. The Epilogue, just before the appendices of this book, recounts these misadventures, but with each situation flourishing with proper maintenance planning.

Bill, Mechanic at Delta Ray, Inc., No Planning

Bill reported to work on time and went straight up to the crew break area. There the supervisor gave out the assignments for the day. Bill received two jobs: one was to take care of a leaking valve on the southwest corner of the mezzanine floor and the other was to check on a reported leaking flange on the demineralizer. The supervisor did not think they would take all day and told him to come back for something else to do when they were finished.

Leaking stuff. Sounded pretty messy, so Bill walked by his locker to put on his older boots. Aaron was at his locker and the two chatted for a moment while they got ready. The first thing Bill did was swing by the jobs. This was always a good idea in case the job needed special tools or something. Maybe the job would not require him to lug his whole tool box there. As he went by the first job, he easily found the deficiency tag matching the tag number on his work order. Bill had the work permit and there were hold cards everywhere so he knew it was safe to work. The valve was at chest level so there would not be any scaffolding or a lift truck needed. The valve was a 4-inch, high pressure globe valve. Bill decided to look over the other job, then go obtain a valve rebuild kit.

At the demineralizer, the area was also cleared and Bill had the right work permit. But Bill was uneasy. The deficiency tag was hung near a pipe flange, but Bill wondered if the line was an acid line or a water line. In either case, Bill knew the operators would have drained the line, but it would not hurt to put on some acid-resistant gear just in case there were drops or anything.

Bill headed for the storeroom for a valve rebuild kit and the tool room for some acid gear. There was a line at the storeroom so Bill changed direction and went toward the tool room first. On the way, Bill had an idea. He knew Aaron was an experienced technician and had worked on the demineralizer many times. Maybe he would know if the flange was on an acid or water line. After asking around, Bill caught up with Aaron at the pump shop. After a few minutes Aaron came to a good time for a break and walked over to the demineralizer with Bill. Aaron was confident that the line was only for water so Bill decided to skip the acid gear. It was now break time so Aaron and Bill headed for the break room.

After break, Bill got in line at the storeroom. The storeroom happened to have a rebuild kit for the 4-inch valve. Bill took the valve kit and his tool box up to the mezzanine floor and got to work. This was an interesting type of valve. Bill was hoping that it could be rebuilt in place. After unbolting several screws on the top of the valve, Bill was able to remove the internals. Bad news. Although Bill had the right kit to replace the valve internals, it was obvious that the valve body was shot. The whole valve would have to be replaced. The only problem was that Bill was not a certified welder and this high pressure valve had welded connections. Bill went straight to his supervisor and explained the situation. The supervisor wanted to complete this job today and called the crew's certified welder on the radio. The welder could come over in about an hour and start the valve job. The supervisor asked Bill to return the valve kit to the storeroom and check out a replacement valve for the welder. Bill waited again at the storeroom to make the exchange, then took the new valve to where the welder was and explained how far he had gotten along. Then Bill took his tool box over to the demineralizer to be ready to go after lunch.

After lunch, Bill took the flanged connection apart at the demineralizer. In order to obtain access to the leaking flange, Bill had to dissemble two other connections as well. All three flanges looked like they had Teflon gaskets, so Bill went to the tool room for material to cut gaskets. Since he was waiting in line at the tool room, it was a good time to call the dentist to make an appointment for next month. With the gasket material in hand, Bill went to his work bench and cut three gaskets using one of the old gaskets as a template. Bill realized that with these gaskets, he could finish up this job in no time. He wondered what the next job would be if he went back to his supervisor. It would probably be cleaning under the auxiliary boiler. He hated that job. Why couldn't he be given a pump job or something important? Well, there was no sense worrying about it. Bill gathered up his gaskets and started toward the job. On the way he passed

Gino cutting out some gaskets at his work bench. After stopping to compare notes for a few minutes, they both noticed it was almost break time, so they decided just to stay in the shop and talk.

After break Bill started reassembling the flanges. Most of the bolts looked in good shape, but a couple looked a little ragged. Bill thought that the plant had a good handle on completing most of the maintenance work. It would probably be a wise use of time to go to the tool room and replace those bolts. The tool room had an open crib for bolts so he did not have to waste any time in line acquiring new bolts. Soon Bill finished the job and he wiped down and cleaned up the area. He then reported to his supervisor so the work permit could be signed off and taken to the control room. By then, there was about an hour and a half left in the work day. It was customary that the crew could use the last 20 or 30 minutes of the day filling out time sheets and showering. Therefore, instead of starting a new job, the supervisor decided to have Bill go assist Jan who was finishing up a job on a control valve. Bill helped Jan complete her job. Then he filled out a time sheet and headed to his car at the end of the day.

On the way out to his car, Bill reflected how you had to keep busy all day long just to finish one or two jobs. He wondered if he did enough work.

Sue, Supervisor at Zebra, Inc., No Planning

Sue considered herself a capable supervisor. She knew that to keep the operations group satisfied, the maintenance crew had to respond to urgent maintenance requests. She worked the crew hard and kept on top of high priority work. Whenever a priority-one work order came in, she assigned it immediately even if it meant reassigning someone from a lower priority job. The crew knew the importance she placed on completing high priority work and was always willing to work overtime when required. In return for their cooperation, Sue did not push the crew when there were few high priority jobs. She was sure that the crew would eventually complete the lower priority jobs, but the operations group really needed the higher priority jobs completed or production would suffer.

Her normal method of job assignment was to assign one job at a time to each technician, putting persons on what they did best. Sometimes this required the art of deciding who would receive which jobs. Since all she had to go on was the work request from the operations group, it was sometimes difficult to tell what craft skill was required and for how long. Her experience came to her aid frequently, but she still preferred to assign one job at a time and trust the individuals to work expeditiously. She knew the crew worked hard because they rarely lounged either on the job or in the breakroom. When they finished the jobs they were on, they would come to her for other assignments. Earlier in the day, Jim came in for another job. After looking through the backlog in the file cabinet, she assigned a pump repair. Jim was great working with pumps. She noticed a higher priority, air compressor job in the file, but Donna knew the most about air compressors and she was on leave for the week. A few moments later, another technician came into the office. This particular technician had not earned Sue's confidence so she assigned that technician to go help Jim.

Lately it seemed that all the work was high priority and production was suffering. She used to feel that sometimes the operations crews would exaggerate the priority of minor jobs just to make sure they were done. However, from looking at the recent work orders, there really were many urgent jobs, some bordering on near emergencies. She knew the crew was beginning to tire of working in a near panic mode and it seemed some of the crew was slowing down. Hopefully, after they completed this recent batch of jobs, things would calm down. In order to keep the crew moving along, Sue decided always to make sure that each technician had a personal backlog of at least two or three jobs to do. Next Sue began to monitor starting and quitting times closely in addition to break time. Nevertheless, things did not seem to be improving.

Juan, Welder at Alpha X, Inc., Has Planning

Juan received two jobs for the day. Planning had planned one, but not the other one. Both were jobs to replace valves that were leaking through. Eli in the predictive maintenance group had used thermography to find the problems. Juan hoped he could finish both jobs before quitting time.

Juan got to the site of the planned job and looked at the work order again. The west economizer drain root valve was leaking through. The work plan called for replacing the valve and gave a detailed plan. The plan gave the following steps.

- Obtain valve from storeroom, 15 minutes.
- Obtain welding machine, welding materials, and chainfall from tool room, 20 minutes.
- Make sure the work area is cleared by the operations group, 3 minutes.
- Clear the immediate area of combustible material since welding is involved, 15 minutes.
- Unwire and set aside equipment tag, 1 minute.
- Support the old valve to be removed with a chainfall from the tool room, 5 minutes.
- Use a cutting torch to cut out the old valve, 30 minutes.
- Prepare both pipe ends, 25 minutes.

- Check new valve for obvious defects and move into place with chainfall, 10 minutes.
- Make root pass, 10 minutes.
- Finish welding, 45 minutes.
- Grind to smooth edge of weld if necessary, 10 minutes.
- Heat treat weld areas for 5 hours or as directed by supervisor, 4 hours.
- (During heat treating, clean up area, 20 minutes.)
- Replace equipment tag with wire, 1 minute.
- Take old valve to scrap and return equipment to tool room, 15 minutes.
- Turn in work permit and fill out paperwork, 10 minutes.
- Total time: 3 hours 15 minutes plus 4 hours for heat treating.

Juan thought the plan was ridiculous. He did not mind having the valve identified and reserved. Nonetheless, Juan felt that the planner must think he was an idiot not knowing how to weld. Juan was a certified welder for which the plan called, after all. Juan also figured that the planner being an apprentice explained why the heat treatment information was all wrong. This type job required preheating with a torch and temperature stick for about 5 minutes. A simple wrapping with an electrical treatment blanket at the end of the job kept the valve from cooling too quickly. Juan could go on to another job and come back in 2 or 3 hours to retrieve the blanket. Juan also seemed to remember working on this valve last year. Did he have to drain the water through the root valve before he could cut out the valve? It just seemed that planning was not all it was cracked up to be.

Jack, Planner at Johnson Industries, Inc.

Jack came in ready to go. As a planner for 20 technicians, he knew that each day he needed to plan about 150 hours worth of work orders. Standard preventive maintenance work orders that needed no planning would add about 50 hours. That would keep 20 technicians busy for a 10-hour shift. He could not afford to become bogged down.

Reviewing the work requests from the previous day, Jack got to work. He decided first to make a field inspection for eight of the most pressing work orders. He hoped to have them planned before lunch and start on another group. He gathered the eight work orders on his clipboard and headed for the door. At the door he met George and Phil. They had just started a pump job and wanted the pump manual. Jack agreed to help them look through the planning files for the book. After some minutes they found a copy in the technical file section with the other OEM manuals.

Jack had no problem finding and scoping most of the jobs, but one job was hard to find. Jack made a trip to the control room and waited a few minutes for an operator to be able to take a look with him. While he waited, he received a radio call from the Unit 1, mechanical crew supervisor. Jack's plan had indentified the wrong valve for a job and the supervisor wanted him to help the technicians at the storeroom pick out the right one. Telling the operator he would have to come back, Jack headed for the storeroom. Once there, Jack agreed that the application called for a globe valve. It was now about 10 A.M. and Jack decided to meet with the operator after break.

After break, Jack and the operator found the elusive job site. Jack made a mental scope of the job and headed for the planner office. Once there, another technician, Jim, caught his attention and pleaded for help. He was replacing some bearings on an unplanned job and needed size information. Together, they searched, but did not find a manual nor any information in the equipment files. This required a call to the manufacturer who was glad to help. Then two more technicians working unplanned jobs came in and asked his help finding parts information. Since their jobs were underway, it was logical that he should stop and help them. After all, he was very adept at finding information and his job existed to support the field technicians.

So it was after lunch that Jack finally sat down to write detailed work plans for the jobs he had scoped. The equipment files had parts information for two of the eight jobs. Jack went ahead and wrote those work plans for about 12 hours of technician work. Looking through the storeroom catalog yielded parts information for three jobs. One job needed no parts and the last two jobs required parts not carried in stock. Jack requested the purchaser to order them. Technicians twice more interrupted as Jack wrote out plans for the six jobs. Near the end of the day, he completed the six job plans totaling about 50 hours. Jack realized he had only completed plans for 62 hours of technician work that day. He had hoped to complete some of the other work orders. Maintenance seemed to be in a cycle where crews would have to work unplanned jobs because there were few planned jobs available. Then crews would need parts help for the jobs already underway, which kept him from planning new jobs. Something was not right.

Maintenance
Planning and
Scheduling
Handbook

The Benefit of Planning

One cannot discuss maintenance planning without first considering an overall perspective of maintenance itself.

Plant capacity is the lifeblood of a company. Plant capacity must be reliable for the company to produce a product to stay in business. Yet Sandy Sutherland and Gordon (1997) of Kemcor Australia point out an astounding conflict in the business models of many companies. These models show reliable plant capacity connected with revenue streams while showing plant maintenance in the fixed cost section elsewhere in the models. The management and financial groups of these companies do not realize that reliable plant capacity is by definition an investment in maintenance. In real life, capacity must be maintained. Capacity is not reliable by itself. Poor maintenance equals poor revenue streams.

Maintenance provides a competitive edge in many companies. In 1993, good maintenance helped operating crews at a large electric power station achieve an excellent 93% equivalent availability (a utility measure of generating capacity), well above industry average. If the capacity were not available, the station could certainly not sell electricity. The significance of higher availability extends even beyond the daily increase of sales and reduction of generation interruption. If maintenance can achieve continued superior availability, then a company can defer construction of new capacity even as annual sales grow. The ability to defer capital construction as a company grows leads to lower company capital cost, a financial blessing. Today's money invested in proper maintenance ensures high capacity and guards against premature future construction. Proper maintenance makes a company cost competitive.

Company Vision

The purpose of maintenance is to produce reliable plant capacity. The company vision for producing a profitable product should understand that effective maintenance provides reliable plant capacity. Some of the most important maintenance decisions are made before a company even builds a plant. Gifford Brown (1993) of Ford Motor Company explains the *1-10-100 Rule*. This rule means that every $1 spent up front during engineering to reduce maintenance eliminates a later $10 cost to maintain equipment properly or $100 in breakdown maintenance. In this sense, Brown says, "The company vision should be how to *prevent* maintenance, *not* how to do it efficiently." Companies should spend more effort purchasing machines that need a minimum of attention. This is preferred first over being efficient at either performing work to keep machines from failing or reacting to repair failed machines. Any company would prefer machines that run constantly without any attention. Phillip Young (1997) of DuPont says industry typically does not involve maintenance intellect up front, a serious fault. By far, the greatest maintenance opportunities exist before the company installs equipment. Therefore, the first step in dealing with maintenance effectiveness involves working actively with engineering and construction departments before installing equipment (Fig. 1.1).

Nevertheless, some maintenance attention will be required after a company installs equipment. Once equipment is installed and operating, the second step in dealing with maintenance effectiveness is to be proactive. *Proactive maintenance* means to act before breakdowns occur. It acts through preventive maintenance, predictive maintenance, corrective maintenance, and project work. Proactive maintenance recognizes and addresses situations to prevent them from ever becoming urgent problems or breakdowns. Urgent maintenance performed under schedule pressure is rarely cost efficient. Breakdowns interrupt revenue producing capacity and destroy components. Maintenance does not want to recover plant capacity by repairing broken components. Proactive maintenance programs stay involved with the equipment to prevent decline or loss of capacity. In this sense, maintenance produces a product which is capacity; maintenance does not just provide a repair service. These concepts of proactive maintenance of John Day, Jr. (1993) play an important part in planning in Chap. 4.

Effective Maintenance Provides Reliable Plant Capacity

Figure 1.1 Management must make this connection in their vision for producing a profitable product.

Why Improvement Is Needed in Maintenance

Effective maintenance reduces overall company cost because production capacity is available when needed. The company makes a product with this capacity to sell at a profit. This explains the reliability–cost relationship: focus on overall cost reduction and reliability gets worse, but focus on reliability improvement and overall cost goes down. Nevertheless, examining the cost of the maintenance operation cannot be dismissed as unimportant. After maintenance effectiveness, maintenance efficiency must be considered. What if the same or better maintenance could be provided for less cost? What if the company could grow by adding new production capacity and maintain it without increasing the current maintenance cost?

Keeping the purpose of maintenance in mind, one may focus on the cost of the maintenance operation. Understanding the details of one's maintenance system provides the information on how it may be improved. Many companies trying to become more competitive change their maintenance budget without any understanding of how their maintenance system works. They may increase the budget to add maintenance personnel when making capital plant additions. They may reduce their budget for an existing plant. They may not increase the budget when making capital plant additions. They may hope that budget pressure will cause the maintenance force to "work harder" or "do what it takes." Nonetheless, to make improvements to the efficiency of a maintenance operation, one must understand the details of the system.

What are the details in the maintenance system? The following case shows a pertinent example of the details involved in a maintenance system.

In the 12 months resulting in 93% availability, the previously mentioned power station spent over $9 million in maintenance. This amount included more than $5 million in wages and benefits for the mechanical, electrical, and instrument and control (I&C) crafts. A study revealed that productivity of maintenance personnel was about 35%. That is, on the average, a typical maintenance person on a 10-hour shift was making productive job progress for only $3\frac{1}{2}$ hours. The other $6\frac{1}{2}$ hours were spent on "nonproductive" activities such as necessary break time or undesirable job delays to get parts, instructions, or tools. The study only included persons who were available for the entire shift so training time and vacation time were not even included. For example, if mechanic Joe Stark had a pump job and a valve job for a 10-hour day, typically he would have physically performed maintenance on the equipment for only $3\frac{1}{2}$ hours. The rest of the time Joe might have done something very necessary for completing the job. He may have stopped to get a gasket or a special wrench, but when he stopped, the job did not progress. If the job did not progress when it otherwise might have, the company lost an opportunity not only to regain plant capacity, but also to have Joe perform another job that day. If Joe had not had to stop, the work would have proceeded much faster. Overall, only 35% or $1,750,000 of the $5 million paid to the employees was for productive maintenance. The company paid 65% or $3,250,000 for unproductive maintenance. Considering that training time and vacation time were included in the $5 million would make the actual amount paid for productive maintenance even lower. The company was surprised to learn that 35% productivity was typical of good traditional-type maintenance organizations. However, the company realized that the average of 6½ hours of nonproductive time per person accompanying the significant cost of maintenance was an opportunity to improve maintenance efficiency.

Understanding the details in the maintenance system leads to improvement opportunities. Understanding what is happening allows selection of maintenance strategies for the specific opportunities to improve. *Maintenance planning is a major strategy to improve maintenance efficiency with regard to unproductive maintenance time.* Implementing proper planning and scheduling can improve productive maintenance time from the 25 to 35% of a typical organization without planning to 50 to 50%, almost doubling the ability to get work completed. Nevertheless, Brad Peterson (1998) of strategic Asset Management, Inc. (SAMI), says that "Planning is a discipline that is difficult to achieve and difficult to maintain. It needs to be nurtured and developed carefully. This is the greatest issue to maintenance improvement in most plants."

What Planning Mainly Is and What It Is Mainly Not (e.g., Parts and Tools)

All plants require some maintenance and planning can help maintenance efficiency. Some of the primary aspects of planning are well known. Maintenance planning involves identifying parts and tools necessary for jobs and reserving or even staging them as appropriate. The common perception of planning is that after someone requests work to be done, a planner would simply determine and gather the necessary parts and tools before the job is assigned. The planner might even write instructions on how to do the job. With this preparatory work done, the craftperson actually doing the job would not have to waste time first getting everything ready. This planning methodology would be thought to increase maintenance productivity.

Figure 1.2 shows the common perception of what a planner would do for parts. The planner would write a job plan that identified parts needed such as specific gaskets and impellers along with their storeroom identification numbers. Then the planner would reserve them in the storeroom to ensure their availability when the job was executed. If any

Parts

✦**Identify on Plans**   ✦**Illustrated Parts Breakdowns and Lists**

✦**Reserve**

✦**Order Nonstock**   ✦**Vendors Lists**

✦**Some Staged**   ✦**QA and QC**

Figure 1.2   It is commonly recognized that planning consists of parts.

needed parts were not carried in inventory, the planner would have them ordered to be on hand when needed. The planner might stage some of the parts by placing them in a convenient location such as the job site before the job starts. With staged parts the technician performing the work would not have to wait at the storeroom. The planner would also provide a bill of materials or an illustrated parts diagram. These documents would help the technician identify parts unanticipated at the time of planning or understand how the parts fit together. The planner would also work with vendors to ensure good sources of material supply. Finally, the planner would be involved in quality assurance and quality control of vendor shipments.

Likewise, Fig. 1.3 shows the common perception of what a planner would do for tools. The planner would write a job plan that identified special tools needed such as a chainfall or even a crane. The planner would reserve or schedule certain items such as the crane so everyone would not be expecting to use it the same time. The planner might even stage the special tool, such as having the crane moved to the job site in anticipation of the work to begin.

Unfortunately, most organizations do not see a significant improvement in maintenance after over 10 years of trying maintenance planning based on getting parts, tools, and instructions ready. It just seems that something is missing. "Why is it that when you are driving and looking for an address, you turn down the volume on the radio?" That something is not working is obvious when studies show productive maintenance time is never

Tools and Vehicles

✦**Identify on Plans**
✦**Reserve and Schedule**
✦**Some Staged**

Figure 1.3   It is commonly recognized that planning also consists of tools.

more than 40%, hardly better than an organization without planning. In these companies, planning also has a bad reputation among the crafts for not offering much assistance anyway.

One must first understand the *system* of planning to use it effectively. Maintenance planning is a system analogous to the bubble in the carpet. If one simply pushes down the bubble, it will only appear elsewhere. Understanding the carpet as a system allows focus on the edge of the carpet, the leverage point, to pull the slack areas and eliminate the bubble. One must understand the planning system with its specific important characteristics. Fortunately, these characteristics are not complicated. This book explains the characteristics of the system with principles, guidelines, and specific techniques so one can work with planning to improve maintenance productivity.

It turns out that identifying parts and tools is not *the* purpose of planning. This concept is important to state. If planning does not increase overall maintenance effectiveness or efficiency, it does not matter that planning expends effort gathering parts and tools. The purpose of planning must focus on the high productivity desired from the application of the planning and scheduling principles.

In the proper planning system, the maintenance process proceeds this way. After someone requests work to be done, a planner plans the work order by specifying job scope, craft and skill level, and time estimate, as well as specifying anticipated parts and tools. The planner does not necessarily specify a detailed procedure. By including the skill levels and time estimates on jobs, scheduling can assign the proper amount of work to the crews. In actual practice, scheduling control contributes more to managing productivity than do parts and tool delays in and of themselves.

Consider a one-person company: the owner works extremely hard, conscious of every job needing completion. If he finishes one more job, the owner makes more profit and pays the house mortgage once again. After several years of prosperity, he hires ten salaried technicians. The owner later senses that less work per person is accomplished than before. The conversation goes like this:

OWNER: How did it go this week?

TECHNICIAN: We did a lot!

OWNER: Well, how much was that?

TECHNICIAN: We turned out 50 jobs which was more than we ever did before!

OWNER: But how much was that compared to how much you should have been able to turn out?

TECHNICIAN: But you don't understand…We really worked hard!

So the next week the owner looks at every job in the backlog and estimates how long each should take. Then on Friday the owner selects 400 hours worth of work and tells the crew, "There are ten of you each working 40 hours next week, so here are the jobs we need to complete." The next Friday, the owner has some basis for knowing how much work should have been done and has another conversation:

OWNER: How did it go this week?

TECHNICIAN: We did a lot!

OWNER: How much of the work that I gave you last Friday is done?

TECHNICIAN: Well, most of it.

OWNER: Let me have the jobs back that you haven't started yet. [Then after a minute] I see you didn't start about 100 hours' worth of the jobs. What happened?

TECHNICIAN: Let's see. On three of the jobs we didn't have the right parts in stock so we had to order them. On one of the other jobs that we did complete, the time estimate you gave us just didn't work out; that job ended up taking George and John twice as long even though the job didn't have any special problems. Then on one of the other jobs Fred completed, the work took extra long because he ran out of solvent and had to run to the supply center and buy some. So overall, we didn't finish all the work you had wanted."

OWNER: Well, that's okay. On some weeks that just happens. I know we were working hard because I was on the shop floor several times this week. But I am concerned a little bit about three jobs not having parts available. When I scheduled them, I didn't think they would require anything special. We probably also need to look at how much solvent we normally carry; that's not something we should be running out of. Also, if we didn't start on three jobs, were we able to work in any other jobs that we didn't think we would start this week?

The owner utilizes a basis for controlling the work force. The word *control* in this context means that the owner can compare the actual amount of work done against something. In this case the something was the amount of work hours the owner had originally assigned for the week. The crew may not have been able to do all the assigned work, but the point is that now there is a basis for questioning and examining the work done. Could the owner gain this information without having assigned a specific amount of work and just by asking if there had been any problems or delays? He could have, but consider if the technician had said "No, it seemed to be a normal work week. We worked pretty hard." The technician may well have presumed it was just part of the job to scramble for parts or supplies here and there. He may accept that jobs sometime seem to run on forever. The owner or manager of a maintenance group cannot accept that delays are normal before any scrutiny. It is a fair question to ask why 40 hours of work are not accomplished in the 40 hours a technician works. But the question cannot be asked if the amount of work assigned and completed is unknown. Planning and scheduling assigns the proper amount of work to the crews and a control tool becomes available for managing productivity.

Maintenance managers greatly need the information just to allow scheduling. If a crew has 1000 person hours available for the upcoming week, a planning system allows 1000 hours of work to be scheduled. In actual practice without such a systematic approach, supervisors typically assign much less work than should be done during the course of the week.

Note that the context of discussion is not major plant outages or turnarounds. The book touches on outage scheduling in Chap. 6, but that is not the book's focus. Outages are very important, but very well managed already. Management gives much attention to the execution and improvement of outage maintenance. On the other hand, consider a week of routine maintenance. How much work should be done? How would one know? And if one did know, how would it be done? The system of planning and scheduling answers these questions. The *Maintenance Planning and Scheduling Handbook* focuses on the planning and scheduling of the routine, day-in and day-out maintenance. This maintenance most affects the reliability of plant capacity and makes up the bulk of the budget, yet so far it has received the least attention. This undeveloped area of maintenance provides the greatest opportunity for leverage.

Proper planning also provides identification of parts and tools, but not as commonly perceived. Planning departments usually maintain a list of parts for each piece of equipment. The planner would send this list out with the planned work order. On the other hand, if parts identification is not readily available the first time the plant works on a machine, the planner does not necessarily create a list. The planner knows the technicians can determine what is needed and provide identification through feedback to help future jobs. The planner accumulates the feedback from completed jobs to establish a parts or tool list over time. So the planner becomes somewhat of a file clerk for the 20 to 30 technicians.

In the broadest view, a maintenance planner gives the maintenance manager information to allow scheduling enough work and gives the field technicians file assistance. The planners do not necessarily place first priority on extensive research to determine possible parts and tool requirements.

Nevertheless, knowing what constitutes planning does not make it happen. One must know the specific results to expect and the principles and practices involved.

How Much Will Planning Help?

Planning provides dramatic, tangible help. The amount of work accomplished rises. The work force is freed up. The extra labor power can be reallocated to added value activities. One can calculate and measure the actual amount of increased productivity.

The practical result of planning:
freed-up technicians

After 1993, the previously mentioned power station examined its position. The station had achieved reliable plant capacity with a year of superior availability, but studies showed an opportunity to improve work force productivity. Management decided to redirect its existing maintenance planning group. They implemented a planning *system* according to the guidelines in this handbook for its mechanical maintenance craft (approximately 30 persons). Less than a year later in 1994 the practical result of planning was 30 maintenance persons yielding the effort of 47 persons. Figure 1.4 is the central statement of this book and the subject of planning: how planning leverages 30 persons to produce as much work as 47 persons. They did not hire anyone new. The benefit received was as if 17 new persons suddenly started helping. These new persons did not cost the company any money because they were free.

Although planning had existed since 1982, planning according to the system principles began in 1994. The start of weekly scheduling began in the middle of May 1994. The maintenance group completed so much work that in mid-June there started to be insufficient backlog to schedule for the entire amount of work hours available for each crew. This happened because in about a month the crews had worked down their entire outstanding backlogs. These backlogs had even included some work orders that were over 2 years old. The power station was thus able to proceed into its Fall 1994 major overhaul of its largest unit with the other units caught up in backlog.

PLANNING
30 Techs Yielding the Effect of 47 Techs

17 Extra Techs

Figure 1.4   The practical result of planning.

Emerging from the overhaul of the unit, the utility included the electrical and I&C crafts as well as two other plants into the planning system. The total of the maintenance force at this point was 137 personnel. With the productivity improvement from planning and scheduling assistance, the utility could expect to free up in effect 78 technicians. These technicians could be available for work to stay ahead of the maintenance backlog. They could do work previously outsourced or given to contractors for outages and projects. They could even build parts in-house. They could also do more preventive maintenance and do maintenance for others at other stations. They could accept attrition without rehiring.

Planning achieves this effect for improvement in maintenance productivity. How would you like to have 17 extra persons for free? How about 78?

The reduction of delays is where planning impacts productivity. Productivity of 25% to 35% is typical of traditional-type maintenance organizations using "wrench time" as a measure. Remember the case of 35% productivity of available maintenance persons where, on the average, a typical person on a 10-hour shift is only making productive job progress for $3\frac{1}{2}$ hours. The other $6\frac{1}{2}$ hours are spent on nonproductive activities such as necessary break time or undesirable job delays such as getting parts, instructions, or tools. Simply implementing a fundamental planning and scheduling system should help improve productivity to about 45%. Then as files and information become developed to allow avoiding problems of past jobs, productivity should increase to 50%. The last improvement to over 55% is attributed to special aids, such as inventory or tool room sophistication or perhaps computerization of certain processes. This last improvement is only possible after the basic processes leading to the first improvements are well utilized. [A mention of computerization is appropriate at this point. Planning is obviously more than computer data collection and research. Planning leverages maintenance productivity. Whether or not a computer is employed, there are certain principles necessary to make planning and scheduling effective to provide leverage (see Fig. 1.5).]

What the Reduction of Delays Does

Figure 1.5   Improvement in productivity.

Paradigm shift: Immediately one may be thinking "There is no way our wrench time is below 80% in the first place because we are always busy. We are working as hard as we can. Sure, there are some delays, but they are unavoidable." See Fig. 1.6.

In reality, productive maintenance time is not nearly that high. Many studies classify work activities differently. For example, should break time be included in the study? How about lunch? Regardless, there is wide agreement across most industries that productive maintenance time is less than 35%. Productive work is commonly reported between 25% and 35%. It may actually be less. Keith Mobley (1997) of ISI says that "Typical maintenance technicians spend less than 25% of their time *actually* maintaining critical equipment. The balance of the time is spent on nonproductive tasks." Yes, one can be busy, but if one is obtaining a part instead of working on the job site, one is in a delay situation that might have been avoided. One would think it would be hard to get only $3\frac{1}{2}$ hours productive work from the average, 10-hour shift mechanic. Where does the other time go? Unfortunately, a minute here and a minute there getting tools, traveling, and the like add up to significant delays. That is why the results of statistically valid studies are so important, because of this great false notion of high productive time.

Statistical work sampling studies properly measure productive time, also known as wrench time. Separate studies done over time indicate if planning is getting better or worse. At issue is not so much the time the technicians spend doing productive work. Figure 1.7 shows there is a significant proportion of time involved with delays. To determine if any of the delay time is avoidable requires analysis of the nonproductive time. For example, how much time is spent waiting for parts? What are the specific delay areas? Can these delay areas be avoided? Analysis of the nonproductive time is one of the most valuable parts of a work sampling study.

A key point has been uncovered. Where there is a great difference between reality and perception, there is great opportunity for improvement. Here is

Figure 1.6   False common perception.

Figure 1.7   Reality of productivity.

an opportunity to improve productivity. Maintenance planning addresses this opportunity to reduce delays and free up technicians for more productive work.

"World class" wrench time

Not only is there a widespread mistaken belief among maintenance practitioners that wrench time is not very low, but many industry consultants believe this as well. There have been numerous consultant surveys passed among maintenance organizations asking about wrench time among other plant workforce questions. A typical wrench time question asks the respondent to score their own wrench time with the following choices: ___30%, ___40%, ___50%, ___60%, ___70%, ___80%, ___90%. The availability of such high choices could be a leading question, in case the respondents have the mistaken belief that wrench time could even be so high. However, the absence of choices below 30% indicates that the consultants who drafted the questions do not understand the severity of typical low wrench time as well. Furthermore, sustained achievement of wrench times above 60% averaged across a workforce is nearly impossible. The *world class* performance target for wrench time is probably keeping a workforce average between 50% and 55%. Does this mean that although low wrench time is a reality, it is too difficult to improve performance? Not at all. This book has its roots in the real world improvement experienced as a good performing workforce moved from 35% wrench time without planning to 55% with planning. In addition, the following section calculates the benefit of planning with a more typical improvement of a plant at 25% wrench time improving to only 50%, a doubling in labor productivity!

Beyond any rigorous calculation of improvement based on wrench time, knowledgeable reliability professionals acknowledge the value of planning. Brad Peterson (1998) of SAMI reports that planned work typically requires only one-third as much labor as unplanned work. Another highly respected reliability manager of an international company told the author that because of the intangible benefits of planning providing better control of maintenance work, if he had only two or three maintenance persons, he would make one a planner!

The specific benefit of planning calculated

How does one measure the leverage of work order planning?

The specific improvement in maintenance can be quantified as shown in Fig. 1.8.

Consider three persons working without the benefit of planning, but placing them at the highest productivity common in such organizations (35%). Their combined productivity (105%) can be thought of as one person always working productively who never has a delay and even gives the company some extra time at the end of the day.

Without Planner:

$$3 \text{ persons at } 35\% \text{ each} = 3 \times 35\% = 105\% \text{ total productivity}$$

Now, take one of those persons away from the work force and make that person into a planner. The planner helps boost the productivity of the remaining two persons up to 55% each. The planner's productivity is considered to be 0% because productive time is defined as time physically working a job. Envision turning a wrench. The planner no longer turns a wrench. The combined productivity of all three persons is now 110%, a little better than all of them working without planning.

With Planner:

$$2 \text{ persons at } 55\% \text{ and } 1 \text{ planner at } 0\% = (2 \times 55\%) + (1 \times 0\%)$$
$$= 110\% \text{ total productivity}$$

How many planners? It is intuitive if a planner could help multiply the productivity of a single craftperson by a factor of 1.57 (55% divided by 35%), that a breakeven point would be to take one of every three craftpersons and convert them to planners. Furthermore, experience has shown that a single planner can plan for 20 to 30 persons. Consequently, there should never be any question that a person cannot be taken out of the work force to become a planner. This area is a big problem for many companies. They fail to provide enough planners for one of two reasons. One, they might select one planner for a group of 50 technicians, a serious underinvestment. Two, they may select two planners, but then dump extra duties on them preventing them from the real work of

The Leverage of Planning

✦Three Technicians without "Planning"
$3 \times 35\% = 105\%$

✦One Planner, Two Technicians
$1 \times 0\% + 2 \times 55\% = 110\%$

✦Ratio Planner to Technicians   1 : 20 - 30

✦$55\% / 35\% = 1.57$ (57% Improvement)

✦30 Technicians $\times 1.57 = 47$ Technicians

Figure 1.8 The mathematics of the leverage of planning.

proper planning. In either case, planning fails because management does not really believe that they could take out one of every three technicians and keep the same productivity.

A 30-person maintenance force is leveraged as 30 persons times 1.57 to yield a 47-person effective work force. Instead of 30 persons working at 35% productive time each, the work force is boosted to be equivalent to 47 persons working at 35% each. This means that one is obtaining the effect of having 17 extra persons on the team without having to hire anyone new; 17 persons for the cost of one planner.

Another way of examining this benefit is that the old, 30-person work force averaged 35% productive time each for a combined total of 1050%. The new, planning assisted, 30-person work force averages 55% each for a combined total of 1650%. The difference of 600% divided by 35% equals 17.1 showing the extra persons.

Seventeen persons at $25 per hour (including benefits) for a year are worth $884,000. (17 persons times 2080 hours per year times $25 per hour equals $884,000.)

Consider a 90-person work force, leveraging to get a 1.57 improvement yields a 141-person work force. The extra 51 persons are worth $2,652,000 for a year.

The above cases are conservative in considering a boost from 35% wrench time to 55%. The above cases are fantastically improved if starting at 25% wrench time and moving to only 50, which is a 50/25 = 2.0 factor improvement. The 30-person work force is increased to 60 persons in effect; the 30 persons are worth $1,560,000 annually. The 90-person work force is increased to 180; the extra 90 persons are worth $4,680,000. What an improvement to the maintenance force!

What do these "extra" persons do? In the company with much reactive work, one leverages or uses them to put out all the fires. In the company with reactive work under control that is focusing on planned work, one leverages them to do more proactive maintenance work avoiding fires. Finally, in those world class companies with preventive maintenance well in hand, one leverages them to invest in training to increase labor skills and in projects to improve equipment or other work processes. Each of these companies has the ability to grow or allow natural attrition without hiring.

While the measure and value of the "extra" productivity can be calculated rather easily in terms of work force, how one uses this extra labor is what matters. Getting more work done and done right leads to other significant savings that are as easy to calculate, but are more difficult to attribute. Planning does not work in a vacuum, but brings the other aspects of maintenance together. Nevertheless, these are important considerations. Beyond just performing enough maintenance work to keep the plant on line and stay in business, there are considerations of increased reliability of existing capacity, improved efficiency, and deferred capital investment for more capacity. The monetary value of these benefits is company and situation specific. For example, for an electric utility with a 1000-MW steam system, each 1% availability improvement might be worth over $300,000/year in power transaction capability. Each 100 Btu/kWh improvement in efficiency might be worth over $600,000/year. A single 1% sustainable improvement in availability means not having to build 10 MW of future power plant capacity. At $1800/kW construction prices, that is $18 million.

One can measure the leverage of work order planning on maintenance to see that it is definitely a good investment.

Why does this opportunity exist?

To understand further why the opportunity of improving productivity through planning exists, this section considers an "organization" (see Fig. 1.9). An organization is a group of coordinated specialized persons with a common goal. Everyone understands a group; everyone involved in the area of the company being considered. Everyone also understands a common goal; in the context of business, the group wants to make a profit with decent wages while being environmentally, socially, and legally responsible and safety conscious. The reason to organize is to allow specialization, but with specialization comes a need to coordinate. Everyone also understands specialization, but no one understands coordination.

Organizing allows specialization. Companies understand that divisions of labor where different persons perform different tasks offer great improvements in overall productivity. For example, hundreds of years ago, Adam Smith published his classic text *The Wealth of Nations* (1776) wherein he studied a plant manufacturing industrial pins. There were 18 distinct tasks involved in making a single pin. The tasks included activities beginning with drawing and cutting the metal to length and continuing all the way to finally inserting the pins into paper and packaging them. However, the plant had specialized by identifying different tasks associated with the pin making and grouping them into 10 different jobs. To each job, the plant assigned a different person to specialize. Management had set up an assembly line of sorts through which each pin passed. Ten persons together could manufacture 48,000 pins in

**An "organization" is
a *group* of *coordinated*
specialized persons with a
*common goal.***

Figure 1.9 Definition of an organization.

Less than 20 Pins per Person

+

Specialization

=

4800 Pins per Person

Figure 1.10   Adam Smith's result in 1776.

a single day. Average daily pin production achieved 4800 pins per person. Upon investigating, Smith deduced that had each employee performed every task necessary to make each pin alone, each employee could have made less than 20 pins in a day. Specialization increased per person pin production from less than 20 to 4800 (see Fig. 1.10). Obviously, industry to this day continues to use specialization to maintain high productivity.

That is why companies train and maintain separate groups of mechanics, electricians, and instrument technicians. Companies have many special needs where it makes sense to have a dedicated group of experts. This may include persons that maintain buildings and grounds, machine shops, storerooms, tool rooms, and a myriad of other specialties (see Fig. 1.11). The only key is that there is enough distinct work to keep the persons occupied.

On the other hand, where there is specialization, there must be coordination. In practice, coordination is much less understood than specialization. A group must coordinate the different areas of specialty. The typical maintenance organization affords an obvious illustration of both specialization and coordination. The specialized groups of mechanics, electricians, instrument technicians, and others need coordination. The maintenance superintendent, maintenance clerk, and supervisors of each specialized group in Fig. 1.11 coordinate the specialized efforts of the craftpersons. Figure 1.12 shows that it is in this area that planning belongs.

Figure 1.11 We understand *specialization* very well.

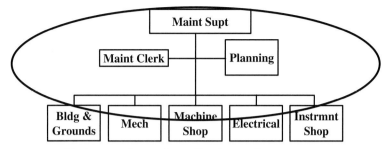

Figure 1.12 We do not understand *coordination* at all.

Specialization cannot function in a vacuum. Coordination must exist over specialization. That companies acknowledge coordination is evident from the typical company reorganization every 5 years or so. However, that they do not understand coordination is evident from their statements that "As soon as the organization settles down, we can lessen the numbers of clerks, secretaries, ..." that were in special need during the reorganization. In fact, many reorganizations combine crews and eliminate supervisors. Simply by specializing tasks, a company reaps an extreme increase of productivity over having all the maintenance employees exist as "jacks of all trades." Specializing is not a competitive edge because all companies do it. Yet companies commonly do not properly coordinate the efforts of maintenance specialization and therefore experience low wrench times. By using planning to coordinate the efforts better, they could further increase the productivity of maintenance, perhaps as much as doubling productivity. Proper coordination could be a competitive edge.

With this recognition of the purpose of a coordinating function, this handbook can properly address maintenance planning. Maintenance planning functions as a coordinating mechanism within the maintenance department. A maintenance planning group helps coordinate many of the specialized activities of maintenance. After a person writes a work order for the maintenance department, maintenance planning takes action. A maintenance planner takes the work order and does preparatory planning for the crew supervisor and craftpersons who will ultimately execute the work. The planner considers the proper scope of work for the job. The work requester may have identified a noisy valve. The planner judges whether the valve should be patched or replaced. The planner also identifies appropriate materials for the specified job and whether they are available or must be specially ordered. In addition, the planner specifies the appropriate craft skills for the job. Having these determinations made before the crew supervisor assigns a job for execution helps avoid problems such as delays stemming from assigning a person with insufficient craft skills or from not having required

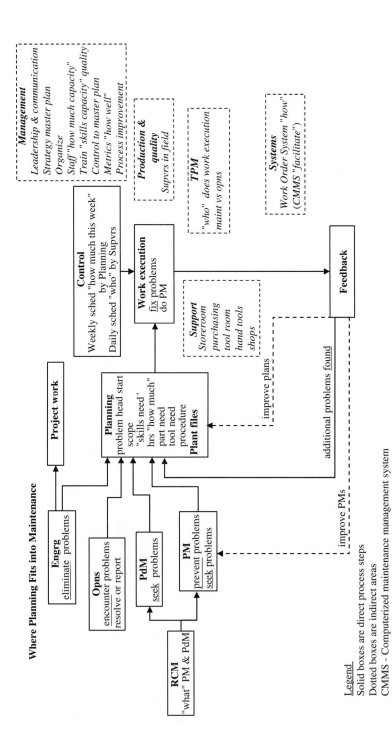

Where Planning Fits into Maintenance

Management
Leadership & communication
Strategy master plan
Organize
Staff "how much capacity"
Train "skills capacity" quality
Control to master plan
Metrics "how well"
Process improvement

Production & quality
Supvrs in field

TPM
"who" does work execution
maint vs opns

Systems
Work Order System "how"
(CMMS "facilitate")

Control
Weekly sched "how much this week"
by Planning
Daily sched "who" by Supvrs

Work execution
<u>fix</u> problems
do PM

Support
Storeroom
purchasing
tool room
hand tools
shops

Feedback

Project work

Planning
problem head start
scope
"skills need"
hrs "how much"
part need
tool need
procedure
Plant files

improve plans

additional problems <u>found</u>

Engrg
<u>eliminate</u> problems

Opns
encounter problems
resolve or report

PdM
<u>seek</u> problems

PM
prevent problems
<u>seek</u> problems

improve PMs

RCM
"what" PM & PdM

Legend
Solid boxes are direct process steps
Dotted boxes are indirect areas
CMMS - Computerized maintenance management system
TPM - Total productive maintenance philosophy

Figure 1.13 An example of overall maintenance process showing planning's coordination role.

materials or tools available. Having time estimates also allows crew supervisors to judge how much work to assign individuals and thus better control their work. Control in this sense means that the planned information gives the supervisors a reference for expected work completion time and thereby helps them more quickly to become aware of any problems that might lessen productivity. Consequently, maintenance planning brings together or coordinates the effort of many other aspects of maintenance.

Figure 1.13 gives an idea where planning in a coordination role fits into the overall maintenance process. Appendix A further describes many of these aspects of a modern maintenance organization and their relationships to planning.

Quality and Productivity Effectiveness and Efficiency

Planning pushes hard for productivity. However, it is very dangerous to push for productivity if there is not a quality focus in the work place. Figure 1.14 shows these sometimes opposing considerations. If one is not effective, it does not matter how efficient one is. If one is not doing the right thing, it does not matter how fast one does it. Presume technician Barbara Smith is assigned to a 5-hour, planned work order. She gets over to the equipment needing attention and soon finds herself staring at a job that is going to take at least two days to do right. One does not want Barbara to say "Well, if they want a 5-hour job, that's what I'll give them!" No, Barbara has to speak up and say to her supervisor, "Look here, this is the situation," so the matter can be resolved. Perhaps the

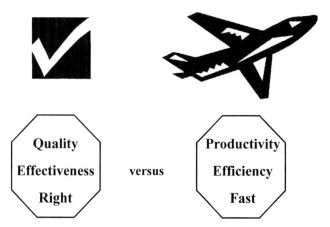

Figure 1.14 Stop! Quality is still most important.

planner was wrong in the time estimate or the technician is reading the scope wrong. The supervisor may point out that the work order calls for a thorough lubrication, not an overhaul. Among other possibilities, maybe the initial scope did only call for lubrication, but after opening up the equipment, it obviously needed an overhaul. The point is that the shop floor understands the need to do the right job in the right way before considering how fast to do it or just blindly following a work plan. If a company does not have a quality focus, it must hold up on pushing for productivity itself. It is wrong to push for productivity if there is not a quality focus present. Craft personnel must have the attitude that work being done in a quality fashion is more important than meeting a production schedule. The individuals on the floor must communicate concerns with the crew supervisor if they need more time to complete work properly.

Nevertheless, while it must be emphasized that quality is more important than productivity, the benefits of planning actually involve quality as well as productivity.

Tangible quality savings come from planning in two ways. First, planning focuses on correctly identifying work scopes and provides for proper instructions, tools, and parts being used on the jobs, thereby facilitating quality work. Second, productivity improvement frees up craft, supervision, and management time to do more proactive work. This proactive work includes root cause analyses on repair jobs, project work to improve less reliable equipment, and attention to preventive maintenance and predictive maintenance.

Planning Mission

When making any decision the ultimate question is, "Will this decision help improve maintenance performance?" The mission statement should guide the planning organization. One must first understand the mission and what planning is trying to accomplish. Then one can further try to understand the system in order to determine how to set up planning.

A planning mission statement might read: "The Planning Department increases the Maintenance Department's ability to complete work orders. Work plans avoid anticipated delays, improve on past jobs, and allow scheduling. Advance scheduling allows supervisors to assign and control the proper amount of work. A work crew is ready to go immediately to work upon receiving a planned and scheduled assignment because all instructions, parts, tools, clearances, and other arrangements are ready. The right jobs are ready to go."

The last sentence really captures its overall meaning. "The right jobs are ready to go" sums up the planning mission statement. Having the "right jobs" involves job priorities, crew schedules, and work type such as preventive maintenance versus breakdown work. Having the jobs "ready to go" involves correctly identifying the work scope, considering the safety

Have the Right Jobs Ready to Go!

✦Schedule - Priority - Worktype

✦Scope

✦Safety

✦Anticipated Delays

✦Parts and Tools

✦Instructions

✦Clearances

✦Other Arrangements

Figure 1.15   Summation of the planning mission.

aspects of the job, and planning to reduce anticipated delays. When a crew gets a planned job, they should be able to go to work. They should not have to ask, "What exactly am I supposed to do? I'm not sure what parts I need. Do I need a crane for this or do I just use come-alongs?" They should be able to receive the assignment and perform the work.

Note that planning is not so much "Information Central," but "Control Central" or "Coordination Central." Planning uses information, but its primary mission is not that of a research function. Planning brings resources to bear on leveraging productivity. See Fig. 1.15.

Frustration with Planning

Disenchantment in implementing a planning organization is frequently due to an attempt to provide detailed work plans on reactive jobs. Since reactive jobs by their nature are urgent, it is frustrating to everyone to wait on a planning group to turn over the work. Once equipment has actually broken down and is interfering with operating the plant, a planning group adds an extra step in the repair process supposedly "to speed the job up." Planners try to write detailed job plans and come up with parts lists from scratch. This effort delays the execution of urgent work and results in frustration. Successful planning organizations concentrate on planning proactive work. By concentrating on work to circumvent later breakdowns, the planning organization can produce good work plans without schedule pressure. Reactive work receives minimal planning attention before crew assignment. At the same time, the entire maintenance organization should be committed to schedule proactive work as well as to give feedback after every job to aid future job plans. In this manner, the overall percentage of reactive-type work should decrease.

Many companies have planning organizations that are sources of considerable frustration to the maintenance effort. Most supervisors do not realize the great value of a planner simply writing down a proper job scope along with craft skill and time requirements. The supervisors feel that "if a detailed job procedure and parts list are not provided, then planning has not done anything." Nevertheless, with the "minimal" planning even on reactive, urgent jobs the supervisor has schedule control and avoids problems such as assigning a mechanic to a job needing a welder. The correct assignment avoids subsequent job reassignment and delay. Planning still has a job function for these reactive jobs and the concept of how planning handles reactive versus proactive work is extremely important for making planning leverage the maintenance productivity. The concept of reactive versus proactive work is not the same as unplanned versus planned work. After the planning and scheduling principles are developed, this important area is discussed in more detail in Chap. 4.

Summary

Effective maintenance is vital to provide reliable plant capacity. The application of maintenance planning makes possible dramatic improvement in maintenance productivity. Moreover, the aspects of planning must be understood in the context of a system in order to avoid the frustrations of many companies that have tried planning without success. The *Maintenance Planning and Scheduling Handbook* explains how the planning system works and the principles and techniques that make the dramatic leverage possible in any maintenance program.

Overview of the Chapters and Appendices

Following the Preface, the Prologue narrates several typical scenarios of maintenance, some with and some without planning. These scenarios all have significant problems which many readers recognize in their own organizations. At the end of the book, the Epilogue recounts these misadventures with every situation flourishing because of proper maintenance planning.

The book progresses directly through the vision, the basics, and the application of planning in the main body of the text. Where does planning fit into maintenance? What principles make it work? And exactly how is planning done? Then, extensive appendices provide additional resources for planners and persons otherwise responsible for maintenance planning.

Chapter 1, The Benefit of Planning, shows the importance of maintenance on plant profitability and the opportunity for better labor productivity. The chapter quantifies the improvement in labor productivity possible through planning. Planning is described as a coordinating role within the maintenance organization.

Chapter 2, Planning Principles, begins with the vision and mission of planning, then presents the principles or paradigms that profoundly affect planning. These principles must be understood to have effective planning. The principles are having planning in a separate department, focusing on future work, having component level files, using planner expertise to create estimates, recognizing the skill of the crafts, and measuring performance with work sampling for direct work time. There is a thorough discussion of what is commonly known as wrench time, a frequently misunderstood measurement. The chapter concludes by showing why scheduling is a necessary part of planning.

Chapter 3, Scheduling Principles, presents first the vision, then the principles or paradigms that profoundly affect scheduling. Effective scheduling is inherent in effective planning. The principles are planning for lowest required skill levels, respecting the importance of schedules and job priorities, forecasting for highest skills available, scheduling for every forecasted work hour available, allowing the crew supervisor to handle the current day's work, and measuring schedule compliance.

Chapter 4, What Makes the Difference and Pulls It All Together, explains the final ingredients necessary to make planning work. Several of these factors make planners do different things for different types of jobs, and several factors greatly influence the overall application of the principles. Lack of appreciating these factors frequently makes planning programs fail. The programs fail because the programs are trying a "one size fits all" approach to different types of jobs, and the programs are not sensitive to the immediate needs of reactive jobs. The chapter distinguishes proactive versus reactive maintenance. It distinguishes extensive versus minimum maintenance. It describes the resulting planning adjustments. The chapter also further discusses communication and management support regarding these adjustments.

With the principles of planning and scheduling well in hand, Chap. 5, Basic Planning, proceeds into the nuts and bolts of exactly what a planner does and how in the context of the preceding chapters. The chapter resolves the question: "We know the purpose of planning, but what exactly does a planner do?" The chapter follows the actual planning process and includes such areas as when and how a planner scopes a job, what the planner writes on the work order form, and how the planner files information.

Chapter 6, Advance Scheduling, continues the nuts and bolts of making the planning system work with regard to a weekly schedule created in the planning department. The chapter shows a specific method to create the schedule.

Chapter 7, Daily Scheduling and Work Execution, goes beyond the duties of the planning department. The craft crew supervisors create the daily schedules and manage daily work execution. Yet because these are critical elements of the overall planning system, this chapter provides specific details pertinent to planning and scheduling.

Chapter 8, Forms and Resources Overview, explains with discussion and examples the types of forms and resources a planner uses and why.

Forms help collect and use data and information. Resources include areas such as the plant files and plant schematics, what they are and how they are used. Reference is made to App. D containing blank copies of useful forms.

Chapter 9, The Computer in Maintenance, speaks to why a computer "might" be used in planning and how. Appendix L contains more information on computerization so as not to distract from the presentation of planning in the main body of the book. Maintenance planning is not simply using a computer.

Chapter 10, Consideration of Preventive Maintenance, Predictive Maintenance, and Project Work, covers the specific interfaces of these important areas with planning for the overall success of maintenance.

Chapter 11, Control, finally gets to the all important issue of how does one make sure planning works from a management and supervisory standpoint? Surprisingly, it is not on the basis of indicators; although two of the twelve planning and scheduling principles describe indicators. It is on the basis of the selection and training of planners.

The summary of the matter in Chap. 12, Conclusion: Start Planning, and the Epilogue help to tie together all the principles and techniques to achieve the vision set forth for planning. Then the appendices provide additional help for the interested reader. The various appendices are helpful to planners, analysts, planning supervisors, and maintenance managers responsible for planning.

Appendix A explains why planning is a "tool" and where it fits into the maintenance picture. Planning does not solve everything, but planning certainly brings together many of the other aspects of maintenance. The appendix describes other necessary maintenance tools and their relationship and relative importance to planning. Other tools needed include a work order system; leadership, management, communication, and teamwork; qualified personnel; shops, tool rooms, and tools; storeroom support; and maintenance measurement. In addition, consideration of reliability maintenance as preventive maintenance, predictive maintenance, and project maintenance is essential.

Appendix B addresses the people side of maintenance planning. The appendix identifies and discusses important soft aspects of maintenance that are critical to mindsets and attitudes for making planning successful. Management must work within the real world of real people.

Appendix C gives a starting point for what to buy and where to buy it when starting a planning organization.

Appendix D provides sample datasheets and forms that can be used directly by the purchasers of this book for maintenance in their organizations. It also illustrates sample completed work orders for planned jobs. These very helpful samples illustrate the proper information included at various stages of the maintenance process including requested work, coded work, planned work, and completed work.

Appendix E overviews the specific duties of a maintenance planner in a more step by step fashion and with less reasoning behind each step than does Chap. 5.

Appendix F overviews the planning-related activities of other employees besides the planner. This appendix covers the job steps of scheduler, clerk, operations coordinator, purchaser, crew supervisor, planning supervisor, maintenance manager, and maintenance analyst as well as a potential project manager to implement a new planning group.

Appendices G and H contain the reports of actual work sampling studies. This type of productivity study is the primary measure of planning and scheduling effectiveness. Consultants typically conduct them. Both sample studies contain complete procedures for conducting an in-house study. Appendix G shows a streamlined, simple study requiring a minimum of effort. Appendix H contains a more traditional study with many more measurement observations. The traditional study contains a section validating the accuracy of streamlined studies.

Appendix I on special factors affecting productivity covers in more detail how the scheduling system affects productivity. For example, what happens with blanket work orders?

The appendix also provides and discusses detailed cause maps that identify the underlying roots of schedule compliance and priority system problems. The practitioner can utilize these maps to help zero in on and overcome the key sources of trouble in their own unique maintenance environments.

Appendix J on work orders is a must. Planning follows a work order system as the most important improvement one can make to a maintenance program. The work order system is a process which maintenance uses to manage all plant maintenance work. The system assists the plant in keeping track of, prioritizing, planning, scheduling, analyzing, and controlling maintenance work. The plant must have a viable work order system as a foundation to planning and this appendix develops a typical system.

Appendix K on equipment, schematics, and tagging sets forth guidelines for equipment identification and tagging. Planning uses the equipment tag numbers in its filing system.

Appendix L on computerized maintenance management systems gives additional information on utilizing and implementing a computer application to assist planning. It provides essential guidance including critical cautions concerning initial implementations, patches, major upgrades, testing, training, consultants, and interfaces with other company software.

Appendix M on how to go about implementing a planning organization is a valuable feature in the book for its practicality. This appendix addresses how to organize the planning department as well as how to select and train the planners. Special topics are covered: traditional versus team environments, older versus newer facilities (even under construction), and centralized versus area maintenance. The appendix also provides an exhaustive aids and barriers analysis for setting up a planning group. It identifies aids and barriers to establishing each of the major functions of a successful planning and scheduling system. The analysis allows the practitioner to develop successful action plans to use these aids and avoid these barriers to ensure success. The analysis further breaks down critical aids and barriers themselves to provide in-depth implementation help.

Appendix N shows an example of a formal job description for planners to give a company that must establish one, a head start.

Appendix O has example training tests and illustrates the type of knowledge a planner should be gaining when becoming familiar and adept with planning techniques.

Appendix P illustrates the type of questions and information a maintenance manager uses when implementing a planned maintenance process and culture.

Appendix Q thoroughly examines contracting out work, one of the most controversial issues in labor relations. Great pressures exist for companies to contract out some or all of their maintenance and these companies typically involve the planning group in any effort to utilize contractors. Therefore, this appendix provides the practitioner of maintenance planning with the information to judge the potential benefits versus risks of different levels and approaches for contracting out plant work. The information of necessity also covers modern arbitration decisions regarding contracting.

To conclude the *Maintenance Planning and Scheduling Handbook*, App. R is a concise recap of the text of the many principles and guidelines pulled together from throughout the book and helpfully put into a single place for reference.

Planning Principles

This chapter recaps the vision and mission of planning and then presents the principles of effective planning. Each principle identifies an important crossroads. At each crossroad, the company has to make a decision regarding alternative ways to conduct planning. The decision the company makes regarding each situation determines the ultimate success of planning. Each principle presents the recommended solution to the crossroads.

Six principles greatly contribute to the overall success of planning. First, the company organizes planners into a separate department. Second, planners concentrate on future work. Third, planners base their files on the component level of systems. Fourth, planner expertise dictates job estimates. Fifth, planners recognize the skill of the crafts. And sixth, work sampling for direct work time provides the primary measure of planning effectiveness. Figure 2.1 shows the entire text of these principles.

The Planning Vision; The Mission

As presented in Chap. 1, the mission of planning revolves around making the right jobs "ready to go." Maintenance management uses planning as a tool to reduce unnecessary job delays through advance preparation. To prepare a job in advance, a planner develops a work plan after receiving a work request. The work plan is nothing more than the assembled information that the planner makes ready for the technician who will later execute the work. Some organizations call the work plan a *work package* or a *planned package*. At a minimum, the work plan includes a job scope, identification of craft skill required, and labor time estimates. The planner may also include a procedure for accomplishing the task and identify any parts and special tools required. With the proper planning or preparation for each job, this effort sets the stage to increase the productivity of the maintenance force.

2. The Planning Department concentrates on future work—work that has not been started—in order to provide the Maintenance Department at least one week of work backlog that is planned, approved, and ready to execute. This backlog allows crews to work primarily on planned work.

Crew supervisors handle the current day's work and problems. Any problems that arise after commencement of any job are resolved by the craft technicians or supervisors.

After every job completion, feedback is given by the lead technician or supervisor to the Planning Department. The feedback consists of any problems, plan changes, or other helpful information so that future work plans and schedules might be improved. The planners ensure that feedback information gets properly filed to aid future work.

3. The Planning Department maintains a simple, secure file system based on equipment tag numbers. The file system enables planners to utilize equipment data and information learned on previous work to prepare and improve work plans, especially on repetitive maintenance tasks. The majority of maintenance tasks are repetitive over a sufficient period of time. File cost information assists in making repair or replace decisions.

Supervisors and plant engineers are trained to access these files to gather information they need with minimal planner assistance.

4. Planners use personal experience and file information to develop work plans to avoid anticipated work delays and quality or safety problems.

As a minimum, planners are experienced, top level technicians that are trained in planning techniques.

MAINTENANCE PLANNING PRINCIPLES

1. The planners are organized into a separate department from the craft maintenance crews to facilitate specializing in planning techniques as well as focusing on future work.

6. Wrench time is the primary measure of workforce efficiency and of planning and scheduling effectiveness. Wrench time is the proportion of available-to-work time during which craft persons are not being kept from productively working on a job site by delays such as waiting for assignment, clearance, parts, tools, instructions, travel, coordination with other crafts, or equipment information. Work that is planned before assignment reduces unnecessary delays during jobs and work that is scheduled reduces delays between jobs.

5. The Planning Department recognizes the skill of the crafts. In general, the planner's responsibility is "what" before "how." The planner determines the scope of the work request including clarification of the originator's intent where necessary. The planner then plans the general strategy of the work (such as repair or replace) and includes a preliminary procedure if there is not one already in the file. The craft technicians use their expertise to make the specified repair or replacement. The planners and technicians work together over repeated jobs to develop better procedures and checklists.

Figure 2.1   The six maintenance planning principles.

The vision of planning is simply to increase labor productivity. The mission of planning is to prepare the jobs to increase labor productivity. As simple as this sounds, when management implements planning, it becomes apparent that the planning system abounds with many subtleties. The inability of many companies to recognize or deal with these subtleties prevents their planning organizations from yielding productivity improvements. The following principles guide planning through these particular difficulties to be effective.

Principle 1: Separate Department

Planning Principle 1 (Fig. 2.2) states

> The planners are organized into a separate department from the craft maintenance crews to facilitate specializing in planning techniques as well as focusing on future work.

The first principle dictates that planners are not members of the craft crew for which they plan. Planners report to a different supervisor than that of the craft crew. The company places planners into a separate crew of their own. They have their own supervisor. With a small number of planners, the planners might report to the same manager who holds authority over the crew supervisors. There may be a lead planner with some responsibility to provide direction and ensure consistency within the planning group.

The problem with giving the crew supervisors authority over their respective planners is that the crew focuses almost exclusively on executing assigned work. The crew members execute work; the planners do not. The planners must be engaged in preparing work that has not yet begun. In actual practice, the crew supervisor receives too much pressure for the supervisor not to use the planner to assist work that has already begun. The crew supervisor must have repairs completed. It is tempting to reassign a planner to a toolbox and say, "The planner is a qualified welder who can come help us." Even in a plant with few reactive jobs, the supervisor should still have significant motivation to keep actively completing an assigned backlog of work to keep the plant out of a reactive maintenance mode. The supervisor has an obligation to complete the assigned work in an expeditious manner with a minimum of interruptions

Principle 1

Separate Department

✦**Planners not on craft crews**
✦**Planners not pulling wrenchs**

Figure 2.2 Separation reduces temptation.

or delays. Once any job encounters delays, the supervisor feels pressure to minimize them. With direct access to the superior craft skills of a maintenance planner, the supervisor would always have significant motivation to take a planner away from planning duties. To the crew supervisor, the present is always more urgent than the future. The work in progress is always more important than the job not yet begun.

Management may contribute to this problem when planners report to crew supervisors. The pressure is especially intense if the maintenance manager has given a specific direction to the crew supervisor, such as "Put that pump back on line today!" How does the supervisor balance this instruction against the manager's admonition last year, "Try not to use the planner on field work unless necessary"? There will always be important work to complete today and the temptation to delay preparing for tomorrow's work.

Not only does the crew supervisor favor assigning craft work to the planner, the rest of the crew members as well place more relative importance on the work in progress than the paperwork of the planner. Such peer pressure encourages the planner to assist on jobs already begun or to take assignments directly for craft work willingly.

The natural inclination of the crew supervisor to place highest importance on assigned work, the unconscious pressure from management to encourage supervisors to give craft work to planners, and the peer pressure from fellow crew members all contribute to taking planners away from planning duties. In actual practice, planners on maintenance crews frequently work craft jobs and devote inadequate time to planning activities. As a result, crews have insufficient work to execute on a planned basis merely because planners do not have time to plan much work. This situation may also lead to another problem that manifests itself in an insidious fashion. Because planning contributes to scheduling, the lack of planning effort may decrease the number of work assignments to crews. The amount of work the company expects from each crew decreases. The work assigned becomes more reactive in nature because the plant executes less proactive work to head off problems. Gradually, the plant returns to a situation in which crews routinely repair equipment under urgent conditions and with little time remaining for maintenance to prevent equipment problems.

A self-fulfilling prophecy occurs for the manager who assigns planners to field crews. Supervisors frequently put planners on their tools to pull wrenches instead of plan. Planners plan less work. Less work is assigned. Work that is assigned is more reactive in nature, needing more on the job assistance. An apparent, but false, validation results showing that planners need to be on crews to help.

The problem is not managers, supervisors, or crew members with inadequate organizational discipline or inadequate understanding of the nature of planning. The problem is poor alignment of the company organization with the company vision. Simply removing the planners out from under the crew supervisors allows the planners to perform planning duties. The problem is not having persons who can resist the temptation to use a planner's craft skills. The problem is creating a situation where the temptation exists. The company avoids this situation by removing the planners from direct control of the maintenance crews. Then when the supervisor presumes it necessary to use a planner as a technician on an emergency job, the maintenance manager makes the call, not the supervisor.

If problems do arise where extra craft help is necessary, the supervisor has several options besides using a maintenance planner. The supervisor may assign more capable technicians to difficult jobs. The supervisor may decide overtime work is appropriate. The supervisor may decide to extend the job duration and not complete the job on schedule. The supervisor may decide to take advantage of an existing contract to provide contract labor assistance. The supervisor may decide to contract the job altogether. Perhaps the supervisor could increase productivity by personally supervising the work. The supervisor might request help from another crew. The labor contract might allow the supervisor to use another craft as a helper. For example, an electrician might be an adequate helper for a machinist on a particular task. Supervisors might also contribute their own hands to the execution of the work. Many options besides using the planner exist to expedite pressing field assignments.

Only after considering other avenues of help might the supervisor request using a planner as a technician through the maintenance manager who applied the job pressure in the first place. It is one thing for a manager to say "Fix that pump today!" and another thing for the maintenance manager consciously to redirect other resources to the task. Because a single planner helps leverage 30 technicians into 47, the planner in effect is worth 17 persons. The planner is the last person the manager would want to pull away for a field assignment. Compare the cost of time and a half overtime paid to a mechanic versus 17 times straight time opportunity lost to the company for using a planner on a field assignment. Even triple overtime does not compare to the economic waste of using a planner for execution of work. Pulling a planner for a field assignment must be the absolute last resort for the manager who understands and believes in the leverage of planning. Making the manager involved in each case for such a decision helps prevent such reassignments.

The manager might expect the crew supervisor to complain that management took some of the best technicians from the work force to create the planner positions. The manager must understand that for each technician transformed into a planner, the work force receives the equivalent of 17 technicians in return. It is in everyone's best interest to make planning work. Time spent in explaining the leverage and benefit of planning to supervisors both at these times of questioning and at the outset of initiating planning is time well spent.

Another reason the company organizes planners into a separate group is to facilitate or help the planners become specialized in planning techniques. Planners need to work closely together to ensure proper execution and consistency of planning work itself. There are ample opportunities to conduct planning in different manners. Planners need the reinforcement of each other's help to plan jobs and follow the planning principles in a common fashion. Preparing work to be accomplished in the future while the field technicians on crews scurry after jobs-in-progress is a new experience and is difficult to master alone.

Illustrations

The following illustrations demonstrate this principle of planning. The first section shows problems occurring as a result of not following the principle. The second section shows success through application of the principle.

Not this way.   Maintenance Manager Scott Smith walked over to the office of the mechanical crew supervisor. Each crew had its own planner who had a partitioned section of the supervisor's office with a desk and computer. Smith did not expect to see the planner necessarily because he knew that planners had to travel quite a bit to go to all the jobs for scoping. So it was not unexpected that the planner was not at the desk. The crew supervisor was not there either, which was appropriate, because Smith likewise expected supervisors to spend time in the field with their crews. However, on the way back to the front office, Smith happened to pass the fuel oil transfer pumps and saw the mechanical crew planner on a scaffolding assisting another mechanic hoist a valve into place. After questioning the planner, it appeared that the crew supervisor wanted to have the valve job completed today. He had directed the planner to help the mechanic who was having trouble managing the bulky valve alone. Smith could understand that the planner was under the direction of the supervisor, but Smith had begun to notice an uncomfortable trend. At least half of the time when he saw a planner, the planner would be working on a crew. This probably contributed to the indicator Smith tracked showing that the crews spent most labor hours on unplanned work. Last week Smith had even seen one of the planners working as a tool room attendant. The supervisor of the tool room had borrowed the planner from one of the crews because the tool room was suddenly short-handed that day. Smith was somewhat reluctant to counsel his supervisor because the supervisors took such great pride in managing their own work. However, in order for planning to work, obviously there had to be some planners doing planning. Smith decided to meet with his supervisors again regarding the matter.

This way.   Maintenance Manager Scott Smith walked over to the office area of the maintenance planners. Each planner had a partitioned office cubicle with a desk and computer. Smith did not expect to see all the planners necessarily because he knew that planners had to travel quite a bit to go to all the jobs for scoping. So it was not unexpected that only two of the four planners were at their desks. One of the planners present appeared to be attaching plan information to a work order and the other planner was going through a file to find equipment information. On the way back to the front office, Smith happened to pass the fuel oil transfer pumps and saw two mechanics hoisting a valve into place. After questioning the mechanics, it appeared that the job plan was helping them expedite the job. The plan had given the valve weight so that the right straps could be checked out of the tool room before the job started. The plan had also advised the supervisor ahead of time that the job required two persons because of the valve's bulkiness. After talking to the mechanics, Smith started again back to his office. As he was crossing the pump yard he noticed one of the remaining planners carrying a clipboard with a stack of work order forms. This planner claimed to be en route from the power house where three jobs had been scoped and was heading toward the chemical waste treatment system to scope four more work orders. Smith was comfortable that the planners were engaged in planning activities as he wanted. Smith knew that the supervisors also knew the importance of completing the planning. This morning he had turned down a request for a crew supervisor to borrow a planner for a field assignment. After discussing the particular work order, Smith had advised that the crew supervisor would have to extend the schedule for its completion.

Managers need to place maintenance planners out from under the control of crew supervisors to prevent the planners from being assigned field work as technicians. The temptation to use planners as field technicians on current jobs is usually too strong to allow the planners time to do helpful planning for future work. A separation arrangement allows the planners to concentrate on planning future work.

Principle 2: Focus on Future Work

Planning Principle 2 (Fig. 2.3) states

> The Planning Department concentrates on future work—work that has not been started—in order to provide the Maintenance Department at least one week of work backlog that is planned, approved, and ready to execute. This backlog allows crews to work primarily on planned work.
>
> Crew supervisors handle the current day's work and problems. Any problems that arise after the commencement of any job are resolved by the craft technicians or supervisors.

Principle 2
Focus on Future Work

Figure 2.3   The snowball of improvement.

After every job completion, feedback is given by the lead technician or supervisor to the Planning Department. The feedback consists of any problems, plan changes, or other helpful information so that future work plans and schedules might be improved. The planners ensure that feedback information gets properly filed to aid future work.

The reason the planners need to be separate is they need to focus on future work. Planners do not become involved in work that is already ongoing. A simple definition of future work is when the crew has not yet been assigned to start on the work order. Once a crew has started working on a job and they find out they need more information, they do not come to the planner for assistance, but work it out themselves. Then after the crew successfully completes the current job, feedback to planning helps avoid similar problems in the future.

The problem with the planner having the duty to help technicians find file information for jobs already under way is that the planner soon has no time left to plan or gather job information to help future work. A vicious cycle is then in place. No jobs receive the benefit of advance planning because there is no time to refer to past feedback or otherwise anticipate problems ahead of time. The question at the crossroads is whether planners are really in the business of planning or are they in place to help technicians quickly find information to help resolve problems for work that has already started. The planners are most knowledgeable about the plant technical documents, and jobs that are underway need help fast when problems arise. Nevertheless, this use of planning is almost as short-sighted as using planners as field technicians.

Think of the circle in Fig. 2.3 as a repeated cycle of maintenance over the life of a piece of equipment. Maintenance does a job to maintain the equipment. During the course of the work the field technicians learn about the equipment or task. For example, they may learn that a certain pump bearing can only be removed from the inboard side because of an almost imperceptible taper in the design. The technicians learned this fact from trial and error and spent most of a morning doing it the wrong way. After the job the technicians give feedback on the work order form about the design and delay. Then the next time that particular pump needs maintenance, the planner can refer to the previous problems and the resolution because the planner filed the previous feedback. The planner reports this information as part of the job plan before the crew starts the task. As a result, previously encountered delays might be avoided on the subsequent maintenance operations. In the example of the tapered bearing, the second time the crew replaces the bearing, they should not have to waste time trying to remove the bearing from the wrong side. The crew avoids an entire morning of wasted time. Each time the crew works on a particular piece of equipment, they might learn something new that could help future jobs.

This cycle of maintenance and planning concept carries some important implicit presumptions. The first and most important presumption is that a planner is available to review feedback from previous jobs and otherwise plan for new work. Another presumption is that feedback is not only obtained, but kept after each job. The final presumption is that equipment is worked on repetitively. These presumptions are not taken lightly.

The first presumption is that a planner is not only willing, but available to plan new work. As planning recognizes the need not to be on the tools (Principle 1), they are still frequently hindered from focusing on future work. As the planners leave their tools and arrive in the office to focus on future work, they meet a new challenge. The problem that arises is that if a planner is planning for 20 to 30 technicians, how many of those technicians are going to want some additional information? Probably at least two or three will do so. So these two or three technicians come to the planning office and ask the planner for help; after all, the technician regards the planner as the information finding expert. With this constant interruption, the planner does not have the time for the filing or work necessary to focus on future work. The planner helps with work-in-progress, not future work. Figure 2.4, Chasing Parts, illustrates what happens.

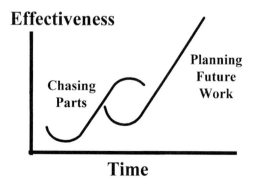

Figure 2.4   Chasing parts for today's jobs cannot help as much as focusing on future work in the long run.

Figure 2.4 presents a variation of the common product life cycle that illustrates the planning effectiveness challenge. As management takes good technicians out of the work force (Principle 1) to be planners, the work force's effectiveness initially suffers. Then as the planners become proficient at finding file information (albeit on work-in-progress), there is overall improvement for the work force. However, the first curved line shows an upper limit to how much help this practice can deliver. The second curved line shows when planners turn away from constantly helping work-in-progress and focus on future work that maintenance effectiveness can improve further. Opportunity for further improvement exists because when the planners only help work-in-progress, they are not helping the crews avoid previously encountered delays. Every job becomes a new job without any history advantage. No wonder so many techs need help with work-in-progress; they have no opportunity to avoid what has happened in the past. It is no wonder the planner cannot focus on future work. Every job in progress runs into problems creating another vicious cycle. The planners become known as "parts chasers" excitedly helping technicians find parts information or solve other problems on most jobs. Every job is urgent once it starts.

This is a very sensitive area for existing planning departments. Management may have started the planning department with the published intent of helping everyone with obtaining information at any time. A planner soon learns the impracticality of planning in advance for 20 persons while at the same time helping with work-in-progress. The best alternative at this point is to try to designate one of the planners for helping all jobs-in-progress to shield the other planners.

It is best to start out with the understanding that "planners will not replace the need for a tech (or supervisor) to find technical information." However, once a technician has found information the planner will save and reissue all job feedback on future work. This arrangement is also necessary for the crew supervisors to maintain their familiarity with the files and also encourages feedback from the technicians. Once technicians have to find technical information for a job, knowing that they will have to find the information again themselves the next time unless the planner can extract the data from the files, encourages feedback.

The future work concept is important. If a crew has already started working on something and they find out they need some more parts, they do not come to the planner to help find those parts. That would be counterproductive overall. Think back to before the company had planning; then the crew supervisors knew how to obtain parts. The crew supervisors knew how to find file information. That previous familiarity should be maintained. Management wants the "added value" of looking at future work. Therefore, after the job starts, the techs or crew supervisors must find any additional information just as they did before planning existed. That lets the planner focus on getting all the jobs planned.

Principle 2 does not accept planning being a highly efficient department of persons to help crews look for parts once jobs start. The craftperson who changes the plan or has problems should write that information down after she finishes the job and give it back to planning for filing. The next time that piece of equipment needs work the planner will take the filed information and insert it for an improved job plan.

Management needs to monitor the time planners spend planning future work versus helping jobs-in-progress. If using a timesheet system, management may consider planners using a one time accounting number when planning and another number when providing technical assistance. A balance should be struck between the use of separate numbers legitimizing "chasing parts" and showing that "chasing parts" is not planning.

The second implicit presumption is that feedback will be received and used. Many companies almost hopelessly damage their planning effort with misconceptions regarding this point. These organizations start their planning groups with the expectations that field technicians would never have to look for information and that planners would always plan perfect jobs from scratch. In other words, their concept is that each planner would pick through the technical manuals every time a job came up to support the planner's 20 to 30 technicians. The field technicians thereby never have to find information because the planner always has it ready. This approach fails for two reasons. The first is that a planner cannot keep up with the work load researching each job from scratch. This is why planning organizations have a difficult time in their first 6 months of existence. In effect, every job is being built from square one before the files slowly become built up and useful. The second reason is that the most valuable information needed on plans is not available from equipment manuals. Information such as potential work permitting problems, the probability that certain parts will be needed, and corrected local inventory stocking numbers are learned from past jobs. A planner must be able to find the helpful feedback on those last three work orders from the last 3 years to help the crew avoid previous problems. For example, if the planner finds that the last time the crew worked on this job they did not have a certain part, the planner makes sure they have that part this time. Each and every job is on a learning curve. Looking to the files helps achieve that improvement opportunity. The correct concept is that the planner to a large degree is essentially a file clerk for the technician. The planner promises that if the technician reports any information, the planner will have that information available for next time. The field technicians must be willing to research and resolve problems as they come up on jobs-in-progress and report feedback to their file clerk. The technicians must not have the false impression because certain information was unknown that the planner failed to adequately plan the job. On the other hand, the planner must understand the importance of saving and referring to this important feedback. The planner does not plan each job from scratch. By using feedback in the plant files, the planner not only has the opportunity of continuously improving job plans, but has time to plan all the work orders.

The last presumption concerns doing work repetitively. Working on equipment repetitively is a reality. One typically thinks of preventive maintenance as the only repetitive work in the plant. Yet the 50% rule says that if a piece of equipment requires work, there is a 50% chance it will require similar, if not the same, work on it again within a year's time. Moreover, the 80% rule says that there is an 80% chance the equipment will be worked on again within a 5-year period. These percentages are not for preventive maintenance. Why are these percentages so high? One reason is "infant mortality." After any work on any equipment, there exists an increased chance of additional maintenance soon being required. Problems from the initial job might include faulty materials or maintenance practices. The feedback from these jobs is especially important for the planner to scrutinize for opportunities to avoid repeated problems. Another reason is that some equipment simply requires more attention than others. Out of 10,000 different pieces of equipment, 300 might continuously need attention while the other 9000 or so never seem to need work. On the other hand, there is a common perception that "Nothing is ever the same" or "It is always something different." These statements reflect a perception that none of the equipment receives repetitive maintenance attention. This perception is false, but understandable. For one thing, the exact same technician might not be involved each time. For another thing, working on a piece of equipment only once or twice a year just does not seem to be very repetitive, especially if the exact same task is not involved. Nonetheless, one must move beyond the horizon of a crew thinking of one week at a time. The 30 plus years of a plant's life mean that the vast majority of maintenance tasks will be executed repetitively. And if the vast majority of jobs are repetitive, each presents the potential opportunity of contributing to increased labor productivity through heeding the lessons of the past. That means there is a tremendous opportunity to improve through avoiding past delays. There is a cycle and a snowball effect. As maintenance crews work jobs, they learn helpful information about delays. Then they give that information to planning as feedback at the end of a job. Planning references this information when the next job comes up for that equipment and the snowball picks up momentum as repeated jobs avoid past delays.

A final comment is appropriate regarding future work. Even without regarding the repetitive nature of maintenance work, there is a serious problem when the plant overfocuses on helping jobs-in-progress. When technicians run into a problem, there is generally a job delay while they resolve the matter. Unless these technicians can quickly move to other work, there will be several technicians standing around wasting time even if the planner rapidly resolves the problem. It is undeniably much better to have the planner anticipate problems ahead of time and spend time resolving them while no one is waiting.

Illustrations

The following illustrations demonstrate this principle of planning. The first section shows problems occurring as a result of not following the principle. The second section shows success through application of the principle.

Not this way.   Sally Johnson was the planner for the mechanical work for Crew A's ten mechanics and ten welders. Since it was Monday, she planned to scope and compile plans for all the jobs that the weekend operating crews had reported. In addition, there were a number of jobs completed last week for which she needed to file the work orders. Before she could complete checking her email, however, two welders came into the office requesting her help to run pick tickets for them to receive a valve out of inventory. Soon after she provided this help, a mechanic called her on the radio for assistance obtaining bearing clearances for the forced draft fans. She knew this would be a problem and she spent the better part of the morning locating and talking to the manufacturer. By midafternoon, the interruptions had kept coming and Johnson still had not scoped the first job. At least she felt a sense of accomplishment that she kept important jobs going through her efforts.

This way.   Sally Johnson was the planner for the mechanical work for Crew A's ten mechanics and ten welders. Since it was Monday, she planned to scope and compile plans for all the jobs that the weekend operating crews had reported. In addition, there were a number of jobs completed last week for which she needed to file the work orders. After checking her email, she began filing. As she started to assemble information for the new jobs, she again returned to the files. Good, she thought, here is a list of parts for the air compressor job. That will help the mechanics when they start that job. On about half the jobs, she found useful information from previous work orders. After compiling the information and making field inspections, she finished the required planning by about midafternoon. That left part of the day to talk to one of the plant engineers from whom she had asked some material selection advice. She felt a sense of accomplishment that she was part of a new service for maintenance that boosted productivity and ultimately company profits. She could feel that her efforts were part of a better process than the old "Just work harder" mind set.

As one can see, the repetitive nature of equipment maintenance provides great opportunity for planners to give technicians a head start in avoiding past problems. Technicians need to be mindful to resolve problems without planner assistance and provide feedback on circumstances encountered and information gained. Planners need to be heedful to their task of keeping and utilizing past work order information to improve jobs being planned. To make the cycle of job improvement work through avoiding past delays, planners must be allowed to focus on future work. Nevertheless, past delays can only be avoided if they are remembered, which leads to the next principle.

Principle 3: Component Level Files

Planning Principle 3 (Fig. 2.5) states

> The Planning Department maintains a simple, secure file system based on equipment tag numbers. The file system enables planners to utilize equipment data and information learned on previous work to prepare and improve work plans, especially on repetitive maintenance tasks. The majority of maintenance tasks are repetitive over a sufficient period of time. File cost information assists making repair or replace decisions.
>
> Supervisors and plant engineers are trained to access these files to gather information they need with minimal planner assistance.

The concept of component level files or "minifiles" is a vital key for successful planning. Principle 3 dictates that planners do not file on a system level or basis, but on an individual component one. A minifile is a file made exclusively for an individual piece of equipment the first time it is maintained. The term minifile helps convey the understanding that the file does not keep information for multiple pieces of equipment together. Planners make new equipment a minifile when it is purchased. Planners label the file with the exact same component tag number attached to the equipment in the field. Planners consult the minifile for each new job to take advantage of the lessons and information gained on previous jobs. This principle takes advantage of the fact that equipment requires repetitive attention over the life of the plant. In particular, cost information available through the files helps planners and

Principle 3

Component Level Files

✦**Paper and Computer**
✦**Work Order and Equipment Databases**

Figure 2.5 Filing so that information can be used.

others make important decisions on replacing or modifying troublesome equipment. The files are arranged in a secure fashion to keep data from being taken away unadvisedly and lost, but are arranged simply enough for other plant personnel to be able to access their information. Engineers and supervisors directly use the files for obtaining information for projects or jobs-in-progress rather than interrupt the planners from planning future work.

The crossroads, so to speak, in this instance is whether to file information by systems or by individual equipment. A simple few files make it easy to put certain information in, but later difficult to find and take out that information. A complex, multiple file arrangement would require more time to find the right file in which to put the information. On the other hand, later it would be easier to find the information again. Starting from the extreme, the easiest arrangement into which to place information would be a single repository or file for the whole plant. Planners would have no trouble filing information because it all goes into one place together. However, later if a planner wanted information saved last year for the clarifier drain valve, it would be impractical to find it amidst the mass of other saved data. Moving to a slightly less simple arrangement, the plant could file information by building or plant area. A planner would file all the waste treatment information together and later might have a less difficult time finding the clarifier drain valve data. Continuing to how many plants do actually file data, a plant could file by equipment system such as the liquid waste system, the high pressure steam turbine system, and the polisher system. This makes the planner have to take a little more care filing the information to place it in the right place. Later the planner has a much easier time finding the information if needed. The next less simple filing system would be filing information by the equipment itself such as the clarifier drain valve. Obviously, the planner would have little trouble later retrieving information, but to begin with the planner would also have to exercise considerable care filing information. The extreme case would be to file information separately even by nearly every discrete subcomponent such as a valve body, a valve actuator, a pump, or a pump bearing. These arrangements become too complicated for filing or retrieving information. Alternately, the plant may file equipment information by manufacturer or vendor. Filing by manufacturer or vendor is common, but generally not favored because manufacturers and vendors change over time for particular pieces of equipment.

Consider a road or street address system in a town or city. Persons might take a multiple lane highway to arrive at the town. They turn off the highway onto a major road to go to the neighborhood. Then they look for particular side streets leading to the street of interest. Once at the specific address of the home of interest, they turn onto a specific narrow driveway. The seekers can locate all of the occupants of this home because they are at a specifically numbered address within the city. One cannot find any of the occupants of the home by simply arriving in the city. Planners likewise cannot find any of the work orders for a piece of equipment if the population of total work orders is significant at all.

Consider a doctor's office. Many physicians have a paper file system directly behind the receptionist. There is a separate paper file for each patient or, at most, family. The physician can easily determine the patient's medical history by looking at the filed information. A patient would also be uncomfortable if the physician did not think any past history was ever important. A patient would also be uncomfortable if the physician filed all history by single neighborhood files. Similarly, planners know that history is important for all equipment and there is not too much trouble in filing by equipment.

The conventional wisdom is threefold for filing. Do not file information that one knows will not be needed in the future. File in fat files what probably will not be needed in the future, but if needed must be found. File in skinny files what will be needed and used in the future.

Maintenance work orders decidedly fall into the last category considering that the majority of equipment maintenance is repetitive over the years.

It turns out that once management takes the planners off their tools (Principle 1) so they may actually focus on future work (Principle 2), a new situation arises. The files where everything has been put for years are not useful unless information is filed by individual pieces of equipment. Say a planner is planning a job on the polisher cation regeneration valve. According to the 50% rule, there should be at least one or two previous work orders from the past couple of years that would help. The problem is that the plant used a single file to place *all* the work orders from the polishers; there must be 250 work orders. The planner does not have time to dig through them looking for the several cation regeneration valve ones if this situation is the norm encountered for every piece of equipment and job planned.

As planning is implemented, it soon becomes evident that it is not feasible to check individual equipment history and technical information if they are kept in system files. System files have too much information to allow quick reference for individual equipment. Once the planner receives job feedback for future reference, it cannot go into a system file. A system might have 20 to 100 or more different components alone with multiple work orders for each. When a file is that large, planners cannot practically find information on a single piece of equipment. Therefore planners use a component level file for each piece of equipment. When the planner receives a work order, the planner consults the specific file to find the previous work orders for that equipment. The filing mirrors the obvious work order arrangement. Normally, planners plan work orders for discrete pieces of equipment. It makes sense to file information in the same manner.

Consider a simple, paper file system. This file system is the equipment database complete with work order history for each piece of equipment. With a minifile, the first thing a planner does when a job comes in is go to the minifile, pull it out and find the previous work orders for the equipment. If the planner finds that the last time the crew worked this job they did not have a certain part, the planner makes sure they have that part this time. The job is on a learning curve.

As discussed in the previous principle, many persons think a crew never works on the same thing over again, that it is always something different. Yet in reality they work on the same things over and over again, just not every day. It might be 9 months to over a year before a crew works on it again and even then with a different technician. So persons just have a feeling that they are working on different things all the time. Notwithstanding popular opinion, if a planner can find those last three work orders over the last 3 years, the planner can help the crew avoid previous problems. Furthermore, if a planner can tabulate the previous cost, the planner can make better repair or replace decisions. For example, "The last two times we worked on that, it cost $1000. I know I can buy a completely different valve for $500 that probably will not need as much maintenance." Looking to the files helps the planner reach that improvement opportunity. In addition, since the majority of jobs have been worked on before, most of the jobs currently in the plant would benefit from a planner being able to review past information through an adequate file system. Filing information by the individual equipment allows that opportunity.

Experience has shown that after only 6 months of conscientious feedback and planning, most jobs in the plant receive a benefit from feedback learned on previous jobs.

The next issue concerns how the planners should physically arrange and number the files.

First, an intelligent numbering system of some sort is preferred. Many plants might have the equipment files labeled by the written names of the equipment. For example, one file might have Polisher Cation Regeneration Valve as its label. The plant may order these files within systems alphabetically or by process location. However, using the filing system becomes somewhat cumbersome as the quantity of equipment rises. For one thing, not everyone may refer to the equipment by the same name. On the other hand, a plant-wide coding system allows better file arrangement through intelligent numbering. For example, from the number N01-CP-005, one could tell that the equipment is a part of the Condensate Polisher system of North Unit #1. This number allows not only a unique, file reference number, but also the grouping of all polisher equipment together. This system is preferred although some thought will have to be spent on developing an appropriate numbering system. Some companies have already tagged their equipment with unique numbers just for the benefit of ensuring maintenance does their work on the correct machines. Planning should use these existing numbers as the basis for the filing system whenever possible. Appendices J and K give practical advice on setting up a numbering and tagging system.

Second, when using a numbering system, the company must make sure to follow through on one action. Not only must they label the files, but it is almost imperative that they hang matching equipment tags on the field equipment. This simple step greatly assists the operators and other writers of work orders tie the equipment number to the work order. This tie helps the planner find the correct equipment files. Some filing programs have failed not because the filing system was somewhat complex, but because there were no corresponding equipment tags.

Third, the planners must set up the files so that the supervisors and plant engineers do not ask the planner to look in the files; they look in the files themselves. The planners intend for these persons still to work with files and information. For this reason, paper files should be open and easy to see with side labels on individual folders. Files that are enclosed within closing file cabinet drawers tend not to be inviting or as user friendly as possible. Large labels should clearly declare the contents of different shelf areas. For this same reason, planning should keep all the files in a common area, not within individual planner cubicles.

Fourth, if other persons have access to the files, management may have some concern for security. Generally, having the file area located so that persons must first pass through the planner area is acceptable. This arrangement strikes a balance between making the files accessible and making the files less prone to wander off by knowing who is there. Supervisors may want to designate that only certain individual technicians may access the files depending on the competence of the technicians in this regard.

The objective of this principle is to create a file system that delivers useful information to the planner and the rest of the plant personnel.

Illustrations

The following illustrations demonstrate this principle of planning. The first section shows problems occurring as a result of not following the principle. The second section shows success through application of the principle.

Not this way.   David needed to plan two jobs. One job required a simple filter change and the other required stopping a drip on the hypochlorite discharge piping. Both jobs were fairly routine. The filter was not on a PM route because varying operating modes caused the filter to plug at different intervals. The operators monitored the pressure differential and wrote a work order whenever the filter was beginning to show signs of clogging. David first skimmed through the thick system files behind his desk for past work orders, FC for Fuel Oil Service System and IR for Intake Chemical Treatment System. He was sure there were at least some for the filter. After several minutes he was able to find one for the filter, but not the piping. David copied down the filter and gasket inventory numbers off the previous work order plan. From his field inspection of the discharge piping, he determined that maintenance needed to cut away and replace the PVC piping. David included PVC piping inventory numbers and a statement to obtain PVC glue from the tool room in the job plan.

As David was finishing up the job plans, Supervisor Juan asked where the equipment information was for the hypochlorite pumps. David explained that all the information from past work orders was together in the system file and waited patiently as Juan shared his cubicle looking through the file.

This way.   David needed to plan two jobs. One job required a simple filter change and the other required stopping a drip on the hypochlorite discharge piping. Both jobs were fairly routine. The filter was not on a preventive maintenance (PM) route because varying operating modes caused the filter to plug at different intervals. The operators monitored the pressure differential and wrote a work order whenever the filter was beginning to show signs of clogging. The operators had written the equipment tag numbers on the work orders so David was able to walk over to the planner file area and immediately locate the two pertinent file folders, N02-FC-003 and N00-IR-008. As he had suspected there were several work orders for the filter and one for the piping.

David noticed that out of the three times the plant had changed the filter, two times the technician had reported having to redo the job because the assembly had leaked upon pressurization. David decided to change the work plan and include a reminder to tighten the strainer cover in a criss-cross pattern. David also included a step to request the operators to pressure test the line before the technicians packed up and left because of past trouble with the lid. David also copied down the filter and gasket inventory numbers off the previous work order plans. From his field inspection of the discharge piping, he determined that maintenance needed to cut away and replace the PVC piping. David included PVC piping inventory numbers and a statement to obtain PVC glue from the tool room in the job plan. David also noticed that the previous job in the file for this piping had recorded a job delay to wait on the operators to drain the pipe. Apparently the pipe was not self-draining as previously thought. David included a note in the plan for the supervisor to remind operations about the potential clearance problem.

As David was finishing up the job plans, Supervisor Juan asked where the equipment information was for the hypochlorite pumps. David pointed to the file area and explained that any information they had from past work orders was in the N00-1R section in several specific pump files. If Juan could not find what he wanted there, Juan might want to try the O&M manuals on another shelf area in the same room. David asked that if Juan found anything useful, to make David a copy and he would file it in an equipment specific minifile.

Caution on computerization

A computer certainly gives more capability to the maintenance effort. For instance, a computerized maintenance management system (CMMS) might allow accessing work order information away from the planning shop (by operators, engineers, and managers). It might allow sorting work orders (such as for specific types of outages). A computer might be able instantly to tabulate previous work order histories with costs and even eliminate a paper file system altogether. However, these benefits are not the specific leverage of planning. They are either additional points of leverage or acceleration of the manual planning operation. Planning itself is not the use of a computer. First one must learn to add, subtract, multiply, and divide before employing a calculator. The calculator simply helps the existing process.

Be cautious in thinking that having a computer system is itself planning. Planning multiplies a work force by 157%; it transforms 30 technicians into 47. Is management properly thinking that the computer system may help reach the top of this percentage increase or is management only thinking in terms of replacing two clerks currently entering work orders or typing PMs? Management needs a sense of perspective. Do not be unnecessarily eager to abandon a paper file system.

Figure 2.6 declares that computerizing a poor maintenance process will not help maintenance. This is especially true of the planning process.

As one can see, having unique numbers for equipment and then filing equipment work orders and information by those numbers make it possible for the planner to file and retrieve information as needed. Planners serve as file clerks to a large degree and need an accurate filing process.

When Using a Computer

1.   If you do not know how to do something without a computer, doing it with a computer will not help.

2.   Doing something wrong is faster with a computer.

Figure 2.6   First learn planning, then computerize.

Principle 4: Estimates Based on Planner Expertise

Planning Principle 4 (Fig. 2.7) states

> Planners use personal experience and file information to develop work plans to avoid anticipated work delays and quality or safety problems.
>
> As a minimum, planners are experienced, top level technicians that are trained in planning techniques.

Principle 4 dictates that the plant must choose from among its best craftpersons to be planners. These planners rely greatly upon their personal skill and experience in addition to file information to develop job plans.

The crossroads that this principle addresses is twofold. First, the plant has to decide what level of skill planning requires. The choices range from using relatively lower paid clerical skill all the way up to higher paid engineering skill. Second, the plant must decide the appropriate method of estimating job time requirements. A wide range of choices also exists for this issue.

It would seem that with the feedback and file system in place, clerks might be utilized as planners. However, as a minimum, planners need to be top level, skilled technicians, so that they can best scope a job or inspect the information in a file for its applicability to the current job being planned. One issue at stake is in whether to have (hopefully) good execution on an excellent job scope or have excellent execution of perhaps the wrong job scope. Identifying the correct job scope is of primary importance. One of the best persons to scope a job is the skilled craftperson who has successfully worked the job or ones similar many times in the past. Even if the planner has not worked the particular task, a skilled craftperson can research or make an intelligent estimate for what the task might require. A second issue involves the files. Planners cannot simply be clerks or librarians in this regard, either. Again as a minimum they need to be skilled craftpersons so that when they review information in a file, they can gather all possible help for the current job. They can look and see if a part used on a previous job was a "one in a million" type of part or whether it really needs to be a part used on most future jobs.

Principle 4

Plans with Estimates Based on Planner Expertise

Figure 2.7 Estimates are easy for planners that are accomplished craftpersons.

Companies have considered apprentices for planner positions. These appointments run into two problems. First, an apprentice rarely has the experience to scope jobs properly simply from a lack of experience. An apprentice has also not had the opportunity to develop a top level of skill. The second problem is that experienced craftpersons receiving a job plan from an apprentice tend to cast doubt not only about the job plan, but management's support of planning as well.

A newly promoted technician rising from the apprentice class has essentially the same weaknesses in the planner position as an apprentice. There is more possibility that an experienced technician may make a good planner, but consider that the planner will be dictating certain job requirements to all of the field technicians. If an existing technician is not a star performer, the technician may not have the skill desired to be scoping all the plant work. The rest of the technicians also have some reason to doubt the specifics of any job plan based upon their perception of the talent of the planner as a technician.

Companies have also used engineers and technologists as planners. However, they typically do not possess the skill to plan most maintenance jobs. Most maintenance jobs consist of routine valve replacements, filter changes, or equipment adjustments that the technical experience of the engineer or technologist does not encompass. Each of these seemingly simple tasks is laden with potential job problems and delays beyond their experience. On the other hand, even if these personnel have actually risen through the ranks of the maintenance force while earning their degrees, they are not cost effective to utilize as planners for routine maintenance. Routine maintenance offers the highest potential for planner contribution to company success because more intricate or unusual maintenance tasks normally already receive help from plant engineering.

Supervisors make excellent choices for maintenance planners because they were typically experienced, top level technicians before promotion. Because planners also must have a high degree of self-initiative, they possess another of the qualities mandatory for supervisors, but possibly lacking in some technicians. Existing company guidelines for selecting supervisors frequently are satisfactory for selecting the best planners. Because companies realize that they must attract the best technicians to make planning work, many companies pay planners at or above the first-line supervisor level. A recent survey indicates this is the case for over half of the electric utilities with maintenance planning. A company might want to consider moving an existing supervisor into a planner role or providing an additional promotion opportunity for its existing technicians. Making the planner position a step toward supervisor may also increase support in maintenance for planning. Another argument for paying planners at the level of supervisors is that the planners deal with the crew supervisors, not the technicians, at a peer level.

Companies not accepting that planners should be supervisor level might have one or two other considerations in mind. The company might feel that responsibility over personnel is more difficult than responsibility over a process. This thought has some merit, but consider that companies typically pay engineers higher than crew supervisors because of market demand. The market might also attract away some of the companies' best technicians if there is not ample room for growth. Paying planners as supervisors offers one solution to keep company strength in technical talent. Another consideration might be that the company does not support planning all the way. The company is keeping open an option to revert the planners back into the work force if planning does not work. The company might also be leaving an avenue to replace one or two planners that do not do well. The company so inclined must be very careful that it is not holding back the support a planning organization must have to succeed. The company might also have a weakness in not being able to remove unqualified supervisors. If the company's strategy does not select the best planners, the company does not follow this principle at the peril of planning.

Appendix M, Setting up a Planning Group, gives more guidance on selecting maintenance planners.

Another issue is the development of time estimates. The opinion of the skilled technician-planner is preferred over strict file information, pigeon holing, or other built-up time estimates.

File information yields historical data about past jobs, but can only offer general guidelines for current estimates. For example, the same job to clean an oil burner gun showed the following actual time requirements. One time the job took one person 20 hours. The next time the job took two persons 4 hours each. The last time took two persons 6 hours each. A planner might be tempted to average the times and plan for two persons at about 7 hours each. However, it is difficult to understand why the past jobs were so different especially if feedback was minimal as in these cases. The longest job might have had an inexperienced technician assigned or the person assigned was given no other jobs or schedule pressure. In the latter case, the person may have simply taken all of two 10-hour days to complete the work. If this was the case the planner might be more inclined to average only the two shorter jobs and plan for two persons at 5 hours each. Alternately why might not the planner insist that the target should be two persons at 4 hours each since that rate had been achieved once? On the other hand, what if the technician feels that from personal experience that, if done properly, the job should take two persons an entire day, 10 hours each?

Perhaps the planner could use the historical time estimates to create job standards for certain repeated tasks. The problem with this approach is first that historical time estimates might not reflect the appropriate time to do the job right. Second, other than for routine PMs, the day-to-day maintenance tasks are typically not repeated often enough or with enough similarity for studied measurements. In addition, management might be reluctant to press for early PM completion where one of the objectives of PM is to take care of all necessary minor adjustments.

Pigeon-holing offers another option for estimating jobs. Pigeon-holing involves estimating a job's time requirements by referring to a table or index of similar jobs and making adjustments for particular job differences. For example, if the job at hand is to rebuild a 25-GPM pump, the planner might refer to a table for pump work. The planner finds a suitable chart showing overhauls for 20-, 50-, 100-, and 200-GPM pumps. The planner figures that a rebuild is probably about the same as an overhaul and adds a little time to the estimate offered for the 20-GPM pump. The problem with this effort is its complexity and the time consumed finding and using the correct tables even if they are available and accurate.

There are industrial engineering estimates available for minute portions of tasks that are generic to many jobs the planner is planning. Times for taking off individual bolts of various sizes, walking certain distances, and particular hand or body motions are given. The planner could build up a time estimate for different maintenance operations using these standards. It is doubtful that the estimates these built-up estimates would yield would be worth the planner's time in creating them.

In certain industries such as maintenance of automobiles, auto shops have available books of standards for almost any maintenance task regarding almost any car. The great numbers of identical cars make these books possible.

The jobs in many industrial plants do not yield themselves as well to such universal standards. These plants use a variety of equipment in a host of different applications. The plants also have unique spatial or geographic layouts and unique maintenance facilities and personnel skills.

In practical application, the estimates that a qualified planner can make based on personal experience supplemented by file information are entirely adequate. The objective in planning is to help boost labor productivity, not create perfect time estimates or meet standards. On the bottom line, maintenance supervisors need estimates to help schedule and control work assignments. The planners' estimates are therefore considered the plant's standards for jobs even though they are not "engineered standards."

This need for an easily determined time estimate that the field technicians will respect is one of the reasons a planner must possess the skills of a top level technician.

Two issues arise after accepting how the planner determines the job estimate. Should the planner plan for a certain skill level and should the planner allow time for delays? The resolution to both of these concerns is that the planner estimates how long the job should take a *good* technician without *unanticipated* delays.

These issues are discussed briefly here and more thoroughly in the Chap. 5 section on estimating work hours and job duration. First, the planner wants to set a standard for performance through the estimate. The planner does not want to set an ambitious target or goal. The planner wants the standard to be met, but at the same time provide for proper maintenance execution of the work. The planner does this by deciding that every job will be done by a good technician. This methodology encourages most technicians on most jobs although it requires the supervisor to shore up weaker technicians on certain jobs. Second, the planner does not allow extra time for delays that the planner does not expect. This keeps the estimate accurate when the technicians encounter no delays, and provides the supervisor a reference time for controlling the work when unexpected delays do occur. The supervisor can judge the appropriateness of the performance taking into account the specific delays dealt with and the time estimated for the job without those delays. Setting time estimates for jobs not to include extra time for unanticipated delays also sets forth the expectation that maintenance should proceed as expeditiously as possible under normal conditions.

At this point, it is appropriate to discuss the accuracy of such estimates determined by the planner with extensive technician experience. Experience has shown that job estimates for individual work orders may be off plus or minus as much as 100%. That means that on the average, a job planned for five labor hours has as much chance of being accomplished in 1 or 2 hours as it might in 10 hours. This is especially true of the smaller work orders that make up the bulk of many maintenance operations. Does this mean that planner estimates are worthless? Absolutely not. On the contrary, the planner estimates are more accurate overall as the work horizon widens out because as many jobs run over as run under the estimate. (Statistically speaking, that means that such estimates are very accurate and give the average of any individual job performed many times under the same circumstances.) A supervisor can use these estimates to assign and control work. Instead of simply assigning a single job at a time, a supervisor should be assigning a day's worth of work. Instead of assigning a single 5-hour job to a person on a 10-hour shift, the supervisor might assign the 5-hour job, a 3-hour job, and a 2-hour job. The 5-hour job might only end up taking 3 hours, but the 3-hour job might take 4 hours and the 1-hour job might take 2 hours. This is an idealistic example, but makes the point that if the supervisor does not hold technicians accountable to a single job, but a day's worth of work, the supervisor can use the planned estimates to control work. In addition, the supervisor does not get too excited about any single day's work being overrun and is interested more about the long-term performance of technicians against job estimates (standards). Experience also shows that over a week's worth of crew labor, the overall estimate of planned hours becomes extremely accurate, only off as much as 5% or less. The objective of planning at this point is to provide the supervisor and technicians useful information for helping control their work, not to produce a perfect estimate. The planner's estimate is entirely adequate for this purpose.

Is there a better way to produce a more accurate estimate for routine maintenance work? Perhaps not. This *is the nature* of maintenance work. Maintenance work is not assembly line repetitive task work. Anyone that has ever worked on their own car in their driveway can attest to this. Has anyone ever planned to do a simple task under the hood for a couple of hours on a Saturday morning and still been engaged after dinner? How about planning to take all day the next Saturday to do something else, but for some reason the job is finished before lunch? This is the nature of maintenance where different technicians are performing different tasks all the time on a very wide range of equipment. Thus, the simple labor estimate of a planner with significant craft experience is preferred over methods that are more complex simply because it yields an estimate that is as accurate as possible. Nevertheless, as a planning program progresses in experience over time, it does produce procedures and checklists that help better standardize and define specific maintenance tasks. These "evolved" job plans help provide better consistency in maintenance practice and time expended, lowering the uncertainty of the accuracy of individual estimates.

Another concern regarding the expertise of the planner involves skills outside the normal experience of the planner. Some jobs require crafts outside the background of the planner. An example might be a requirement of electrical work on a mostly mechanical job. The mechanical planner has several options. The planner might ask an electrical planner for input. If there is no established planner for electrical work, the planner might also consult an electrical technician or leave it up to the electrical craft supervisor to coordinate the electrical input at the time of work assignment. The planner might also be able to provide basic file information from previous jobs that might be helpful to the electricians. A mechanical planner might even have difficulty planning certain mechanical tasks. Many pieces of equipment have become so specialized that not all technicians within the same craft might be familiar with them. In these cases the planner simply consults with the specialists who have knowledge. The planner attempts to provide useful information regarding scope, schedule, and file data even on these jobs to help the later scheduling and execution efforts. In certain plants planners may become specialists in planning different work and do not attempt to plan all the jobs. Jobs requiring the expertise of another planner are referred appropriately.

Two final concerns regarding planner training include maintaining a planner's craft skills and developing skills in specialized planning techniques. First, experience has shown that a planner retains practical knowledge of craft skills even when not applying them in the field. This is because of the close association to the actual maintenance through the planning duties. These planning duties allow the planner continually to develop strategies for jobs and review feedback from actual execution. The planner also spends significant time in the field talking to technicians and supervisors. Second, there are formal courses available for training planners in planning techniques, but on-the-job training provides the most effective training of planners. An experienced planner guides the new planner through the processes. The first planning principle to keep the planners in a separate group together facilitates this learning.

Illustrations

The following illustrations demonstrate this principle of planning. The first section shows problems occurring as a result of not following the principle. The second section shows success through application of the principle.

Not this way.   The planner sat down to estimate ten jobs. Lynn was by classification an apprentice who had completed all of the requirements necessary for promotion to technician and was waiting for a technician job to become available. He had been one of the few persons interested in the job as planner when it became available. The first job was a pump alignment. He had been trained and done several alignments, but never on a pump of this size. He looked in the file and was able to find a previous alignment work order for this very pump. The previous work order had estimated 10 hours for the task and the actual field technician had reported taking 10 hours. Lynn therefore used 10 hours as the job estimate. The second job required rebuilding a fan and there was no previous information available. Fortunately, Lynn had personally been involved in two rebuilds of either this same fan or its redundant spare nearby in the same service. He felt very confident that the job should take two persons a total of 2 days. However, just in case something came up, Lynn put an extra half day into the estimate. Lynn continued to estimate times for the remainder of the jobs.

Later the mechanical supervisor who was about to assign several of the jobs looked at the pump alignment and fan rebuild work orders. Brittany had not had a chance to see the jobs in the field and was inclined to accept the estimate of the planner who had. Still she wondered why the alignment procedure should take so long.

The technician received the pump alignment work order and knew right away that the alignment would only take 4 or 5 hours. Dana decided she would spend the morning setting up for the job and complete it in the afternoon. That would ensure a quality job. After completing the alignment, she reported to her supervisor an hour before the shift ended. The job had only taken 9 hours instead of the estimated 10.

Meanwhile, Scott and Fred had received the fan rebuild assignment. Surprisingly, the total job lasted exactly $2\frac{1}{2}$ days as estimated even though there had been several unexpected delays. Fred had been temporarily reassigned for several hours at one point. One bearing had also been damaged beyond repair and a new one had been obtained from inventory.

Several days later Lynn received the completed work orders for both jobs for filing. The alignment had only taken 9 hours Lynn observed and the fan rebuild had apparently gone off exactly as planned since no unusual feedback was reported.

This way.   The planner sat down to estimate ten jobs. Lynn had been a certified mechanic with over 15 years of experience. He had competed for the job of planner when it became available since it was a promotion. Lynn had been able to pass the test and interviews successfully. The first job was a pump alignment. He had aligned most of the pumps in the plant in his 15 years including this one. He looked in the file and was able to find a previous alignment work order for this very pump. The previous work order had estimated 10 hours for the task and the actual field technician had reported taking 10 hours. There did not seem to be any unusual reasons the alignment had taken so long for the last person. Lynn thought that most good mechanics ought to be able to align the pump in about 5 hours. Lynn used 5 hours for the estimate. The second job required rebuilding a fan and there was no previous information available. Fortunately, Lynn had personally been involved in two rebuilds of either this same fan or its redundant spare nearby in the same service. He felt very confident that the job should take two persons a total of 2 days. Lynn used that for the estimate. Lynn continued to estimate times for the remainder of the jobs.

Later the mechanical supervisor who later was about to assign several of the jobs looked at the pump alignment and fan rebuild work orders. Brittany had not had a chance to see the jobs in the field and was inclined to accept the estimate of the planner who had. She had confidence in Lynn's ability to estimate the jobs.

The technician received the pump alignment work order and knew right away that the alignment would take 4 or 5 hours. Dana spent the morning setting up and aligning the pump. No unusual delays came up and she reported to her supervisor an hour after lunch. The job had taken 6 hours instead of the estimated 5.

Meanwhile, Scott and Fred had received the fan rebuild assignment. The total job had run over about a half day because there had been several unexpected delays. Fred had been temporarily reassigned for several hours at one point. One bearing had also been damaged beyond repair and a new one had been obtained from inventory. Scott, the lead technician, carefully explained the delays on the work order after the job was completed.

Several days later Lynn received the completed work orders for both jobs for filing. The alignment had taken an extra hour Lynn observed and the fan rebuild had run into problems according to the feedback. An extra hour shorter or longer was not unusual nor was a problem for most jobs since estimating was not an exact science. The bearing damage was a concern, however, and Lynn knew that it would be advisable either to have the bearing inventory number available or stage the bearing the next time the crew rebuilt the fan.

The experience of the planners makes a big difference in the success of planning. Planners must have the skills of a top level technician to create timely, useful estimates necessary for increasing labor productivity.

This discussion has concentrated chiefly on the general scope and time estimates of the job plans. The following principle addresses the specific content of the job plans regarding maintenance procedures and specific details. Although top level technicians should be utilized for planners, there is still a great reliance on the craft skills. The utilization of superior skilled planners does not mean that unskilled technicians are acceptable in the work force.

Principle 5: Recognize the Skill of the Crafts

Planning Principle 5 (Fig. 2.8) states

> The Planning Department recognizes the skill of the crafts. In general, the planner's responsibility is "what" before "how." The planner determines the scope of the work request including clarification of the originator's intent where necessary. The planner then plans the general strategy of the work (such as repair or replace) and includes a preliminary procedure if there is not one already in the file. The craft technicians use their expertise to make the specified repair or replacement. The planners and technicians work together over repeated jobs to develop better procedures and checklists.

Principle 5

Plans Recognize the

Skill of the Crafts

♦ **What, Why, before How**

♦ **Evolving Standard Plans**

♦ **Coordination of Engineering**

Figure 2.8 The planning department's guidelines on level of detail.

This principle dictates that planners count on the workforce being sufficiently skilled so that the planners can get all the work planned through putting a minimum level of detail into initial job plans. Strict adherence to the job plan is not required of technicians as long as feedback is received at job completion.

The crossroads encountered regarding this principle is primarily a choice between producing highly detailed job plans for minimally skilled crafts or producing less detailed job plans for highly trained crafts. An associated issue involves whether all the work should be planned or are there only certain jobs that would benefit from planning. Another issue is whether strict adherence to a job plan by the technicians is required. The resolution of these questions regards considering the company's desire for productivity and quality.

Planning promotes productivity by examining work for potential delays and scheduling work. Planning and scheduling more work increases labor productivity. Nearly all work has potential for delays and learning from past history and so most work merits planning attention. The plant has better control over work that is scheduled and so most work merits some schedule control. To assist the plant in completing work, planners need to plan most of the plant's work. Planners have to be careful not to put so much detail in each plan that they cannot plan most of the work. A general strategy for 100% of the work hours is preferable to developing a detailed plan for only 20% of the work hours. How much detail should planners put into plans? If there is a procedure already in the file or the persons who worked on the equipment previously wrote down some things that are important, the planner should include those items in the work package. If no file information exists, planners might only have time to develop a rudimentary procedure, in some cases only one or two steps. The planners must respect that the craftpersons know how to work within their expertise. The planner may in a sense develop a "performance spec" on certain jobs. That is, the plan describes the intent of what needs to be done, not necessarily how best to accomplish it.

In addition, there are frequently different ways to do the same job. Classical industrial engineering sets forth that there is one best way to perform each job. However, engineered standards help productivity for jobs that are repeated twice per day, not twice per year or less. Maintenance planning seeks more to avoid past delays and provide scope and scheduling assistance than to minutely examine each welder's technique on any individual job. In addition, individuals sometimes have perfected their individual methods of accomplishing some routine tasks. Requiring a technician to perform a particular minor task in a way less familiar, though not necessarily superior, may lead to lower quality simply from unfamiliarity. It is the supervisor's job to help promote good work practices, not the planner's job to dictate consistency among them every time.

Even when including a minimum of detail on an initial job plan, the planner must be cognizant to include certain information. First, a planner should include information as to "why" the planner chose a certain job strategy, especially when the file history helped make the decision. For example, "This valve is being replaced since patching it in the past has not worked well" (the planner knows the file history). The technician needs this information to avoid making unwise field decisions. A planner at one company reviewed the history file and recommended a valve be replaced because of past unsuccessful repair attempts. The planner did not mention the history leading to the replacement decision on the job plan. Consequently, when the technician finished the job, he returned the completed work order with the following feedback, "I saved the company money by repairing the valve instead of replacing it." Second, the planner should include known legal or regulatory requirements if adherence to a particular procedure is necessary and not commonly known by technicians.

Nevertheless, the plant is also vitally interested in the quality of maintenance. (Remember Chap. 1's admonition that effectiveness comes before efficiency.) Jack Nicholas (2005), a long-time proponent of procedures-based maintenance, reminds us that we cannot always count on sufficiently trained craftpersons.

Nicholas reported that the U.S. Navy moved to a procedures-based maintenance effort after it found causes of some disasters rooted in relying too heavily on the field technicians. In particular, the military moves persons from base to base and duties change upon promotion. "Word of mouth and on-the-job training were simply too unreliable to ensure safety and consistency in maintenance practices." And there was not enough time for formal training to train sufficiently. In the submarine program, maintenance policy required detailed procedures for critical equipment and safety systems. The policy allowed "skill-of-the-craft maintenance practices" for less essential systems. The submarine programs in the 1970s and surface ships in the 1980s greatly reduced their infant mortality failures of equipment, very likely from maintenance and operations personnel having and complying to detailed procedures. They repaired equipment correctly "the first time."

Nicholas also reported that after the U. S. Nuclear Regulatory Commission mandated that nuclear powered electric stations use detailed procedures and checklists, overall reliability and capacity improved.

Keith Young (2001), an associate of Jack Nicholas at Maintenance Quality Systems LLC, very neatly sums up the level of detail versus craft skill trade-off with several statements. "As a general rule, the greater potential consequences of performing the task incorrectly, the greater degree of formality required in the execution of the procedure." "The higher the level of knowledge of the craft which the worker possesses, the less detail required within the procedure. The 'skill of the craft' is a necessary starting point from which procedures grow, however the requisite 'skills' should be defined and not left to chance."

If detailed procedures and checklists contribute to better reliability, performance, and safety, how does a company move toward this objective? The preferred method to become a procedures-based organization might be for a company to assemble teams of experienced technicians, planners, engineers, and even vendors to create detailed procedures for most of the common maintenance tasks on critical equipment and safety equipment. In lieu of using crew technicians, the plant could bring back recent (or soon to be retired) technicians for the express purpose of developing procedures. A facilitator familiar with the standard format for proper procedures and other plant and personnel characteristics might assist with much of the paperwork or administrative tasks of such an effort.

Although such efforts by a plant are desirable, the *Maintenance Planning and Scheduling Handbook* presumes a typical plant situation wherein the planners have no such help and much of the large amount of work to plan is reactive. In this scenario, the planning department establishes a system that utilizes the skill of the crafts to improve job plans and their procedures over time.

First, within a constraint of planning most of the jobs, the planner chooses where to put more time. The planner should give more attention to critical equipment and safety areas. The planner should help describe job circumstances where misunderstandings may be common. Other jobs perceived as less complicated or on insignificant systems might get only minimal planner attention. (Chapter 4 deals much more extensively with the issue of planning reactive work versus more proactive work of lesser urgency, but more plant importance.)

Second, the planner calls for a minimum craft skill on a job plan. The plan dictates the skill set necessary to accomplish the work given the state of the job plan.

Third, having planners with experienced craft backgrounds and skills, allows a planner to leverage that expertise into job plans to make helpful initial job plans. And remember, this is only for initial job plans. There may be a procedure already in the file or at least, important information that persons who worked on the equipment previously recorded. The planner would include those filed items in the planned package.

Fourth, the plant does strive to maintain technician skills with proper hiring and training practices. At some level, planners must count on the skill of the crafts. Supervisors must shore up technicians with deficient skills rather than the planners planning every job for the lowest skill level of the workforce.

Fifth, the plant counts on experienced supervisors to help ensure that field technicians get proper support and guidance.

Finally, the planning system counts on the skilled technicians giving feedback on job plans so that their expertise and the planner's expertise both contribute to adding information to future plan.

Thus, the planning system is designed to move toward a procedures–based organization over time by a systematic method of improving job plans. In the case of larger, more complex jobs such as a large pump overhaul, the planners and supervisors should work together if a carefully prepared plan cannot be developed in advance. The plan should help guide the technicians toward relevant O&M manuals and other references. At the conclusion of such large tasks, the supervisors should allow the technicians to take time to assemble important information together to form the basis of a more standard job plan.

Planning concentrates on adding value. Before there was a planning function in existence, the technicians had to decide how to accomplish work requests. Planning does not take over this function, but rather adds a new function of value. The planners give the technician a head start from scoping the field situation and reviewing the history file review, and planners give the supervisor information for scheduling control. The planners also use their own expertise with that of the technicians to formalize best practices on individual job plans.

There are different schools of thought for ownership of work orders. In one, technicians must execute jobs exactly as planned. In the other, there is freedom for technicians to change the plan.

One school believes that technicians must execute the job precisely as planned for two reasons. One reason is that the planner had access to the necessary information including specifications, history, and engineering to develop the proper job plan. Any recommended deviations from the job plan must be approved by planning before execution. A second reason is that restriction promotes consistency of job execution. A third reason is that restricting execution to the plan ensures reliable history records without having to count on accurate job feedback. One can sometimes recognize this school by work order forms or computer systems that have limited or no space for reporting job feedback. An example area where this may be appropriate would be an automobile repair shop. One would like to approve any work done to one's car before it is begun. This type of arrangement normally has a larger planning staff because of the iterations sometimes necessary before a job can commence. So in this school where the planning department essentially owns all jobs throughout the work process, a more substantial planning investment is required and less emphasis is placed on technician competency for determining the job scope and procedure for execution.

On the other hand, this book follows the other school of thought. While a methodology of strict adherence to job plans may be necessary for some industries (nuclear power comes to mind), it could be counter-productive. In the first place, planners do not possess perfect information. Without an internal inspection, planners "guess" at what is wrong with equipment based on file information and personal experience. Frequently, they ask others for opinions. Nevertheless, they do not "know." They plan the job around their guess and let the technician go to work. The technician later finds out the real source of trouble and must modify the plan based on actual conditions encountered. In the second place, planners do not have the time to review these technician reasons for changing the job plan for improvement during the course of each job while in progress. The technicians and their supervisors must take ownership of the job after they receive it to make necessary decisions. Finally, the planners simply do not have the time to develop a detailed step-by-step procedure for every job the first time. As discussed previously, the planner must divide time in favor of jobs on more critical equipment when there is time to add more details to job plans.

The first school of thought keeps ownership in the hands of the planners for control of the work. On the other hand, the second school of thought keeps moving the ownership of the job to the current holder of the work order. This second school is the accepted model for this book. When the job is being planned, the planner owns and controls the work order. Later, after assignment, the field technician owns the work order and is responsible for it. When the job is being planned, the planner uses field scoping, file information, and personal experience to develop a good job scope with the best procedure possible under the circumstances and pressure of time. Planning gives the technician a head start. Scheduling gives the technician a time requirement. When the technicians receive the work order in the field, it is their job. They own it. The technician is part of a team in the process, however, and this process requires good feedback for file history to help future work. In the second school adopted by this book where the ownership actually passes to the field technicians, a leaner planning effort requires competent field technicians. A higher reliance is made on receiving good feedback to make history records accurate and allow avoiding future job delays. After planning the job, the planner no longer "owns" the job. The planner gives the technician a head start on the job, but the technician now "owns" the job. After beginning the job, the technicians are free to accomplish the job scope as they see fit. They may have a closer intimacy with the job than the planner had time to develop. The technicians must give feedback on any job changes or delays encountered so that future plans can benefit.

Keep in mind that recognizing the skill of the technicians does not mean that "anything will do." The principle requires that skilled technicians employ appropriate practices for their craft specialty. This may involve their being able to follow a provided technical manual correctly. A certified welder will know how to perform simple weld heat treatment.

A skilled mechanic will be able to follow and perhaps improve upon a guide to rebuild a boiler feed pump. The planner may have to provide particular standards for particular jobs such as unusual safety precautions or machine tolerances. Warren Riggs (1995) of Eastman Kodak correctly notes that empowerment is in direct conflict with standards and that some jobs require standards.

This ownership passing to technicians could be a stumbling block for the planning group that feels they "own" all the jobs from start to finish and are responsible for making sure the crews execute the jobs properly. Explicit advice is necessary for these planners to reorient their thinking to the team concept.

Once planning accepts this principle, planned coverage can take a big leap as shown in Fig. 2.9. Planned coverage is the percentage of all work hours spent on planned jobs. One hundred percent planned coverage would indicate that the company spends all labor hours on work assignments only on planned jobs. Fifty percent planned coverage would indicate that the company spends half of the labor hours on planned work. This company was able to move from work crews spending only about 45% of their work hours on planned work to about 65% of the work. The company made the improvement simply by changing its approach to allow more dependency on the skill of the technicians. Planners were able to plan more work for the crews by spending less time specifying unnecessary details. Craft satisfaction with the work plans also increased as technicians felt more responsible for making decisions in line with their craft expertise and more a part of the system for helping improve the job plans. Not only were planners planning more of the work, but also they were no longer insulting the technicians with excessive details.

Two final issues include technicians desiring the extra detail every time and jobs requiring engineering.

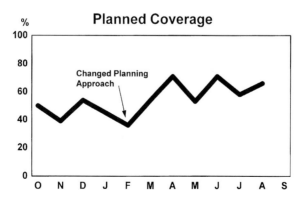

Figure 2.9   Putting less detail on initial plans.

A sensitive area for an existing planning group is when the previous management direction and craft expectation had been having total information on every plan. It does not matter that the planning group was only planning about one in five jobs. The plans that they did complete were often favored by some technicians. The less skilled technicians may complain when fewer details start to appear on job plans. Communication and management commitment to the program must focus here on the purpose of planning. One of the problems is that unless they are informed, technicians and supervisors either may not understand how helpful a simple job plan is or may not understand the cycle of improvement. A simple job plan may have a good job scope, craft identification, and time estimates along with a knowledge of previous job delays to avoid. The supervisor must accept the responsibility to assist weaker technicians on certain jobs.

The crew supervisor still has an option regarding work plans deemed unsatisfactory. The supervisors can return job plans to planning for additional detail or information as long as they have not yet assigned the work. Once the work has been assigned or has commenced, the crew owns it and must resolve problems and give appropriate feedback to improve future planning efforts.

Engineering assistance also merits some comment. Planners should plan work within their level of expertise. Planners should recognize, but not become bogged down with design considerations beyond their expertise. The planner is responsible for coordinating work requests to plant engineering where appropriate. The planner still owns the job at this point and should request a quick turnaround of answers to routine questions. If the questions point to an extended effort on the part of engineering, the planner should take other steps. The planner should formally assign the work order to the engineering group or otherwise request that a project be initiated. A few plants have an engineer assigned under the planning supervisor to provide easy access to engineering support. This engineer would answer uncomplicated questions and coordinate questions requiring more extensive research or determination. Utilize caution when mixing a staff engineer into the production environment of the planning department. The planners must not become staff assistants to the engineer gathering file information. Planners must not become distracted from their planning chores.

The bottom line is that planners must try to plan all the work. Not planning all the work allows some jobs to go into the field without the benefit of *any* field scoping or file review. In this case, planners must put fewer details into some or all of the job plans. Planners might choose jobs on noncritical equipment to rely more heavily on skilled craftpersons and write fewer procedural job steps so that they have relatively more time for jobs on more critical equipment. If the planners are planning all of the jobs, they can put more details into some of them.

Illustrations

The following illustrations demonstrate this principle of planning. The first section shows problems occurring as a result of not following the principle. The second section shows success through application of the principle.

Not this way.   Typically it seemed the crews worked only about one out of five jobs on a planned basis. This distressed Hosea, the supervisor of planning. The problem was not so much that the supervisors did not want the planned work, but that planning simply could not get to the jobs before the crew had run out of planned work. In these cases the crew naturally turned its attention to the unplanned work backlog. There were ample planners. There were five planners for only 100 technicians. The planners were busy as well. The planners continually worked to provide very detailed procedures on every plan.

The problem with the crews working unplanned work was that they were simply not able to take advantage of parts lists or other information the planners had available from past work. Supervisors also had inadequate information to control schedules. That brought up another problem. With the planners being so busy, they were not filing all of the completed work orders. So even on planned jobs, the files were not as helpful as they might be.

There were also some indications that particular members of some of the crews thought planning was a "waste of time," in their words. Hosea had talked to one electrician who told him flat out that he did not need to be told how to run a conduit. This electrician had felt irritated at the thought that he had to be baby sat.

One of the planners had also expressed irritation recently, but not for the same reason. This planner was upset that the crew supervisor had not taken the plan's advice to rewind a motor in-house. Instead the supervisor had agreed with the technician to send the motor out to a local motor shop. The planner wanted to know why the supervisor did not understand that in-house work could provide better quality. The planner asked if Hosea would bring the matter to the plant manager to resolve.

This way.   Typically it seemed the crews worked about four out of five jobs on a planned basis. This was acceptable to Hosea, the supervisor of planning. The problem was not so much that the supervisors did not want the planned work, but that sometimes the supervisors directed technicians to unplanned work. The unplanned work was pressing and did not appear to require much planning. Hosea knew that after becoming more used to planning, they would want even more of their jobs reviewed by planners before starting them. There were ample planners. There were five planners for only 100 technicians. The planners were busy as well. The planners continually worked to provide adequate job scopes, time and craft estimates, file parts information, and other notes to help avoid previous job delays. The planners were able to provide planning for all the work orders that the supervisors had not immediately written up and started themselves.

The advantage of the crews working mostly planned work was that they were able to take advantage of parts lists or other information the planners had available from past work. Supervisors also had adequate information to control schedules. The planners were busy, but still filed all of the completed work orders. So to improve all of the planned jobs, the files were becoming ever much more helpful.

There were still a few technicians that did not understand how helpful the scoping and file information were to them or the scheduling information was to their supervisor. Some technicians thought that without an extremely detailed, step-by-step procedure, planning was a waste of time, in their words. Hosea had talked to one electrician who told him he did not receive a diagram on how to run some field conduit. Hosea carefully explained to the technician that the planner had considered this to be a field decision. On the other hand, the planner had reserved 60 feet of conduit to avoid a parts delay, enough to satisfy any layout.

The planners had accepted their roles of giving the technicians a head start and the planner duty carefully to save any feedback on actual job performance. One of the planners had recently received feedback that a plan to rewind a motor in-house had been contracted. The planner made sure to record the contract motor shop's address and warrantee information for the files. The planner also checked with the supervisor to see if future plans should consider such an option or if this was just a one-time event.

Planning provides the *what* and the technicians provide the *how* for many initial job plans. This ensures that the company best leverages the skill of the technicians. The company wants the technicians to do what they were trained to do. At the same time, this allows the planners to ensure planning all the work so that every job can have the benefit of advance planning. This principle presumes the company invests in the acquisition and training to produce and maintain a staff of skilled technicians. Planning gives skilled technicians a head start. However, the best quality and consistency of work comes from a procedures-based maintenance effort. Planners and technicians work together over repeated jobs to improve and formalize procedures.

Principle 6: Measure Performance with Work Sampling

Planning Principle 6 (Fig. 2.10) states

> Wrench time is the primary measure of workforce efficiency and of planning and scheduling effectiveness. Wrench time is the proportion of available-to-work time during which craft technicians are not being kept from

productively working on a job site by delays such as waiting for assignment, clearance, parts, tools, instructions, travel, coordination with other crafts, or equipment information. Work that is planned before assignment reduces unnecessary delays during jobs and work that is scheduled reduces delays between jobs.

Principle 6 ordains that measuring how much time craft technicians actually spend on the job site versus other activities such as obtaining parts or tools determines the effectiveness of the maintenance planning program. This principle holds that delays are not simply part of a technician's job and should be avoided. Figure 2.10 shows an example of the distribution of technician time. Only category 1 is productive time on the job. All of the other categories identify delay time.

The mind of management must resolve two crossroads considerations. (1) Does management have a specific mission for planning to keep technicians on job sites or does management have a more vague idea of planning somehow contributing to effectiveness? (2) Is working in a delay area such as obtaining parts or tools merely part of the job or is it a delay to be avoided? Does management's strategic vision involve moving technicians out of delay areas and onto job sites or does the vision only have technicians working hard to do everything?

The purpose of planning is to help put everyone on their tools in front of a job instead of traveling, waiting for parts, or otherwise being delayed. The purpose of planning does not include making sure persons are productively working once they are in front of a job and not being delayed. The issue of productively working once on a job is important, but it is not centrally associated with planning (other than the planner setting an informal time standard through the estimate). Nevertheless,

Principle 6

**Measure Planning Performance
by Analysis of Delays with
Work Sampling**

Figure 2.10   This company's time on the job is only 35%.

consider that whether or not time in front of a job is as productive as possible, simply increasing the proportion of time so spent by a workforce should increase the number of jobs completed by maintenance. That improvement is the purpose of planning. Similarly, planning is not concerned with administrative time spent for activities such as training, meetings, or vacation. Planning concerns itself with the time technicians do have available to work under the control of their supervisors.

Work sampling to determine wrench time gives this measure of how much planning helps. The time the employees are at their job sites working is called direct or productive work. At issue is not so much the time the technician spends doing productive work. What is actually important is the analysis of the nonproductive time. For example, how much time is spent waiting for parts; how much time for tools; how much time for instruction? If the technician is obtaining a part, instruction, or tools, the job is actually not progressing. Separate studies done over time indicate if planning is becoming better or worse with regard to reducing these delays. Has the time waiting for parts gone down; has time waiting for tools gone down; has time waiting for instruction gone down? Interestingly, measuring the technicians tells about the planning function, not the technicians. The planning tool should have an effect on the technicians.

The interesting thing about this principle is that it does not make planning work per se, it only measures how well planning is working. A company could believe in planning and successfully implement planning according to the other planning principles without ever conducting a wrench time study. Similarly an automobile could function flawlessly without a speedometer. Nonetheless, measuring wrench time does tell directly if the objectives of planning are being met. The objectives of planning are to reduce delay times and put technicians on their tools. Measuring wrench time thus also gives an overall indication of how well the other principles have been implemented or accepted. The other principles must be in place for planning to succeed. Wrench time analysis is an indicator, not the control of planning or the work force. Chapter 11 deals exclusively with the control of planning.

While management might not use wrench time measurement to conduct or control planning, it might use it to demonstrate the need for planning. Maintenance planning effectively helps improve labor productivity exactly because there is such a great misunderstanding of the current level of direct work time. That is why analysts present the results of work sampling studies to management, supervisors, and technicians. The realization that delays consume over 70% of workforce time and direct work is less than 30% generates extremely beneficial dialogue toward accepting the concept of planning and productivity improvement. An important issue is that everyone understands while technicians are being paid by the hour to handle delays, the company

is not receiving any benefit from such activities. The company benefits when productive maintenance keeps equipment in service to make
a product for market. The company does not benefit from avoidable
activities that consume over 70% of its workforce labor hours. Such a
discussion time is a marvelous opportunity to explain that delays are
undesirable. The technicians view the results of the initial wrench time
studies as even more remarkable when they realize that during the
course of the study, they had made a special effort to be productive.
That means the observation effect of the study showed the results to
be even more confirming that at best the productivity had been less
than 30%.

Simply conducting a wrench time study to illustrate what planning is
all about and why the company employs technicians (to work on equipment) could be worth more than the results of any study itself. The measuring of wrench time does not yield planning improvement, it only
quantifies it. A properly structured planning system within a maintenance organization yields the improvement whether or not it is measured.

It is difficult to agree with industry claims that productive time could
possibly be so low without the results of a valid study. One supervisor
submitted a scenario showing how hard it would be for an employee to
try to have such a low wrench time. This supervisor showed a theoretical technician through an average day. The tech first took 30 minutes
to start going in the morning. During the course of the day the tech spent
45 minutes receiving instructions from the supervisors and 60 minutes
waiting at either the tool room or storeroom. 45 minutes were consumed
traveling. The tech took a total of 90 minutes in breaks and 30 extra minutes for lunch. The tech also took 90 minutes for showering and otherwise getting ready to go home at the end of the day. With all this wasted
time, the tech had only 210 minutes left out of the 10-hour shift for
work. This time arrangement netted the tech a 35% wrench time and
65% delay time. Incredible as it seems, the typical wrench time reported
in industry ranges between 25% and 35%. While some employees at each
plant are in more productive situations than others, studies show overall productivity measurements are in this range. A few minutes here and
there add up to a productivity problem with significant delays.

Wrench time is accurately measured with a properly structured, statistical observation study. The study sets up statistical procedures to
ensure proper observation techniques. Generally, a study conducts observations over several weeks or months to ensure a time period representative of the workforce's normal activities. An observer has a list of
maintenance employees at the plant each day of the study and has a
methodology for selecting a sample of employees to locate each half
hour or other time period. The first moment the observer locates a
selected employee, the observer categorizes the activity as a type of
work or delay. The observer does not merely follow an employee around

to gain observations. The observer also does not locate jobs instead of persons because some persons may not be even assigned to work. At the end of the study, the study reports the proportions of observations in each category. Appendices G and H present actual work sampling studies conducted at an electric utility.

Other less formal methods of measuring wrench time have been explored. One method has been to have several individuals in the work force carry special scorecards. A clerk pages these individuals at specified random times during the day. When a person's pager goes off, that person records the appropriate category on the scorecard. The problems with this method are several. First, there is not a single person deciding the appropriate category to use. Second, there tends to be great reluctance on the part of any but the most productive employee to participate and carry a scorecard. Third, this method requires extreme integrity on everyone's part instead of on a single observer. Fourth, there is also extreme "observation effect" in that the person being measured is continually aware of the ongoing measurement. As might be expected, studies using this method have recorded average wrench times about 20 to 25% higher than what a normal study would show on the same work force. That means when the actual work force wrench time was probably about 35%, there would be reports of 55 to 60%. On the other hand, studies such as this can often be conducted with good humor and effectiveness, not to find out wrench time, but to help educate the work force of the importance of direct work versus delay activities.

Similarly, efforts to have entire crews where everyone keeps track of their daily time in the different categories have resulted in reported wrench time hardly ever below 80%. These studies with everyone participating even if just to raise awareness are probably not a good idea. They seem to degenerate into a "liars' club," damaging the integrity of everyone and everything, including the wrench time concept. It is about impossible for an individual to keep track of the minute-to-minute delays that impact one's work on a continual basis. This factor combined with the often disbelief that wrench time could be fairly low anyway leads everyone to guess high. Consider this point applicable to work order or time sheet systems that expect everyone accurately to quantify all their delays during a job or time period.

Nearly everyone has apprehensions that conducting a wrench time study could be taken by supervisors and technicians in a mean-spirited way. That does not have to be the case. Communicate the reasons before, during, and after the study. After the study, report the results to everyone. It is difficult to imagine too many persons objecting to a program designed to boost productivity only to 55%. Also, after some studies workforces were able to demonstrate the need for new tool boxes, a better storeroom, and even go-carts. During the study consider using a familiar, agreeable person as the observer.

A further mention of administrative time is appropriate in the discussion of wrench time. The wrench time study observations do not include any employees not available for work. If employees are scheduled for training all day, those employees are not observed. This administrative time is time the company has decided to invest other than for immediate work. On the other hand, consider the implications about wrench time. Consider if employees are only available for work 80% of the time because of administrative time. A wrench time of only 35% is only a measure of the percentage of time available to work that the employee was directly working. The percentage of time paid that the employee was directly working was a mere 28% (35% × 80%). Looking at the cost to the company another way, say that the technician is paid $25 per hour. Because the employee is only working 28% of that time on the average, the company actually pays $89 for work that the employee accomplishes. This is why contracted repair persons charge a seemingly high rate for time spent at the company's location. The work force needs to understand its own high cost to the company and join forces with management to raise productive time and lower the rate of company labor cost. While planning can help with the productive portion of available time, the company cannot take the impact of the other administrative time lightly. The company must balance among providing competitive company benefits, investing in training, and making technicians available to work.

Illustrations

The following illustrations demonstrate this principle of planning. The first section shows problems occurring as a result of not following the principle. The second section shows success through application of the principle.

Not this way.   Management could not understand why reliability continued its slow decline. From discussions with the planning department, nothing seemed to be out of the ordinary. The crew supervisors claimed to have their hands full, but were able to stay on top of things.

This way.   Management could not understand at first why reliability continued its slow decline. From discussions with the planning department, nothing seemed to be out of the ordinary. The crew supervisors claimed to have their hands full, but were able to stay on top of things. However, from observing the general state of the work force, management suspected a lower than desirable productivity. Management had noticed lines at both the tool counters and storerooms. In addition, it appeared that breaks were somewhat excessive. Management decided that direct work time on the jobs needed to be improved and that meant there was a problem with the planning and scheduling process.

Planning has the responsibility to help move personnel onto jobs and out of delay situations. Even without making formal measurements, understanding this concept of wrench time as valuable time and delay time as waste leads to improvement. Properly conducted studies can quantify the direct work time, help educate the work force on the need for improvement, and demonstrate improvements. The wrench time is not so much a measure of the work force's performance, but that of the success of the leverage being employed by the planning process. Planning takes direct aim at reducing the causes of job delays.

Summary

So far the planning effort has mainly focused on making individual jobs ready to go by identifying and planning around potential delays. Consideration of six basic principles greatly boosts the planning program efforts toward success. Each principle resolves a crossroads decision that affects the planning effort. At each crossroads, the company has to make a decision regarding alternate ways to conduct planning. The decision the company makes regarding each situation determines the ultimate success of planning. Each principle presents the recommended solution to the crossroads. While a plant must incorporate or consider all of the planning principles to be successful, ignoring a single one can often spell the ineffectiveness of the entire planning effort.

The principles are having planning in a separate department, focusing on future work, having component level files, using planner expertise to create estimates, recognizing the skill of the crafts, and measuring planning performance with work sampling for technician direct work time. Having planners separate from the control of crew supervisors avoids the temptation of using planners for field work instead of for planning. Planners also need to avoid continually being interrupted to resolve problems for jobs already under way. Planners need to focus on future work not yet begun. Because most jobs are repetitive, file history can help technicians avoid previous problems encountered. Only when planning keeps a separate file for each piece of equipment is it practical to retrieve information when needed. Planners must possess the experience of top level technicians in order to scope jobs, utilize files, and estimate times adequately. Engineered standards or other sophisticated time estimating techniques are unnecessary to accomplish the specific objectives of maintenance planning. At the same time, craft technicians must also demonstrate considerable skill during job execution. Planners count on technician skill and the planners focus on providing adequate job scopes for first time jobs rather than on providing an abundance of job procedure details. During actual job execution, technicians decide how best to accomplish job scopes and later give adequate feedback for planner files. Finally, wrench time measures whether the objectives of planning are being met, that of reducing job delays.

So utilizing planned work packages increases the maintenance department's ability to complete work orders effectively, efficiently, and safely. With maintenance planning based on the six planning principles, will the planning effort "work"?

Here is what one utility discovered. They had only a marginal planning program. The planning department consisted of apprentices tasked with developing very detailed job plans on lower priority work orders. The crews worked very few of the planned jobs and primarily worked only on unplanned higher priority work as soon as operations wrote the work orders. With only this planning program under way, management commissioned a work sampling study. Wrench time was only 37% and an analysis of the delay areas indicated that the plant could do a better job with parts and tools. This was either symptomatic of tools and parts availability problems or planning problems, or both.

Considering this and other information, the company placed a renewed emphasis on planning. Management replaced the apprentices with technicians for planners. (However, there was no compensation program to make planner pay competitive. In fact, because the plant did not allow planners to work much overtime, the real pay of planners ended up lower than that for most field technicians.) The company also purchased separate hand tools for each craftperson to reduce sharing problems. The company also virtually doubled the number of parts categories carried by the storeroom to reduce ordering needs. A follow-up work sampling study revealed that wrench time was still at only 37%.

Since analysis of the last wrench time study showed travel time was at 22%, management purchased bicycles and golf carts to help reduce travel time. At the same time, however, management overhauled the planning program and adopted the six planning principles. The company took the planners out from under the control of the crews. The company encouraged the technicians not to seek planner assistance for problems on jobs already started. The company adopted an equipment numbering system to begin creating specific equipment files and filing by system ceased. The company again replaced the planning personnel. This time management selected technicians who had all passed the supervisors test, but were yet not promoted due to a lack of positions. These new planners managed to plan all the work though in many cases they provided less detailed job plans. With these principles in place, certainly planning would succeed. The third wrench study revealed only a 35% wrench time. See Fig. 2.11. How surprising since analysis showed travel time had dropped to 15%.

The analysis of this last study revealed a very interesting phenomenon. Large delay times did not exist for parts, tools, instructions, or

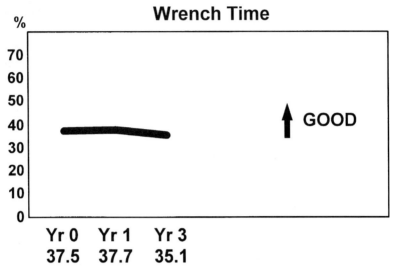

Figure 2.11   Different studies over time.

travel categories. Those were the areas that planning on individual jobs might help to avoid. Large delay times did exist for excessive startup, break, lunch, and shutdown categories. Despite these delay times, to their credit, the technicians had consistently been able to complete all the work assigned them.

Even so, a review of the wrench time for each hour of the day indicated a scenario of how technicians completed their work. When receiving their work for the day, the technicians would scope out the jobs and begin work intermingled with social time and some parts gathering. Then after lunch an incredible burst of activity would see all the work completed where upon the technicians could ease up until the end of the day. Over the years, supervisors had apparently become accustomed to how much work the crews could execute during a day and continued to assign that amount of work every day. The only problem was that now with several systems in place to allow doing more work, supervisors needed to assign more work. Obviously, management needed to consider scheduling of planned work in the planning picture. Maintenance needed some methodology to ensure assigning enough work. This leads to the next chapter on scheduling principles.

3

Scheduling Principles

Effective scheduling is inherent in effective planning. This chapter explains the reason why routine maintenance needs scheduling and then presents the principles of effective scheduling. Together, these principles create a framework for successful scheduling of planned maintenance work. Each principle sets guidelines on how maintenance should handle a different portion of the scheduling process.

Just as for planning, six principles greatly contribute to the overall success of scheduling. First, planners plan jobs for the lowest required skill levels. Second, the entire plant must respect the importance of schedules and job priorities. Third, crew supervisors forecast available work hours one week ahead by the highest skills available. Fourth, the schedule assigns planned work for every forecasted work hour available. And sixth, schedule compliance joins wrench time to provide the measure of scheduling effectiveness. Figure 3.1 shows the entire text of these principles.

Why Maintenance Does Not Assign
Enough Work

Aids such as planning good job scopes and having parts identified and ready make it easier to complete maintenance jobs, but they do not ensure that more work will be done. Adopting all six planning principles from Chap. 2 does not ensure that more work will be done. The reason why is because these aids and principles make it easier to complete individual jobs. That is, each job assigned should be easier to complete than it would have been without such help. If a particular job that used to take about 6 hours now takes 4 hours, that does not mean more work was done. Why? The simple reason is that still only a single job was done. Figure 3.2 explains that productivity cannot increase if supervisors do not assign additional work.

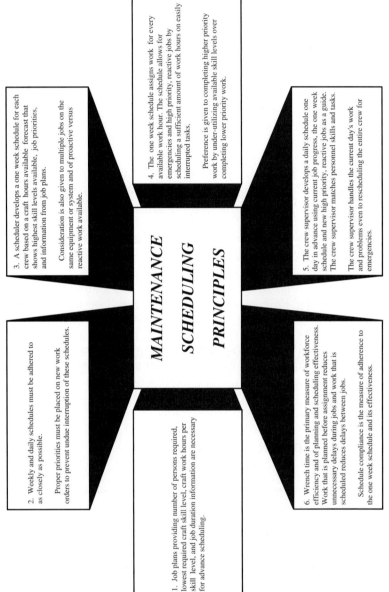

The following content appears within the figure:

1. Job plans providing number of persons required, lowest required craft skill level, craft work hours per skill level, and job duration information are necessary for advance scheduling.

2. Weekly and daily schedules must be adhered to as closely as possible.

Proper priorities must be placed on new work orders to prevent undue interruption of these schedules.

3. A scheduler develops a one week schedule for each crew based on a craft hours available forecast that shows highest skill levels available, job priorities, and information from job plans.

Consideration is also given to multiple jobs on the same equipment or system and of proactive versus reactive work available.

4. The one week schedule assigns work for every available work hour. The schedule allows for emergencies and high priority, reactive jobs by scheduling a sufficient amount of work hours on easily interrupted tasks.

Preference is given to completing higher priority work by under-utilizing available skill levels over completing lower priority work.

5. The crew supervisor develops a daily schedule one day in advance using current job progress, the one week schedule and new high priority, reactive jobs as a guide. The crew supervisor matches personnel skills and tasks.

The crew supervisor handles the current day's work and problems even to rescheduling the entire crew for emergencies.

6. Wrench time is the primary measure of workforce efficiency and of planning and scheduling effectiveness. Work that is planned before assignment reduces unnecessary delays during jobs and work that is scheduled reduces delays between jobs.

Schedule compliance is the measure of adherence to the one week schedule and its effectiveness.

MAINTENANCE SCHEDULING PRINCIPLES

Figure 3.1 The six maintenance scheduling principles.

Wrench Time Cannot
Improve If Crews Are Not
Given More Work

Figure 3.2   The reason planning includes scheduling.

Supervisors are typically responsible for assigning individual work orders to technicians, and there are a number of reasons why supervisors might usually assign an insufficient amount of work. In concert, these factors perpetuate a powerful culture to maintain the status quo. This is not a problem of the personalities of the supervisors. It is a system problem encouraged by how plant management has arranged the processes of maintenance.

First, crew supervisors develop a feel for how much work persons should complete in a day. During the past years that seasoned the supervisors, no planning function existed. The plant also may not have had an adequate storeroom, tools, or other resources now becoming available. It used to take all day for a few technicians to complete one or two work assignments. The technicians had to work hard and stay busy rounding up parts and tools. Frequently they had to clarify instructions and job scopes during job execution. They persevered and completed their one or two jobs. Now, however, with it easier to complete those one or two jobs, the maintenance supervisors may not be assigning more work. Habits are hard to change.

Perhaps the supervisors do assign more work. Perhaps they assign two or three jobs to the two technicians. The supervisors would thus feel very supportive of the company mission. But why two or three jobs? Why not four or five?

Now shift to explore another phenomenon. Consider a scheduled outage such as a major overhaul, sometimes called a turnaround. A maintenance schedule dictates the completion of certain jobs, often at certain times. This is true even for many short unscheduled outages for emergency repair. Everyone also shares a sense of urgency. The maintenance group completes a lot of work. Schedule pressure drives the outage. A consideration for doing quality work and doing the work right may alter the schedule, but the maintenance group still completes a lot of work in a short amount of time. However, that is not the phenomenon being considered here. After the outage, the crew supervisors know that they have just accomplished a lot. They have restored production capacity to full availability, and it is time to relax. What? The phenomenon encountered is that the supervisors may think they are rewarding their crews by not pushing for completing a lot of work *every* day. The supervisor thinks, "How could I expect to work my crew like dogs around the clock during such a critical time and then 'press them' the next day?" The supervisor may feel the outage where everyone works so hard justifies not working so hard later.

In addition, many supervisors feel that the company really does not have quite enough persons during an outage, but that during a regular, nonoutage work day it is a little overstaffed. The supervisor reasons incorrectly that the company has to carry extra persons so that it can be ready for the outages. This reasoning is faulty because there is much work that needs to be done on a normal work day for the competitive company. Outages exhaust maintenance personnel because crews work hard, but they always need to work hard to be competitive. One reason they can still work hard without an outage is that normally there should not be an inordinate amount of overtime when there is not an outage situation. Maintenance personnel can work hard for 40 hours each week without being too exhausted.

The crew supervisor may also feel that there is not enough work for the crews on nonoutage days because they are only working on the urgent or high visibility jobs. They may be ignoring the lower priority jobs to prevent future failures. The crews keep somewhat busy fixing those things that break or fail. The high priority jobs give an enormous sense of satisfaction because technicians can directly relate their completion to plant availability. The lower priority jobs' link to availability is less clear. Extra time exists (remember they can now do a 6-hour job in 4 hours) for performing other maintenance jobs to head off failures. Supervisors just do not seem to assign those lower priority tasks. To make this situation even worse, crews try to make the backlog of satisfying jobs last so that they do not run out of work.

A related practice is a technician receiving a single job assignment at a time with the understanding to come back for a second job when he or she finishes the first. Three things occur. First, the technician feels that the first job is *the* job for the day unless it is very obvious it should only take an hour or two. So nearly every job becomes an 8- or 10-hour job depending not on the job details but on the hourly shift duration. Second, the psychology of the arrangement encourages the technician to presume the next job is somehow a worse job. The fear of the unknown gives appreciation for the current job. "Why rush through it to go to the next job? In fact, I bet the next job is the worst job in the plant, shoveling out the boiler." Third, if the technician does return for the next job, the crew supervisor "cherry picks" through the backlogged work orders in the order of what is urgent and not necessarily by what is serious. If there is nothing urgent in the backlog, the supervisor may well assign the technician to help someone else on an urgent job currently in progress.

Similar to the manner in which many jobs are assigned or executed as 8- or 10-hour jobs, the practice of assigning two persons to each and every job may exist. True, many jobs require the safety consideration of an extra set of hands, but this practice could become a bad habit. Supervisors as well as planners might always assign two persons, needed or not.

Many of the circumstances just noted support a powerful counterproductive culture of peer pressure. Ample reason exists for not productively completing jobs quickly. Very little reason apparently presents itself otherwise. To try to counter this, many facilities do not even write on the technicians' copy of the work orders how many hours the jobs should take. These facilities fear the technicians will slow down if they know they can beat the time estimate. This is not a recommended practice. The technicians are part of the team and the time estimates help them understand the expectations of the job plan. Maintenance management needs a tool that helps supervisors know how much work to assign.

Thus planning is a maintenance manager's valuable tool. Having the estimates of how long a job should take and the number of persons of each skill required is a simple, overwhelming powerful addition to the situation. If a job plan expressly requires a single welder for only 4 hours, two persons for the entire day is obviously not acceptable. A planned estimate may have reduced a task otherwise consuming two persons, 10 hours each, to a 4-hour task. Real labor savings are available to assign elsewhere. Planning has introduced an element of accountability. This is not to say that the crew supervisors were intentionally mismanaging their resources, but planning provides a helpful tool to counteract the natural tendencies.

On the other hand, remember that only a single job has been completed. Even with individual jobs having time and personnel estimates, the proper application of planning provides an allocation of work for a period for the entire crew. This establishes crew accountability in the form of a check and balance system. The principles of scheduling implement this reasoning. Therefore, planning's primary task is not to provide advance information on parts and tools. The most vital application of planning gives the manager the necessary tool to manage how much work an entire maintenance crew should accomplish.

The utility at the end of Chap. 2 had planning without scheduling. Wrench time studies indicated that planning had freed time from earlier delay areas, but overall productive time did not increase. This was because the maintenance group did not assign more work.

Modern maintenance planning considers advance scheduling as an intricate part of planning. Scheduling is necessary for maintenance improvement. The basics of scheduling are centered on giving enough work to the crews to fill up the crews' forecast of work hours available.

Advance Scheduling Is An Allocation

The basics of scheduling involve giving enough work to employees to fill up a forecast of crew work hours available whether for a day or a week. *Advance scheduling* is actually more of an *allocation* of work and not a detailed *schedule* of exact personnel and time assignments (see Fig. 3.3).

Advance Schedule - Why?

✦Sets Goals
✦Ensures a Sufficient Amount of Work
✦Staging
✦Intercraft Coordination
✦Ensures Sufficient Proactive Work

Figure 3.3 Reasons why advance scheduling helps.

It is not a complicated process of making a complex schedule, but rather a simple process of making a list of jobs.

Advance scheduling provides enough work for an entire week *sets goals* for maximum utilization of available craft hours. It helps ensure assignment of a *sufficient amount* of work. Advance scheduling also helps ensure that *sufficient proactive work* to prevent breakdowns is assigned along with reactive work. It also allows more time to coordinate resources such as *intercraft notification* and *staging* of parts. There is also more time to coordinate doing all the work on a system once the operations group clears the system for maintenance.

The planning department can make the advance schedule. Creating the advance schedule in the planning department involves the serious responsibility of selecting the optimum mix of work for the best interest of both the short- and long-term operation of the plant. The scheduler might consult with an operations coordinator to achieve this optimum mix. The craft crews have the responsibility to execute and complete the selected work. This arrangement changes the perceived status quo of the decision previously made by the maintenance crew supervisor about what work maintenance should be performed. Now the scheduler decides what work maintenance should be performed, and the crew supervisor is responsible only for performing it. The crew supervisors see this check and balance system as an unnecessary loss of their control. However, the plant priority system that sets priorities for individual work orders remains the primary driver regarding the order in which crews begin different jobs. The schedule has merely provided the supervisors a service by reviewing the entire plant backlog of work and selecting enough work orders for the crews for the coming week. The supervisor no longer has to pick through an entire plant backlog each time to select individual work orders. The supervisor now has a small week's worth of backlog from which to choose.

The vision of planning is simply to increase labor productivity. The mission of planning is to prepare the jobs to increase labor productivity. The mission of scheduling is to allocate the jobs necessary for completion. Scheduling forms an integral part of planning. Just as outages benefit from having set schedules, routine maintenance benefits as well.

The following principles provide a framework to accomplish effective scheduling.

Principle 1: Plan For Lowest Required Skill Level

Scheduling Principle 1 (Fig. 3.4) states

> Job plans providing number of persons required, lowest required craft skill level, craft work hours per skill level, and job duration information are necessary for advance scheduling.

Maintenance cannot schedule work without some idea of the number of persons and time frames required. Maintenance job plans provide this information in a manner that allows the efficient scheduling of work.

Maintenance job plans first tell what craft specialties are required. Does a particular job require a welder, a painter, or both? Does the job require mechanics or machinists? Does the job require two mechanics or just one? Does the job require three helpers to assist a certified electrician? How many persons are required?

Consider a job that required a certified welder, but the job plan did not specify the number of persons or craft at all. The supervisor would be limited to assigning persons based solely on an interpretation of the job description. The supervisor might err in sending two mechanics to perform the work. In this case, both mechanics would later return to the supervisor explaining their need for welding assistance. Similarly, if a job requires a highly skilled, certified welder, the job plan cannot specify a mechanic with light structural welding abilities. The supervisor needs the information to assign enough welding expertise to the work order.

On the other hand, the essential part of Principle 1 is that job plans identify the lowest skill necessary to complete the work. By identifying the lowest skill necessary, the crew supervisor has even more capability when assigning individuals to execute each job plan. For example, the job plan should specify one mechanic and one helper if a job requires

Scheduling Principle 1

Plans with <u>Lowest</u> Required Skill Level

✦Identify Skills
✦# Persons, # Work Hours, Duration

Figure 3.4   Scheduling requires job plan information.

two persons, but only one needs to be a skilled mechanic. The job plan should not specify two mechanics in this case. The correct specification allows the supervisor who has only a single mechanic to assign the work, presuming the supervisor has other personnel that could be helpers. If the plan incorrectly required two mechanics, the supervisor could not assign the work. Consider a job that requires only light structural welding. The plan should not specify a highly skilled, certified welder. Specifying too high of a skill would severely restrict the supervisor who may see a backlog of mostly certified welding jobs but who may have only one certified welder. The supervisor may have several mechanics that were trained to do light welding. Job plans must specify the lowest qualified skill level to give the supervisors the most flexibility.

Another consideration is if a job could be done equally well with different combinations of persons and hours. Perhaps one person could do the job in 10 hours where two persons would require only 5 hours each. How should the planner plan the job? In these circumstances, the planner does not need to go to great lengths to determine the absolute optimum strategy. The planner's feel for the crew supervisor's preferences usually guides these decisions. The supervisor may normally work technicians in pairs or as individuals. However, the planner should not plan the job example just discussed for two persons with 10 hours each.

Job plans also specify the work hours for each craft skill and the total job duration hours. Work hours are not the same thing as job duration hours. Work hours normally differ from job duration hours for a job. Work hours are the individual labor hours required by each technician. Job duration is the straight calendar time the technicians work on the equipment. Each is necessary for scheduling. Consider a job requiring one mechanic and one helper for 5 hours each to rebuild a pump. The job duration is 5 hours, but the work hours total 10 hours. If the job plan called for an additional 5 hours afterward for painting the equipment, the work hours would total 15. There would be 5 hours each for the mechanic, the helper, and the painter. The job duration would be 10 hours since the painter would have to work after the pump was rebuilt.

The schedulers and crew supervisors need to know how many persons each work order requires and for how many hours each. The job plan specification of persons, craft skills, and labor hours gives this information. The schedulers and crew supervisors also need to know when to send or expect back the appropriate persons on each job. The job plan specification of job duration gives this information.

The operations group also needs to know the duration that equipment will be unavailable for production. The additional time necessary for the operations group to clear up or prepare a piece of equipment for maintenance activities or restore it to service are not included in the time estimates for individual jobs. The estimates are primarily for the use of the maintenance group to schedule maintenance resources. The operations group does their own allocation and arrangement of personnel. Advance coordination keeps technicians from sitting around waiting for the operations group to ready equipment.

For outages, the overall outage schedule addresses where the operations group requires time to prepare and restore equipment, but the estimates for individual job plans do not include this information.

Planners should avoid two common traps when estimating the job requirements on plans. One is always assigning two persons. The other trap is setting the time by using half or whole increments of a shift.

First, planners err when they always presume two technicians must work together. Legitimate reasons to assign more than a single person include operating downtime, task logistics, safety, training, and security. As discussed above, having more persons might decrease the time the operations group has to do without the equipment. Be careful though since this relationship is not always linear. Twice as many persons might not cut the time by as much as a half. Simple logistics might also necessitate extra persons. In this case, assigning extra persons saves time overall. For example, valves might be too heavy or awkward for a single person to manage even with mechanical aids. Two technicians might be able to do a certain job spending 2 hours each where a single technician would take 10 hours. Furthermore, the job might not require the extra person full-time. It might be possible to plan fewer hours for the assisting person than the primary person. In addition, some situations do require two persons for safety reasons. Even work not inherently dangerous might justify needing two persons if located in the midst of an industrial setting away from other personnel. Training is also an issue. The plant strategy might include technicians frequently working in pairs to cross-train each other on a regular basis. Instrument and control crews particularly work in pairs to avoid a problem of over-specialization. Nevertheless, planners should not always plan for more than a single person. The planners should work out expectations for necessary assignments with the crew supervisors.

Second, planners make a mistake if they always round off work hours to shift increments. For example, one might see most jobs requiring either 4 or 8 hours for crews that happen to work 8-hour shifts. Likewise, one might see most jobs requiring either 5 or 10 hours for crews that happen to work 10-hour shifts. This practice damages the scheduling effort. Many jobs require only a couple of hours and many jobs do not require an entire shift to complete. Consider a 2-hour job and a 6-hour job. Both of these jobs could be completed in a single 8-hour day. However, maintenance would incorrectly assign them if one job had been planned for 4 hours and the other for 8. In correct actual practice, planners plan jobs for their true expected time requirements. Then scheduling is able to fit jobs together to improve overall productivity.

Planners also frequently need to address other situations peculiar to specific jobs. These are not usually too difficult to handle. Perhaps insulation has to be removed and replaced. Perhaps the operations group could restore the pump to service before painting if painting could be done on-line. The important point to note is that both job duration and work hour estimates are necessary for scheduling work. The job plans provide this information.

One question that companies ask is whether plans or schedules consider a "high or low wrench time." Usually, job plans and schedules account for technicians having a high wrench time. Job plans do this because the plan time estimates do not allow for unanticipated delays. Moreover, the job plan attempts to avoid or minimize anticipated job delays that the planner feels could occur during individual jobs. Similarly, the weekly schedule attempts to minimize delays that could occur between individual jobs such as excessive idle time, break time, or assignment time. The weekly schedule does this by providing enough work that is ready to go so that crews do not have to waste time receiving new assignments. Because these planning and scheduling efforts aim to reduce delays, they also aim for relatively high wrench time. Remember that high wrench time consists of having technicians on jobs doing productive work rather than being in delay situations.

Illustrations

The following illustrations demonstrate this principle of scheduling. The first section shows problems occurring as a result of not following the principle. The second section shows success through application of the principle.

Not this way.   Paul planned five jobs during the morning before break. Each job required two technicians. The first job required replacing a high pressure steam valve and needed two certified welders for 10 hours each, an entire day. The second also required two certified welders to construct a work bench for the maintenance shop. Paul planned it to take 5 hours for each. The third job was a simple request to move several barrels of waste oil. He planned this job to take two mechanics with a forklift and barrel attachment only 2 hours. The fourth job required replacing a check valve. This was planned to take two certified welders 5 hours. The fifth job required working on a leaking critical control valve. Paul planned this job to require two mechanics an entire day. Before taking his break, Paul figured that he had already planned 64 labor hours' worth of work for the crew.

Later the crew supervisor began to assign work orders to various members of the crew. James had two certified welders, three mechanics, an electrician, and three mechanical apprentices. In addition to the other jobs available to work for the next day, the backlog included the five jobs Paul had planned. There was a significant quantity of mechanic work and, as usual, more work requiring certified welders than the crew had available. Frequently, James had to second-guess the planner and use the apprentice mechanics for some of the mechanic work.

This way.   Paul planned five jobs during the morning before break. Most of the jobs required two technicians. The first job required replacing a high pressure steam valve and needed one certified welder and a helper for 6 hours. The second also required welding to construct a work bench for the maintenance shop. Since mechanics could handle light structural welding, Paul planned it to take one mechanic and a helper 4 hours. The third job was a simple request to move several barrels of waste oil. He planned this job to take one helper alone with a forklift and barrel attachment only 2 hours. The fourth job required replacing a high pressure check valve. This was planned to take a certified welder and a helper 3 hours. The fifth job required working on a leaking critical control valve. Paul planned this job to require one mechanic and a helper 8 hours. Before taking his break, Paul figured that he had already planned 44 labor hours' worth of work for the crew.

Later the crew supervisor began to assign work orders to various members of the crew. James had two certified welders, three mechanics, an electrician, and three mechanical apprentices. In addition to the other jobs available to work for the next day, the backlog included the five jobs Paul had planned. James usually had confidence in the planner's estimate of skill required and knew when apprentices could be sent on jobs as helpers. James first assigned the certified welder and an apprentice to replace both the high pressure steam valve and the check valve in 1 day. James assigned a mechanic and an apprentice to the light structural welding for the work bench to help maintain the mechanic's welding skills. After assigning all the other work, there simply was no electrical work. Although not usually done, James decided to use the electrician as the helper to a mechanic on the critical leaking control valve.

As one can see, the planning function gives the crew supervisor or scheduler the craft skill and time requirements for scheduling work. A job plan tells how many persons the job requires and the minimum skill level.

By not unduly restricting the skill requirements, the planner increases the maintenance crew's flexibility for using different persons for the work.

Principle 2: Schedules and Job Priorities Are Important

Scheduling Principle 2 (Fig. 3.5) states

> Weekly and daily schedules must be adhered to as closely as possible. Proper priorities must be placed on new work orders to prevent undue interruption of these schedules.

The originator of a work order first picks an appropriate priority for the work based on established plant guidelines for setting work order priorities. Depending on the particular plant, the priority may then be reviewed and adjusted by the originator's supervisor, an operations coordinator, planners that code work orders, and at a daily meeting of plant managers or supervisors. The resulting priority should reflect not only the work's level of importance for achieving the plant's objectives but its importance relative to other backlogged work. Therefore, the plant priority system should play a large role in creating the schedule of the work the maintenance group will assign and complete. Management must treat the proper use of the priority system as a serious matter. The plant must expect maintenance crews to work on the jobs that the priority system through the schedule dictated. Management must treat working on scheduled work as a serious matter.

It might seem unnecessary to mention that schedules and job priorities are important, but they cannot be overlooked nor presumed. This is a common area of failure in maintenance management. Advance scheduling enough work for an entire week sets goals for maximum utilization of available craft hours. It helps ensure that a sufficient amount of work is assigned. Together with the priority system, it also helps ensure that the right work is assigned.

A significant source of inefficiency in the maintenance group is the interruption of low priority jobs when more urgent jobs arise. If a true emergency arises, it is always appropriate to delay another job. However, the maintenance group should recognize that interruptions on any particular job add extra time putting away tools, securing the job site, and

Scheduling Principle 2

Schedules Are Important

Job Priorities Are Important

Figure 3.5 Two essentials that management cannot overlook.

later refamiliarizing oneself with the job. An urgent job that is not an emergency should be worked as the next job rather than interrupt any job-in-progress. A nonurgent job should wait until the next day or week altogether so that the job can be scheduled into the overall priority of importance for the plant. Later, parts and tools might be staged to make executing the job more productive at a more appropriate time.

Jobs with priorities falsely set too high improperly interrupt work or cause work to begin without proper preparation. The end result is that the maintenance group completes less work overall. Then a vicious cycle begins. Higher priority work must interrupt lower priority work because there is not enough productivity to complete all the work plus the interruptions. Quite possibly, the maintenance group could complete all the work with more organizational discipline in setting initial job priorities. This would lower the incidence of job interruptions and lowered productivity. Management commitment is important in this area. Conscientious management attention to enforcing adherence to the priority system helps maintenance.

If everyone assigned a high priority to their work just to ensure its completion, then improperly prioritized jobs would also make it hard to recognize true instances of when schedules or work should be interrupted. They might delay starting true high priority jobs even if they did not interrupt them.

Setting false priorities is so serious that Appendix I thoroughly explores causes of false priorities.

In addition, inadequate confidence that crews will execute scheduled jobs hurts the staging program. Staging, as discussed in Chap. 6, can help increase crew productivity by having a job's planned parts and tools ready to go. They are already withdrawn from inventory or storage and ready for the technician to utilize. Planning stages the material before the anticipated execution of the job begins. Technicians avoid delay areas that they might otherwise encounter if they had to gather the parts themselves. Inadequate confidence that crews will execute scheduled jobs may discourage planners from staging parts. On the other hand, if planning continued to stage parts, the staging area might become overflowing with staged parts for jobs that did not start. In this case, the storeroom might run into stockouts for other jobs that maintenance chose instead to start. The stockouts might occur because of parts that were withdrawn for staging. These circumstances significantly diminish the potential for staged parts to expedite jobs.

Illustrations

The following illustrations demonstrate this principle of scheduling. The first section shows problems occurring as a result of not following the principle. The second section shows success through application of the principle.

This plant has the following priority system:

0 – Emergency.

1 – Urgent.

2 – Serious.

3 – Noncritical Maintenance on Production Equipment.

4 – Noncritical Maintenance on Nonproduction Equipment.

Not this way.   Mike finished his operator rounds and wrote work orders for problems he had noticed. Although most were not yet serious, Mike wanted to make sure maintenance completed them. Therefore he set a priority of 1 on the most important ones and 2 on the rest.

Nearly all the jobs in the maintenance backlog had been prioritized as 1's or 2's. They were either urgent or serious. This made it difficult for the crew supervisor, Abby, to select which jobs maintenance should work the next day. Abby selected all twelve priority-1 jobs and three priority-2 jobs to assign.

Near the beginning of the next day, the plant manager asked that Abby immediately assign a few technicians to correct a dripping flange on the installed backup feed pump. Abby interrupted two technicians on one of the priority-2 jobs. These technicians first hastily put their ongoing job in a state where they could leave it. Then with the operations group clearing the pump and themselves having to find suitable gasket material, they worked the rest of the day to replace the flange gasket and correct the leak.

This way.   Mike finished his operator rounds and wrote work orders for the problems he had noticed. Most were not yet serious and Mike set a priority of 3 or 4 on them. He set a priority of 2 on a couple of serious ones.

Mike's supervisor afterwards had Mike change the priority of both of the serious work orders. They changed one to priority 1 and the other to priority 3.

The backlog had work orders with a variety of priorities. Priorities ranged from 1 to 4. This made it fairly easy for the crew supervisor to select which jobs maintenance should work the next day. Abby selected all five priority-1 jobs and eight priority-2 jobs to assign. She also assigned two priority-3 jobs.

Near the beginning of the next day, the plant manager wrote a priority-2 work order for Abby's crew to correct a dripping flange on the installed backup feed pump. Planning went ahead to plan and stage the gasket material. Abby included the flange job with the assignments she was making for the following day. She was also able to assign most of the backlog priority-4 work orders as well. Abby requested the operations group to clear the pump in time for her crew to begin work on it the following morning.

The following morning two of the assigned technicians picked up the staged gasket material and began the flange work order. They completed it within a couple of hours and began another job.

Honoring the schedule.   The following additional illustration shows how an empowered crew supervisor can make a difference.

Victories are won on the shop floor everyday by honoring the schedule even when a new reactive work arises. The operator and the maintenance supervisor are teammates and both have to work together on this issue. Neither one must lie down and get rolled over.

At 8:00 in the morning, the maintenance supervisor answers the "emergency" phone call this way.

"Hello? ...'Send some mechanics right now?' Well, for an emergency, I certainly will, but can it wait until next week? Then the planner can plan the job and schedule it for next week's schedule. I'm already working on this week's schedule. We made a commitment to operations for the work we would try to accomplish this week to set a productivity goal. ...You didn't know about this job last Friday? That's okay. That's why we have operators to know when things happen. Nevertheless, do you think this job can wait until next week?"

"It can't? Well, can it wait until tomorrow? Then the planner can plan it and I'll work it into tomorrow's schedule. I've already assigned everyone on my crew enough work for today to ensure each person does a full day's work. That's our productivity key. I'd sure hate to start reassigning folk. Can it wait until tomorrow?"

"It can't? No problem. But can it at least wait until this afternoon? Then the planner can still plan it by looking in the equipment file to see what we did last time and make this job run smoother. Also, the planner can take a quick look at the job site and see if we need a special skill set. I'd hate to assign a mechanic if the job requires a certified welder. The planner can also estimate how long the job should last so I can coordinate this job with all the other work. Can it wait until this afternoon?"

"It can't? I understand. Well, how about if I start it at 10:00? A couple of mechanics already working on jobs now should finish about 10:00. Otherwise, interrupting a job-in-progress means spending extra time putting away parts and tools so they won't be lost and then later remembering what went where. Then no one's work gets done. Look, can this job wait until 10:00?"

"It *can*? That's great! Okay, 10:00 it is. Give me the work order number. ...What...Of course you have to write a work order for everything, even a 'come-in-the-middle-of-the-night' emergency. I guess if you radioed me from the field about a fire, I would enter a work order for you while

I was radioing my crew to scramble. But you're in the control room. Go ahead and call up the work order module, press 'insert,' and tell me the work order number. Then you can fill out the request while I go and tell the mechanics. ... Oh yes, we need the work order even if we don't plan or schedule it. This work order will allow the mechanics to record feedback. Inventory parts and anything else we learn about the job will be useful next time we work on this equipment. We don't want to reinvent the wheel for anything. Plus, you can't do any kind of equipment analysis if you don't collect the information during the year on work orders. ...Okay, got it. We'll take care of it."

"And listen, by the way, I don't mean to give you a hard time about this emergency and work order thing. I want you and anyone else to call me immediately anytime for an emergency or other problem. I'd be glad to reassign my entire crew at a moment's notice if I have to in order to handle an emergency. But if every week we drive seriously toward completing a week's worth of work, we can usually get everyone's work done in 2 or 3 weeks. And if we ever drift back into simply waiting for operators to call with urgent work, we tend to take care of just that work and then sit back on our heels feeling we've 'done our job.' Then productivity drops and anyone that wants anything done in a reasonable amount of time has to call and say that he has an urgent job."

The operator wins a victory insisting on assurance of the proper response to the true priority of the work. The supervisor wins a victory for keeping the crew working as productively as possible under the circumstances. The plant wins a victory furthering a culture of setting proper priorities and protecting schedules.

Maintenance should avoid interrupting scheduled jobs or jobs-in-progress. Maintenance should also place great importance on the plant following the plant priority system.

Principle 3: Schedule from Forecast of Highest Skills Available

Scheduling Principle 3 (Fig. 3.6) states

> A scheduler develops a one week schedule for each crew based on a craft hours available forecast that shows highest skill levels available, job priorities, and information from job plans. Consideration is also given to multiple jobs on the same equipment or system and of proactive versus reactive work available.

The first two principles set the prerequisites of scheduling. These next three principles introduce the concepts of the foundations of the advance scheduling process.

Principle 3 establishes a 1-week period as the advance schedule of allocation of work time frame. It also presumes that a person apart from

Scheduling Principle 3

Schedule from Forecast of <u>Highest</u> Skills Available

✦One Week

✦Consider Multiple Jobs on Same System

✦Consider Proactive Work

Figure 3.6 The basics of the advance schedule.

the crew supervisor will be the scheduler. The scheduler selects the week's worth of work from the overall plant backlog. The scheduler uses a forecast of the maximum capabilities of the crew for the coming week. The scheduler also uses priority and job plan information. The scheduling process also looks at performing all the work available for a system once maintenance begins work on that system. This includes proactive work.

First, the advance schedule selects a 1-week period for making an advance allocation of the work. Advance allocation means the schedule will select all the work that the crew should be able to finish in a single week. The schedule selects the work from the overall plant backlog. The scheduler does not assign the work orders to individual crew members. The scheduler also does not set specific hours or even certain days on which the work on each work order should start or end. The scheduler merely specifies a block of work as a list or package of work orders. *Advance scheduling* is an allocation of work for maintenance and not a detailed schedule of exact personnel and time slots.

A 1-week period strikes a balance between creating set goals and allowing for gradually changing plant needs. On one hand, a 1-week period is long enough to allow establishing a set block of work for a crew goal. This set block of work also allows planners enough time to stage parts for scheduled work. On the other hand, the plant is constantly writing new plant work orders. The new work orders gradually change the relative importance of all the work in the plant backlog. A 1-week period is short enough for the schedule normally not to need significant alteration due to this new work identification. This may be less true in a plant with more than a moderate amount of reactive work. These plants may normally experience a significant deviation from the set schedule. The 1-week schedule also covers a short enough time period to allow supervisors enough certainty in knowing which of their individual crew members will be available for work.

In addition, as discussed by planning Principle 4 in Chap. 2, a curious phenomenon appears regarding the accuracy of job estimates for individual work orders. Experience has shown that the best job estimates for individual work orders may be off by plus or minus as much as

100%. This is especially true of the smaller work orders that make up the bulk of many maintenance operations. Yet, the planner estimates are very accurate overall as the work horizon widens out to as much as a week. Over a week's worth of crew labor, the overall estimate planned hours becomes extremely accurate, only off by as much as 5% or less. That means that practically as many jobs run over as under due to the myriad of special circumstances surrounding individual work orders assigned to individual technicians on individual days. This confirms that a week is the appropriate allotted time period for advance scheduling. Remember that the objective of scheduling is not to produce accurate time estimates. It is to accomplish more work by reducing delays.

The scheduler publishes this schedule to give to maintenance crews, the operations group, and management. The crews receive the schedule as allocations of goals for the coming week. Supervisors of different maintenance crafts receive the schedules to have an idea of upcoming coordination needs. The operations group receives the schedules to have an idea of what equipment will eventually need clearing. The operations group may also be able to give the maintenance group timely advice of maintenance redirection needed. The operations group as well as management receives the schedule as an indication that maintenance is making progress on work orders. Many times, areas apart from the maintenance group see it as a "black hole" into which work orders enter, but never emerge. Tangible proof of work order schedules increases cooperation from the operations group.

Second, having a person separate from the crew supervisor allows a system of checks and balances. A person separate from the crew determines how much work the crew should be able to accomplish. The question is not necessarily: Which work orders should be done? The plant priority system drives that. The question is: How many work orders should the crew complete? The scheduler is best included as part of the planning department because this person uses planning as well as crew information. Many times it is appropriate for a supervisor of a planning group to perform the duties of scheduler. This allows the planning supervisor routinely to review job plans.

Third, the scheduler receives a labor forecast from each crew supervisor. This forecast tells how many labor hours each crew has for the next week. The scheduler needs this information. The scheduler intends to allocate hours of planned backlog on the basis of the labor hours available for each crew. The crew supervisors are in the best position to forecast the available labor hours on their crews. The crew supervisor may tell the scheduler that the crew will have 1000 labor hours for the next week. The scheduler then has a basis for knowing how many hours of planned work to allocate.

Fourth, the crew supervisor must make the labor hour forecast in terms of the highest skills available. By identifying the highest skills available, the scheduler has more latitude when actually determining which job plans could be executed the next week. Highest skills available means that if a crew has two certified machinists and seven mechanics available for the next week, the supervisor would not just forecast that the crew has nine persons or nine mechanics. The latter forecast would reduce the flexibility of the scheduler who would not be able to assign any complex machining jobs. The scheduler has more flexibility when knowing that there are two certified machinists. The scheduler can then assign complex machining jobs. The scheduler might also decide to assign routine mechanic jobs to the machinists. There is more freedom in what jobs can be assigned.

Fifth, the scheduler will use information from the individual job plans and a feel for the overall priority of plant systems. The scheduler looks at the priorities of the backlogged work to help select jobs. The scheduler looks at the labor hours planned to select enough jobs. Chapter 6 will discuss the actual steps the scheduler follows in this process.

Sixth, the scheduler also considers plant equipment and systems when selecting work. When selecting work for an entire week, the scheduler is able to group work orders for the same equipment. The scheduler may override some individual work order priorities to accomplish this. For example, a priority-2 and priority-3 work order may be both assigned because they are on the same piece of equipment. This might be preferred over assigning two priority-2 work orders on two separate pieces of equipment. Schedulers can also exercise flexibility by initiating certain preventive maintenance (PM) work orders early to take advantage of equipment downtime for other work. This allows improved overall efficiency because the operations group can clear the equipment a single time and the maintenance crew can work on a number of jobs together.

Finally, it is easier for the scheduler to include preventive maintenance or other work to head off failures on a weekly basis. On a daily basis, there is often sufficient justification to put off these seemingly lower priority work orders. On the other hand, when combining a week's worth of work, it becomes clear that PM cannot be delayed. The weekly schedule includes this type of work to encourage the supervisor not to put it off forever, 1 day at a time.

Illustrations

The following illustrations demonstrate this principle of scheduling. The first section shows problems occurring as a result of not following the principle. The second section shows success through application of the principle.

Not this way.   As maintenance manager, George felt that maintenance could increase its productivity. Lately, he had seen more and more technicians heading home early. This was a problem since reliability seemed to be slipping at the plant. He knew that there was a considerable backlog of work, but the supervisors had assured him that they were assigning as much work as the technicians could handle. George was also concerned that supervisors had a habit of putting off PM work orders.

George felt that advance scheduling of some sort was the answer, but the last attempt had been disastrous. Planning had first scheduled hour by hour what work maintenance should accomplish for an entire week. However, by the end of the very first day, the schedule was in shambles. Half of the scheduled jobs could not start at their target times because other jobs had run over their expected completion times. By the middle of the second day, the actual work-in-progress bore no resemblance whatsoever to what the advance schedule had predicted. At this point, the plant had abandoned the concept and gone back to assigning work 1 day at a time. George felt that now was the time to implement a gate carding procedure to make sure employees worked their entire shifts.

This way.   As maintenance manager, George felt that maintenance was increasing its productivity. Reliability seemed to be gaining at the plant. He knew that there was a manageable backlog of work and the scheduling process was helping the supervisors to assign as much work as the technicians could handle. George was also pleased that supervisors were not putting off PM work orders.

George felt that advance scheduling had been a great success. Planning had first developed a list of all the work orders that maintenance should accomplish for an entire week. The amount of work was determined by the labor hours that the crews would have for the week. At the end of the week, George discussed with each supervisor the results of what had actually been accomplished. Although no crew had completed all the allocated work, most crews had finished more work than they had thought possible. By the end of the second month, crews had a firm idea of the amount of work they were responsible for and were becoming more productive. As a result, maintenance crews were executing more work and the plant was increasing its reliability.

The proper period for an advance schedule is normally a single week. This time frame allows setting a goal that can stay relatively fixed as the plant continues to identify more work. The week's worth of work is not an hour-by-hour schedule of work orders, but a bulk allocation. The crew labor forecast is an important part of the scheduling process. Not only should the supervisors forecast how many labor hours are available, but how many in each specialty.

The following principle discusses two concepts relating how the scheduler compares the labor hours available with the planned hours in the backlog.

Principle 4: Schedule for Every Work Hour Available

Scheduling Principle 4 (Fig. 3.7) states

> The one week schedule assigns work for every available work hour. The schedule allows for emergencies and high priority, reactive jobs by scheduling a sufficient amount of work hours on easily interrupted tasks. Preference is given to completing higher priority work by under-utilizing available skill levels over completing lower priority work.

Principle 4 brings the previous scheduling principles together. The first part of this principle is that the scheduler assigns work plans for the crew to execute during the following week for 100% of the forecasted hours. This means that if a crew had 1000 labor hours available, the scheduler would give the crew 1000 hours worth of work to do.

Overassigning and underassigning work are also common and acceptable in industry. However, each causes unique problems that could be avoided.

For example, consider the case of assigning work for 120% of a crew's forecasted work hours. This would mean that the crew that had 1000 labor hours would receive 1200 hours of assigned work. This strategy may seem to be a way to provide enough work for the crew in case operators could not finish some of the jobs. It would also seem to be a way to encourage the crew to stay busy. This is because it sets a more ambitious goal for work completion. This strategy also creates several problems. It becomes difficult to gauge the performance of the crew. Maintenance management has a more difficult time comparing what the crew did accomplish to what it should have been able to do. This is because now there are three factors to compare: what labor the crew had available, what the crew was assigned, and what the crew actually accomplished. In the 100% case favored by this book, the first two factors are identical. The 120% method's three factors makes it more complicated for management to question a crew's performance. If a crew did not accomplish all its scheduled work, management would normally want to know why. However, management might be hesitant to question why a crew only

Scheduling Principle 4

Schedule for Every Work Hour Available

✦100%, not 120%, not 80%
✦Work Persons Down

Figure 3.7   How planned hours and forecasted hours become scheduled hours.

accomplished 1100 hours worth of work with the 1000 work hours it had available. Nearly any source of confusion in communication regarding crew performance is not in management's best interest. Management needs to lessen opportunities for misunderstandings whenever possible. In addition, maintenance coordination with plant operators and other crafts may be more difficult with the 120% arrangement. This is because there is less confidence that jobs will be worked.

Conversely, assigning work for only 80% of forecasted work hours may seem to provide a way to handle emergencies or other high priority work that may occur. However, the maintenance force is trying to eliminate emergencies altogether. Planning significant resources to handle emergencies that may or may not occur is counterproductive. It might also encourage work order originators to claim false emergencies knowing the availability of the resource. In reality, assigning work hours for 100% of a crew's forecasted work hours nearly always inherently includes some jobs that can be easily interrupted in case emergencies arise. A 100% scheduling strategy encourages originators to understand that for every emergency, other work is delayed. The 80% scheduling strategy also makes it difficult to gauge crew performance. Maintenance management also finds it difficult to ask a crew to improve if the crew completed all of its assigned work. A self-fulfilling prophecy is possible. Every week that emergencies do not occur, the crew might complete less work than possible. If the crew completes less work than possible, the work left undone might be work to head off emergencies. Consequently the plant experiences emergencies that justify leaving labor forces unscheduled each week. On the other hand, the 80% arrangement may be preferred in certain situations where maintenance crews must work within an overall time limit. Perhaps an outage with a critical time constraint might meet this criteria. The 80% arrangement might also be justified if the maintenance group has a particular credibility problem with the operations group. The maintenance group could publicize the work that it plans to accomplish and give regular reports to the operations group of its success.

Principle 4 prefers the 100% strategy primarily for accountability and clarity of communication. The 100% rule also keeps the crew busy accomplishing a practical goal. Maintenance handles any emergencies through interrupting jobs-in-progress. Maintenance management should not plan for regular emergencies in this regard.

The second part of this principle, "working persons down," is somewhat more subtle. On a major construction project requiring 20 welders and 20 helpers, the project would simply hire 20 welders and 20 helpers. However, in normal maintenance, the most beneficial jobs requiring completion rarely match the exact skill composition of the standing maintenance force. As a simple illustration, see Fig. 3.8. Consider a planned backlog consisting of 100 hours of high priority work requiring

"Working Persons Down"

Planned Backlog ①   Resource Forecast ③

✦100 Work Hours   ④
② of High Priority   ⟵⎯   ✦100 Hours of
Helper Work   Skilled Machinists
✦100 Work Hours of ⟋✗   Available
Low Priority
Skilled Machinist
Work

Figure 3.8 Doing work most profitable for the plant.

only helpers and 100 hours of low priority work requiring machinists. If there were only 100 hours of machinists available, then the plant should assign them all to the high priority work even though it requires only helpers. The principle has the scheduling process recognize that machinists can do helper work and allows assignment of persons to higher priority work in the plant. Otherwise, think of a not-so-extreme case where there was no machinist work in the backlog and machinists could not "work down." Would a company have high priority helper work sitting in the backlog and machinists sitting in the break room? This is a problem with the automatic scheduling logic of some computerized maintenance management system (CMMS) systems.

Consider what type of multicraft or work agreements are necessary to take advantage of the opportunities in this area.

See also how the note numbers in Fig. 3.8 illustrate the scheduling principles discussed so far. The backlog work is planned by lowest skill level (Principle 1). The backlog is ordered by priority or importance of work (Principle 2). The resources to work the jobs are forecasted by the highest skill level available (Principle 3). Principle 4 shows the correct assignment of technicians to jobs.

Craftpersons typically should not mind working outside of their primary specialties for work that is obviously in the best interest of the plant. It does become a source of resentment when the plant abuses the priority system. Consider management assigning a first class electrician to be a helper for a mechanic. If it is obvious that the mechanical work is much less important than backlogged electrical work, there is a problem.

Principle 4 establishes a methodology in the planning office to assign enough work. In addition, it is worthy to note what actually happens in the field on a day-to-day basis. Because many jobs run over or under, the crew supervisor frequently does not ever have to assign persons outside of their normal crafts. On a day-to-day basis, the supervisor is usually able to assign work from the weekly allocation by craft. There are more occasions where technicians may be used as helpers. For example, a job planned for one mechanic and a helper may be assigned to two mechanics. The next principle describes the basis for the crew supervisor instead of the scheduler making the daily work assignments.

Illustrations

The following illustrations demonstrate this principle of scheduling. The first section shows problems occurring as a result of not following the principle. The second section shows success through application of the principle.

Not this way.   Fred examined the plant's backlog of planned work and selected the work for the maintenance crew for the following week. The crew had forecasted 400 hours' worth of total labor for all the various craft specialties. Normally Fred only scheduled for 80% of the crew's forecast to allow for emergencies. This meant that sometimes he was not able to schedule all of preventive maintenance due on the equipment. This week he was able to schedule 60 hours of PM. At one point when allocating work out of the backlog, it became difficult to match the jobs needing attention with the remaining available electrical skills. Therefore, Fred assigned 20 hours of lesser important priority-4 work. This work required first class electricians and the first class electricians had hours available. The resulting advance schedule was an allocation of 320 hours of planned work for the crew. During the next week, the maintenance crew did not experience any emergencies and completed all 320 hours of work.

This way.   Fred examined the plant's backlog of planned work and selected the work for the maintenance crew for the following week. The crew had forecasted 400 hours' worth of total labor for all the various craft specialties. Fred was able to schedule about 80 hours' worth of preventive maintenance into the schedule. At one point when allocating work out of the backlog, it became difficult to match the jobs needing attention with the remaining available electrical skills. Therefore, Fred put in 20 hours of work requiring only a third class electrician even though the crew had only first class electrician labor hours still available. The third class work was priority-3 work, whereas all of the first class electrical work left in the plant backlog was less important priority-4 work. The resulting advance schedule was an allocation of 400 hours of planned work for the crew. During the next week, the maintenance crew did not experience any emergencies and completed 360 hours of the work.

Scheduling Principle 4 dictates that the scheduler should match the advance allocation of work to the number of hours a crew has available.

To accomplish this task, the advance scheduling process considers working persons out of their strict classifications or below their level of expertise. This methodology allows the scheduler to select the best combination of work orders to achieve plant goals such as reliability and efficiency. The combination of work orders is one in which the crew does possess the skill required to accomplish the work.

Principle 5: Crew Leader Handles Current Day's Work

Scheduling Principle 5 (Fig. 3.9) states

> The crew supervisor develops a daily schedule one day in advance using current job progress, the one week schedule and new high priority, reactive jobs as a guide. The crew supervisor matches personnel skills and tasks. The crew supervisor handles the current day's work and problems even to rescheduling the entire crew for emergencies.

Once the week has begun, obviously some jobs will run over and some will run under their planned work hours. Experience shows that although individual jobs show a wide variance between planned and actual times, over the course of a week there is remarkable agreement between the sums of the planned and actual times. That is the first reason that daily scheduling is best done by the crew leader or supervisor who is close to the field situation of job progress. Equally important is the ability of the crew supervisor to assign particular jobs to individuals based on their experience or even their need to learn.

Each day the crew supervisor assigns the next day's work to each technician. If working 10-hour shifts, each technician would receive assignments totaling 10 hours of work for the next day. The supervisor

Scheduling Principle 5
Crew Leader Handles Current Day's Work

✦**Daily Schedule**
✦**Matches Names to Tasks**
✦**Coordination of Resources & Clearances**
✦**Emergencies**

Figure 3.9 The crew supervisor is in the best position.

intends for each technician to complete 10 hours of planned work each day. The technicians may be continuing on a single job that spans several days or working several smaller jobs in a single day.

During the course of the day, the supervisors are out in the field assessing job progress. If a job runs over the planned hour estimate, the supervisor may have to schedule additional time for the next day. If a job runs under the planned estimate, the supervisor may have to assign additional work to begin a day earlier than expected.

The supervisor normally assigns new work orders out of the work allocation. The supervisor is also free to assign urgent jobs that come up during the course of the week. Ordinarily, the supervisor has the planning group quickly assess urgent jobs. Then the supervisor assigns them as soon as qualified technicians complete current jobs in progress. Because emergency jobs are begun immediately, the supervisor handles them by interrupting jobs in progress. Emergency jobs do not receive planning attention. They are handled entirely as jobs in progress from a planning standpoint.

Because jobs may finish earlier or later than expected, it is not practical to schedule work order assignments more than a day in advance. Because the crew supervisors keep abreast of individual job progress, they are in the best position to create the daily schedule. The crew supervisor creates the daily schedules and works the crew toward the goal of completing all the work allocated in the advance schedule.

The second reason the crew supervisors need to make the daily schedule is they understand the specific abilities of their various technicians. There also might be various personalities making a crew supervisor favor pairing certain technicians together and keeping certain others apart. Some technicians might also work better alone on jobs, while others might work better as a team. A crew supervisor is also best aware of daily personnel concerns, such as persons that call in sick.

To meet the goal of the weekly schedule allocation, the supervisor may also have to challenge some of the technicians. In the past, the supervisor may have allowed certain technicians to accomplish less work or less challenging work than others. Faced with a goal amount of work orders to complete, the supervisor may now be more encouraged to help technicians rise to the occasion. The supervisor approaches these considerations carefully. The situation may be a benefit to technicians who have been "frozen" at their current level of expertise because they only received jobs they could handle.

Because the supervisors create the daily schedule, maintenance also gives them the responsibility to coordinate other daily activities. These may include requirements for another craft to assist on a job. The supervisor makes timely requests from the operations group. Many plants accomplish this type of daily coordination with a brief daily schedule meeting each afternoon. All the craft supervisors attend with the key operations supervisors.

Illustrations

The following illustrations demonstrate this principle of scheduling. The first section shows problems occurring as a result of not following the principle. The second section shows success through application of the principle.

Not this way.   The maintenance planning scheduler sat down to make the weekly allocation of work. This was done by developing a series of daily schedules for a week. After the schedules were complete, the scheduler sent the operations group a list telling which systems and equipment to have cleared at different times each day for work.

As the crew supervisor visited the various job sites during the day, he had a good idea of which jobs would finish early or late. This required constant communication with the operations group, which generally voiced displeasure about the situation. The operations group expected maintenance crews to be able to work on the jobs to which the planning schedule had committed them. Operators generally wasted time clearing systems when the maintenance group did not have personnel ready. He had done the operations group a favor, however, when he was able to immediately put two persons on a fan problem at their request.

The maintenance supervisor did not think that the new scheduling system was any improvement over the past. In the past, the maintenance supervisor had assigned each technician one job at a time after he had checked with the operations group regarding clearances. The operations group could then count on maintenance personnel being ready to work on the cleared equipment.

This way.   The maintenance planning scheduler sat down to make the weekly allocation of work. This was done by developing a list of work orders for a week. After the allocation was complete, the scheduler sent the operations group the list showing which systems and equipment the maintenance group planned to work on sometime during the week.

As the crew supervisor visited the various job sites during the day, he had a good idea of which jobs would finish early or late. The crew supervisor knew that in order to complete the weekly allocation of work, he would have to assign each crew member a full day of planned work for the next day. After making a preliminary daily schedule, he attended the daily scheduling meeting. The operations group said it could clear up all the requested work for the next day. They also said they had earlier written a work order for a fan problem that probably could not wait until next week. The crew supervisor said that he would check with planning to see if they had started planning it. Depending on the craft skills needed, he would probably be able to start it the first thing in the morning. He had several persons who were ready to start new jobs. After the meeting, he called planning. They had just planned the job for two mechanics. The crew supervisor called the operations group, who said they would have the fan cleared for work. He made the necessary changes on his schedule and went to the crew meeting area to post the assignments for the next day.

The supervisor is in the best position to make the daily schedule. This person has the latest information on field progress and can judge when operations should clear equipment. This person has the responsibility of working toward the weekly allocation of work. However, the crew supervisor is still responsible for breaking the weekly schedule when necessary to take care of urgent problems.

Principle 6: Measure Performance with Schedule Compliance

Scheduling Principle 6 (Fig. 3.10) states

> Wrench time is the primary measure of workforce efficiency and of planning and scheduling effectiveness. Work that is planned before assignment reduces unnecessary delays during jobs and work that is scheduled reduces delays between jobs. Schedule compliance is the measure of adherence to the 1-week schedule and its effectiveness.

Work sampling or wrench time is considered the best measure of scheduling performance. However, maintenance management also tracks schedule compliance.

The bottom line is whether or not planning and scheduling have improved the work-force's efficiency. Planning and scheduling aim to do this by reducing delays that otherwise keep technicians from completing work orders. Planning individual jobs can reduce delays such as waiting to obtain certain parts, tools, or technician instructions. However, other than setting an individual job time standard, planning does nothing to reduce delays between jobs. These delays include such circumstances as technicians not receiving an assignment after completing their current work. In addition, not having a sufficient amount of work assigned may encourage technicians to take excessive breaks or have lengthy mobilization and shut down periods at the beginning and end of

Scheduling Principle 6

**Measure Performance by
Analysis of Schedule Compliance**

Case	Sched	Start	Finish	Compliance
①	10	10	9	100% not 90%
②	1	1	0	100% not 0%
③	1000	900	850	90% not 85%

Figure 3.10 Making schedule compliance acceptable to supervisors and practical to calculate.

each day. Scheduling aims at reducing these type delays. Work sampling or wrench time studies quantify both of these type delays. They give the primary measures of planning and scheduling effectiveness.

Schedule compliance is also an important indicator. John Crossan (1997) says that weekly schedule compliance is the ultimate measure of proactivity. When the maintenance force has control over the equipment, the maintenance force decides when to take certain actions to preserve equipment. When the equipment has control over the work force, the equipment drives the efforts of maintenance. A more reactive plant environment has more circumstances of the equipment experiencing problems and causing the maintenance force to break the weekly schedule. The proactive maintenance force in control of its equipment experiences few circumstances of a sudden equipment problem that interrupts scheduled work. Schedule compliance is merely a measure of how well the crew kept to the scheduled allocation of work for the week. Supervisors who adhere to the schedule as much as possible ensure accomplishing as much preventive maintenance and other timely corrective work as possible.

Schedule compliance provides a measure of accountability. It guards against crews working on pet projects or other jobs that are not more important than the allocated work. Yet if other more urgent or serious work arises, crew supervisors must redirect their crews to handle them. The schedule compliance provides a standard against which to discuss those actions. A supervisor may explain a low schedule compliance by telling what other work had to interrupt the schedule. A supervisor may have a low schedule compliance and no other interrupting work. This might indicate there may be a problem such as storeroom performance that needs to be identified and resolved. The schedule compliance scores facilitate discussion and identification of plant problems between maintenance managers and supervisors.

Similarly, a technician's performance measured against the planned estimate of a single job helps facilitate discussion between the supervisor and the technician. The technician must ignore the planned estimate when the actual dictates of the job demand otherwise. The technician and supervisor may need to send job feedback to the planning department to prevent certain problem areas from hindering future work.

Schedule compliance is not a weapon to hold against supervisors. Maintenance management and supervisors want to use schedule compliance as a diagnostic tool. Therefore, it is expedient to measure schedule compliance in a way to give the crew the benefit of any doubt. Figure 3.10 illustrates this approach. Consider if a crew is given 10 jobs and the crew starts all 10 but only completes 9. The crew receives a score of 100% schedule compliance rather than 90%. The second case explains this reasoning where a crew receives only one job, works it all week without interruption, but does not finish. It is not fair to grade the crew as having 0% schedule compliance. Again, the crew receives a score of 100% schedule compliance. In actual practice, case 3 shows how maintenance measures schedule compliance. Schedule compliance actually tracks the *planned work hours* delivered to the crew for the following week's work (1000 work hours). At the end of the next week, the crew returns all work they did not even start (100 work hours). Maintenance calculates the schedule compliance as 90%, which is (1000 – 100)/1000 times 100%. Giving the crew credit for jobs only started in the calculation accomplishes two results. First, the measure gives the crew the benefit of any doubt. This avoids supervisors feeling the calculation gives an unfair poorer-than-actual view of their performance. Second it makes the score very easy to calculate. Otherwise consideration would have to be made for the estimated remaining planned hours of jobs-in-progress. That adjustment would be very subjective and again possibly not seen to the supervisors' advantage. Third, one should remember that the objective is to encourage supervisors to work on scheduled jobs, the objective is not to have a scientifically accurate correlation between an indicator and field performance. The preferred method of calculating schedule compliance is expedient in all of these three regards. Instead of the term "schedule compliance," some companies prefer to call this measure "schedule success" to indicate the plants' attempt to gain control over the equipment rather than over the supervisors.

That the crew in case 3 may have only actually completed 850 work hours is not a problem as long as carryover hours the next week are monitored. For example, there would be a problem if the crew consistently claimed that it had about 200 hours of carryover work each week when the crew only had 200 available labor hours. Carryover hours are part of the crew forecast the supervisor makes each week to determine available labor hours.

Earlier, Scheduling Principle 3 stated that a 1-week period is short enough normally not to need significant alteration due to new work identification. This may be less true in a plant with more than a moderate amount of reactive work. These plants may normally experience a significant deviation from the set schedule. These plants especially should continue to schedule and track schedule compliance. This indicator would determine what improvement maintenance has been able to make in overcoming the reactive situation.

Illustrations

The following illustrations demonstrate this principle of scheduling. The first section shows problems occurring as a result of not following the principle. The second section shows success through application of the principle. Chapter 10, Control, shows an example of the actual calculation of schedule compliance for a crew.

Not this way.   Three plants considered schedule compliance. It made no sense at Plant Shelton to track schedule compliance. The plant simply had too many reactive work orders. However, the crews had become very efficient at taking care of the plant. It was never a problem for maintenance expeditiously to resolve most circumstances encountered.

Plant Bains had made a commitment to track schedule compliance. The plant had assigned an analyst almost full time to the task. Rather than only give the crews credit for completed jobs, each week the analyst would also give credit for some of the work hours for jobs-in-progress. The analyst carefully recorded the actual work hours that technicians had already spent on jobs not completed and added them to the total of the planned hours for completed jobs. There was some concern that the calculation was mixing actual work hours for uncompleted jobs with planned work hours for completed jobs. One alternative was having the planners give an estimate of the planned hours left on each partially completed job. Another alternative was having the supervisors give an estimate of the percentage of each job remaining and proportioning the original planned hours. The analyst doubted there was adequate time to fine tune the calculations each week using either alternative.

Plant Calvin used the schedule compliance indicator as a hammer. The most important task for any supervisor was to finish allocated work. Management used schedule compliance scores as the major part of each supervisor's periodic evaluation. This ensured that crews accomplished all the scheduled preventive maintenance and other work to keep the plant reactive work to a minimum. Supervisors never failed to take charge of emergencies, but they were understandably reluctant to resolve otherwise urgent situations before they became emergencies. Management knew that this was the price to pay for concentrating on proactive work. In the long run, they felt this strategy would provide the plant with superior reliability.

This way.   Three plants considered schedule compliance. It made sense at Plant Shelton to track schedule compliance. Plant Sheldon called it "schedule success." The plant had many reactive work orders. The crews had become very efficient at taking care of the plant. It was never a problem for maintenance to resolve most circumstances encountered expeditiously. On the other hand, the maintenance crews scored fairly low on schedule success each week. The schedule success indicator gave maintenance management one of its few tools to assess the plant's situation. Management knew that somehow they needed to reduce the amount of reactive work at the plant. As management implemented various solutions, they examined the schedule success scores to see if there was any improvement.

Plant Bains had made a commitment to track schedule success. At the end of each week, the planning supervisor gathered back all the work orders that the crews had not been able to start. Then the planning supervisor would sum the planned hours on the work orders separately for each crew. Subtracting these sums from the amount of planned hours the crews had originally been allocated allowed a simple measure of schedule success. This procedure consumed about 2 hours of the supervisor's time near the end of each week, primarily for gathering back the work orders that the supervisors knew that they would not be able to begin. The supervisor reflected that the work orders not started would have to be gathered each week in any case because the scheduler needed them to add back to the plant backlog. The scheduler would then begin the process of allocating work for the coming week.

Plant Calvin used the schedule success measure as an important indicator. It was important for any supervisor to concentrate on allocated work. Management used schedule success scores as one part of each supervisor's periodic evaluation. This ensured that crews understood the importance of accomplishing scheduled preventive maintenance and other work to keep the plant reactive work to a minimum. Supervisors never failed to take charge of emergencies and were also quick to resolve otherwise urgent situations before they became emergencies. Management ensured that supervisors understood their role to keep the plant out of trouble. In the long run, management felt this strategy would provide the plant with superior reliability.

As one can see, the plant's objective is not to have a high schedule compliance. The plant's objective is to have a reliable plant. A low schedule compliance indicates opportunities for management to address other problems in the plant to increase the plant reliability. The schedule compliance score facilitates discussion and investigation of problems. When supervisors are appropriately following the advance schedule and reacting to urgent plant developments, the schedule compliance score indicates the degree to which the plant is in a reactive or proactive mode. A plant cannot bring itself out of a reactive mode by insistence on blind obedience to the advance schedule. If it did, the consistent neglect of urgent developments might put the company out of business. Once it occurs, reactive maintenance needs cannot be ignored.

Summary

Maintenance planning will not increase labor productivity if it only concentrates on planning individual work orders. Making it easier to accomplish individual work orders does not necessarily mean that supervisors will assign more work. A number of system problems discourage crew supervisors from assigning more work orders for completion. Maintenance management must consider scheduling in the maintenance planning strategy to avoid these problems.

Six basic principles form the foundation of successful scheduling. These are using job plans providing time estimates, making schedules and priority systems important, having a scheduler develop a 1-week advance schedule, assigning work for all available labor hours, allowing crew supervisors to make daily schedules, and tracking schedule compliance. When setting craft and time requirements, job plans must plan for the lowest required skill level. This increases later flexibility in choosing jobs. Adhering to schedules is important because interrupting jobs leads to overall inefficiency. The priority system must properly identify the right jobs to start. A separate scheduler from the crew provides a check and balance. A 1-week period strikes a balance between a set goal and changing plant needs. In addition, a 1-week period is long enough to smooth out differences between planned estimates and actual times on single jobs. Knowing of the lowest skills required for jobs and the highest skills available in the labor pool allows developing a schedule with the proper work for the week. The uncertainty of actual job progress and the incidence of unexpected reactive work place the crew supervisor in the best position to create the daily crew work schedule. Finally, schedule compliance joins wrench time as an important indicator of maintenance performance.

Principles 1 and 2 are prerequisites for scheduling. Principles 3 through 5 establish the basis of the scheduling process. Principle 6 sets the overall indicators for scheduling control.

So utilizing planned and scheduled work packages increases the maintenance department's ability to complete work orders effectively, efficiently, and safely. Will the planning effort work with maintenance planning based on the six planning principles and the six scheduling principles?

Here is what the utility discussed at the end of Chap. 2 discovered. The utility established a weekly allocation of work based on all six scheduling principles. The plant management and crew supervisors quickly became extremely frustrated. The frustration was not due to supervisors having a set goal of work. Management and supervisors both accepted the responsibility of the crew to work toward the allocated goal and also respond quickly to urgent plant problems. Management and supervisors understood the balance of both responsibilities. The frustration was caused by the inability of the planning department to adapt the role of the planners for urgent plant needs.

The planners had recognized the supervisors had to deal differently with urgent, reactive work. The problem was that the planners did not recognize that the planners themselves had to deal differently with urgent, reactive work.

The planners insisted on developing significant job plans for reactive work. This delay kept supervisors in a state of frustration having either to wait on planning or proceed without any planning. The former case frustrated the planners who had to hurry. The latter case frustrated the planners whose eventual job plans were ignored. Supervisors realized the need of meeting the urgent needs of the plant, but the planners did not.

Obviously, management needed to consider urgent, reactive work in the planning and scheduling picture. Planning needed to make some adaptation of its work for reactive jobs. This leads to the next chapter on what makes the difference and makes it all come successfully together. The next chapter presents the final consideration necessary for the planning and scheduling strategy to succeed. Planning must not plan reactive jobs in the same manner as proactive work.

4

What Makes the Difference and Pulls It All Together

This chapter explains the final concepts necessary to make planning work. These concepts make planners do different things for different types of jobs and greatly influence the overall application of the principles. Lack of appreciating these factors frequently makes planning programs fail. The programs fail because they try a one-size-fits-all approach to different types of jobs. Primarily, the programs are not sensitive to the immediate needs of reactive jobs. This chapter distinguishes between proactive and reactive maintenance. Likewise, it distinguishes between extensive and minimum maintenance. Most importantly, this chapter describes the resulting planning adjustments. This chapter also discusses communication and management support regarding these adjustments.

The preceding chapter's second illustration of Plant Calvin depicted a fundamental maintenance concept. While the plant should actively engage in activities to prevent problems, problems must be dealt with quickly once they arise.

At Plant Calvin, maintenance crews understood the importance of accomplishing scheduled preventive maintenance and other work to keep the plant reactive work to a minimum. Furthermore, crews also never failed to take charge of emergencies and were quick to resolve otherwise urgent situations before they became emergencies. Management ensured that supervisors understood their role to keep the plant out of trouble.

On one hand, maintenance supervisors must change their past philosophy of executing mostly reactive work. Supervisors must assign more proactive work to head off reactive work. Advance scheduling helps facilitate this change. On the other hand, planners must change their past philosophy of planning all jobs as proactive work. Planning must adapt to an alternative method of planning reactive work. Making several adjustments to the planning department's process removes the last barrier to having an effective system.

Proactive versus Reactive Maintenance

The recognition of the existing maintenance culture helps management change maintenance crews to focus on proactive work. Proactive work heads off problems before they occur. John E. Day, Jr. (1993) has done excellent work developing the concept of proactive maintenance. He points out the standard definitions of maintenance:

Repair: To restore by replacing a part or putting together what is torn or broken: *fix, rejuvenate*, etc.

Maintenance: The act of maintaining. To keep in an existing state: *preserve from failure or decline, protect*, etc.

He explains that, "The key paradigm is that the maintenance product is capacity. Maintenance does *not* produce a service."

Day points out that initial disenchantment in implementing the planning system is primarily due to an attempt to provide detailed work plans on reactive jobs. Since reactive jobs by their nature are urgent, it is frustrating to everyone to wait on a planning group to turn over the work.

Figure 4.1 shows that when something has already broken, the job of maintenance becomes fixing it as soon as possible. "As soon as possible" means the sooner the better. Theoretically, reducing the time to fix it approaches zero (instantaneous fix) as maintenance achieves perfection. When something breaks, to suggest interrupting the crew with notions of waiting to plan the job would not be appreciated. Waiting would only add time and hinder maintenance's quest for perfection on

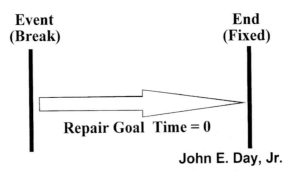

Figure 4.1   The goal for executing reactive maintenance.

that individual job. The concept of keeping the equipment from breaking in the first place actually achieves the zero repair time because the reactive event never occurs. This is not possible once something has already broken.

There are three different schools of thought on how maintenance planning should handle planning and scheduling for reactive work. One school holds that once something breaks, planning does not become involved and leaves the resolution entirely to the pertinent crew supervisor. The second school holds that planning treats all jobs alike. The third school of thought espoused by this book requires planning to become involved in all the jobs, but treat reactive jobs differently from proactive ones. None of the schools recommends planning involvement in true plant emergencies.

The first school concentrates only on proactive work which makes considerable sense for a plant that is in specification condition. That is, all of the equipment is either new or has been maintained well so there are not many reactive situations. Adopting this planning philosophy for an existing plant that has a considerable amount of reactive maintenance forces management to consider two options. Option one is to invest capital to bring the plant into a specification condition. Option two is to only plan and schedule the proactive work. The advance schedule would not include reactive work since there are no time estimates planned for those jobs. Instead, the advance allocation would consist of a small, manageable amount of proactive work to head off future reactive work. Gradually, the proportion of crew reactive work should subside relative to a growing proportion of planned, allocated work.

The second school insists on always planning information to head off probable job delays. If there is not file information available, planners must find and research equipment manuals, even for reactive work. This school counts on files quickly becoming developed and the incidences of having to plan jobs from scratch diminishing. Adopting this philosophy also makes sense for a specification plant where there is not much reactive work. In a plant with considerable reactive work, this philosophy might have planners working quickly to supply information to jobs about to start. Having extra planners at first could help.

There are difficulties seen with the above approaches. In the first school, a plant with much reactive work would not begin doing much planned work. In the second school, planning might develop a bad reputation early on because of the initially underdeveloped files. Planners might be trying to slow the start of jobs they have to research and the technicians might be expecting too much from the job plans.

A third school of thought attempts to resolve these difficulties. Management begins the planning effort primarily as a filing service for the technicians and the maintenance group and understands the technician's role in gathering information that might later be helpful. Therefore, when reactive jobs are first worked, there simply is little information expected from planning. Planning's job is to file the reactive job feedback to help a future job. The scheduling effort is begun to help encourage supervisors to assign more work, especially more proactive work. This book favors this approach for several reasons. There are a great number of plants that have considerable amounts of reactive work. These plants are unable or unwilling to invest in immediately upgrading the plant to specification conditions. These plants could still benefit from planning most of their work. Another reason is that experience has shown that planning usually has a very difficult time becoming established. This is mostly due to early false expectations from supervisors and technicians expecting perfect, complex job plans instead of simply helpful information. Finally, one of the greatest contributions planning makes for improving maintenance productivity is through advance scheduling. This approach allows planning enough detail on job plans to accomplish advance scheduling even while files are becoming developed near the infancy of the program. *Above all, this school (as well as the first school) advocates not holding up reactive work.*

As planning organizations become more mature and plants become more reliable, the differences in these schools of thought become less relevant. For one thing, the plants experience less reactive work. For another thing, files have become fully developed. The schools seem to go apart, but then come together.

In actual practice planning becomes successful when it begins to concentrate on planning proactive work. By concentrating on work to circumvent later breakdowns, the planning organization is able to produce good work plans without schedule pressure. Reactive work still receives planning before crew assignment, but the planners rely more on the technicians in the field researching a job for parts information if there is currently no file information. *For every job, the planner still provides a job scope, craft requirements, and time estimates.* However, the planner treats file information much differently for reactive jobs than for proactive jobs. The planner will always look in the minifiles for information. If there is no helpful file information on a proactive job, the planner will investigate other sources. These sources may include vendor or O&M manuals, consultations with more experienced personnel, or any other avenue thought to yield sought-after information. On a reactive job, however, the planner will not look beyond the specific minifiles. If there is no file or no helpful information in a file, the technicians are on their own for a reactive job. Not only does this methodology allow all the work to be planned to allow scheduling, but it reinforces Planning Principle 2 for feedback.

The challenge is to keep planning and scheduling proactive work while a significant amount of reactive work orders are still being written and planned. Enough personnel resources exist to perform all the reactive and proactive work, but only if all the work is planned so that schedules can be created to set goals for getting it all done. Planners must develop the work plans for all the reactive jobs to show the craft skills and estimated times required.

The objective of proactive maintenance is to stay involved with the equipment to prevent decline or loss of capacity. Planning and scheduling a sufficient amount of proactive work reduces the number of urgent problems and breakdowns. Reactive work receives minimal planning attention beyond a field inspection and minifile check before it is made available to be worked into crew schedules. Crews may have to look up technical information themselves on reactive jobs if the information is not available in the minifiles. Nevertheless, because the repetitive nature of maintenance work continually enhances minifiles with crew feedback, planners are soon able to give complete information and even procedures on reactive jobs.

Deciding to plan differently for proactive and reactive jobs requires definitions for the two types of work when first received by planning. Recommended definitions follow below.

Reactive maintenance is:

1. Where equipment is actually broken down or fails to operate properly.

2. Priority-1 jobs are defined as urgent and so they are reactive.

Proactive maintenance is:

1. Work done to prevent equipment from failing.

2. Any preventive maintenance (PM) job.

3. Work orders initiated by the predictive maintenance group when the need is not otherwise readily apparent.

4. Project work to upgrade equipment.

The essential determination for proactive maintenance is that work is done now to save additional work later. Proactive work heads off trouble. Once reactive situations develop, the operations group is already suffering. Reactive work is where equipment has failed and the plant is reacting to the equipment situation. Reactive work does not include where a specific device or component on a piece of equipment has failed, but the equipment is delivering its intended service satisfactorily to the operations group. For example, a slightly leaking flange on a pipeline might not be considered reactive if the drip is not causing a problem even though the flange itself has failed. (Alternative definitions for reactive versus proactive might be made on the basis of the customer, the operations group. Any job requested by the operations group is reactive because maintenance wants to produce plant capacity for operations, not react to operations problems. Operators should not have problems that they notice. Any job written up by maintenance would therefore be proactive. Maintenance wants to find all the plant deficiencies and correct them before they are noticed by the operations group.)

The practical result of implementation of these definitions should be the completion of all reactive work plans before lunch time for new jobs received that morning. Chapter 5 illustrates the step-by-step methodology planners follow for different types of jobs. Suffice it to say for now that on reactive jobs the planner scopes the job in the field (maybe), checks the file, estimates craft and hours, and puts the job into the waiting to be scheduled file. The crew supervisor then has the option of assigning the job if desired or waiting for the next week's schedule to include it if appropriate in the overall priority of plant needs.

Examples of proactive work include a job to replace the coating on a condenser tube sheet because maintenance has noticed some peeling; a predictive maintenance request to overhaul a pump; changing a filter at a set routine time; changing a filter that has a moderate pressure drop but is not bothering operations; noticing a small noise from a pump, moderate corrosion, painting, a dripping flange, or a sump pump running rough that would not cause an immediate plant problem if it failed; noticing a potentially inaccurate pressure gauge; or a project to replace a troublesome pump.

Examples of reactive work include a condenser tube leak, changing a filter at operations' request, a loud noise from a pump, a dripping acid flange, an operator report of a frozen valve, a clogged filter causing an operations problem, a failed sump pump even if not reported, a dead or obviously wrong pressure gauge, or a work order to restore a pump to service.

Extensive versus Minimum Maintenance

Following the line of reasoning that not all jobs should be planned the same way, it is also not cost effective to spend much time planning certain small jobs. This work is considered minimum maintenance.

This is a different consideration than that of reactive versus proactive. A proactive job may be minimum maintenance or extensive maintenance. A reactive job might also be minimum maintenance or extensive maintenance.

The following definitions are recommended for defining the complexity of maintenance.

Classification of Work Orders

✦ **Reactive versus
Proactive Maintenance**

✦ **Minimum versus
Extensive Maintenance**

Figure 4.2 Classification allows different planning treatment.

Minimum maintenance work must meet *all* of the following conditions:

1. Work has no historical value.

2. Work estimate is not more than 4 total work hours (e.g., two persons for 2 hours each or one person for 4 hours).

3. While parts may be required, no ordering or reserving is necessary.

Extensive maintenance is defined as all other work.

Figure 4.2 indicates the different classifications of work that require different planner treatment. The practical result of implementation of these definitions should be the reduction of maintenance planner time spent on certain jobs. Appendix E illustrates the step-by-step methodology planners follow for different types of jobs. Suffice it to say for now that on minimum maintenance jobs, the planners may put less effort into developing the job plan than they would if the work were extensive.

Examples of minimum maintenance work include hanging a bulletin board, moving barrels, cleaning the shop, tightening valve packing, replacing deck grating (maybe), replacing a 1-inch drain valve (maybe), replacing a frayed electrical cord, washing a fan (maybe), painting (maybe), posting a sign, adjusting dampers, replacing a filter on a PM basis.

Examples of extensive maintenance work include overhauling a pump, changing seals on a pump, changing bearings on a pump, troubleshooting or inspecting a pump, replacing a valve over 2 inches in size, replacing a valve critical to a process, replacing valve packing (maybe), repairing structural steel, welding boiler tubes, or replacing a filter on special request.

Communication and Management Support

Communication among the maintenance groups is especially important regarding these issues. Management support is necessary to keep planning involved and effective.

With an existing planning organization, trying to have a planner reduce the amount of planning that goes into an individual work order is difficult for two reasons. First, the planner may have a hard time accepting Planning Principle 5, to recognize the skill of the crafts. Second, the planner must understand that even with nothing more than a limited field scope and file check, the job is still adequately planned. A field technician's viewpoint on the latter case is similar. When a planned job was received in the past, it had quite a bit of detail.

However, in the past, the crew did not receive *all* its work as planned. Now it does. In the past, the crew did not want to wait on *any* planning for an urgent job. Now, the urgent jobs at least start off with the benefit of the crew supervisor knowing which skill to assign, for how long, for exactly what scope, and with readily available file information, all without waiting. Crews and planners take these things for granted and insist that a job plan without an extensive parts list and set of instructions is not really a plan. Nothing could be further from the truth. The problem stems from a lack of recognition of the value of what technicians and supervisors do receive. Technicians receive all the work as planned taking advantage of previous delay information. A supervisor receiving a week's worth of jobs even with only correct scopes and skill assignments is a tremendous boost toward superior wrench time. Remember that the vision of planning is to leverage productivity, not necessarily to provide "A, B, and C" on any particular job plan even as the plan moves toward having more of a procedure.

This is a sensitive area for the existing planning group that did not come into existence doing it this way. The technicians claim that planning used to provide detailed plans (on the few planned jobs). So communication to the work force with management commitment to understand and explain what is going on is certainly required to avoid derailing planning at this point.

Another point requiring communication and management support, of course, is helping the technicians understand their role to gather information and send feedback to the planners. This support allows the few planners to plan 100% of the work and the many technicians doing a lot of job research in the early days of planning. This is a serious controversy regarding who should do the initial research that management must not take lightly. For every one planner there are 20 to 30 technicians. The planners simply cannot research jobs from scratch and keep up with the workload. One should remember that before planning, the technicians did this anyway. Management does not want to transfer their duties to a specialist group. Management wants to create a value added group, namely planning for filing information to use on future jobs and gradually build ideal plans.

Supervisor buy-in.   The definition of stress is telling supervisors that they are responsible for meeting the weekly schedule and even tying it to their pay. This is not right. Rather, the management should hold supervisors accountable for starting each week with a goal of work and using some manner of a daily scheduling process. The management should hold supervisors responsible for knowing why they did not meet schedules. When the management assures supervisors that they can quickly react to emergencies and urgent work, even if it means breaking the schedule or not waiting for planning, supervisors are supportive of the weekly scheduling effort.

One supervisor neatly summed up this acceptance and even appreciation of the weekly schedule. He was an old sailor. He had retired with 20 years in the U.S. Navy as a Chief Petty Officer and the local electric company had hired him as an apprentice mechanic. The company eventually promoted him to journeyman mechanic and then crew supervisor. Upon his first week as crew supervisor, the planner had helped him develop a crew forecast so that he could receive a weekly schedule of work.

The next week, he said, "You know. When I first had to go along with making the forecast and receiving the stack of work orders, I thought to myself that this was bureaucracy at its finest. I thought that this was a total waste of time, but I was willing to go along if that was what it meant to be a supervisor here. Then a funny thing happened. As I sat down at the table with my crew at the end of the week and I spread out the work orders on the table, I said, 'Look here. This is the work we need to get done next week.' A light came on in my head. I realized that *this* was our purpose for being here. In the past, I had always thought that our purpose was to respond as quickly as possible to operator requests. However, I realized that our purpose was to do as much work as possible to keep operations from having to make requests. Oh sure, if operators tell us about an emergency, we need to respond. Yet at the same time, we need to do as much work as possible so they won't have emergencies. The work on the table represented that work. I'll have to tell you that it was pretty exciting to have that vision."

One Plant's Performance (Example of Actual Success)

When the recognition of reactive versus proactive and minimum versus extensive work planning addresses the maintenance culture, the planning and scheduling principles can deliver the planning mission as shown in Figs. 4.3 and 4.4.

The utility discussed in Chaps. 2 and 3 revamped its planning process to accommodate the abbreviated planning required by reactive and minimum maintenance work orders. These changes allowed planners to plan all the work and also accommodate crew supervisors who wanted to work on urgent, reactive work almost immediately.

Figure 4.3 shows that when the planners were able to schedule all the work, an effective weekly scheduling program helped improve maintenance productivity. Prior to February the maintenance crews were

Figure 4.3 Productivity accelerates.

completing less than 150 work orders each month. In February the planners had changed the approach to begin putting less detail in job plans. This enabled planning more work, and the crew completion of worked climbed consistently over 150 work orders each month. After a few fitful starts at scheduling, the planners began planning reactive and minimum maintenance work in an abbreviated fashion. This enabled them to plan all the maintenance work in the plant backlog and create a meaningful advance schedule. Crew supervisors still broke the weekly schedule to resolve some reactive work without planning. However, the crews began completing over 250 work orders per month. After 2 months of this completion rate, the maintenance crews totally cleared the plant's backlog of work. Without enough work identified to complete, the plant was able to send a portion of its work force to a sister plant to assist with

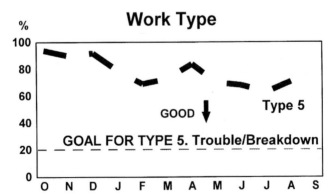

Figure 4.4 Getting more work done decreases the concentration solely on reactive work.

its backlog of work. In addition, the plant backlog reduction justified the plant beginning a major fall outage with a minimum of contract personnel. The plant had less need for a regular maintenance staff to maintain the other steam unit at the plant not involved in the outage.

Figure 4.4 is an excellent illustration of another effect of completing more work. With scheduling, there is more time for proactive maintenance work. The utility's increased rate of work order completion allowed it to complete an increased proportion of preventive maintenance and project work to upgrade equipment. The proportion of the utility's reactive work went from 95% to 65%. (The reactive work in this chart is not necessarily just equipment that has actually failed, but also equipment requiring corrective maintenance or not operating properly. This utility made no distinction for corrective maintenance, which is really proactive work since it heads off later trouble.)

Desired Level of Effectiveness

With this success, the utility decided to expand the planning program beyond the initial mechanical craft at its largest station. The utility decided to add the electrical and I&C (instrument and controls) crafts. It also included two other stations. The result of this expansion would bring 137 technicians under the influence of planning (Fig. 4.5). The resulting productivity should yield the effect of having 78 extra technicians.

Figure 4.6 shows the value of 78 extra technicians for completing new work at this utility. This is new work done with essentially free labor because of improved productivity possible with planning. Insourcing means using in-house resources for providing services such as making spare parts. The high cost of labor sometimes prohibits providing some services in-house, but free labor may make these services worthwhile. A final note is that planning requires a good degree of cooperation among planners, supervisors, management, and technicians. As with any new program, if management intends to lay off or dismiss persons because of productivity improvements, the very programs designed to improve productivity may be destined for failure.

**Add Two More Plants
Add Electrical and I&C Crafts**

**137 Techs Yielding the
Effect of
215 Techs**

78 Extra Techs

Figure 4.5   The leverage of planning on 137 technicians.

78 Extra Technicians to:

✦ Do Outage/Projects/Contractor Work
✦ Add New Fuel Capability to Existing Units
✦ Build & Maintain New Unit
✦ Insource
✦ Maintain New Cogeneration Facility
✦ Do Maintenance Services for Others
✦ Not Lay Off - High Morale

Figure 4.6   Utilization of resulting free labor.

Figure 4.7 shows the reasonable objectives of maintenance that planning may help accomplish. Availability of 95% is not an unreasonable goal. Typical electric utility availability of steam units is in the 85% range. Wrench time from 50 to 55% is desired. This may seem to encompass a lot of delay time, but consider that typical industry wrench time is only around 30%. In addition, scores above 60% are rarely seen workforce-wide. Higher numbers are obtained in specialized crafts such as machinists that have all their work together in a shop environment. Planned coverage represents the percentage of all labor hours spent on jobs that were on planned jobs. Some work will always be done by crews on an unplanned basis. The 80% rule may suggest that expecting greater than 80% planned coverage may not be worth the effort. Maintenance management also desires the continual identification of work to prevent breakdowns. A plant should have at least a 3-week backlog of such work. Concentration on this type work normally takes care of reactive work and overtime needs. The maintenance group should be able to work a normal weekday shift at many industrial plants without experiencing

"Success"
Level of effectiveness

♦ >95% Availability
♦ >50% Wrench time
♦ >80% Planned coverage
♦ >3 Week backlog and equipment NOT breaking
   ■ Reactive work <20% and OT < 3%
♦ Contractor work only on specialty items

Figure 4.7   Typical company maintenance objectives.

off-hour problems. Finally, an in-house maintenance force should prefer maintaining its own equipment for quality reasons. The use of contractors is justified where there is not enough in-house work to justify maintaining necessary qualifications or experience.

Summary

After establishing fundamental principles for planning and scheduling, a few final concepts become apparent for making planning work. Planners must plan different types of jobs differently. This is primarily due to the immediate needs of reactive jobs. Planners put less effort into planning reactive work to accommodate crews that must soon begin work. This also allows the planners time to plan all the work and concentrate more on important proactive jobs to head off failures. Planners also abbreviate their efforts on small tasks that do not justify much planning effort. These tasks are called *minimum maintenance jobs*. These planning adjustments require communication and support from management because of their effect on the plans that crews receive. Crews that previously received detailed job plans may now receive less information on individual jobs without appreciating its value. These concepts make the difference and pull it all together.

5

Basic Planning

The previous chapters have described the principles underlying effective planning and scheduling. This chapter describes exactly what a planner does in the context of the preceding chapters. The chapter follows the entire actual planning process including areas such as how a planner scopes a job, what a planner writes on a work order form, and how a planner files.

More magazine articles have described the concepts of maintenance planning than have described exact steps a planner might take to fulfill those concepts. This may be because there are often many different options for exact steps. Nevertheless, many programs with the right concepts have failed due to difficulty in determining how to execute the concepts. Therefore, this book undertakes the obligation to describe exact steps to clarify the role of a planner to fulfill the concepts. The following sections describe specific planner actions. After understanding what is necessary to execute the principles of planning, readers may implement alternative steps than the ones prescribed. The following sections address some of the considerations involved to allow readers to tailor their own systems appropriately. In this manner, this chapter covers the question: "Exactly what does a planner do?"

Before examining the basics of hands-on planning, observe a planner through a normal work day. This company has a correctly established planning organization. Read the following illustration with the principles and concepts affecting planning identified in parentheses.

A Day in the Life of a Maintenance Planner

Maintenance Planner David Clemons came in to work on Wednesday morning looking forward to another routine day of helping the maintenance department boost its effectiveness and efficiency (**mission of planning**).

After checking his electronic mail for important bulletins, he went to the waiting-to-be-planned file to select work orders (**work order system**) to plan for Steam Unit 1. There were not any reactive-type work orders (**reactive versus proactive**), so David returned to his desk to close work orders already completed by maintenance. David filed information (**Principle 2, feedback**) on repairs made (**history**), delays encountered (**path to improve productivity**), and parts (**parts lists**) and tools (**special tools**) used for each job. The closing included totaling the cost for each work order to help guide future repair or replace decisions (**Principle 5, overall strategy of job**). On one job it was not clear what extra part had been used by the technicians. He made a note to ask them later after the break so the plans for future jobs could have the part number available (**Principle 2, feedback and future work**).

Now it was about 9:30 and the new work orders had come from the supervisors' morning meeting (**other tool, communication**) to the waiting-to-be-planned file. As usual, each work order had pertinent information recorded by the originator (**organizational discipline**). David started with the reactive work orders. David made a copy of each for note taking and placed the originals in the planner active file by his desk. The first job was obviously a simple welding job and required only minimum maintenance attention. The other two jobs needed extensive maintenance consideration (**minimum versus extensive maintenance planning**).

All of the equipment involved in the jobs had component tag numbers (**Principle 3, equipment numbers**) identified on the work orders (**organizational discipline**). Only one piece of equipment did not have a corresponding minifile (**Principle 3, component level files**), so David quickly made a minifile for it. He then entered the work order number and problem description in each minifile. The entry would also later enable any duplicate work orders to be caught by a planner before planning. David also checked the files for each extensive maintenance work order to see a total job history for the equipment. For all of the work orders, David then made a field inspection. Afterward, from his personal experience (**Principle 4, planner skill**) and the minifile information, he made a planned package for the work order. He did this by explaining the work needed on the original work order forms and attaching available technical information from the minifiles. As in the case of most reactive plans, he was careful to plan the general strategy of the job and not spend time including how-to details unnecessary for a competent technician (**Principle 5, technician skill**). David was pleased that for one work order he was able to identify a special tool that had slowed down the last job when it was not available. He then finished the planning for each job by putting the planned package in the waiting-to-be-scheduled file and updating the work order status on the computer. As was normal, all the reactive work was planned before lunch (**reactive versus proactive planning**).

After lunch, David concentrated on the proactive work orders in the waiting-to-be-planned file. Two jobs required extensive maintenance planning and two jobs required only minimum maintenance planning (**extensive versus minimum maintenance planning**). On the first extensive job, a thermography route (other tool, predictive maintenance) had shown a slight leak for a valve. A check of the minifile showed that this valve had a history of leaking (**Principle 2, past job helping future job**). The second extensive job, for a pump, had no identified component tag number because schematics and coded tags were still being developed for that section of the plant (**Principle 3, equipment numbers for files**). There were no minifiles for the pump and the equipment for the minimum maintenance jobs. A computer check for each job showed no additional information.

David put on his hard hat and safety glasses. Then he went out for a field check to scope the proactive work orders. He noted that although the valve was in high pressure service, it had flange connections and would not require a certified welder (**Scheduling Principle 1, lowest skill required**). David decided to include scaffolding in the plan. Since the pump job had no component tag number, David attached a temporary tag directly on the pump by the nameplate (**Principle 3, equipment number for files**). One of the minimum maintenance jobs was as expected, but he had to clarify the other one with the originator. David then returned to the office (Principle 1, separate planning department).

He first finished the minimum maintenance work plans making a minifile (**Principle 3, files**) for each piece of equipment and putting the planned packages in the waiting-to-be-scheduled file. Then David turned to the extensive maintenance jobs. The valve's minifile history showed that the seat and disk had been reconditioned as well as replaced without too much improvement in its time between failures. David decided the present valve was marginal for the service and planned the job to replace the valve with an upgraded valve from the warehouse (**Principle 2, past job helping future job**). For the pump, David made a minifile from the temporary tag number he had installed. Since the job was proactive, David took the time to research the technical and vendor files for certain clearance and parts information (**further research for proactive jobs**). After David found the information he needed, he copied it to the new minifile and finished the work plan. The next time the pump was worked on, that information would be readily available (**Principle 2, past job helping future job**).

With the time remaining in the day, David reviewed the feedback from several preventive maintenance routes (**other tool, PM program**). Each preventive maintenance (PM) route covered multiple pieces of equipment within a single plant system. These inspections usually uncovered most of the problems for which work orders were written in the plant. David changed the frequency of one route from every week to every other week because the route had been run for several months without identifying any adjustments or situations needing correction.

At the close of the day, David walked to the parking lot. He thought about the part he played in the high availability (**objective of planning**) that Steam Unit 1 enjoyed. The backlog of planned work (**Principle 2, future work**) allowed the scheduling of planned work to match (**Scheduling Principle 4, schedule for 100% of hours**) the forecasted available craft hours for the next week (**Scheduling Principle 3, schedule one week from forecast of highest skills**). The weekly schedule set a work goal and made the advance coordination of other crafts and parts staging possible (**objective of scheduling**). These basics would normally boost work force wrench time (**Principle 6, wrench time**) beyond the 35% typical of industry to about 45%. Keeping the plant on a constant learning curve by using information gathered in the minifiles actually increased wrench time to 50%. Technical data was available and previous job delays were avoided. As the computer system became more developed, wrench time was slowly creeping up to 55%. At 55% the productivity of 25 people for which David planned would be the same as for 39 people working at only a 35% wrench time. The benefits of planning actually involved productivity and quality savings. The productivity savings came from reducing delays during and between assignments. The quality savings came from correctly identifying work scopes and providing for proper instructions, tools, and parts to be used. The productivity improvement also freed up craft, supervision, and management time. This allowed them to focus on troublesome jobs requiring more attention and an opportunity to do more proactive work. This proactive work included root cause analyses on repair jobs, project work (**other tool, project work**) to improve less reliable equipment, and attention to preventive maintenance and predictive maintenance. David felt good that his work in planning contributed (**other tools needed, no silver bullet**) to a cycle of continuous improvement.

The previous account shows that the steps job planners take reflect the important principles and concepts of planning and scheduling. The account also demonstrates that the planning system resides as a process within the work order process. The next section describes the overall process of completing work with the work order process from job origination to job closure and notification of the originator. After this section, the chapter focuses on the planning process activity in the work order process.

Work Order System

As explained in App. A, the work order system is the most valuable tool for improving maintenance effectiveness and productivity. Basically, the system helps maintenance personnel obtain necessary origination information and control all the work. The work order system avoids an inconsistent utilization of verbal statements, electronic mail, Post-its, and phone calls. The foundations of the work order system are a consistent format for information and a designated flow for work to proceed. The information needing to have a consistent format (whether on computer or paper) is origination information, plan information, and feedback information. The work order system prescribes the use of specific forms, codes, and work processes. Appendix J provides a complete, sample work order system manual, and also presents a flow chart showing the steps maintenance takes to complete emergency work without planning assistance.

Figure 5.1 shows the steps of the work order process for a typical company using a paper work order system. *Paper* means that the work request is written on a physical paper form. This same physical form passes to the planning department and then to the field for work execution. The form is then returned to the planning department with job feedback. This company also uses the term *work order* to refer to the document when it starts as a work request as well as after it is authorized and becomes a literal order to do work. The form is known as the *work order form*. The company uses a computer, but only to track the work order forms. This book's illustration of a company with a paper-work order system allows the best explanation of the planning process. This typical company process will be used for the remainder of this chapter and the next two chapters to illustrate the planning and scheduling steps. A computer may be used to add value, but it cannot replace the basic process. After the basics of planning are mastered, Chap. 9 explores the possible employment of a computer system.

The first step in Fig. 5.1 shows the origination of the work order. This person may be an operator, a maintenance technician, or anyone in the plant. The originator obtains a work order form and fills it out with the required information. The originator describes the problem or work requested including identification of the involved equipment and its location. The originator makes an estimate of the work's priority. The originator also gives an opinion, if possible, as to which crew or craft the work would be assigned and whether or not the work must be done during an outage. The originator also provides any other information dictated by the particular work order form as well as any other information that might be helpful. This person also hangs a deficiency tag on the equipment if applicable. The originator's supervisor, if required, then reviews the work order, makes any adjustments required, and places the form in a designated collection place. The particular work order form utilized by the company may have multiple carbonless copies to provide copies or a copy machine might be utilized. The collection places are the preferred locations that maintenance planning has established to help speed the work order flow. The collection places might be simple boxes in the control rooms, by the elevators, in the front office, or near other work locations. Maintenance planning collects the work orders placed

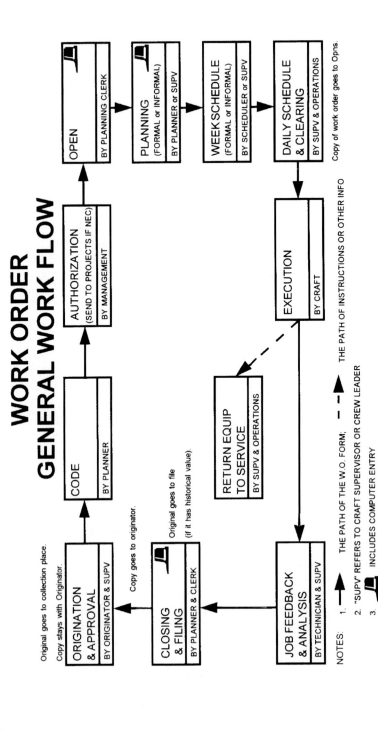

Figure 5.1 Example maintenance work flow diagram.

in these boxes at regular times. The intent is to avoid work orders being placed in the interoffice mail or otherwise delayed or lost through some unusual means of transmission.

After collection of the work orders, the second clockwise box in the flow diagram shows coding. Planners go through the new work orders to code the work orders. When the planners code the work orders, they are placing appropriate codes from the plant coding system on every work order. These codes include designation of work type, whether the work is reactive or proactive, whether the work is minimal or extensive, and other codes. The planners also designate which crew would receive the work order.

The third box shows a step where the work orders are brought to the morning meeting where managers and certain supervisors may swiftly review them to see what is going on in the plant. Plant engineers attend these morning meetings. Any work order may have its priority changed, may be canceled, or may be referred to a project group. The concept of a morning meeting originated before the existence of planning and used to be the place where new work orders were passed from operations to maintenance every day. The use of the morning meeting now is usually a meeting to review a list of work orders and not the work orders themselves. The work orders themselves remain in the planning department where planners may begin planning them and clerks may finish entering them into a computer system. The next step in the flow process allows the maintenance planning clerk to enter any data that any computer system may need. For example, this company uses a CMMS, but just enters the work order information into the computer to allow having a backlog list of work orders. The company also uses two other separate computer systems for inventory and payroll time sheets. The clerk must enter work order numbers to authorize inventory transactions and payroll accounting. After entering the work orders, the clerk puts the work orders in a waiting-to-be-planned file for the planners.

The planners plan work in accordance with the steps discussed in this chapter. Informal planning refers to crafts that do not have planning. Many plants only implement planning for mechanical maintenance when it has the bulk of the work. Electrical and I&C craft supervisors or technicians plan their work on an informal basis. After planning, the planners place the work order forms and all associated pieces of the planned package together in a waiting-to-be-scheduled file.

A scheduler then schedules the work in accordance with the process steps recorded in later sections of Chapter 6. Informal scheduling refers to crafts that do not have planning. After scheduling, the scheduler delivers the scheduled work order forms and planned packages to the appropriate crew supervisors.

The crew supervisors then work the scheduled work into daily crew assignments. The crew supervisors obtain equipment clearances from the operations group, with a copy of the work order if required.

The next step has the technicians executing work on cleared equipment. After job execution, the technicians report job completion to the crew supervisor. The supervisor soon reports job completion to the operations group. The supervisor does not wait until the shift end to report multiple job completions all at once. Giving the operators timely notice allows the operators time away from their own shift change periods and reduces confusion in restoring equipment. Timely notice also allows the operators to restore equipment to service while technicians are still at the plant in case problems are encountered.

The next step requires the technicians to carefully record helpful feedback on the work order form. The technicians and supervisors should go ahead and fill out this information while their memories are fresh. A later section in this chapter thoroughly covers required and desired information for feedback since feedback is essential to the planning improvement process.

After receiving feedback, the last step is for planners to assess the completeness of feedback and file appropriate information including the work order form. The planner may proceed to update future plans even before the plans are required. The planning clerk enters designated information to close work orders in various computer systems. The clerk may also have filing duties. A later chapter section prescribes possible clerk duties. Finally, Fig. 5.1 closes the circle of the work process by sending notification of work completion to the original requester. This notification could be accomplished by copy of the work order, electronic mail, or spoken communication. Only in organizations where the operations group has complete confidence in the maintenance group should the notification of work completion be skipped. Alternately, a computer system where everyone has free and easy access to check work order may suffice for operations to see whether maintenance has completed their work.

Planning Process

In the planning process, the planners take new work orders and add necessary information to allow more efficient scheduling and execution of the work. Figure 5.2 expands the single box that called for planning in the overall work order process. All of the activity shown in Fig. 5.2 occurs within the single planning step of Fig. 5.1. Figure 5.2 shows the general sequence of the planning operations. Understanding this sequence will help explain the discussions that follow in this chapter.

The first box shows where planners place the work orders after coding, computer entry, and morning meeting adjustments have taken place. The planners put all unplanned work orders in the waiting-to-be-planned file. Normally this file has only a single day or two of unplanned work in it because the planning principles and concepts allow for keeping

Planning process flow chart

*According to guidelines for type of work order and plan.

Figure 5.2   General flow of planning activities.

up with the work on a daily basis. However, in a plant just starting to implement planning, there may be a substantial amount of unplanned work in the backlog. In those cases, the file will also have to allow access to crew supervisors who may have to assign jobs before planning has time for them. The work orders are placed in the file to allow easy identification of their planning needs. Normally, the work orders are placed by plant unit. Within the work for each unit the work is arranged by priority code. If there are multiple planners, each planner may be responsible for certain crews or crafts and the file arrangement may reflect this order.

The next box shows the planners taking work orders from the file. The planners generally take the reactive work orders out ahead of the proactive ones. These normally carry a higher priority. The planners want to plan such work orders quickly in case the crew supervisor requests them.

The next box shows a step to provide the planner with a draft copy of the work order to allow note taking while in the field scoping work. The planner does not take the original work order form out into the field. The planners leave the original work order forms out on their desks and go into the field with copies. Alternately, there may be a set location for a "planner active file" in each planner's office or cubicle. Leaving the original in the office also allows the supervisor of planning to help crew supervisors find certain work orders if necessary.

The planners note the type of work orders they are planning. They will plan each work order differently depending on its type: reactive or proactive, minimum or extensive.

The next box shows that a planner will nearly always check the files to see if previous job information will be helpful.

Next the planners scope the job according to the guidelines of the type of work. After scoping the jobs and understanding what they require, the planners will sit down to develop an actual job plan. The planner may update the job's status as planned on the computer. The planners finally file each completed job plan in a waiting-to-be-scheduled file. This file is kept in the planning office area.

Work Order Form

Many maintenance departments use separate work request and work order forms. In such a system, an originator fills out a work request describing the work requested and turns it in to the maintenance department. After planning or other processing, the maintenance department issues the crew or technician a new form, the work order form, which describes the actual work to be done.

This book prefers a single form. Having a single form simplifies the maintenance process and also may avoid the loss of originally attached information. A separate form may also lead to inaccurately copied information to the second form.

After origination, maintenance planning uses the same form to record nearly all planning information.

The scheduler handles the work order forms for scheduling. The form helps the technician and supervisor include vital feedback helpful information.

After job completion, the planner can easily check the structured form for its completeness and any need for the planner to seek additional feedback or initiation of additional work orders. The planner files this form to help future jobs.

When planning a new job, a review of past jobs recorded on consistent work order forms enables the planner to recognize and avoid previous job delays and problems.

Another issue involves work order numbers. Each work order having a unique number facilitates discussion of work orders. Having unique numbers helps persons understand if they are discussing the same work. Sequential, unique work order numbers are easily printed on the top of preprinted work order forms. Unique numbering is almost mandatory for computer systems.

Figure 5.3 shows an example of a typical, structured work order form for a company that employs planning. This example helps to visualize its use in the maintenance process. Many companies have much more complex work order forms with a multitude of specific boxes requesting information. Structured forms provide for consistent information input for all

WORK ORDER

REQUESTER SECTION   Priority ___ ___

Equipment _____ Tag # _____

Problem or Work Requested: Def Tag #_____

Outage Req? Y/N   Clearance Req? Y/N   Confined Space? Y/N

By: Date & Time: **APPROVAL:**

PLANNING SECTION   Assigned Crew:_____   Attachment? Y/N

Description of work to be performed:

Labor requirements:

Parts requirements:

Special tools requirements:

By: Date & Time: Job Estimate: Actual:

CRAFT FEEDBACK (Modify plan sections above:actual labor, parts, & tools)

Work performed including equipment changes & any problems or delays:

Date & Time Started:_____   Date & Time Completed:_____

By: Date :   **APPROVAL:**

CODING

Figure 5.3   An example of a work order form to help guide input.

phases of a job. These phases include origination, planning, job feedback, and coding. Figure 5.4 points out the different main areas of the work order form. For the purposes of illustrating the planning process, this book uses the simple work order form shown in Fig. 5.5.

Consider now an example work order. An operator notices a strainer starting to plug and originates the work order shown by Fig. 5.6. The pressure drop is high, but not too severe. The operator uses the priority codes found in App. J to code the work order as an "R2." "R" means that the

WORK ORDER #001234

REQUESTER SECTION

PLANNING SECTION

CRAFT FEEDBACK SECTION

CODING SECTION

Figure 5.4   The main areas of a work order form.

```
┌─────────────────────────────────────────────────────┐
│           WORK ORDER  #001234                        │
│  REQUESTER SECTION                                   │
│                                                      │
│                                                      │
│                                                      │
│                  APPROVAL:                           │
├─────────────────────────────────────────────────────┤
│  PLANNING SECTION                                    │
│                                                      │
│                                                      │
│                                                      │
│                                                      │
│                                                      │
│                                                      │
│                                                      │
│                                                      │
├─────────────────────────────────────────────────────┤
│  CRAFT FEEDBACK                                      │
│                                                      │
│                                                      │
│                                                      │
│                  APPROVAL:                           │
├─────────────────────────────────────────────────────┤
│  CODING                                              │
└─────────────────────────────────────────────────────┘
```

Figure 5.5   Example simple work order form used for illustration.

situation involves plant reliability and "2" deems it serious. The operator fills in other information required by the plant.

Coding Work Orders

Two planners go through the new work orders early every morning to code work orders. When the planners code the work orders, they are placing appropriate codes from the plant coding system (such as found in App. J) on every work order. These codes include designation of work

```
┌─────────────────────────────────────────────────────┐
│           WORK ORDER  #001235                        │
│ REQUESTER SECTION                      Priority R2   │
│ #2 Control Valve B Strainer (N02-FC-003) has high    │
│ pressure drop.  Def. Tag #010304.  No outage required.│
│ Needs mech crew. Clearance Req.  No Confined Space.  │
│ J. Smith 4/21/98 3:00pm.APPROVAL:  S. Brown 4/21/98  │
├─────────────────────────────────────────────────────┤
│ PLANNING SECTION                                     │
│                                                      │
│                                                      │
│                                                      │
│                                                      │
│                                                      │
│                                                      │
│                                                      │
├─────────────────────────────────────────────────────┤
│ CRAFT FEEDBACK                                       │
│                                                      │
│                                                      │
│                                                      │
│                            APPROVAL:                 │
├─────────────────────────────────────────────────────┤
│ CODING                                               │
└─────────────────────────────────────────────────────┘
```

Figure 5.6 Work order after the originator completes the information.

type, whether the work is reactive or proactive, whether the work is minimum or extensive, and other codes. The planners also designate which crew would receive the work order. If a number of work orders have been received in the middle of the day or if there are fairly urgent work orders, the planners may not wait until the usual morning period to code them. If the company utilizes an operations coordinator, this person reviews the work order priorities and has the authority to adjust them as needed. The operations coordinator might also be responsible for gathering work orders to bring to planning. Frequently, an operations coordinator is a member of the operations group on loan to planning. Less commonly, planners have the authority to adjust work order priority.

Coding could also take place in a morning meeting to verify priorities and assigned crews.

Plants code data to increase their ability to use information. For one thing, it clarifies communication. It provides for a consistent designation for each type of information to reduce confusion. Once planners code a work order, persons using the work order later might not have to digest the entire document to get necessary information. For example, once the planners code a work order as reactive, later interpretation is not necessary when the planner is reviewing many work orders to select several to plan. For another thing, plants code their data to allow better analysis of information. For example, coding all breakdown work as work type 5 allows the plant later to evaluate the percentage of plant work it does on equipment after it has failed. Coding all pumps as equipment type 01 may help find pump work orders or plans later on a computer database.

There might be a number of codes the plant develops to enable various uses of plant data. Appendix J shows typical codes. Examples are work type, priority, outage, department, craft, crew, plant, unit, building, plant area, process group, process system, equipment type, manufacturer, proactive versus reactive, and minimum versus extensive.

The planners code the work orders rather than everyone in the plant coding his or her own work orders. This maintains better consistency of data. The planners maintain the plant's filing system and the coding allows access to this critical information. Because the codes lend themselves to interpretation at times, it is best to use as few persons as possible to code them. This is similar to a library. Once a book is returned to the wrong shelf location, it becomes lost for all intents and purposes. A computer especially requires consistency of data. The exception to this guideline is having the originators place the initial priorities on work orders because they have first hand knowledge of the work situation. The plant consistently using an intelligent, equipment numbering system helps ease the problem of selecting the wrong codes.

The planning department uses at least two planners for coding because the work orders must be processed without fail. Two persons allow the absence of one planner not to delay the coding.

Each morning, designated planners code spaces at the bottom of the work order form for each new work order. This means that planners are the first persons in the maintenance department to see new work orders. Sometimes the planners notice insufficient data supplied by the originator. The planner may be able to obtain this information later when scoping the job. However, if the originator supplies all the readily

```
┌─────────────────────────────────────────────────────┐
│              WORK ORDER  #001235                      │
│ REQUESTER SECTION                      Priority R2    │
│ #2 Control Valve B Strainer (N02-FC-003) has high     │
│ pressure drop. Def. Tag #010304. No outage required.  │
│ Needs mech crew. Clearance Req. No Confined Space.     │
│ J. Smith 4/21/98 3:00pm APPROVAL: S. Brown 4/21/98    │
├─────────────────────────────────────────────────────┤
│ PLANNING SECTION                                      │
│                                                       │
│                                                       │
│                                                       │
│                                                       │
│                                                       │
│                                                       │
│                                                       │
│                                                       │
├─────────────────────────────────────────────────────┤
│ CRAFT FEEDBACK                                        │
│                                                       │
│                                                       │
│                                                       │
│                     APPROVAL:                         │
├─────────────────────────────────────────────────────┤
│ CODING  Plan Type RE   Group/Syst FC   Crew    1-2    │
│         Work Type 5    Equip Type 18   Outage  0      │
└─────────────────────────────────────────────────────┘
```

Figure 5.7   Work order after the planner completes the coding.

available information, maintenance planning has a much easier job planning the work.

Receiving new work orders with insufficient data is a very sensitive area. On one hand, the plant desires all persons to be able and willing to write work orders as they see areas in the plant needing attention. On the other hand, poorly written work orders hinder the maintenance effort (going back to the days of illegible Post-its). This is a management commitment area to stress how timely information helps maintenance support the entire plant's reliability. Planners should not simply complain of wide-spread problems with origination, but help management understand the problem with specific work order examples.

Figure 5.7 shows the example work order after morning coding. The planner used the coding system in App. J. The planner could have made a case for coding the work order as P (proactive) since at first glance this important plant process had not yet been affected. This logic was furthered by the operator setting the priority as a 2 (serious) rather than 1 (urgent). However, the planner realized that normally the operators blew down strainers into the oil tank without a need for maintenance attention. The strainer had failed and needed manual cleaning. Therefore, R (reactive) was the proper code. For similar reasoning, the work type was 5 (trouble and breakdown) rather than 9 (corrective maintenance). The work was also not minimum maintenance since planning desired to keep track of strainer work in history. The planner coded it as E (extensive). The equipment group was F (fuel) and the specific system was C (service pump). The equipment involved was an 18 (strainer). Crew 1 to 2 would be assigned the work and the proper outage code was 0. No outage was necessary.

Using and Making a Component Level File

The planner begins the planning process by selecting work orders from the plant's unplanned backlog. The waiting-to-be-planned file holds this work. The planner normally picks several work orders at the same time. This may allow scoping several jobs at once in a common plant area. After selecting the proper work to start planning, the planning process dictates that the planner should first consult the information already filed for the equipment in question. The planner should consult the files before scoping the jobs in the field. The files may yield information on past events that would help the planner know what to look for. The planner may also not find certain information that could be gathered in the field. The planner would then know to gather that information when inspecting the equipment on site.

Checking the files simply involves simply walking to the file section and looking for the appropriate minifiles. Component level files are called minifiles because they do not contain information for more than a single, specific piece of equipment. A minifile was made the first time the planning department ever planned for the involved equipment. If the planner finds that a minifile exists for the piece of equipment, the planner scans the included information inside. If there is no component level file for that equipment, the first thing a planner would do is make one. Chapter 7 describes how a planner makes a minifile.

Planners should not underestimate the value of making a minifile for nearly every piece of equipment on which maintenance performs work.

If certain equipment in the plant is important enough to have a separate identification number, it is important enough to have a minifile.

The planner gains what useful information is available in the minifile for each piece of equipment before scoping the work orders.

For our example work order, there was no file for that piece of equipment, so the planner created a new file using the equipment number N02-FC-003. Then the planner proceeded to scope all the selected work orders.

Scoping a Job

Scoping simply means identifying all the work required. Scoping is a subset of the planning process. Scoping refers to the overall work scope and not other planning tasks such as identification of parts and time estimates. Scoping is necessary even though the person who requested the work provided descriptive information. Work requests sometimes have only a description of the problem itself. If the work request states that "The boiler feed pump is running hot," obviously the work needed to remedy the problem has not been defined. The process of defining that work is called scoping. At other times, work requests come with a description of the work desired. A work order may state, "Replace the gasket for the leaking flange." Although the work appears to be defined well enough to continue planning, scoping by the planner is still beneficial. The planner scoping the job looks through the eyes of a skilled maintenance technician, rather than those of an operator or less skilled technician. The skilled technician planner may see a pipe hanger 10 feet away has come loose causing the flange to leak. The planned job scope should include attention to the pipe hanger in addition to the flange. Instead of the pipe hanger, the planner may realize that a nearby valve is leaking instead of the flange where the drip appears. Whether the valve packing needs tightening or the valve needs replacing becomes the job scope decision. The scope may include a need for scaffolding after the planner sees the height of the valve. The planner ensures that the plan specifies the right job through scoping work orders.

During scoping, the planner may alter specific information contained in the work request. Consider the example of a leaking valve causing a flange to appear to be leaking. The originator requested the flange gasket be replaced. The planner would mark over the request and explain the actual problem, the valve. The planner would specify that the valve be replaced if that is the proper course of action. The planner would correct any equipment tag information in the request. The planner would have to change the equipment tag number specified from the pipeline to the valve.

The work involved in correctly scoping a job increases in complexity from simple minimum maintenance, reactive work to extensive proactive work.

Beginning with work that is both minimum maintenance and reactive, the key is personal planner experience for scoping. The planner should not need to consult any minifiles or the computer. A field inspection may not even be necessary, but is usually recommended. A work request might state, "Replace the broken window in the field office." In this case, a field inspection might be necessary to ensure the correct job is planned. Close inspection of the window job might reveal that the broken window has a rotten wooden casing so its replacement must also be put into the job scope. The consequence of missing the rotten casing could be the wasted hour or two of an assigned technician. The technician would return and declare the window cannot be replaced before a carpenter is called. That is one reason why planners should have top level technician-level skills. An experienced technician has run into many of these type problems and knows what to inspect.

For jobs that are reactive but also extensive the planner would always consult the minifile and the computer. The planner would always make a field inspection. However, because of the urgency in processing reactive work, only if there is great uncertainty would the planner consult operators or otherwise proceed further in investigation.

For proactive work, urgency is not critical. The objective is to avoid future reactive work by planning excellent proactive work. For proactive work, the planner may take time to consult any minifiles and the computer even for minimum maintenance work to develop a good job scope. For proactive work that is extensive, the planner develops the best job scope possible, not hesitating to consult operators or other knowledgeable sources, if needed. Sources to be researched for proactive work may require looking beyond the minifiles and include equipment technical files and vendor files. These files are described in Chap. 7, Forms and Resources Overview. A root cause analysis may also be advisable. Operation of the equipment may be necessary to observe the problem in some proactive cases.

Troubleshooting

How should planners scope a job that requires extensive troubleshooting? The answer lies in planning Principle 2, the repetition of maintenance work. Planners cannot entrap themselves trying to make *this* job perfect. They will get another shot at it later (and later again), each time armed with more information. Management must not sell planning as a perfect job agency. It must sell planning as a library service that should be able to give technicians a good head start.

First, the plant wants to have all the jobs planned so it can attempt scheduling (where the productivity gains exist). That means that the planner cannot make *this* job perfect. He has to make his guess from his experience and a quick look at the files. However, the mechanic has the luxury of actually working on the job and finding what was truly wrong.

Later, the planner has the opportunity to file that information in a component specific file and make a better guess for the *next* job. Faithfully, this same equipment will need service again in a year or so and probably in the same general way. The planner must present himself to the crews as "I'm your file clerk. Tell me what happened and I'll have this information for you next time." The planner heads off those delays caused by "What was that part we used last year? Didn't this machine have a special lug? Did we have to heat the bearing last year?" This notion is especially important for electrical planners. Many times the troubleshooting lasts 2 days and the fuse replacement lasts 10 minutes. The first time, the planner might suggest a half-day "troubleshoot and repair" job scope with the admonition to report what happened. The next time, the planner might suggest a half hour troubleshoot and repair job scope mentioning the last time a 10 amp fuse blew in panel N00-RA26-001.

Yes, it would be better for the operators to be TPM oriented and take care of a lot of troubleshooting. They should be able to chase down fuses from reviewing drawings and even fix what they can. However, do not think that planning cannot make a great contribution in another environment. If operators do not perform any troubleshooting or maintenance, planning should still create an improving file system the best it can. The next section discusses engineering, but consider: Should things be allowed to fail repeatedly? Planners can take some measures as they see history files showing obvious plant weaknesses. The planners can write some work orders to make some things better. Nevertheless, planners should not become bogged down in changing the design of the plant. They must plan all the work orders at hand. Let the engineers do their jobs to recognize and improve weak areas in the plant, perhaps with some suggestions from the planners. Furthermore, concerning job plans that let the technicians perform the bulk of troubleshooting, a word is appropriate about jobs-in-progress. Say even that the planner cannot make a decent guess (guessed wrong and the wrong parts got ordered). Once the mechanics have opened up the equipment and figured out exactly what they need, do not let the planner become entangled in a job-in-progress. The mechanics should have access to order parts whether it be the crew supervisor or some kind of expediter in the planning department. From a planner's perspective, the key is to collect feedback for the next time.

Rotating spares form a special kind of troubleshooting situation. In these cases, management has decided that troubleshooting and repairing certain equipment should not be done at the expense of unit downtime. The plant has purchased entire assemblies that maintenance can swap out and then troubleshoot the failed piece without hurry. Presume that just because the primary equipment failed, the rotating spare will not immediately fail. The plant has the technicians take the rotating spare apart on a nonemergency basis and recommend to the planner what to order for a new work order to repair it. This can be done in a shop environment where things do not disappear. Nevertheless, this is "opened-up equipment." The planner should order needed parts delivered quickly as for a job already in progress.

Performance testing or engineering

However, any time proactive or reactive work requires performance testing or engineering, the planner refers it to the planning supervisor for possible reassignment. Reassignment is desirable for a couple of reasons. One reason is not to bog down planners with a few unusual jobs and neglect the rest of the backlog. Another reason is to get special expertise when needed.

First, the planner cannot afford to become bogged down on one or two jobs and not plan the other 20 jobs. Most of the work performed by maintenance is routine. That means it does not require extraordinary measures to scope the jobs. Jobs such as valve replacements, flange leaks, and loose linkages far outnumber situations where a skilled technician cannot make a fairly rapid determination of the job requirements. The planner needs to make sure all of those routine jobs are planned so that crews can schedule their work and avoid unnecessary delays, such as having to find part numbers already available in the files.

Second, some jobs merit specific expertise. Resources for planning to consider might include engineering such as plant engineers, predictive maintenance, performance testing, and the plant controls group. Planners must involve these groups when necessary.

The planning supervisor or planning department may have guidelines for when the planners should request assistance. The guidelines for what jobs to reassign and whether to have an engineer in the planning group itself depend on what type of jobs would slow the planners. A very skilled planner group may be able to handle a wider variety of complex jobs than a less experienced group without becoming bogged down. One of the supervisory responsibilities of the planning supervisor would be to monitor what jobs are troublesome and cost too much time in planning for the planning group.

Planning groups must also maintain an awareness of plant procedures to coordinate certain actions with outside agencies. The plant may have an agreement with its insurance carrier to notify it whenever a fire prevention system is turned off for maintenance. Normally, the primary responsibility for this type of notification would fall with the operators who perform the actual clearing and shutting off of the system. However, the planners may be able to use their file systems to provide helpful reminders. Some plants also place a burden on specific plant engineers to remember and fulfill these requirements.

The plant also benefits from having some mechanism to allow cancel-ing of work orders. The best scope of work may be not to do the requested work and the planner sometimes coordinates these decisions. The work orders may not be thought necessary because of the broader scope of knowledge of plant operations that planners may possess. Certain work orders may not be needed because of a soon-to-be-executed project to resolve the situation. The plant may be able to cancel certain work orders to repair portions of an old demineralizer beyond immediate repairs if the project group intends on installing a new demineralizer. The plant can reconcile work orders to planned projects in different ways. Plant engineers may have the responsibility of reviewing new work orders with this in mind. Planners may have the primary responsibility to be aware of these projects to turn back unnecessary work orders.

Illustrations

The following illustrates the actions of a planner to scope different types of work orders. The planner selects four work orders from the waiting-to-be-planned file.

In the first case, the planner has to scope a reactive, minimum main-tenance job. A plant engineer has written a work request to replace a pressure gauge. The planner knows there are many of these gauges in the storeroom. The planner walks out in the field to make sure which gauge the engineer is referring to. The planner also checks to see if the gauge has an isolation root valve or if the system has to be cleared.

In the second case, the planner has to scope a reactive, extensive maintenance job. The operators have reported a control valve that is leaking through. The planner consults the minifile for the valve and inspects the valve in the field. Since this valve has no information in the minifile to indicate otherwise, the planner decides that maintenance should replace the valve.

In the third case, the planner has to scope a proactive, minimum maintenance job. The plant environmental engineer has written a work request to make and place a "No Swimming" sign by the percolation pond. The planner reviews the area and decides the sign should be attached to an existing fence. The planner asks the engineer if this would be acceptable.

In the fourth case, the planner has to scope a proactive, extensive maintenance job. The predictive maintenance group has reported another control valve leaking through. The planner consults the minifile for the valve and inspects the valve in the field. The planner decides that maintenance should try to replace the valve. Since this valve has no information in the minifile, the planner spends some time to research several technical manuals and talk to a supervisor to see if this valve might be rebuilt in place.

Note how scoping the jobs varies in complexity because of the need to move reactive work quickly to maintenance and do careful analysis on proactive work. The overview of duties for a maintenance planner in App. E provides a more formal, step-by-step type checklist of these activities for the different type work requests received.

Engineering Assistance or Reassignment

Most of maintenance is routine and can be handled by planners. The planners must get all the work planned so that effective scheduling can consider all the plant's work. Planning must not become delayed on the exception work orders that could prevent the other work from being planned.

The common maintenance task is not an engineering concern. Adjust a valve here, replace a gauge there, overhaul a pump here. These ongoing maintenance tasks make up the vast bulk of maintenance. This work could arguably include some desire for engineering scrutiny. However, the main task of maintenance is maintaining or preserving a performance level at a previously engineered specification level. This engineering was completed at the original installation of the equipment long ago. Let the engineers concentrate on analysis in projects designed to make things better, but that does not include most of what is going on in maintenance.

Occasionally jobs are received in planning that need engineering solutions or designs. An example would be an access platform where structural integrity is important or where detailed drawings are necessary. Another example would be where a repeated failure indicates a need for a different type valve or special material. In general, a reactive work order would get less of this type scrutiny and be left up to the maintenance crews to resolve, but a proactive work order's objective is to head off future reactive work, so engineering help may be appropriate. A plant engineer should be able to advise whether PVC piping could be used to replace a chronic failure situation for carbon steel. A job plan that requires sizing a pump could be handled quickly by a plant engineer. Some jobs obviously need to become turned into projects such as adding a new demineralizer. The planners need to make sure these jobs do not come through planning, but are handled entirely by a project group or the plant engineers. If a plant intends to use maintenance labor to install the equipment, the scheduling aspect may come into play. However, planners do not need to become involved in the design. Other jobs not worthy of being called outright projects still may not need to go into planning until after a design is completed. Overhauling a steam trap system may need to go through plant engineering for trap selection and piping sizing and routing. Then planning can take the completed design to coordinate parts and assign estimates for labor and schedules.

To reassign work or receive engineering assistance, the planner consults with the planning supervisor or planning department guidelines. Some plants have a plant engineer as part of the planning group to handle quick questions without reassignment and to coordinate the jobs that do become assigned to the plant engineer group. Otherwise, the planner is responsible for keeping up with planned packages that are waiting for engineering.

A word of caution to planning groups with a strong engineering presence: The engineers should not be allowed to redirect the planners' focus. Engineers may be tempted to use planners to collect equipment data rather than allow them to plan routine maintenance work.

The planner should indicate on the final work plan any engineering input received and why. For example, the planner would not write simply "Replace with staged new valve." A better plan would be: "Replace with staged new Teflon valve per engineer Brown's recommendation to avoid previous corrosion problems." The failure to include the "why" of a job plan has resulted in many field technicians coping within their own experience to a fault. If the technicians thought they could patch the existing valve to save the company time and money, the expanded job explanation from the planner would help them see the big picture. The idea is to explain the "what" as well as possible so that the technicians can react with the best information to the actual job circumstances. Planners should always pass along technical information.

Developing Planned Level of Detail, Sketching and Drawing

Similar to scoping a job, the planner spends the least time describing what work needs to be done on reactive, minimum maintenance jobs and the most time on proactive, extensive jobs. Only on a proactive, extensive job would the planner consider searching files other than the minifiles to find new information. The planner must adhere to Planning Principle 5 and avoid telling an experienced crew "how" to do the work on initial reactive plans. The plan must emphasize the "what." The unnecessary effort takes up valuable planning time.

For example, a planner might say, "Repair" or "Replace" a valve, but never, "Repair or Replace." It may be acceptable in certain circumstances to say "Attempt to repair the valve unless internal valve inspection requires valve replacement."

The planner might say in the case of repair that the "Internals probably need replacing," but might not give steps on how to dissemble the valve if an experienced craftperson should know how to do it already. On the other hand, if the minifile already had an easy-to-copy procedure available, the planner might attach it as a reference. A copy of any standard plan should be attached if there is one for the job. Also, on more critical plant systems, a planner might provide more steps and rely less on the skill of the field technicians.

The description of "what work is needed" goes in the planner section on the work order form. The planner should write down the scope of the job and the results of any research to give the technician helpful information. For example, "Replace the valve. History file indicates repeated failures and patches." Another example, "The equipment could not be run, but information from operator J. Smith indicates a failed bearing. Disassemble to inspect and replace the bearing or other corrective action necessary." Another example, "Replace the entire pump because cost information indicates repairing is not cost-effective."

The planner may feel that certain steps should be given to help clarify the intent of the plan. The steps may also be needed to help coordinate resources. When giving steps, the planner might number them. For example:

1. Erect scaffolding

2. Replace valve

3. Remove scaffolding

One intent of the job plan work scope and plan detail is to provide a good technician with enough information to reduce the incidence of the technician having to delay the job seeking additional instructions on what to do. The planners reduce potential job delays when the work order plan identifies enough information so that the technician does not have to make extra trips to seek help from the supervisor. As with any trip, the technician's leaving the job site causes a job delay. Any trip away from the job might not only consist of a momentary delay finding the crew supervisor. Frequently, trips allow other distractions to hinder the technician's prompt return to execute the work.

Another intent of the scope and plan detail is to provide enough information for any technician that might receive the work order to avoid mistakes and provide a consistent maintenance result. This is best achieved through detailed procedures and precise checklists. However, the planner guidelines dictate that it is more important to plan all the work before achieving this objective. Planning all the work allows all jobs to receive file information to avoid significant delays of the past and allows the scheduling process (which needs craft and time estimates) to function. This is preferred over having a few "perfect" jobs, many unplanned jobs, and ineffective scheduling.

All the same, as presented by Chap. 2, Principle 5, having detailed maintenance procedures is a worthy objective. Detailed procedures enhance reliability, performance, and safety. Maintenance planning aims to accomplish this by gradually adding more details and helpful job steps as jobs repeat over time. In this manner the expertise of the crafts assists the skilled planner develop procedures over time. The planners with extra time should first concentrate on putting more detail into proactive work to head off problems and into critical (or safety) systems or equipment to best serve plant reliability objectives. Planners can provide fewer details on work that is more reactive or on less critical plant systems or equipment. As a minimum, planners call out the appropriate minimum skill needed to work on the equipment regardless of the procedures included on the job plans. Chapter 8 presents more advanced procedures that could be called "standard plans" although even these plans should continue to evolve.

Many plants use experienced technicians, supervisors, or even recent retirees in a structured way under the facilitation of a person experienced in procedure writing to create initial procedures for common maintenance actions on critical equipment. These groups develop the procedures for the planners to use on real jobs. These groups often also follow up the initial execution of the work "walking down" the actual maintenance to provide realistic first job plans.

On the other hand, a point of diminishing returns exists for providing information. For example, consider an electrical planner sketching a conduit run. The planner had been sketching for an hour. If the planner did not do the layout, then a field technician would have to spend about an hour doing one. The crew could simply field run the conduit because system constraints did not mandate any certain route. The planner was doing the layout in order to know how much conduit to reserve from the storeroom. However, the planner could just guess that between 100 and 200 feet of conduit would be needed without doing a layout. In this case, the planner should simply reserve 200 feet of conduit and send the plan on its way. Think about it. The one electrical planner was supposed to be planning for 20 to 30 electricians. There is no way the planner can justify spending an entire hour to save a technician only a single hour in the field. A planner is worth 17 technicians because the planner can help leverage the work of 30 persons into the work of 47 persons. To plan the specific details of the conduit run in this case would be to violate Planning Principle 5. The planner violates Principle 5 in not recognizing the skill of the crafts. The planner also violates Planning Principles 1 and 4. The planner violates Principle 1 when the planner does field work the crew should do. The planner violates Principle 4 because the planner is not quickly using a planner's expertise to estimate a job. Consequently, the planner might not plan all the backlogged work. The domino effect starts. Not getting all the work planned leads to not being able to schedule enough work for a crew for a week. Thus the crew does not have a goal of work to complete. In addition, an excessive amount of unplanned jobs causes a multitude of problems. It frustrates the crews if the crews are not allowed to have it.

If they are allowed to work on unplanned work, it is not only less efficient, but there is an excuse not to do planned work.

One of the reasons the electrician example ended up as a problem may be a false perception that planning takes care of all paperwork. This includes the time before and even after a job. This reasoning is false because planning is supposed to add value to the maintenance process, not just do something that someone else used to do. There is nothing wrong with a technician doing some paperwork before the job such as figuring how to field run conduit. This particular job might not require a drawn conduit run at all. Technicians filling out paperwork for feedback after a job is essential to the whole planning concept. It provides the basis for improvement on subsequent jobs. Paperwork is not a criteria involved in deciding how much detail to put into a job plan.

Sketching or drawing is a particular concern. Planners may need to provide plant schematics or other engineering drawings as attachments to jobs. However, planners should rarely have to draw diagrams themselves unless they have a gift for sketching. Some planners can make sketches to illustrate job needs faster and better than they could describe the job in words. On the other hand, many planners have felt required to provide sketches on jobs when the planners were not talented in drawing and the sketches simply were unnecessary.

The planners must respect the skills of technicians. This allows planning all of the work. The planner must also consider other plant specialists when planning. This permits utilizing the proper expertise on certain jobs.

Attachments

If the planner needs any more room than the space provided on the work order form, the planner might have to attach additional pages for writing. The planner should always identify any attachments on the first page of the work order form. This allows the schedulers and crews to make sure they have all of the planned information. It also allows the planner to check its return after job completion.

A planner might attach a single page or so of text by stapling it to a paper work order, but more extensive attachments are possible. Sometimes planning departments create small booklets of important equipment information that they keep in the minifiles. These booklets contain especially critical information culled from the more massive O&M manuals. Planners bind these booklets with report covers so they can be sent out on jobs.

Such larger attachments may be attached by paper clamps to work orders or they may be kept in files in the planning department. The job plan should state that attachment so and so is in the equipment minifile in the planning department. Alternately, the planning department might keep the attachment in a special "Attachments" file in the planning department. This file keeps attachments in the planning department filed by work order number (they have a copy of the work order attached) where technicians can come get them when ready.

On the other hand, not that infrequently jobs benefit from having the entire O&M manual available. Such jobs should refer to the manual and planners should allow technicians to access the file system. Technicians should be able to view the manuals in the planning office where there should be ready access to tables for study and copiers. Technicians should even be able to check out manuals through an accountability process.

English 101

Next to consider when writing job plans is simple English. The following is added for the planner's consideration:

1. Use the second person imperative form. This means saying, "Do this" or "Do that" as if you are telling the reader what to do.

2. Use active, not passive voice. Say, "Do this" instead of "This should be done." "Do this" clearly indicates the technician should do something whereas "This should be done" might lead the technician to think that someone else is supposed to do something.

3. Use action verbs that clearly tell the technician to complete some action

4. Avoid abbreviations when not absolutely clear. Usually, the first time a plan mentions an abbreviation, it should explain what the abbreviation means.

5. Avoid wordiness. Tell the technician what to do in as few words as possible. As a minimum, tell the technician straightforwardly what to do before explaining why if necessary.

6. Be very careful with person pronouns such as "it," "they," or "them." The procedure may be much clearer if the planner repeats naming the items intended. "Clean the positioning screw" is better than "Clean it."

7. Use standard conventions for warnings, cautions, and notes. Standard industry convention dictates that "WARNING" indicates possible safety concerns that could lead to injury or death; "CAUTION" indicates possible equipment concerns where mistakes could lead to equipment damage; "NOTE" indicates additional information that might help the technician understand a step or item. Use these words at the beginning of appropriate lines on the job plan to help the technician better understand the job plan. The job plan should place these

statements as close as possible to the job plan steps to which they are associated.

8. Do not be afraid to use spell check. Use Microsoft Word or any other word processing program to check spelling and grammar. These programs not only detect poor spelling, but grammatical problems such as passive sentences and the like as well.

Figure 5.8 shows the earlier example work order after the planner decides on the scope of the job and adds the necessary level of detail to describe the steps of the job.

Craft Skill Level

Planners designate craft skills needed on job plans. This allows schedulers to select the appropriate jobs matching the skill levels possessed by each crew. It also allows crew supervisors to determine to which persons to assign the jobs. With the crafts specified by the job plan, schedulers and supervisors do not have to thoroughly read each work order and decide for themselves which crafts the jobs require.

The planner also plans each job to allow the lowest qualified skill level. This increases the flexibility of the schedulers and supervisors in selecting or assigning the work. Different jobs require a different minimum craft skill level. If a certain job could be done either by a junior mechanic or a certified mechanic, the job plan would not want to specify a certified mechanic. That would limit the choices for who could do the work. On the other hand, the planner could not specify a trainee who would not have the minimum skill levels necessary to be qualified for the job.

The designation of crafts and skill levels provides for communication from the planners to the schedulers and supervisors. The planners, schedulers, and supervisors must have a common understanding of the terms used.

The following designations have been adopted for the purposes of illustrating the selection of crafts and skill levels in this book. All of the maintenance departments at the plant understand what each designation means when it appears on a work order from planning.

Trainee: Any person newly hired in the maintenance, but not yet enrolled in an apprenticeship program.

Mechanical trainee: A person newly hired in the mechanical maintenance craft, but not yet enrolled in the apprenticeship program.

Electrical trainee: A person newly hired in the electrical maintenance craft, but not yet enrolled in the apprenticeship program.

I&C trainee: A person newly hired in the instrument and controls maintenance craft, but not yet enrolled in the apprenticeship program.

```
┌─────────────────────────────────────────────────────────────┐
│            WORK ORDER  #001235                                │
│ REQUESTER SECTION                          Priority R2        │
│ #2 Control Valve B Strainer (N02-FC-003) has high             │
│ pressure drop. Def. Tag #010304. No outage required.          │
│ Needs mech crew. Clearance Req. No Confined Space.            │
│ J. Smith 4/21/98 3:00pm APPROVAL: S. Brown 4/21/98           │
├─────────────────────────────────────────────────────────────┤
│ PLANNING SECTION                                              │
│ Clean strainer positioned in front of the control valve.      │
│ Remove strainer element, clean, and replace.                  │
│ Replace gasket if needed                                      │
│                                                               │
│                                                               │
├─────────────────────────────────────────────────────────────┤
│ CRAFT FEEDBACK                                                │
│                                                               │
│                                                               │
│                                                               │
│                       APPROVAL:                               │
│ CODING  Plan Type RE   Group/Syst FC   Crew  1-2             │
│         Work Type 5    Equip Type 18   Outage 0              │
└─────────────────────────────────────────────────────────────┘
```

Figure 5.8 Work order after the planner completes scoping and adds the appropriate level of job step detail.

Apprentice: Any person enrolled in the apprenticeship program for one of the maintenance crafts. These persons are normally not assigned work by themselves when the primary objective of the plant's apprenticeship program is for them to learn alongside higher skilled technicians.

Mechanical apprentice: A person enrolled in the apprenticeship program for the mechanical maintenance craft.

Electrical apprentice: A person enrolled in the apprenticeship program for the electrical maintenance craft.

I&C apprentice: A person enrolled in the apprenticeship program for the instrument and controls maintenance craft.

Helper: Anyone. This designation is used if the job does not require any special skills outside that of an adult maintenance employee. This person could possess only the skills of a new trainee, but may also possess the skills up to any certified technician. The planner will use this classification alongside that of a more specific craft skill to indicate that the primary need is for an extra set of hands. The planner may also require only helpers on a job. This does not mean the supervisor would assign two apprentices, but that the job does not require any specific skills.

Technician: Thus far, this book has used this term to mean anyone in the work force. This has allowed the illustration of the principles of planning without unnecessary details of craft skills. However, in the use on particular job plans, this term means a person that has passed the apprenticeship programs in one of the craft areas. The term also implies a measure of responsibility and accountability. The plant would normally hold a technician accountable rather than an apprentice for the results of a particular job.

Mechanic: A mechanic technician. In this plant, the mechanic possesses a certain amount of structural welding skills and light machining skills.

Welder: A welder technician able to do high pressure welding.

Machinist: A machinist technician able to perform most machine work for the plant.

Painter: A painter technician.

Electrician: An electrical technician.

I&C technician: An instrument and controls technician.

Lab technician: A technician that works in the plant laboratories.

Certified mechanic: A mechanic technician who has gained special experience, knowledge, and skills in the mechanical craft. The plant has given this technician the extra privilege and responsibility of being able to stop certain jobs in the field whether or not assigned to them for quality concerns.

Certified welder: A welder technician who has gained special experience, knowledge, and skills in the welding craft. The plant has given this technician the extra privilege and responsibility of being able to stop certain jobs in the field whether or not assigned to them for quality concerns.

Certified machinist: A machinist technician who has gained special experience, knowledge, and skills in the machining craft. The plant has given this technician the extra privilege and responsibility of being able to stop certain jobs in the field whether or not assigned to them for quality concerns.

Certified painter: A painter technician who has gained special experience, knowledge, and skills in coating and corrosion. The plant has given this technician the extra privilege and responsibility of being able to stop certain jobs in the field whether or not assigned to them for quality concerns.

```
┌─────────────────────────────────────────────────────────┐
│              WORK ORDER #001235                          │
│ REQUESTER SECTION              Priority R2                │
│ #2 Control Valve B Strainer (N02-FC-003) has high        │
│ pressure drop. Def. Tag #010304. No outage required.     │
│ Needs mech crew. Clearance Req. No Confined Space.        │
│ J. Smith 4/21/98 3:00pm. APPROVAL: S. Brown 4/21/98      │
├─────────────────────────────────────────────────────────┤
│ PLANNING SECTION                                         │
│ Clean strainer positioned in front of the control valve. │
│ Remove strainer element, clean, and replace.             │
│ Replace gasket if needed.                                │
│ Labor: 1 Mech                                            │
│        1 Helper                                          │
│                                                          │
├─────────────────────────────────────────────────────────┤
│ CRAFT FEEDBACK                                           │
│                                                          │
│                                                          │
│                                                          │
│                      APPROVAL:                           │
│ CODING   Plan Type RE   Group/Syst FC   Crew   1-2       │
│          Work Type 5    Equip Type 18   Outage 0         │
└─────────────────────────────────────────────────────────┘
```

Figure 5.9 Work order after the planner determines the craft skill level required.

Certified electrician: An electrical technician who has gained special experience, knowledge, and skills in the electrical craft. The plant has given this technician the extra privilege and responsibility of being able to stop certain jobs in the field whether or not assigned to them for quality concerns.

Certified I&C technician: An I&C technician who has gained special experience, knowledge, and skills in the instrument and controls craft. The plant has given this technician the extra privilege and responsibility of being able to stop certain jobs in the field whether or not assigned to them for quality concerns.

Certified lab technician: A laboratory technician who has gained special experience, knowledge, and skills in the laboratory. The plant has given this technician the extra privilege and responsibility of being able to stop certain jobs in the field whether or not assigned to them for quality concerns.

Using specific craft and skill designations provides a common ground for discussions and reduces misunderstandings. Commonly used terms can be agreed on if they do not represent specific labor classifications used by payroll.

Figure 5.9 shows the addition of craft skills to the work order. The planner has determined that the job requires two persons. One person must possess at least the skills of a mechanic technician. The other person could be anyone to assist the mechanic.

Estimating Work Hours and Job Duration

Work hours (or labor hours) are the actual craft personnel hours that technicians later enter on personal time sheets and job duration is how many hours the job lasts. Consider a pump job that lasted 2 entire days and had two persons working 10 hours each day. The work hours would be 40 hours and the job duration would be 20 hours. Note that for job duration, only calendar time when someone was working on the pump applies. The estimated work hours and job duration are necessary to schedule the work efficiently.

Operators need the duration information to know how long to expect equipment should stay cleared and unavailable. The job duration is also useful when considering what work could be completed in a short outage situation. Consider a short outage that suddenly occurred and had an estimated length of 24 hours. Any backlogged outage job with an estimated job duration of equal to or less than 24 hours should be considered.

The planner uses the work order form to write the estimated work hours for each job. The planner specifies the number of persons and the work hours for each person estimated for the work. Then the planner totals all labor hours on the work order form. The planner also specifies the estimated job duration hours on the form.

The planner develops the work hours and job duration estimates from personal judgment and consultation of the minifile in most cases. The previous jobs in the minifile are helpful, but the planner is not restricted to using previous estimates or actual hours reported. The planner is attempting to estimate the hours reasonably required by *experienced craftpersons without unexpected delays.*

Experienced craftpersons means that the planner should plan the job time estimate for a good technician, not the average technician, not the slowest technician, and not the fastest technician. There is no way that the planner knows to whom the supervisor will assign the job. The supervisor may assign the job to the worst technician, an average technician, or the fastest technician on the crew. For whom should the planner plan? This entire consideration is essentially the same as for technician skill level where Planning Principle 5 recommended that the planner respect the skill of the crafts.

Estimating the time of each job for the slowest technician means adding a lot of extra time to every job. When every job is planned this way, the planner allows too much time that most technicians do not need. Instead of planning a job with the slowest person in mind, the planner plans knowing that although the slowest person may need extra time, the help may not come from the plan itself. These technicians might get extra help from being assigned to jobs where a more capable technician is also assigned to the same job. The supervisor may spend more time on the job with a technician in a coaching role. The supervisor may choose to give this technician the extra time that the technician requires.

On the other hand, estimating each job's time for the average technician has problems. *Average* means probably as many technicians in the same classification need more time as need less time. In addition, one might think there is a small standard deviation away from this average or mean. That is, on the whole, most of the technicians in the same classification possess about this mean level of capability. This is a very dangerous presumption that may not be true. In many organizations, the skill level widely varies within a single classification. There is no large, single aggregate of persons with a similar capability. Out of a group of 20 technicians, the more realistic case may be two or three near the bottom, three a little faster, a couple above them in speed, four above them, three considered decent, three considered good, and two or three technicians who can do almost anything quickly. To complicate matters, the general hierarchy changes with respect to different types of work within the classification. The absolute slowest technician with respect to pumps may be quite decent with respect to valves. Two of the decent technicians with respect to valves and piping may be quite deficient with regard to rotating equipment. So the fastest technicians in some respects could be the slowest technicians in other respects and there may be only an illusion of what average means.

One problem of planning all jobs for the average technician is similar to planning all jobs for the slowest technician. Although the planners can estimate less time than if they were planning for the slowest technician, the planners are still allowing more time than some persons need.

The most significant problem is that there is no standard of what time a good job should take if all jobs are planned with average in mind. If maintenance assigns a job, the company has some profit motive in mind. The company wants good work and so does the technician. What time should a good job take? Jobs planned for good technicians mean that technicians pull their weight and a little more. They are not just keeping up with the sled. Jobs planned for good technicians set a time standard for all technicians by which they can judge their skills. Planning for the level of a good technician gives the crew an idea of how good it is.

This concept of a *standard* is very important. Although the plan itself is a standard in several ways (what skill should be assigned, what should be done, and so on), it is the time estimate that is of most interest to the productivity issue. The standard states, "This is the time the job should take." This is of such importance that complex methods exist to determine meaningful time standards. Chapter 2 discussed engineered built-up standards and standards derived from techniques such as slotting or pigeon-holing. Unfortunately, for the practitioner of maintenance planning, these techniques are not much more accurate than that "guesstimated" by an experienced craftsperson after reviewing the job and past history. In addition, these more formal techniques generally take more time to employ. Maintenance in a modern industrial plant encompasses different technicians working on different equipment in different environments constantly. There is little assembly line, repetitive motion activity that might benefit from an intense industrial engineering scrutiny for an exact time estimate. On the other hand, the standard is not so much needed to grade an individual technician's performance, as to enable the scheduling process. The standard of the planner's simple estimate is entirely adequate. Chapters 2 and 3 discussed that these estimates have a wide variation for accuracy. Yet as many jobs are overestimated as are underestimated and grouping jobs allows making meaningful schedules with much less overall variation of accuracy.

Why should planners not plan all jobs for the fastest technician? There are some problems associated with planning jobs for the fastest technician, namely schedule realism. For one thing, the supervisor needs some accuracy to schedule work. The estimates also set time standards. The time standards should allow average technicians to rise to the challenge to do a good job. These challenges and the scheduling are not practical if all jobs were planned for the fastest technician.

How do planners know how much time a good technician needs on any particular job? Planners cannot know unless they are at least good technicians themselves. By putting a best technician in planning, management assures itself that the planners can easily judge different jobs and assess what time frames and details should be needed by a good technician.

In addition, having a superior technician as the planner gives the job plan credibility. This helps when the supervisor finally assigns a planned job in the field. The field technician presumes the job plan to be practical and does not second-guess the standard.

Another consideration is that of job delays with wrench time. Although proper planning and scheduling helps produce high wrench time, the planners do not consciously consider wrench time when estimating jobs. The planners simply estimate each job without unexpected delays.

Without unexpected delays means the planner should consider including break time, but not time to find unexpected parts, tools, or instructions that were not identified by the planner. Jobs that take longer or shorter than estimated should be identified by the craft personnel reporting what happened on the work order form. These comments help guide the planner reduce delays on future jobs. The planner may be tempted to put in an extra hour in case problems arise. If the planner automatically included time in the estimate for unexpected delays or inexperienced craft personnel, actual delays or problems may not be reported if the overall time estimates were met. The preceding discussion addressed slower technicians, but not other type delays. Even the best planned jobs can run into unexpected problems. If a job runs over or under the estimated time, it is not as important as if crews become better each time they do the job. Crews become better each time doing the job with planning around expected delays. The planner anticipates some delays from personal experience, but the best source of delay information on any particular job for any particular equipment is from its equipment file. The planner can look in the file and readily see that previous jobs required an extra gasket. So the planner includes having an extra gasket in the job plan. This improvement would not be possible if the use of an extra gasket had never been reported by the technician. Having the time estimated for a smooth job (no unforeseen problems) obligates the technician to explain and record unusual circumstances that slowed the job. Therefore, planners should not include miscellaneous extra time in job estimates.

There are also other reasons to include only the amount of time planners actually estimate the job should take. The time estimate helps explain the job scope. For instance, giving a technician a 10-hour estimate for a job that obviously should take no more than 4 hours may confuse the intent of the job plan. Consider a job scope that calls for changing the lubricant of a fluid drive, but the planned estimate includes far too much time. The technician may wrongly conclude the job includes changing all the filters as well. The estimate of only the actual time expected also gives the technician a reasonable target to shoot for and keeps the work moving. In addition, many jobs go exceptionally well and end up taking less time than the planner imagined, so any extra time included on estimates should be kept to a minimum.

Planners should be mindful of avoiding the practice of estimating job durations by shift hours. This practice might cause many jobs to be planned as 4- or 8-hour jobs when crews work 8-hour shifts or 5 or 10 hours when crews work 10-hour shifts. This needlessly adds hours to jobs and reduces the schedulers' ability to put small jobs in schedule gaps. A 3-hour job should be planned for 3 hours, not 4.

A special word concerning breaks and startup or wrapup time is appropriate. The planners' estimates do not have the precision to be unnecessarily worried about such details. The planners figure that 10 hours of work can be accomplished in a 10-hour day. Planners usually consider there would be no breaks in short jobs, but one or two breaks in longer jobs during a single day. Similarly, the planner might realize there is a certain amount of time lost in starting up or wrapping up at the beginning and end of shifts. Plans for a short job do not make special provision for these times, but plans for longer ones might. On the other hand, job plans should recognize that time might be required for cleanup at the end of certain jobs regardless of size.

Finally, one must remember that even before management ever initiates planning and scheduling, a quality focus must be in place. Technicians must not feel they must meet a limited time estimate rather than do good work. Planning pushes so hard for productivity that the work force must have adequate concern for quality.

Figure 5.10 shows that the planner of this example work order estimated 5 hours for each person. The planner estimated the total job to require 10 labor hours and 5 duration hours.

Parts

The identification and coordination of parts or material is an area where the planner can greatly help improve craft productivity. Although scheduling information provides the greatest planning help to maintenance, planning's help with parts is the most visible. This is the reason most organizations begin planning departments. The offer of help with parts usually encourages technicians to accept planning.

On the other hand, the advertised idea that planning will identify all future parts severely hinders the accomplishment of the planning mission. Planning cannot gather all parts information before all jobs. The technician's idea of that purpose which planning does not fulfill gives planning a poor reputation that hinders later cooperation from technicians. The vital role that planning fulfills is to save and retrieve parts information that the technicians have previously gathered. This is how planning helps future jobs. Management should first introduce planning properly to avoid later misunderstandings.

The intent of planning with regard to parts is to identify them and ensure their availability before the job begins. The planning department may also stage certain parts to reduce delays in technicians having to gather them. As with job instructions, the planners reduce potential

WORK ORDER **#001235**

REQUESTER SECTION
Priority R2

#2 Control Valve B Strainer (N02-FC-003) has high
pressure drop. Def. Tag #010304. No outage required.
Needs mech crew. Clearance Req. No Confined Space.
J. Smith 4/21/98 3:00pm APPROVAL: S. Brown 4/21/98

PLANNING SECTION

Clean strainer positioned in front of the control valve.
Remove strainer element, clean, and replace.
Replace gasket if needed.
Labor: 1 Mech   5hr   Total labor 10hr
       1 Helper 5hr   Job duration 5hr

CRAFT FEEDBACK

APPROVAL:

CODING   Plan Type RE   Group/Syst FC   Crew   1-2
          Work Type 5   Equip Type 18   Outage 0

Figure 5.10   Work order after the planner estimates the time required.

job delays when the work order plan identifies parts because the technician does not have to make extra trips to procure them later after the start of a job.

To identify the parts a job requires, a planner first consults the equipment's minifile to review previous job requirements or problems. The plans and feedback from previous jobs contain most necessary parts information. Other material in the minifile may contain lists of possible parts. If the necessary information is not in the minifile, the planner may be able to easily identify the parts necessary in the inventory catalog or listing. The planner might do this even for reactive jobs if there is time. On proactive jobs, the planner might spend a considerable amount of time searching the plant's other technical files or vendor files. The planner makes sure to make a copy of any helpful information found to put in the minifile.

The planner writes down the information for needed parts directly on the work order form. The planner should identify only anticipated parts within the plan itself on the work order form. By only identifying anticipated parts within the work plan itself, the planner helps clarify the job scope. However, many jobs may need additional parts other than that first anticipated. The planner should include the identification of other parts that there may be a fair chance of using. This allows the technicians to obtain them more quickly if needed. The planner may identify these parts lower on the work order form and not necessarily within the plan of the anticipated work.

Even if there have not been delays on previous work, including a parts list in the planned package helps the technician understand the equipment and be ready for procuring unanticipated parts. The planner might attach a list of all parts the equipment uses if the list is readily available. The planner attaches a copy of this list to the back of the work order form. The planner never sends the original out in the field. The planner also makes a note on the work order form identifying attachments. This note helps the technician know there should be attachments in case something gets lost. It also helps the planner account for returned items at the end of the job. The planner would not attach a complete parts list for a simple job that would definitely not need any of the parts on the list. This would simply encumber the technician with extra paperwork.

Once a job begins the planner is no longer responsible for helping identify any necessary parts. The responsibility then falls on the crew supervisors and technicians. These persons may use any available resources including all the files in the planning department. They must record parts information uncovered during job execution as feedback on the work order form to help future work.

Equipment parts list

Information on parts that equipment might require comes in many forms. Lists of parts are known by different names such as part breakdowns and bills of material. Manufacturer "exploded view diagrams" or illustrated views showing how the different parts fit together are extremely useful.

The specific work orders from past jobs contain parts information either in the plans or in the feedback. Numerous repetitions of previous jobs develop a thorough listing.

The planner might use a parts form to record in one place the identification collected from previous work orders. Such a form is shown in Chap. 7, Forms and Resources Overview, for planners to include in minifiles when they create them.

If there is a standard plan as shown in Chap. 7 for a particular piece of equipment, it might contain a thorough listing of parts.

The storeroom inventory might be arranged by equipment number or contain a comparable sort. In addition, the CMMS computer system inventory module might automatically record the previous use of parts by equipment. This would provide an ongoing development of a parts list for each piece of equipment as maintenance works on jobs. Unfortunately, many CMMS inventory modules do not provide this automatic feature because they presume the plant has a complete set of parts lists for all its equipment. The modules primarily concern themselves with tracking quantities on hand and reservations for specific work order numbers.

When the company purchases new equipment, the equipment normally comes with an O&M manual showing a breakdown of parts. In addition or instead, the O&M manual might contain a list of recommended spare parts. Planners must insist on receiving these manuals when new equipment arrives. Planners should then create the appropriate minifiles to facilitate their use. If the planners merely place the manuals on the technical file shelves without sorting the information into minifiles, later efforts will be hindered. The planner may later find the manual to contain information for all the manufacturer's models and the planner will then be faced with an additional identification task.

Equipment sales proposals or negotiation notes may contain listings of proposed spare parts.

The planner might obtain lists of parts from vendors or manufacturers.

Management might initiate a project to gather parts lists from vendors and manufacturers. This effort could involve the identification of desired equipment, the collection of the lists, and the insertion into the proper minifiles. The project might involve plant engineers or other technical specialists. Management might choose to involve contract labor. Management might also choose to use field technicians to gather the data. Including technicians would facilitate their acceptance of the planning file concepts.

Purchasing

Sometimes a planner finds that a part necessary to execute a job is unavailable. The planner is responsible for all planned packages that are waiting for material. This material may be out-of-stock inventory items being ordered by the storeroom or nonstock items to be ordered by the planner. For nonstock items, the planning department is responsible for procuring the part.

The planner identifies on the work order form nonstock items that the planning department procured. The work order identification would normally name the item and describe where it was placed or staged after the plant received it.

The planning department has to include a purchasing capability or some method to procure parts. This purchasing capability may involve close coordination with a company or plant purchasing department. (In some companies, planning itself may be a department within the plant purchasing department.) Planning may also be able to purchase some material through direct vendor contact and purchase orders. The planning department might control several blanket purchase orders set up by the purchasing department to allow planning to buy certain material with a minimum of paperwork and administration.

When nonstock purchased material arrives at the plant, the planning department verifies that the shipment contains the proper material and makes the work order available for scheduling.

Many planning departments employ a separate purchaser or expediter to handle most of the purchasing coordination duties. This allows one or two persons to develop more familiarity with the company's purchasing requirements. This familiarity helps the person push through urgent work requests faster through the system if needed. In addition, this person might contribute to the planning department by having a special talent for finding parts that are difficult to locate. On the other hand, the planning department might prefer that the planners locate the desired parts and initiate the purchase. The purchaser would be more of an expediter to complete the details of the transaction and ensure the timely arrival of the material. Having a separate person might take an administrative burden from the planners.

In addition, once a job is in progress, the craft supervisor is responsible for procuring any parts that were unanticipated during planning. If the storeroom does not have these parts, they must be purchased. The planning department does not want supervisors to interrupt planners, but recognizes that a specialist might best handle purchasing. Having a separate person in planning that the supervisors can access for purchasing might allow resolution of both concerns. The supervisors could receive help, but not interrupt the planners.

A planner also needs to spend considerable time in the field. A planner spends time in the field to scope work as well as to inspect jobs-in-progress. Looking at jobs-in-progress allows the planners to increase their feel for the degree that jobs proceed according to plans and the completeness of feedback. On the other hand, a purchaser frequently needs to be available to accept return phone calls from vendors. Having the purchaser separate from the planner may facilitate these different planning department needs.

A plant might consider having a separate person to handle some purchasing duties under the following circumstances:

1. There is a lot of bureaucracy or complexity in the purchasing procedures.

2. Maintenance has a person available with the special talent required.

3. The planning department is fairly large and has more than three or four planners.

4. There is a continual receiving of shipments that must be inspected and staged.

5. There is a continual need to spend time verifying that parts will be compatible with the system or existing standards.

6. Planners are each planning for 30 technicians rather than 20.

In general, the purchasers would either be responsible to or report to the planners. Depending on their duties, the purchasers may not necessarily have to be technicians. Clerks may be qualified to serve as purchasers. Appendix F includes an overview of these purchaser or expediter duties.

Storeroom, reserving, and staging

In addition to naming any parts on the work order form, the planner also records the storeroom identification or stock numbers. This precisely identifies the material to avoid possible misunderstanding or delays at the storeroom. The technician presumes any planned part with a stock number is kept in the storeroom. The planner identifies which storeroom stocks the material if the plant has more than one storeroom.

Next the planner reserves anticipated parts in the storeroom and marks the planned items as "Reserved" on the work order form. *Reserving parts* means that the planner has placed a reservation on an item with the storeroom. This action ensures that the storeroom will not run out of the items before maintenance executes the particular job that had a reserved part. The storeroom uses the work order number to identify the reservation requirement. Normally the planner reserves parts that the plans require from the storeroom even if the storeroom has an ample supply of the items. This allows the storeroom more timely notice of the consumption of parts and helps it prepare for replacing inventory. The planning department also develops good work habits by reserving most parts. There are certain parts that the storeroom keeps in limited supply and the reserving of parts on a routine basis reduces the possibility of overlooking their reservation. Otherwise, a technician may start a planned job and find that a planned part is unavailable.

One common problem plants encounter is not experiencing a high stockout rate when, in fact, the storeroom frequently does not have enough parts the planner or technicians seek. Stockouts measure how many times the storeroom is out of an item that the plant currently requires.

Stockouts do not necessarily measure that a storeroom is either out of a material or has a less than desirable quantity on hand. The discrepancy is caused by the planners or technicians not reserving parts or requesting pick tickets for items that are out of stock. When a planner sees that there is no desired item available, the planner will often plan the job another way. For example, a planner may change a plan to replace a valve bonnet to replacing the entire valve because the storeroom is out of bonnets. A technician may change the execution of a job in the same manner. Because no one ended up requesting the deficient part, the plant never experiences a stockout. Therefore, the planner and technicians should always reserve the parts to cause a stockout even though the planner and technicians alter the jobs. In this manner, the jobs are planned and executed expeditiously and the storeroom receives the appropriate signal to have enough quantities on hand in the future.

Another related problem is the storeroom having parts reserved for jobs that maintenance has already completed. This is usually caused by technicians requesting parts at the storeroom without informing the storeroom of a previous reservation. The technician receives the requested item, but the storeroom does not reduce the reservation quantity. The storeroom handles this problem by checking work order numbers for previous reserved parts when issuing parts to technicians. The storeroom should also check completed work order numbers against outstanding part reservations occasionally and canceling the reservations.

Next the planner stages any appropriate parts and marks on the work order form which items are "staged" and where. If items will not be staged until later, the planner does not mark on the work order form. Those items will be marked later when staged.

Staging is done for the same reason the operations group clears equipment ahead of time: to avoid delays. A plant does not desire for a crew to stand around waiting for operators to clear a piece of equipment. Similarly, the plant prefers a crew not to stand around waiting for a part or tool to begin a job if the item could have been provided ahead of time. The provision of the tool ahead of time is referred to as *staging*. Staging goes beyond reserving items in the storeroom or tool room. Staging places the items in convenient locations for the technicians. These convenient locations reduce the need for crews to make more time-consuming arrangements or side trips to gather the items.

The planning department always stages special purchased parts that inventory did not carry. The planner must note on the job plan where these parts were placed after their receipt.

It may be more appropriate for supervisors or schedulers to determine whether parts should be staged based on scheduling concerns. Therefore, the next chapter on scheduling discusses guidelines to assist planners determine the need to stage parts. These guidelines are summarized in App. R.

Figure 5.11 shows an example work order after the planner has determined the necessary parts. In the equipment's minifile, the planner found that a previous job identified the gasket stock number. Because the job was reactive, the planner would not have otherwise spent much time trying to determine a gasket number if it had not been readily available. The planner noted the gasket's price when reserving it and wrote the necessary information on the job plan.

WORK ORDER #001235	

WORK ORDER #001235

REQUESTER SECTION   Priority R2

#2 Control Valve B Strainer (N02-FC-003) has high pressure drop. Def. Tag #010304. No outage required. Needs mech crew. Clearance Req. No Confined Space. J. Smith 4/21/98 3:00pm. **APPROVAL:** S. Brown 4/21/98

PLANNING SECTION

Clean strainer positioned in front of the control valve.
Remove strainer element, clean, and replace.
Replace gasket if needed.
Labor: 1 Mech   5hr   Total labor 10hr
          1 Helper 5hr   Job duration 5hr
Parts: Strainer lid gasket   GSK-RR-130* Qty 1 Cost $10ea

                    *RESERVED

CRAFT FEEDBACK

                APPROVAL:

CODING	Plan Type RE	Group/Syst FC	Crew 1-2
	Work Type 5	Equip Type 18	Outage 0

Figure 5.11   Work order after the planner determines the parts required.

Special Tools

Similarly, special tools is an area where planners can help boost crew productivity by reviewing past jobs in the minifile. A *special tool* is any device that would not ordinarily be carried in a craft tool box. Examples are come-alongs, cranes, and shim packs.

The planner's intent is to allow the technicians to gather all the tools they should need before they first go to the job site to avoid extra trips later. As with parts and job instructions, the planners reduce potential job delays when the work order plan identifies special tools because the technician does not have to make extra trips to procure them later after the start of a job. As with any trip, the technician's leaving the job site delays the job and might not consist of merely traveling to the tool room and returning with the proper tool.

Special tools might be kept in the tool room or other places. Certain tools may be available in one of the craft shops. Other highly specialized tools may be kept at the site of the equipment where it is normally utilized. For example, one plant has a special bar that is used to apply enough torque to unbind a particular control valve. The plant keeps the bar next to the valve. One of the planner's duties when scoping this job is to ensure the bar is there. Another plant keeps a locker full of special tools to work on burner parts on the burner deck of its unit. The planner should remind the technicians to retrieve the locker key from the tool room when they perform work on the burners.

The primary source for special tools is the planner's personal experience and information from past jobs in the minifile. There also may be lists of special devices recommended by manufacturers or vendors in O&M manuals. These lists should be kept in the minifiles once they are found. The tool room might also employ a "job tool card" to keep track of tools issued to jobs rather than individual technicians. This allows the tool room to issue tools over a period of time to larger jobs that might be conducted by several crews of technicians working alternating shifts. In these jobs and other jobs, it might be easier for the tool room to issue the tools to jobs rather than individuals. On these jobs the planner should encourage the technicians and tool rooms to provide copies of the cards after job completion to keep in minifiles.

Infrequently, the planner may also be able to help the technician avoid carrying an entire tool box to the job site. For instance, a certain job may only require a flathead screwdriver. In these cases the planner would note that "job only requires" the particular few tools prescribed.

The planner writes special tools on the work order form. The planner writes any special identification numbers if the tool room has such a system. (As discussed later, the planner also uses this section of the form to indicate if the job needs any contractors such as insulators.) The planner also identifies any special tools that the planner staged in accordance with the plant's guidelines. Staging of special tools often helps crews avoid delays. Chapter 6 discusses staging.

Figure 5.12 shows an example work order after the planner finished writing in tool information. The technicians would take out the strainer and place it in the plastic bag. The technicians would then transport it and clean it in the shop steam cleaning room. The technicians would then replace the strainer and refasten the assembly. The technicians would clean up the area and put the rags in the special hopper in the

---

WORK ORDER #001235

REQUESTER SECTION   Priority R2

#2 Control Valve B Strainer (N02-FC-003) has high pressure drop. Def. Tag #010304. No outage required. Needs mech crew. Clearance Req. No Confined Space. J. Smith 4/21/98 3:00pm. **APPROVAL:** S. Brown 4/21/98

PLANNING SECTION

Clean strainer positioned in front of the control valve.
Remove strainer element, clean, and replace.
Replace gasket if needed.
Labor: 1 Mech   5hr   Total labor 10hr
          1 Helper 5hr   Job duration 5hr
Parts: Strainer lid gasket GSK-RR-130* Qty 1 Cost $10ea
Tools: Rags, Can of degreaser, Plastic garbage bags.

                              *RESERVED

CRAFT FEEDBACK

                    APPROVAL:

CODING	Plan Type RE	Group/Syst FC	Crew 1-2
	Work Type 5	Equip Type 18	Outage 0

Figure 5.12 Work order after the planner determines the tools required.

fuel oil room. The planner needed only to write down the rags, degreaser, and plastic bag as special tools.

Job Safety

Job planners never take job safety for granted. The planner first makes sure the origination section of the work order specifies whether the operations group must clear the equipment. The planner contemplates if there are conditions on the job site that will affect the safety of personnel. For each special safety concern, the planner describes the necessary safety issues on the work order form and attaches or references any pertinent information. The planners always ensure that any developed or researched information is copied to the minifile. The planner may find useful information already in the minifile. This is one circumstance in which a planner might do more extensive research on a reactive or minimum maintenance job.

The planner also considers whether the job will be in a confined space or involve special chemicals. As usual, the minifile becomes a repository to help scope jobs in this regard and retain useful information.

Confined space

A confined space is an area with a potentially dangerous environment regarding respiration. The planner considers several questions regarding possible confined space work. Is it possible this is a confined space? Is it possible the space requires continuous air monitoring and a hole guard? Is there work to be done at intervals that will require special monitoring? Will the job scope change during the repair affecting the conditions as permitted?

If the work involves confined spaces, the planner ensures that the work order origination information and work plan reflect this requirement to follow the established plant procedures. The planner adds "entry supervisor" or "hole person" to the persons required section of the work order form.

Material safety data sheets

If there are hazardous chemicals present on the job site, the planner writes or attaches the necessary MSDS information on chemicals present. The planner will write the process necessary to protect maintenance personnel if necessary.

The Internet is becoming a useful source for accessing MSDS sheets for certain substances. A CMMS might also be helpful in this regard.

Individual countries, states, and plants have unique requirements for what information and safeguards must be available and what must be included in individual job plans. The planning department should meet with the designated company safety authorities to clarify particular requirements for work orders.

Estimating Job Cost

At one plant the planning manager had the planners stop totaling up the cost for each work order because no one was tracking it. This rationale misses the point of planning. The planners are one of the major users of this information. The planners should be able to see how much past work on a piece of equipment has cost in order to make intelligent maintenance scope decisions. If the past maintenance costs have been high, the planner can properly present a case for expensive replacement equipment without resorting to such statements as "We don't know how much it's costing us, but it's really hurting us." Management cares about bottom line cost. Maintenance professionals must learn the language of financial dealings. Management desires realistic cost comparisons over opinions. Maintenance would rather consider a case of "The old pump seals have cost $2000 per year to maintain, but we can eliminate that cost with a new seal design for only $3000."

Any cost collection is better than none. It is not legitimate to wait on the eventual installation of a computer or CMMS to collect cost information. The cost information provided by work plans is the first step in having useful data even if actual field costs are not gathered. This information is accurate enough to guide the decisions that will be made. Whether a job cost $200 or $400 is not as important as whether it cost $400 or $3000. The minifiles will show the trend and magnitude of maintenance costs accumulating from multiple work orders as the equipment is maintained. The minifiles cannot do this practically if the information is not estimated at the time of the work. The information needs to be in a state where the planner can easily glance at previous job costs as replacement equipment is considered. Engineers can go through minifiles and spend time occasionally to total up job costs for a few projects. However, planners must be able to see this cost data without strenuous effort as they plan routine maintenance jobs daily.

Another reason actual costs may not be critical to maintenance assessments is due to a philosophy of what the job cost. Consider a job that planning estimated for a single day, but through poor scheduling the job took 2 days to complete. However, the technician put 3 days on a time sheet because the crew supervisor did not assign any more work. Did the equipment cost the company 1 day, 2 days, or 3 days of labor and unavailable equipment? The equipment probably cost the company 1 day and poor maintenance practices cost the company 2 more days. The reason the equipment cost only a single day was that the job's standard was established by the planner. That was what the job should have taken.

On the other hand, most accounting systems and computer programs might attribute all 3 days to the equipment. This is because they do not attribute any cost to an inefficient maintenance organization. In addition, it is very difficult to establish on a job-by-job basis the cost to the company for lost availability. Therefore most maintenance cost systems do not account for them on individual work orders. Whether or not all of these differences are meaningful is beside the point. The differences should not delay the collection of maintenance information. The rapid accumulation of helpful job information will begin when it is started.

Another reason to show cost information on the job plan is to guide technicians when working with parts. Certain low cost parts may not be practical for the technician to return to the storeroom when unused. Other small parts may surprisingly cost thousands of dollars. These parts should be handled with more care than others. The technician can only make these determinations when the planner includes price information on the job plan.

The planner first calculates labor cost. The planner first uses a standard rate for all labor hours. This allows the planner quickly to add up the total cost of labor. It also reduces jealousy among different crafts and skill levels not to have actual wage information thrown in their face. However, the use of some labor rate shows how much labor does cost the company and encourages everyone to become as efficient as possible. Using a standard rate for all labor hours is justifiable. Maintenance decisions require an accuracy showing whether certain equipment costs hundreds of dollars each year or thousands. The wage difference between a trainee and a certified electrician will not skew data used for this purpose. In addition, the planner has planned the job for the minimum skill level without any certainty who will actually be assigned. Therefore the planning department uses a standard labor wage rate. (A CMMS might easily allow using exact labor rates but caution should be used considering different crafts may not appreciate seeing exact wage differentials.)

This book uses a standard $25 per labor hour to illustrate the concept of estimating labor hours. This figure includes all wages and benefits that the company pays per labor hour. It does not consider wrench time or administration time. It merely represents what the company would pay to an employee each year in benefits and wages divided by the usual number of hours worked each year on straight time. Job plans do not consider that jobs may be worked on overtime. Overtime is more of a cost of scheduling or maintenance practice than planning. This figure may not be appropriate for many industries or geographical areas.

The planner uses a standard rate of $25 per hour for each work hour and writes the labor cost estimate directly on the work order form.

The planner then calculates the cost of parts. First, the planner only needs to include the cost of parts that the planner anticipates the craft will use. Second, the planner determines the cost of the items. The planner notes the inventory price when the planner reserves the part in the inventory computer system. If the computer lists several prices for an item, the planner uses the last purchased price since that is the actual cost to the company as a whole. If there is no price information, the planner makes an educated guess. The planner may also have the benefit of items identified with a price on a bill of material or other equipment breakdown list. All of these sources are usually accurate enough for the purposes of overall cost accumulation for equipment. The planner does not need to include items not of significant value such as a few dollars worth of bolts. The planner should not tie up the time of a purchaser or parts expediter to determine routine cost information. The planner should use the parts expediter to provide the value for parts needing special purchasing when the expediter orders the part. The planner includes the cost for all anticipated parts on the work order form. The planner writes the individual cost for each part and includes the total cost of parts in the total estimate for the job.

Next, the planner considers the cost of special tools. The planner does not include a cost for a special tool unless the item is not available in the tool room and a special cost will be incurred. The cost for contracting out work would be included as a special cost. This chapter discusses contracting work in the next section.

The planner then totals all the cost estimates for labor, parts, and tools at the bottom of the planning section of the work order form.

Finally, the planner informally consults the planning supervisor if the estimated cost is over a certain amount established by a planning department guideline, $5000 for example.

The planning department has a guideline to scrutinize more expensive job plans to determine if another strategy is advisable. Some persons may feel that this is an unnecessary precaution. Their reasoning suggests that because the plant already exists, it must be maintained. They reason that all jobs must be executed. However, the planner consultation is not necessarily checking to see if the plant will execute the job. The planner consults to see if a more prudent alternative exists. For expensive jobs, it never hurts to get a second opinion. In addition, perhaps the plant should not execute the job after all. This plant exists, not another plant. Sometimes employees request improvements to the plant that are simply not advisable for a number of reasons. Many times the proposed task exceeds the economic point of diminishing returns. There would be a benefit, but the benefit would not outweigh the cost of the job. Identifying projects that would modify the plant is one reason the plant classifies jobs according to work type. Project work adds capability that was not had before; the plant or equipment is better than before. Projects must be carefully weighed to see if they are good users of the company's funds. For major projects, most companies have a project proposal and approval process, usually not involving the planning group. However, maintenance considers smaller project work continually. Sometimes the classification is not very clear; a repair job to overhaul a broken pump may be planned to include a better design impeller. These considerations occur daily in planning at the work order level, and mechanisms need to be in place to guide routine maintenance work with regard to expensive tasks.

Figure 5.13 shows the example work order after the planner finishes the job plan by completing the job estimate.

WORK ORDER #001235

REQUESTER SECTION

Priority R2

#2 Control Valve B Strainer (N02-FC-003) has high pressure drop. Def. Tag #010304. No outage required. Needs mech crew. Clearance Req. No Confined Space.

J. Smith 4/21/98 3:00pm **APPROVAL:** S. Brown 4/21/98

PLANNING SECTION

Clean strainer positioned in front of the control valve. Remove strainer element, clean, and replace. ·
Replace gasket if needed.

Labor: 1 Mech 5hr   Total labor 10hr
       1 Helper 5hr   Job duration 5hr

Parts: Strainer lid gasket GSK-RR-130* Qty 1 Cost $10ea

Tools: Rags, Can of degreaser, Plastic garbage bags.

                                        *RESERVED

Planner D. Lee 4/22/98 Job estimate: $260

CRAFT FEEDBACK

                          APPROVAL:

CODING	Plan Type RE	Group/Syst FC	Crew 1-2
	Work Type 5	Equip Type 18	Outage 0

Figure 5.13 Work order after the planner estimates the total plan cost.

Contracting Out Work

A *contractor* is a company that the plant hires to do specific tasks. Different companies have different strategies regarding contracting out work. Some companies prefer to use contractors as little as possible. Other companies regularly contract work. Some companies are themselves contractors using planning principles to increase their productivity. Normally other company divisions manage work done by contractors either as projects or general contractor work. Occasionally planning must coordinate outside contractors for ordinary work orders such as the setting of a safety valve. This section presents information for a planning department that has some interaction with contractors. Appendix Q thoroughly reviews the issues of contracting maintenance work.

A word of caution advises management that planners and technicians sometimes fear that the establishment of the planning department promotes contractor work. Their reasoning suggests that the primary reason management implemented planning was to create work plans for contractors less familiar with plant equipment. Management might try to ease this worry by pointing out two things. One, planning should improve in-house efficiency so that contractors are less competitive. Two, Planning Principle 5 specifies that planners create plans for technicians who are familiar with plant equipment.

The planner identifies and writes the cost for contractor work in the special tools portion of the work order plan.

Insulation

Presume for the purpose of illustration that a plant routinely contracts all insulation work. This plant feels that the concern for asbestos and the need for special tools and materials to work with insulation makes using a contractor advisable. The contractor also has the ability to ramp up and down personnel levels faster than the plant. This is useful for periods when the plant does not require much insulation work. The plant has the insulation contractor under a special contract that pays a specified rate for insulation work.

Insulation work lends itself to a minimum of craft interference because the contractor can remove the insulation with the equipment still in service before maintenance work commences. The insulation contractor can later replace the insulation after maintenance completes its work and the plant returns the equipment to service. Scaffolding presents a similar situation.

When insulation must be removed, the planner puts the work order in the waiting-for-insulation-work file. The planning supervisor (or another designated individual) coordinates the insulation contractor work and returns the work order to the planner after the contractor removes the insulation. The planning supervisor gives the contractor's cost estimate for removing and replacing the insulation to the planner at this time. The planner then proceeds to finish scoping or otherwise planning the work, if needed, and passes the work to the waiting-to-be-scheduled file. The planner must exercise caution that necessary equipment tags or job markings remain in place or are replaced in order to help the technician later executing the job.

When the craft completes its work and the planner receives the work order back, the planner then makes a copy of the work order form. The planner places the copy of the form in the waiting-for-insulation-work file for the planning supervisor to coordinate the contractor for replacing the insulation. The planner uses the estimated insulation cost for the actual cost also. The paperwork of determining the actual cost from the blanket work order used for insulation is not practical. However, in unusual circumstances, the contractor informs the planning supervisor of the actual cost. This information is returned to the planner for updating the minifile.

Other contracted out work

The planner handles on-site contractor work that is not routine or does not have special contracts in place the same way as purchasing nonstock parts. The planner has the planning department purchaser determine the cost and have the contractor ready to mobilize. The planner then puts the work order in the waiting-to-be-scheduled file with a note describing what coordination is necessary for using the contractor. The maintenance scheduler initiates the coordination if advance notice is required. The crew supervisor makes any coordination requiring less than a few days. The crew supervisor supervises the contractor on-site.

The planner does not supervise contractors because that would interfere with future job planning and otherwise engage the planner in an activity that does not leverage maintenance. Anyone from the crew would be spending an hour to supervise when the planner would have to spend an hour as well. There is no leverage from the planner. On the other hand, there is a leverage from the crew supervisor's standpoint. The crew supervisor should be out in the field anyway so that the crew supervisor supervising a contractor may not take any time. The crew supervisor should supervise the contractor, not the planner.

Plant engineers or technical specialists normally supervise or inspect contract work performed off-site. However, some small jobs such as getting parts rebuilt might involve the maintenance force. For getting parts rebuilt off-site for a job-in-progress, the crew supervisor coordinates this work through the tool room or through the purchaser (or expediter). For rebuilding used parts after maintenance completes a job, the crew supervisor writes and submits a new work order to rebuild and have the parts placed in stock. The planners would then coordinate the work. Placing the responsibility to initiate follow-up work directly on the crew supervisor lessens misunderstandings resulting in parts not being rebuilt.

Closing and Filing Feedback after Job Execution

The planner now performs one of planning's most important tasks. To move each future job up the learning curve, a planner must place information used or discovered during a job into the minifile. The planner completes the work's actual cost directly on the work order form and updates any necessary minifile sheets. The planner places the original work order form in the minifile. When filing the work order form, the planner sends a copy to the planning clerk for updating the computer for job closure. The clerk subsequently forwards that copy to the originator or otherwise notifies the originator of job completion.

If the technician or planner has indicated on the work order form that drawing or equipment technical data has changed, the clerk sends an extra copy to the plant engineering department. If the plant keeps a single CMMS database for equipment design information, the engineering department might not need information other than to revise drawings.

The planner needs to ensure that the necessary details of the actual job execution are clear enough to maintain the equipment database and help future work. Occasionally, the planner must dig and dig to get good job feedback. The planner might need to consult with the technicians or supervisors to clarify job details. A routine failure of a crew to report feedback must be brought by the planner to the crew supervisor or planning supervisor's attention. In addition, if the work order feedback indicated that maintenance made only a temporary repair, the planner may need to ensure that the necessary follow-up work orders have been written to address the situation.

The following provides guidelines for adequate job feedback the craft technicians should provide.

1. Identify quantity of persons and specific craft and grade of each person. Identify the names of the persons.

2. Identify labor hours of each person. Give start and finish times of job. Explain any variance from the plan estimates if greater or less than 20%.

3. Thoroughly describe the problem if not accurately specified by the plan.

4. Thoroughly describe the action taken if the job did not proceed according to the plan. Report any special problems and solutions.

5. Identify actual quantities of parts used and report stock numbers if not given by the plan.

6. Identify actual special tools used or made if not given by the plan.

7. Return the original work order and all attachments provided by planning. Include any field notes and return any datasheets that the technician filled out whether or not planning provided them.

8. Return updated drawings.

9. Note any changes to equipment technical information such as new serial numbers and model numbers and names. Return any manufacturer's information or literature that was received with any new parts being installed. This information is especially vital and often cannot otherwise be determined to help future maintenance.

10. Include any other information such as bearing clearances (radial and thrust), wear ring clearance, shaft runout clearance, bearing to cap clearances, coupling condition, and gap clearance.

11. Make any recommendations to help future plans.

Technicians generally want to give good feedback on job plans, especially when encouraged by proper management support and communication regarding the role of planning. Four particular strategies further help promote receiving good feedback. First, the simple notion that they are supposed to give feedback surprises some technicians. The plant might want to conduct a short, 1- or 2-hour training class for the maintenance workforce. The emphasis of this class would describe the technician's role in improving job plans and procedures as part of the planning program. Second, a planner should attend regular crew meetings at least once every week or two. The planner would emphasize the help expected from the technicians. The planner might say, "Listen. I'm not Superman with x-ray vision when I scope out a job. I give the plan my best effort, but I expect you folks to tell me what the equipment actually needed and what you did. I may not be able to give you a perfect plan, but I surely promise to be a faithful file clerk. I can attach those part numbers you find to future jobs, so we don't have to 'reinvent the wheel' on every job in the future." Having the planner attend some crew meeting greatly reduces "us versus them" restriction of information and ideas, and encourages the players to work together as a team. Planning as a system requires working together. Third, regardless of whether the company employees a CMMS, paper work orders should accompany technicians in the field. It is on this document that the technician can precisely mark areas of the plan needing improvement. A CMMS that merely allows a check next to a step or the technician filling in a long description field at the end of the plan is inadequate. Experience shows that on the same job, technicians who make extensive comments all over a paper work order form tend to enter much fewer comments on the computer. This reduction in helpful information is due to the difficulty in describing the exact parts of the plan needing attention and the slowness of typing in the information. The paper work order in the field also allows actual input in real time before the technicians forget exactly what they meant. Tasking the technician to copy the paper comments into a computer later is duplication of effort. Providing the technicians with computerized field devices means small handheld devices where data entry is difficult and is not geared toward individual job steps. Larger laptop field devices allow easier data entry, but at some point pass the point of diminishing company returns for replacing simple paper with expensive rugged electronics for every field technician. Technicians can take paper work orders directly to job sites and mark them up. The technicians hand these papers back through their supervisors to the planning department. The planners directly view the marked up papers and file them in minifiles to help future jobs. In a pure paper environment, the planner would simply review the feedback and file the completed work in the minifile to help future plans. In a CMMS situation, the planner would review the feedback, update any electronic standard plan, and then allow a clerk to manually enter the feedback as best as possible into the standard CMMS data entry fields and file the paper work order. Even with a CMMS, making minifiles to hold paper work orders is advised for better future research. It is easier to scan paper files than scan computer files in some cases and for different persons who may need to consult files. In any case, be aware that the quality of job feedback may diminish as a plant abandons paper in favor of computer data entry for field technicians. A similar issue arises with paper or computer feedback. Should there be a survey type form asking specific questions about the job? Could parts be improved? If so, how? Could safety be improved? If so, how? And so forth. These types of questions tend to make work order forms too long and actually discourage some technicians from marking up the plan in specific places. Experience shows that a general statement at the end of a work order to "Mark up the plan where needed for improvement either on the pertinent section of the plan, on the space below, or on the back of the work order" may be more beneficial. This type of statement might be recommended most for new PMs or entirely new plans in a plant where the technicians are not used to giving feedback. Fourth and finally, plans having a plan history section at the end of the work order greatly encourage feedback because technicians see that the plans use their recommendations. This is not so practical in a pure paper environment, but a CMMS that prints out work orders for technicians can easily list this type of information. The history should include the date, the change, and the planner making the change. This section should always be in the same area of the work order for easier reference (usually at the end). (In addition, adding reference sections for other than history is a good idea even if not for feedback reasons, especially ones identifying O&M manuals.) Thus although technicians usually support giving good feedback when they understand the planning program, four strategies help encourage better feedback. Giving a short feedback expectation class, having planners visit crew meetings, using paper work orders in the field, and including a procedure history section all help facilitate better feedback.

Figure 5.14 shows an example work order after the crew has executed the work and given feedback to planning. Figure 5.15 shows the closing notes the planner made to the work order to update the work

---

WORK ORDER #001235

REQUESTER SECTION
Priority R2
#2 Control Valve B Strainer (N02-FC-003) has high pressure drop. Def. Tag #010304. No outage required. Needs mech crew. Clearance Req. No Confined Space. J. Smith 4/21/98 3:00pm APPROVAL: S. Brown 4/21/98

PLANNING SECTION
Clean strainer positioned in front of the control valve. Remove strainer element, clean, and replace. Replace gasket if needed.
Labor: 1 Mech   5hr   Total labor 10hr
        1 Helper 5hr   Job duration 5hr
Parts: Strainer lid gasket GSK-RR-130* Qty 1 Cost $10ea
Tools: Rags, Can of degreaser, Plastic garbage bags.

                                    *RESERVED
Planner D. Lee 4/22/98 Job estimate: $260

CRAFT FEEDBACK
Cleaned strainer. Replaced gasket. Used 2" combination, 2" impact socket, and impact wrench. 2 mech-7 hours ea. Job Started 4/23 7am. Finished 4/23 2pm.

C. Jones 4/23/98
                APPROVAL: L. Vincent 4/23/98

| CODING | Plan Type RE | Group/Syst FC | Crew 1-2 |
| | Work Type 5 | Equip Type 18 | Outage 0 |

Figure 5.14 Work order after the crew executes the job and provides feedback.

```
WORK ORDER #001235
REQUESTER SECTION                         Priority R2
#2 Control Valve B Strainer (N02-FC-003) has high
pressure drop. Def. Tag #010304. No outage required.
Needs mech crew. Clearance Req. No Confined Space.
J. Smith 4/21/98 3:00pm APPROVAL: S. Brown 4/21/98

PLANNING SECTION
Clean strainer positioned in front of the control valve.
Remove strainer element, clean, and replace.
Replace gasket if needed.
Labor: 1 Mech   5hr    Total labor 10hr    Actual  14
       1 Helper 5hr    Job duration 5hr    Actual   7
Parts: Strainer lid gasket GSK-RR-130* Qty 1 Cost $10ea
Tools: Rags, Can of degreaser, Plastic garbage bags.


                          *RESERVED
Planner D. Lee  4/22/98  Job estimate: $260  Actual: $360

CRAFT FEEDBACK
Cleaned strainer. Replaced gasket. Used 2" combination,
2" impact socket, and impact wrench.  2 mech-7 hours ea.
Job Started 4/23 7am. Finished 4/23 2pm.


C. Jones  4/23/98
                    APPROVAL: L. Vincent 4/23/98
CODING  Plan Type RE    Group/Syst FC    Crew   1-2
        Work Type  5    Equip Type 18    Outage 0
```

Figure 5.15 Work order after the planner writes in the actual field cost for the history file.

order form totals for time and cost. Figure 5.16 shows a new work order an operator wrote 5 months later with a similar problem. Figure 5.17 shows how the planner was able to use feedback from the job completed previously to improve the new work order's job plan. The planner was able to identify several special tools to help the craft technician avoid an extra trip to gather an impact wrench and sockets after arriving at the job site. Notice that the planner did not change the plan for craft skill or time required. The planner still felt that the job required only a single

```
┌─────────────────────────────────────────────────────────┐
│          WORK ORDER  #002107                            │
│ REQUESTER SECTION              Priority R2              │
│ Unit 2 Cntrl Vlv B Strainer (N02-FC-003) high          │
│ differential, needs attention. Def. Tag #037114. No outage. │
│ Mech crew.  Clearance Required.  No Confined Space.    │
│ F. Balder 9/11/98 2am   APPROVAL: S. Brown 9/11/98     │
├─────────────────────────────────────────────────────────┤
│ PLANNING SECTION                                        │
│                                                         │
│                                                         │
│                                                         │
│                                                         │
│                                                         │
│                                                         │
│                                                         │
├─────────────────────────────────────────────────────────┤
│ CRAFT FEEDBACK                                          │
│                                                         │
│                                                         │
│                                                         │
│                    APPROVAL:                           │
├─────────────────────────────────────────────────────────┤
│ CODING  Plan Type  RE   Group/Syst  FC   Crew    1-2   │
│         Work Type  5    Equip Type  18   Outage  0     │
└─────────────────────────────────────────────────────────┘
```

Figure 5.16 New work order later on same equipment after the originator completes the information.

mechanic with a helper for 5 hours. Finally, Fig. 5.18 shows a more sophisticated work order with reference and history sections.

Summary

Seeing explicit descriptions of the steps a planner takes helps one understand how the company actually conducts maintenance planning. The chapter described first the work order process and then the flow of planning

WORK ORDER **#002107**

REQUESTER SECTION   Priority R2

Unit 2 Cntrl Vlv B Strainer (N02-FC-003) high differential, needs attention. Def. Tag #037114. No outage. Mech crew. Clearance Required. No Confined Space. F. Balder 9/11/98 2am   **APPROVAL:** S. Brown 9/11/98

PLANNING SECTION

Clean strainer positioned in front of the control valve. Remove strainer element, clean, and replace. Replace gasket if needed.

Labor: 1 Mech   5hr   Total labor 10hr
       1 Helper 5hr   Job duration 5hr

Parts: Strainer lid gasket GSK-RR-130* Qty 1 Cost $10ea

Tools: Rags, Can of degreaser, Plastic garbage bags,
       2" combination, 2" impact socket, and impact
       wrench.

                                        *RESERVED

Planner D. Lee 9/11/98 Job estimate: $260

CRAFT FEEDBACK

                          APPROVAL:

CODING   Plan Type RE   Group/Syst FC   Crew 1-2
             Work Type 5   Equip Type 18   Outage 0

Figure 5.17 New work order after the planner improves the job plan with feedback from previous work on equipment.

activity within the work order process. As the discussion unfolded, a planner scoped and planned an example work order and the reasoning behind each step was explained. This chapter should allow companies to tailor their own systems to implement effective planning. The next chapter presents the same level of detail in describing the specific activities of advance scheduling and daily scheduling. Appendix E narrates duties through step-by-step activities for the maintenance planner and App. F does the same for many of the other persons involved in the planning process.

Work order		**#002107**

Requester section   Priority R2

Unit 2 CntrlVlv B strainer (N02-FC-003) high differential, needs attention. Def. Tag #037114. No outage. Mech crew. Clearance required. No confined space.

F. Balder 9/11/98 2 am   **Approval:** S. Brown 9/11/98

Planning section

Clean strainer positioned in front of the control valve.
Remove strainer element, clean, and replace.
Replace gasket if needed.

Labor: 1 Mech   5 hr   total labor 10 hr
         1 Helper   5 hr   job duration 5 hr

Parts: Strainer lid gasket GSK-RR-130* Qty 1 cost $10ea

Tools: Rags, can of degreaser, plastic garbage bags,
         2" combination, 2" impact socket, and impact
         wrench.   *Reserved

Special reference and procedure history section*

Reference (not needed on job): ACME Mnl#7 pg 14

job plan/procedure history:

04/25/98 Added 2" combo/socket/impact per fdbk. Lee

planner D. Lee 9/11/98 job estimate: $260

Craft feedback

                        Approval:

Coding	Plan type RE	Group/syst FC	Crew 1–2
	Work type 5	Equip type 18	Outage 0

Figure 5.18   Add reference and history sections to better explain plan and encourage feedback.

6

Advance Scheduling

This chapter continues the nuts and bolts of making the planning system work with regard to scheduling. The chapter shows exactly how to do a practical method of scheduling.

In actual practice, it may be helpful to note that persons may consider scheduling a somewhat vague term. To be more precise, *advance* or *weekly scheduling* means a scheduler allocating an amount of work orders for a week without setting specific days or times to begin or complete individual work orders. Likewise, *daily scheduling* means a crew supervisor assigning specific work orders to specific individuals to begin the next day. A maintenance group uses both weekly and daily schedules. This chapter describes the activities to accomplish weekly and daily scheduling. In addition, the chapter covers how maintenance personnel stage material and tools. Although this book focuses on routine maintenance, the book also explains key scheduling concepts behind successful outages. Finally, the chapter compares and contrasts the concepts of scheduling with the concepts of quotas, benchmarks, and standards.

Weekly Scheduling

The scheduler performs most of the tasks of advance scheduling. The scheduler first gathers jobs from the waiting-to-be-scheduled file and any work returned from the previous week's schedule. The scheduler then allocates them into each crew's work hour forecast for the next week. The scheduler allocates jobs by work order priority, then number of work hours, but also makes other considerations per the scheduling principles. The scheduler utilizes scheduling worksheets for assistance. The end product is a package of jobs that the crew should be able to complete the next week. The scheduler then delivers the jobs for each crew to the crew supervisor. The scheduler also sends a copy of each work order that will need intercraft coordination to the supporting craft.

Each day the crew supervisor makes the daily schedule mostly from the work orders allocated in the weekly schedule. The crew supervisor assigns work to individual crew members based on the current day's activities and progress on work. The supervisor attends a daily scheduling meeting with other craft supervisors and the operations group representative to coordinate work for the next day. Near the end of the week, the crew supervisor returns to the scheduler the jobs that he or she does not expect the crew to start that week. The scheduler then considers them for inclusion in the advance schedule then being prepared.

Here are the step-by-step actual activities for weekly scheduling. This is the type of physical process which must be understood and which a computer (CMMS) would mimic.

Forecasting work hours

Each crew supervisor forecasts the crew's available work hours as the first step in the advance scheduling process.

This activity takes place near the end of the work week, usually at the beginning of the last shift for the period already in progress. For example, this would be Friday morning for a crew that works Monday through Friday, 8-hour day shifts or Thursday morning for a crew that works Monday through Thursday, 10-hour day shifts.

Near the end of the work week, the scheduler takes a Crew Work Hours Availability Forecast worksheet (shown in Fig. 6.1) to each crew supervisor. Taking the availability forecast worksheet to the supervisors impresses upon the supervisors that next week they will be responsible for completing the amount of work for which they have labor. They are involved in the process. The scheduler is only helping them determine how many and which jobs should be selected from the backlog.

The Crew Work Hours Availability Forecast worksheet has blanks to guide the crew supervisor in determining how many hours each craft level has available. This section and the following sections illustrate the crew supervisor's use of this form.

The crew supervisor fills out and returns the availability forecast worksheet as soon as possible after receiving it from the scheduler. The supervisor receives the forecast worksheet from the scheduler at 8 AM. The crew supervisor should already have an idea of who will be in training and who has requested vacation for the next week. On the other hand, the supervisor may need to check on jobs currently in progress to determine which ones the crew will probably not finish this week. Those jobs will need carryover hours reserved for them next week. In addition, the supervisor needs to assess which new jobs the crew will start that day. Some of these jobs might be finished or might also run over into next week. One thing the supervisor does not plan for is unexpected absences. Scheduling is not based on the unexpected. The supervisor may later use unexpected absences that occurred to explain why the crew did not meet

CREW WORK HOURS AVAILABILITY FORECAST

For week of: __/__/__ to __/__/__
For crew: _____ By:_____ Date: __/__/__

Craft	# Persons	Paid Hrs	Leave	Train	Misc	Carry-over*	Avail Hrs
_____	____ x 40 =	____ -	____ -	____ -	____ -	____ =	_____
_____	____ x 40 =	____ -	____ -	____ -	____ -	____ =	_____
_____	____ x 40 =	____ -	____ -	____ -	____ -	____ =	_____
_____	____ x 40 =	____ -	____ -	____ -	____ -	____ =	_____
_____	____ x 40 =	____ -	____ -	____ -	____ -	____ =	_____
_____	____ x 40 =	____ -	____ -	____ -	____ -	____ =	_____
_____	____ x 40 =	____ -	____ -	____ -	____ -	____ =	_____
_____	____ x 40 =	____ -	____ -	____ -	____ -	____ =	_____
_____	____ x 40 =	____ -	____ -	____ -	____ -	____ =	_____
_____	____ x 40 =	____ -	____ -	____ -	____ -	____ =	_____

Totals ____ x 40 = ____ - ____ - ____ - ____ - ____ = _____

*Carryover work is any work which has been
physically started in the current period, but will not be
finished and will run over into the forecast period.

Figure 6.1 Worksheet to assist crew supervisors forecast how many labor hours are available for scheduling work the following week.

a schedule. When the crew supervisor finishes these determinations, he or she gives the scheduler the completed Crew Work Hours Availability Forecast worksheet. The crew supervisor also hands over any physical work orders which the crew received to do this past week, but will not start. The crew supervisor makes all these determinations by 10 AM and returns the worksheet. This gives the scheduler time to create the weekly schedule by 1 PM after lunch. (The scheduler must finish the weekly schedule by early afternoon so that the crew supervisor can make a daily schedule for the first day of the coming week. The crew supervisor must attend a late afternoon daily scheduling meeting to give the operations group information on clearances for the beginning of the following week and initiate coordination with other crafts, if necessary. The supervisor may also begin to make individual technician assignments.)

The maintenance group can easily computerize the availability forecast worksheet. Many CMMS packages contain work calendars for the crew supervisor to update crew member availability for the coming work days. Then the scheduler can access the data whenever needed. Of course, the crew supervisor must keep the data current. Even without a CMMS system or CMMS labor calendar module, the company can use a computer network. The scheduler can make a form on email or attach a spreadsheet program or word processing document with a representation of Fig. 6.1. By exchanging this form back and forth by email, the crew supervisor and scheduler produce the necessary forecast information.

Without getting current information from the crew supervisor, the scheduler might use a standard forecast of how many work hours were normally available. In other words, the scheduler would presume for a given crew that a certain number of labor hours were available every week. This standard forecast might presume that for a 10-person crew, one person would be unavailable for one reason or another. So the scheduler would schedule for nine persons each week. The scheduler and the crew supervisor would meet together occasionally to assess any needs to adjust the standard. The problem with using a standard forecast is that crew supervisors may not take the resulting schedule seriously. Whether or not they finish all their scheduled work, they will get the same amount of new work the next period. There is not much attention to carryover work which could be a significant problem. This approach figures that carryover work from the previous work remains about the same each week. Carryover work would be work that the crew started the previous week but did not complete. One of the main reasons to base weekly scheduling on a precise weekly forecast is to facilitate communication about performance. Scheduling to a standard forecast hinders achieving this purpose. Nonetheless, this approach may be necessary in situations of extreme crew reluctance and minimal management support.

Figures 6.2 and 6.3 illustrate the use of the Crew Work Hours Availability Forecast worksheet for a mechanical maintenance crew working a 10-hour shift. B. Jones, supervisor of A Crew, has just received the availability forecast worksheet from the scheduler. The crew consists of five persons: one skilled welder, one apprentice, one painter, and two mechanics. Jones feels that one of the mechanics possesses a very high degree of mechanical skill, but the other mechanic is significantly less capable at the present time. Jones knows that none of the crew will be in training any of next week, but the painter will be off 2 days for vacation.

CREW WORK HOURS AVAILABILITY FORECAST

For week of: ___/___/___ to ___/___/___

For crew: _____ By: _____ Date: ___/___/___

Craft	# Persons	Paid Hrs	Leave	Train	Misc	Carry-over*	Avail Hrs
Mech _____	___ x 40 =	___ -	___ -	___ -	___ -	___ =	___
Welder _____	___ x 40 =	___ -	___ -	___ -	___ -	___ =	___
Machinist _____	___ x 40 =	___ -	___ -	___ -	___ -	___ =	___
Painter _____	___ x 40 =	___ -	___ -	___ -	___ -	___ =	___
Tech _____	___ x 40 =	___ -	___ -	___ -	___ -	___ =	___
Helper _____	___ x 40 =	___ -	___ -	___ -	___ -	___ =	___
_____	___ x 40 =	___ -	___ -	___ -	___ -	___ =	___
_____	___ x 40 =	___ -	___ -	___ -	___ -	___ =	___
_____	___ x 40 =	___ -	___ -	___ -	___ -	___ =	___
_____	___ x 40 =	___ -	___ -	___ -	___ -	___ =	___

Totals ___ x 40 = ___ - ___ - ___ - ___ - ___ = ___

***Carryover work is any work which has been
physically started in the current period, but will not be
finished and will run over into the forecast period.**

Figure 6.2 Input of normal craft and skill level designations.

Jones considers the current jobs in progress. The welder's current job
will not be finished today and requires about 5 hours next week to
finish. The higher skilled mechanic has been working a job for the past
2 days and claims it will take 5 hours next week as well. The other
mechanic will finish one job and start another job today that will also
take about 5 hours next week to finish. The other crew members should
finish their current work today as well. Jones plans to have them start
and finish a new assignment. With this information, Jones estimates the

---

CREW WORK HOURS AVAILABILITY FORECAST

For week of: __5/10/99__ to __5/13/99__

For crew: __A Crew__   **By:** __B. Jones__   **Date:** __5/ 6/ 99__

Craft	# Persons	Paid Hrs	Leave	Train	Misc	Carry-over*	Avail Hrs
Mech	1 × 40 = 40	-	-	-	-	5 =	35
Welder	1 × 40 = 40	-	-	-	-	5 =	35
Machinist	× 40 =	-	-	-	-	=	
Painter	1 × 40 = 40	- 20	-	-	-	=	20
Tech	1 × 40 = 40	-	-	-	-	5 =	35
Helper	1 × 40 = 40	-	-	-	-	=	40
	× 40 =	-	-	-	-	=	
	× 40 =	-	-	-	-	=	
	× 40 =	-	-	-	-	=	
	× 40 =	-	-	-	-	=	

Totals __5__ × 40 = __200__ - __20__ - ___ - ___ - __20__ = __165__

*Carryover work is any work which has been
physically started in the current period, but will not be
finished and will run over into the forecast period.

Figure 6.3 Completed forecast for the A Crew.

total carryover work to be 5 hours of skilled welding, 5 hours of highly
skilled mechanic work, and 5 hours of lesser skilled mechanic work.

Figure 6.2 shows that the supervisor begins completing the availability forecast worksheet by starting with a craft skill level listing for
this specific crew. First, note that the worksheet considers apprentices
simply as helpers. This plant does not regard apprentices as strictly in
training for their craft and may utilize them where the high priorities
of the plant lie. It is expedient to follow this philosophy in the forecasting phase of the scheduling, but to use the daily scheduling assignments

to try to keep apprentices within their craft specialties. Second, note that the scheduler and the supervisor use the term mechanic to designate a fairly well-skilled mechanic technician. They use the term technician to designate a less skilled technician in the primary craft of the crew, in this case mechanical. This use of the terms allows planners, schedulers, and supervisors to communicate regarding skill level even within a standard classification. Although the plant does not have a certification program, significant differences between the skill levels of the mechanics exist. The plant needs to ensure not to allocate too many jobs requiring highly skilled mechanics at the same time. Normally a company could distinguish overall skill level through some certification process or a progression of rank such as third, second, and first class mechanics to identify the better mechanics. However, the subject plant has only mechanic, apprentice, and trainee formal designations. Therefore, the planners and schedulers informally address the needs for jobs by using the terms technician and mechanic on the job plans. When a planner uses the term technician, the job plan does not require a more capable mechanic.

Figure 6.3 shows the availability forecast worksheet after the supervisor enters all the quantities for persons and hours. Jones only forecasts for the 2 days of vacation approved for the painter. Supervisors do not presume there will be unexpected absences due to personal illnesses or sudden vacation day requests. The advance schedule sets a goal based on current knowledge and encourages everyone to meet the schedule. Typically, management above the supervisor directs training and special meetings. Management decides and sends various persons to different training classes or schools as well as coordinates special meetings such as safety or outage planning. Management does the crafts a great favor by scheduling these types of special events at least a week ahead of time. Once maintenance has set a weekly advance schedule, management assists maintenance in building the plant's confidence in the schedule by not encouraging deviation. Jones's management has not scheduled training or meetings for anyone. Jones's estimate of carryover hours is important because the scheduler must allow the crew time to finish jobs already in progress. The scheduler must not allocate new work for these labor hours. The supervisor's estimate of the amount of time required to finish carryover work is adequate. Finally, the supervisor completes the total's line for each type of work hour. The total's line helps in several ways. First, it helps to check the entries for accuracy of addition and subtraction. It also draws attention to how many hours exist in the various categories. The total magnitude of paid hours available to the crew and the effect of lost hours due to training or carryover work often are unappreciated. In this case, 200 labor hours represent a significant company expense. Out of this 200 paid hours, the supervisor forecasts 165 available for work next week in the shown crafts and skill levels. Finally, the totals lend themselves to tracking areas for improvement.

Figure 6.4 illustrates the use of the Crew Work Hours Availability Forecast worksheet for another mechanical maintenance crew. J. Field, supervisor of B Crew, has just received the availability forecast worksheet from the scheduler. The crew consists of 15 persons. The crew has two welders, two machinists, and six mechanics. Field considers three of the mechanics to be significantly more capable than the others. Two trainees and three apprentices make up the remaining five employees.

CREW WORK HOURS AVAILABILITY FORECAST

For week of: __5/11/99__ to __5/14/99__
For crew: __B Crew__   By: __J. Field__   Date: __5/ 7/ 99__

Craft	# Persons	Paid Hrs	Leave	Train	Misc	Carry-over*	Avail Hrs
Mech	3 x 40 =	120 -	-	-	3 -	10 =	107
Welder	2 x 40 =	80 -	20 -	-	2 -	20 =	38
Machinist	2 x 40 =	80 -	-	-	2 -	10 =	68
Painter	x 40 =	-	-	-	-	=	
Tech	3 x 40 =	120 -	-	-	3 -	10 =	107
Helper	5 x 40 =	200 -	-	30 -	5 -	30 =	135
	x 40 =	-	-	-	-	=	
	x 40 =	-	-	-	-	=	
	x 40 =	-	-	-	-	=	
	x 40 =	-	-	-	-	=	
Totals	15 x 40 =	600 -	20 -	30 -	15 -	80 =	455

*Carryover work is any work which has been physically started in the current period, but will not be finished and will run over into the forecast period.

Figure 6.4   Completed forecast for the B Crew.

Field will forecast the crew available work hours from the following information. One of the welders requested 2 days of vacation next week. All B Crew apprentices must attend an entire day of classroom training. B Crew will have a 1-hour safety meeting on Wednesday. After checking on jobs in progress, Field makes an estimate for carryover work. Carryover work will consist of 2 days of welding needing a welder and a helper, 1 day of machine work, 1 day of skilled mechanic work, and another day of less demanding mechanic work.

As before, the supervisor classifies the mechanics according to skill describing them as three mechanics and three technicians. The supervisor forecasts five helpers including the three apprentices and both trainees. The classroom training makes only 30 hours unavailable since out of the five helpers, only the apprentices must attend. The safety meeting on Wednesday makes 1 hour for each person unavailable for scheduling in the Miscellaneous column. The specific day of the week is irrelevant to both the forecast and the weekly allocation. Only the available hours for the entire week matter. The daily scheduling routine later will take this into account. Finally, out of the 600 paid hours, 455 are available for new work.

Sorting work orders

In preparation of allocating work orders into the crew availability forecast, the scheduler sorts the plant's backlog. If the crew is responsible for only a certain area of the plant, the scheduler will only sort those work orders. Later after the scheduler allocates work for every crew, the scheduler might be able to recommend that certain crews assist other crews. The scheduler sorts the backlogged work orders in order to preferentially select work orders to allocate. If there are more job hours in the backlog than the forecast, obviously the scheduler cannot expect the crew to complete all the work. If the backlog has 1000 hours of jobs, then a crew expecting to have 455 labor hours cannot do all the work in a single week. The scheduler must select 455 hours worth of work to allocate for the week. The scheduler sorts the backlog of work orders into an overall order that will help determine which particular work orders are appropriate. Note that this presumes Scheduling Principle 4, proceeding with a preference to schedule 100% of the forecasted crew work hours, not more (120%) or less (80%). The procedure for allocating the work hours into the backlog proceeds with assigning the higher priorities before the lower priority work. The procedure also makes allowance for jobs on the same system and proactive work. In addition, the process shows how to select among jobs of equal priority and work type.

The scheduler returns to the planning office after delivering the Crew Work Hours Availability Forecast worksheet for the crew supervisor to complete. The scheduler collects all the planned work orders for a particular crew from the waiting-to-be-scheduled file and organizes them on a conference room table. The scheduler sorts them into separate piles, one pile for each priority. The waiting-to-be-scheduled file might already have separate folders for each priority to facilitate this step. Appendix J describes the priority codes for the work order system. Then, the scheduler sorts each pile into a particular order.

The scheduler first sorts the highest priority pile, priority 1 (urgent). Note there are no priority-0 (emergency) jobs in the backlog. Emergency jobs would already be under way and so are neither planned nor scheduled. By definition, all priority-1 work is classified as reactive work. Why is this reactive work sorted ahead of any proactive work? The ideal situation would be to have only proactive work in a plant. The plant prefers doing lower priority, proactive work to head off emergencies and urgent work. However, priority-1 work is urgent. It must be addressed, usually to restore lost capacity or remedy an immediate threat to production. The purpose of proactive work is to head off any reactive work, but when proactive action has failed to prevent a reactive situation, the situation must be addressed. Maintenance must first schedule urgent work to restore the plant. However, lower priority, reactive work should wait until after equal priority proactive work. In giving general preference to proactive work in this manner, the incidence of reactive situations diminishes. Therefore, the scheduler considers priority-1 work, which is all reactive and urgent, first.

The scheduler sorts the priority-1 jobs into order with jobs requiring the highest total of work hours on top. When the scheduler later allocates the work orders, jobs with more total work hours are put in the schedule first. This will allow fitting smaller jobs of equal priority into gaps of time remaining during the allocation process. This gap fitting would be more difficult if the scheduler allocated smaller jobs first. The scheduler would have to allocate only portions of larger jobs unnecessarily. The scheduler makes an exception to this sorting for smaller jobs encountered if they belong to the same system as larger jobs already higher in the group. In that case, the scheduler physically removes the smaller work order from its natural order and staples it to the work order for the larger job in the same system.

Scheduling work together for the same system is important for several reasons. Productivity increases if technicians can move from one job to a nearby job on the same system. They do not need to lose time familiarizing themselves with a different system at the start of each job. They avoid having to demobilize, travel, and move personal tool boxes to a different site to set up again. Sometimes they can use the same scaffolding or insulation clearing to get to the work. There is also a psychological boost to remaining on the same system. Frequently the time between jobs is taken advantage of as being a logical time to take breaks or "rest a moment" even when the jobs only last an hour or less. When jobs are scheduled on the same system, there is a tendency to look at the entire system as a single job through which to proceed with minimal delay. On the operations or production side, combining same system jobs also helps improve plant operations. An operator prefers to clear up a single system a single time for several jobs. A less organized scheduling effort might have the operator clear up the demineralizer on Monday and Wednesday for two jobs that could have both been done on Monday. Then on Friday, when the maintenance group requests the operator to clear up the demineralizer a third time, the operator must refuse. The operator must explain that the plant is in jeopardy of not having enough water in its storage tanks. This scenario frequently occurs when no advance scheduling exists. The supervisor then assigns work by picking through the entire plant backlog for each next job. The scenario also can occur if the scheduler does not place the jobs together in an advance allocation. Unnecessarily clearing up a system multiple times wastes time and frustrates the operators.

Then the scheduler similarly sorts the priority-2 (serious) work orders by work order size, biggest jobs on top, with three exceptions. The first exception is that the scheduler puts all, preventive maintenance (PM), jobs at the top of the pile. PM jobs are always prioritized as priority-2 jobs. They are considered serious and not simply routine maintenance with a lower priority. The scheduler sorts PM jobs by job size within themselves, larger ones first. The second exception is similar as for priority-1 work orders. If a work order is encountered that belongs in a system for which a larger or higher priority work order has already been sorted, the work order is moved up and attached to the previously sorted work order. This is done even if it means moving a work order from the priority-2 pile to the priority-1 pile to attach it to the other work order. The third exception is that for work orders of approximately the same size with the same priority, the scheduler places a proactive work order ahead of a reactive one.

Then the scheduler sorts the pile for priority-3 work orders with larger work orders on top and proactive work getting preference for work orders of approximately the same size. The scheduler physically moves any work order for a system already encountered and attaches it to the other work order, even if it is in a higher priority pile.

Similarly, the scheduler sorts each pile of same-priority work orders. Finally, there is a finished group for each priority arranged from top to bottom by size. There are exceptions for same-system work orders stapled together, exceptions for similar-sized proactive work ahead of reactive work, and PM work orders at the top of the priority-2 stack (unless a PM was moved to the priority-1 stack to be with a same system work order).

Note that the advance scheduling process does not consider aging of work orders. The concept of aging is that an older work order should get higher attention than a similar work order only recently written. That is, a low priority work order written 9 months ago might justify more attention than a serious work order written only yesterday. Aging might help a workforce that does not have weekly scheduling. If a crew is only completing the high priority work with low productivity, increasing the relative priority of an older, low priority job might encourage the crew to include it as well. Aging is not as helpful if the maintenance group allocates and expects a crew to complete a proper amount of work each week. A properly sized maintenance crew is capable of handling all the work that comes up, not just the high priority tasks. That means that the crew can complete all the work so aging is not necessary to bring older jobs to the top. Scheduling keeps the crew from lowering its productivity to handle only the high priority work. Scheduling Principle 2 states the importance of correct priorities. Working lower priority jobs ahead of clearly more important jobs leads a crew to doubt its leadership. Doing the most important work first gives the plant more benefit by definition. So aging is really a tool to increase a crew's low productivity. Aging would interfere with an already highly productive crew. Aging figures that a crew is less productive than it should be. Aging figures that a crew is only doing the high priority work by choice. So aging simply raises the priority of some of the work into the higher priority work to which the crew will give attention. On the other hand, a crew already getting as much work done as it should can only give attention to another job by not doing a job already intended. So aging for the highly productive crew has just made the crew complete a lower priority job instead of a higher priority job to the plant's detriment.

Claiming that less important work should not be done ahead of more important work does not say work should never be reprioritized. If an older, low priority work order for some reason merits more importance recently, then the priority should be increased. With the same reasoning, perhaps a higher priority work order is now less important than previously thought. The planning system should allow for changing the priority of each of these types of work orders. One of the helpful additions to a planning group in this respect is an operations coordinator. This person or any knowledgeable operator can benefit a planning group by a monthly review of an extensive backlog. The operations coordinator can determine if some work orders should be reprioritized. This person has the authority to change the priority of any work order on the spot.

The scheduler receives two things from the crew supervisor a few hours after the last shift starts. The scheduler receives the completed Crew Work Hours Availability Forecast worksheet and any work orders which had been scheduled, but are now not going to be started this week. The supervisor has had time to assess the crew projected attendance for the next week as well as the status of current jobs-in-progress. In the case of an entirely paper-driven system, the crew supervisor physically hands over the work orders that the crew will not start. (In the case of an entirely computer driven system, the maintenance group might only print out the physical work orders when assigning them to the field technicians, if at all. The supervisor updates the computer changing each job to "in-progress" at the beginning of the shift or the end of the previous shift for the coming day. Therefore, the scheduler may consult the CMMS each week to determine which allocated jobs will not be started.) The scheduler takes the jobs not to be started and places them into the priority piles of work orders if they have already been arranged. The scheduler places these new work orders into the piles where they would belong if they had already been in the backlog. The actual allocation sequence can now take place.

Tables 6.1 through 6.3 illustrate sorting a plant backlog. The backlog belongs to mechanical maintenance A Crew. An earlier illustration used this crew as an example for forecasting. Table 6.1 shows the backlog

TABLE 6.1   Plant Backlog for the A Crew Listed by Work Order Number

WO No.	Unit	System	Priority	Work type	Outage	No. of persons and craft	Est. hours	Est duration
001	N00	ZE	4	5	0	1 welder	35	35
002	N01	CP	1	5	0	1 tech	10	10
003	N01	CV	3	9	0	1 painter	40	40
004	N01	FC	1	5	0	1 tech, 1 helper	7, 7	7
005	N01	CP	2	7	0	1 tech, 2 helpers	6, 12	6
006	N01	FC	3	5	0	1 tech, 1 helper	15, 15	15
007	N01	JC	2	5	0	1 mech, 1 helper	20, 20	20
008	N01	JX	3	9	0	1 painter	4	4
009	N01	CP	4	5	0	1 mech, 1 helper	2, 2	2
010	N01	IF	3	9	0	1 tech, 1 helper	3, 3	3
011	N01	CD	2	8	0	1 mech, 1 helper	10, 10	10
012	N01	BV	1	5	0	1 welder, 1 helper	8, 8	8
013	N01	IF	4	9	0	2 tech	6	3

TABLE 6.2 Plant Backlog for the A Crew Grouped by Work Order Priority, Size, and PM

WO No.	Unit	System	Priority	Work type	Outage	No. of persons and craft	Est. hours	Est. duration
012	N01	BV	1	5	0	1 welder, 1 helper	8, 8	8
004	N01	FC	1	5	0	1 tech, 1 helper	7, 7	7
002	N01	CP	1	5	0	1 tech	10	10
005	N01	CP	2	7	0	1 tech, 2 helpers	6, 12	6
007	N01	JC	2	5	0	1 mech, 1 helper	20, 20	20
011	N01	CD	2	8	0	1 mech, 1 helper	10, 10	10
003	N01	CV	3	9	0	1 painter	40	40
006	N01	FC	3	5	0	1 tech, 1 helper	15, 15	15
010	N01	IF	3	9	0	1 tech, 1 helper	3, 3	3
008	N01	JX	3	9	0	1 painter	4	4
001	N00	ZE	4	5	0	1 welder	35	35
013	N01	IF	4	9	0	2 tech	6	3
009	N01	CP	4	5	0	1 mech, 1 helper	2, 2	2

TABLE 6.3 Plant Backlog for the A Crew Adjusted for Work on Same Systems and Other Proactive Work

WO No.	Unit	System	Priority	Work type	Outage	No. of persons and craft	Est. hours	Est. duration
012	N01	BV	1	5	0	1 welder, 1 helper	8, 8	8
004	N01	FC	1	5	0	1 tech, 1 helper	7, 7	7
006	*N01*	*FC*	*3*	*5*	*0*	*1 tech, 1 helper*	*15, 15*	*15*
002	N01	CP	1	5	0	1 tech	10	10
005	*N01*	*CP*	*2*	*7*	*0*	*1 tech, 2 helpers*	*6, 12*	*6*
009	*N01*	*CP*	*4*	*5*	*0*	*1 mech, 1 helper*	*2, 2*	*2*
007	N01	JC	2	5	0	1 mech, 1 helper	20, 20	20
011	N01	CD	2	8	0	1 mech, 1 helper	10, 10	10
003	N01	CV	3	9	0	1 painter	40	40
010	N01	IF	3	9	0	1 tech, 1 helper	3, 3	3
013	*N01*	*IF*	*4*	*9*	*0*	*2 tech*	*6*	*3*
008	N01	JX	3	9	0	1 painter	4	4
001	N00	ZE	4	5	0	1 welder	35	35

arranged by work order number. The backlog consists of 243 total estimated work hours as planned. Because A Crew has only 165 work hours
forecasted available, the scheduler must select the proper 165 hours to allocate for the next week. A plant also generally prefers to have 2 to 3 weeks
of backlog available. This plant has less than 2 weeks of backlog. If this
shortage is the normal case, the plant might not be identifying enough
corrective maintenance situations to head off later breakdowns. The
plant might also not be creating enough PM tasks. On the other hand,
the maintenance crew might be overstaffed for the work area. Of the
work orders in the backlog, work order codes as defined in App. J define
certain information necessary for scheduling. First, the unit code N01
shows that most of the work orders are for North Unit 1, which is A
Crew's primary responsibility. Second, within Unit 1, the system codes
show that a variety of different systems need work. Third, work type
codes indicate the nature of the work. Code 5 is breakdown and failure.
Code 7 is PM. Code 8 is work recommended by predictive maintenance.
Code 9 is corrective maintenance that can head off failure and breakdown. Fourth, outage code 0 illustrates that the scheduler considers
only work not requiring an outage of the entire unit for the normal
work week.

Table 6.2 shows the backlog after the scheduler has physically grouped
the work orders into different priorities. In addition, the largest work
orders have been placed ahead of smaller jobs within each priority
group. Note the scheduler places WO (work order) no. 012 requiring 16
total labor hours ahead of WO no. 004 requiring only 14 hours. Similarly,
WO no. 004 is ahead of WO no. 002, which requires only 10 labor hours.
Each priority group is similarly arranged with the only exception being
for the priority-2 work orders. The scheduler must place PM work orders
first within the priority-2 work orders. Work type code 7 defines the work
as preventive maintenance. Therefore, the scheduler places WO no. 005
at the head of the priority-2 group even though it has the fewest hours.

Table 6.3 shows the backlog after the scheduler has adjusted the
groups considering same-system work and other proactive work. First,
the scheduler takes WO no. 006 from the priority-3 stack and attaches
it behind WO no. 004 in the priority-1 group. These work orders are both
in the same system, FC. Second, the scheduler takes WO no. 005 from
the priority-2 stack and WO no. 009 from the priority-4 group. The
scheduler attaches both work orders behind WO no. 002 in the priority-1
stack. All three are in the same system, CP. Finally, the scheduler scans
the work orders to see if any of the proactive work orders besides PM
should be moved up in the allocation preference. Proactive work type 8
is work recommended by predictive maintenance and reactive work
type 5 is work to restore something that has already failed. WO no. 011
(work type 8) is currently behind WO no. 007 (work type 5). If WO no. 011
were closer in size to WO no. 007, say 35 to 39 labor hours, the scheduler

would move it ahead. However, in this case, the relative size dictates that preference be given to the reactive work. In the priority-3 group, all of the work is proactive, work type 9 so no adjustments can be made. The scheduler will use the order presented by Table 6.3 to select work orders to allocate into the A Crew work hours availability forecast.

Tables 6.4 through 6.6 illustrate another example of sorting a nonoutage, plant backlog for B Crew at the same plant. This crew has

TABLE 6.4   Plant Backlog for the B Crew Listed by Work Order Number

WO No.	Unit	System	Priority	Work type	Outage	No. of persons and craft	Est. hours	Est. duration
021	N02	BS	1	5	0	2 tech	6	3
022	N32	UA	3	9	0	2 tech	14	7
023	N00	FO	4	9	0	1 tech, 1 helper	20, 20	20
024	N00	HA	2	7	0	1 tech, 1 helper	20, 20	20
025	N00	HC	2	5	0	1 tech	17	17
026	N00	FO	2	5	0	1 welder	3	3
027	N00	HD	2	7	0	1 mach	2	2
028	N00	HP	1	5	0	1 mach, 1 helper	8, 8	8
029	N00	FW	4	9	0	2 helpers	12	6
030	N00	HD	1	5	0	1 mech, 1 helper	1, 1	1
031	N02	DA	3	3	0	1 mech, 1 helper	20, 20	20
032	N02	DO	3	9	0	1 helper	40	40
033	N00	FO	2	9	0	1 mach, 1 tech	14, 14	14
034	N00	FO	2	5	0	1 mach, 1 tech	6, 6	6
035	N00	KD	1	5	0	1 tech	20	20
036	N00	FW	1	5	0	1 mech, 2 helpers	6, 12	6
037	N00	FO	2	7	0	1 mech, 1 helper	4, 8	8
038	N31	UZ	2	5	0	1 tech, 1 helper	3, 3	3
039	N00	FO	2	8	0	1 welder, 1 helper	2, 2	2
040	N02	FD	3	8	0	1 welder, 1 helper	10, 10	10
041	N00	FO	1	5	0	1 tech, 1 helper	4, 4	4
042	N02	FD	1	5	0	2 tech	8	4
043	N31	UA	3	9	0	1 mech, 1 helper	20, 20	20
044	N31	UZ	4	9	0	2 helpers	2	1

TABLE 6.5   **Plant Backlog for the B Crew Grouped by Work Order Priority, Size, and PM**

WO No.	Unit	System	Priority	Work type	Outage	No. of persons and craft	Est. hours	Est. duration
035	N00	KD	1	5	0	1 tech	20	20
036	N00	FW	1	5	0	1 mech, 2 helpers	6, 12	6
028	N00	HP	1	5	0	1 mach, 1 helper	8, 8	8
041	N00	FO	1	5	0	1 tech, 1 helper	4, 4	4
042	N02	FD	1	5	0	2 tech	8	4
021	N02	BS	1	5	0	2 tech	6	3
030	N00	HD	1	5	0	1 mech, 1 helper	1, 1	1
024	N00	HA	2	7	0	1 tech, 1 helper	20, 20	20
037	N00	FO	2	7	0	1 mech, 1 helper	4, 8	8
027	N00	HD	2	7	0	1 mach	2	2
033	N00	FO	2	9	0	1 mach, 1 tech	14, 14	14
025	N00	HC	2	5	0	1 tech	17	17
034	N00	FO	2	5	0	1 mach, 1 tech	6, 6	6
038	N31	UZ	2	5	0	1 tech, 1 helper	3, 3	3
039	N00	FO	2	8	0	1 welder, 1 helper	2, 2	2
026	N00	FO	2	5	0	1 welder	3	3
031	N02	DA	3	3	0	1 mech, 1 helper	20, 20	20
032	N02	DO	3	9	0	1 helper	40	40
043	N31	UA	3	9	0	1 mech, 1 helper	20, 20	20
040	N02	FD	3	8	0	1 welder, 1 helper	10, 10	10
022	N32	UA	3	9	0	2 tech	14	7
023	N00	FO	4	9	0	1 tech, 1 helper	20, 20	20
029	N00	FW	4	9	0	2 helpers	12	6
044	N31	UZ	4	9	0	2 helpers	2	1

a different backlog because it is responsible for a separate section of the plant. Table 6.4 shows the backlog arranged by work order number.

Table 6.5 shows the backlog after the scheduler has sorted the priority groups and placed PM jobs at the top of the priority-2 work orders.

Table 6.6 shows the backlog after the scheduler has grouped same system work orders. This grouping should help maintenance and operations concentrate on the most needy systems in an organized manner. Notice this grouping also allowed the scheduler to move a number of

TABLE 6.6   **Plant Backlog for the B Crew Adjusted for Work on Same Systems and Other Proactive Work**

WO No.	Unit	System	Priority	Work type	Outage	No. of persons and craft	Est. hours	Est. duration
035	N00	KD	1	5	0	1 tech	20	20
036	N00	FW	1	5	0	1 mech, 2 helpers	6, 12	6
029	*N00*	*FW*	*4*	*9*	*0*	*2 helpers*	*12*	*6*
028	N00	HP	1	5	0	1 mach, 1 helper	8, 8	8
041	N00	FO	1	5	0	1 tech, 1 helper	4, 4	4
037	*N00*	*FO*	*2*	*7*	*0*	*1 mech, 1 helper*	*4, 8*	*8*
033	*N00*	*FO*	*2*	*9*	*0*	*1 mach, 1 tech*	*14, 14*	*14*
034	*N00*	*FO*	*2*	*5*	*0*	*1 mach, 1 tech*	*6, 6*	*6*
039	*N00*	*FO*	*2*	*8*	*0*	*1 welder, 1 helper*	*2, 2*	*2*
026	*N00*	*FO*	*2*	*5*	*0*	*1 welder*	*3*	*3*
023	*N00*	*FO*	*4*	*9*	*0*	*1 tech, 1 helper*	*20, 20*	*20*
042	N02	FD	1	5	0	2 tech	8	4
040	*N02*	*FD*	*3*	*8*	*0*	*1 welder, 1 helper*	*10, 10*	*10*
021	N02	BS	1	5	0	2 tech	6	3
030	N00	HD	1	5	0	1 mech, 1 helper	1, 1	1
027	*N00*	*HD*	*2*	*7*	*0*	*1 mach*	*2*	*2*
024	N00	HA	2	7	0	1 tech, 1 helper	20, 20	20
025	N00	HC	2	5	0	1 tech	17	17
038	N31	UZ	2	5	0	1 tech, 1 helper	3, 3	3
044	*N31*	*UZ*	*4*	*9*	*0*	*2 helpers*	*2*	*1*
031	N02	DA	3	3	0	1 mech, 1 helper	20, 20	20
032	N02	DO	3	9	0	1 helper	40	40
043	N31	UA	3	9	0	1 mech, 1 helper	20, 20	20
022	*N32*	*UA*	*3*	*9*	*0*	*2 tech*	*14*	*7*

proactive work orders up into the priority-1 group. There are seven proactive work orders moved into this group, two PM work orders, three corrective maintenance work orders, and one predictive maintenance work order. Without this grouping, the workforce may have had a tendency to concentrate solely on the high priority reactive work for the week. A sanity check may also be needed after the forecast work hours are compared with the sorted backlog preference order. WO no. 023 is a fairly large priority-4 work order that the scheduler has moved into the priority-1 group. After reading exact job descriptions, the scheduler or operations coordinator may prefer holding it until the plant has addressed some of the priority-2 work. An obvious case might be the inadvisability of attaching a 40-hour, priority-4 job to a single 3-hour, priority-1 job. However, in the case of WO no. 23, there are already 67 hours of work to be done for system FO.

Allocating work orders

With the crew work hour forecast in hand and the work orders stacked on the table, the scheduler uses the Advance Schedule Worksheet, Fig. 6.5, to allocate the right work orders for the week. This worksheet allows the scheduler logically to connect the backlog with the crew time available. The worksheet form essentially consists of a blank sheet of paper with horizontal lines. The horizontal lines lie on the form at the same heights as the Crew Work Hours Availability Forecast worksheet (Fig. 6.1). The scheduler may tape or staple the schedule worksheet side by side to the availability worksheet. The scheduler places the worksheets together such that there is a long blank line immediately following the available work hours for each craft skill level. For the ease of this discussion, the scheduler does not physically attach the worksheets together, but writes the craft level and forecast available hours on the extreme left of each line. The long blank lines allow for tabulation of remaining crew hours as the scheduler places each job into the schedule allocation for the next week.

After writing the available hours on the left of each line, the scheduler selects the top work order from the pile of highest priority work orders. The scheduler uses the long blank line beside each craft skill level to write the remaining craft hours available after the scheduler subtracts the hours required by the selected work order. The scheduler then puts the selected work order into the week's worth of work being allocated. This work constitutes the weekly schedule.

The scheduler physically places the paper work order itself into a set of folders to deliver to the crew supervisor. There is a folder for each craft skill level which is the lead for the work order. For example, if the work order requires two helpers, the scheduler places the work order into the helper folder. If the work order requires a mechanic and a helper, the work order also goes into the mechanic folder since the lead person is a mechanic. If the work order requires two welders and an electrician, the work order goes into the welder folder since the bulk of the work (and probably the lead) is welding. Later these folders will help the crew supervisor find work while assigning to the different crafts and skill on the individual crew. The craft skill level folders not only allow sorting of the work by trades or classifications, but they are also the vehicle that physically transports the scheduled work orders over to the crew supervisor.

ADVANCE SCHEDULE WORKSHEET

For week of: _____ **to** _____
For crew: _____ **By:**_____ **Date:** _____

Forecast   **Available Hours Left**

Totals _____   _____

Instructions: Subtract job work hours from available line total until balance reaches zero for each line or backlog runs out.

Figure 6.5   Worksheet to assist the scheduler determine which work orders to allocate for the week.

Later at the end of the scheduling period, the supervisor returns the folders to the scheduler with work orders that were not started. The scheduler places these work orders back into the backlog for possible inclusion in the next weekly schedule as the scheduler follows the advance scheduling routine.

The scheduler then repeats the work order selection process to allocate work orders into the available craft hours. The scheduler selects the next work order from the top of the highest priority work orders. If the selected job requires more hours than the hours left on a particular craft line, the scheduler must make a decision. The scheduler first tries to "work persons down," such as using a mechanic as a helper. The scheduler might want to schedule only part of a job, such as scheduling 30 hours of a 60-hour job if only 30 hours of a particular skill are available. The scheduler might also decide the job cannot be scheduled because there are insufficient hours available. The scheduler places subsequent work orders behind any work orders already in each folder. The scheduler continues the allocation process until either the crew runs out of available work hours or the backlog runs out of work orders. The backlog might run out of work orders altogether or just run out of work orders for which the crew has qualified labor.

After making the initial allocation grouping, the scheduler makes a final consideration of proactive work and consults operations. The scheduler considers if he or she ought to place any more proactive work into the schedule to replace low priority, reactive work. This might be advisable for the allocation in which there is almost no proactive work whatsoever. There will never be a reduction of reactive work if there is never any proactive work performed. Both of the previous examples of backlogs contain a modest amount of proactive work and so need no adjustment. Next, the scheduler consults the operations coordinator giving this person a chance to replace any of the allocated work with work that the scheduler did not choose. This might be done in a formal weekly schedule meeting. Maintenance and production schedules must be integrated even when not considering outage work. The operations coordinator understands overall constraints of operations being able to clear or release certain equipment at the present time. The operations coordinator may also make final adjustments for the best benefit of the plant. In both of the example allocations, the operations coordinator decides not to make adjustments.

The following examples use the Work Order Allocation Worksheet to combine the previous forecast examples and previous backlog sorting examples into a week's worth of work, the weekly schedule allocation.

Figures 6.6 through 6.12 illustrate using the scheduling worksheet to allocate the nonoutage backlog for A Crew. The scheduler first copies the craft levels and forecasted hours from the Crew Work Hours Availability Forecast worksheet for A Crew (Fig. 6.3). Figure 6.6 shows the resulting Advance Schedule Worksheet after this first step.

Then the scheduler selects work order no. 012, the first work order from the highest priority group for A Crew (shown in Table 6.3). WO no. 012 requires a welder for 8 hours and a helper for 8 hours. Therefore, the scheduler subtracts 8 hours from the 35 hours available for welders leaving 27 hours available. The scheduler writes down "27" on the welder line indicating the hours now available. Similarly, the scheduler subtracts 8 hours from the 40 helper hours available and writes down "32"

ADVANCE SCHEDULE WORKSHEET

For week of: **5/10/99** to **5/13/99**

For crew: **A Crew**   By:**C. Rodgers**   Date: **5/ 6/ 99**

Forecast   Available Hours Left
                       (Work Order #)
                       #12

Mech	35		
Welder	35	27	
Machinist			
Painter	20		
Tech	35		
Helper	40	32	

Totals **165**   _____

**Instructions: Subtract job work hours from available
line total until balance reaches zero for each line or
backlog runs out.**

Figure 6.6   Input of original labor forecast and first work order for the A Crew.

on the helper line to indicate there are only 32 helper hours now available. The scheduler places work order no. 012 into a folder labeled "Welder" for eventual delivery to the crew supervisor. Figure 6.7 shows the resulting Advance Schedule Worksheet.

Figure 6.8 shows the Advance Schedule Worksheet after the scheduler selects work order nos. 004, 006, and 002. The scheduler places all of these work orders into a folder labeled "Technician" because that is the lead skill required on each.

The scheduler then selects work order no. 005 from the backlog. This work cannot be allocated into the available hours as simply as the

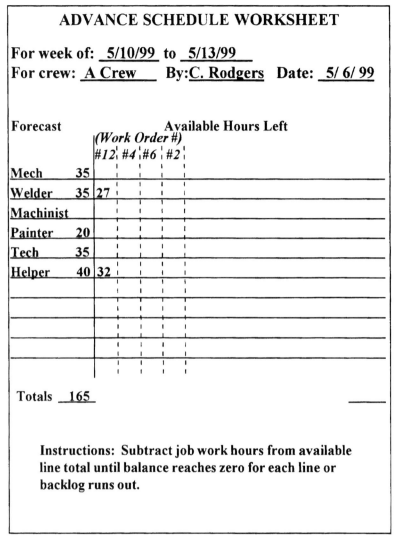

Figure 6.7 Setting of columns to illustrate labor calculations with next three work orders.

preceding work orders. This work order requires 6 technical hours, but there are only 3 technical hours available. The scheduler takes the 3 technical hours available and then takes 3 welder hours to use as technician hours. The scheduler considers that a welder can perform the less complex mechanical tasks required by a technician. The scheduler does not wish to use a more skilled mechanic for the work because the backlog contains a significant amount of priority-2, skilled mechanic work. The backlog has more welding work, but it is priority-4 work. Next, the work order requires 12 helper hours, but only 10 are available. Therefore

---

ADVANCE SCHEDULE WORKSHEET

For week of: _5/10/99_ to _5/13/99_
For crew: _A Crew_   **By:**_C. Rodgers_   **Date:** _5/ 6/ 99_

Forecast   **Available Hours Left**
                     (*Work Order #*)
                      #12 #4 #6   #2

		#12	#4	#6	#2	
Mech	35					
Welder	35	27				
Machinist						
Painter	20					
Tech	35		28	13	3	
Helper	40	32	25	10		

Totals _165_   _____

Instructions: Subtract job work hours from available line total until balance reaches zero for each line or backlog runs out.

Figure 6.8   Labor calculations of available labor hours remaining.

the scheduler takes the 10 helper hours available and takes 2 more welder hours to use as helper hours. Thus the scheduler subtracts a total of 5 welder hours from the 27 available leaving 22 available welder hours. See Fig. 6.9. These decisions of where to take hours require judgment on the scheduler's part. The exact choices are not critical. What is critical is that the scheduler realizes the ability to allocate work to other than the exact craft and skill specified by the job plan. The scheduler must remember that the job was only planned for the minimum skill level required.

```
ADVANCE SCHEDULE WORKSHEET

For week of: _5/10/99_ to _5/13/99_
For crew: _A Crew___ By:_C. Rodgers_ Date: _5/ 6/ 99_
```

Forecast		Available Hours Left (Work Order #)					
		#12	#4	#6	#2	#5	
Mech	35						
Welder	35	27				22	
Machinist							
Painter	20						
Tech	35		28	13	3	0	
Helper	40	32	25	10		0	

Totals __165__   ____

**Instructions: Subtract job work hours from available
line total until balance reaches zero for each line or
backlog runs out.**

Figure 6.9 Allocating first work order to use other than the minimum labor skill
that was planned.

Another question arises regarding splitting hours from different areas.
Is the scheduler requiring different persons to perform fractions of jobs?
No. As the actual week later progresses, some jobs will run over and
others under allowing the crew supervisor to assign whole jobs to specific individuals. The crew supervisor may assign a welder as a helper
for an entire job, but probably not for only half of a job. Actual experience has shown that this method of allocating work for the week results
in a quantity of work for which the crew possesses the appropriate labor.

ADVANCE SCHEDULE WORKSHEET

For week of: 5/10/99 to 5/13/99

For crew: A Crew   By: C. Rodgers   Date: 5/ 6/ 99

Forecast   Available Hours Left
                 (Work Order #)

		#12	#4	#6	#2	#5	#9	#7	#11
Mech	35						33	13	3
Welder	35	27				22	20	0	
Machinist									
Painter	20							10	
Tech	35	28	13	3	0				
Helper	40	32	25	10		0			

Totals   165   _____

Instructions: Subtract job work hours from available line total until balance reaches zero for each line or backlog runs out.

Figure 6.10 Allocating next three work orders. Each uses other than the minimum labor skill that was planned.

As the scheduler selects more and more of the backlog, more of these type decisions are made. Figure 6.10 shows the results of the scheduler allocating work order nos. 009, 007, and 011 into the weekly schedule. For both WO no. 009 and WO no. 007 the scheduler has the mechanic hours needed, but has to use welder hours for all of the helper hours. For WO no. 011, the scheduler again has the needed mechanic hours available. However, the scheduler must again use another skill level for the helper hours required. This time there are no more welder hours

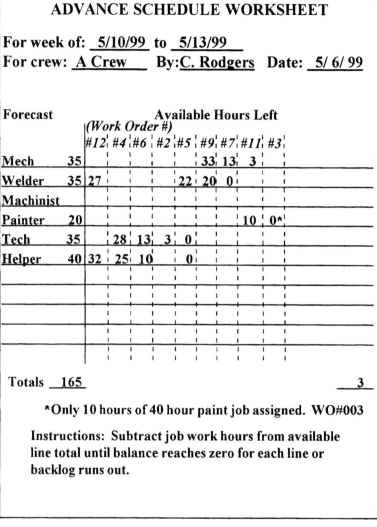

ADVANCE SCHEDULE WORKSHEET

For week of: __5/10/99__ **to** __5/13/99__

For crew: __A Crew__   **By:**__C. Rodgers__ **Date:** __5/ 6/ 99__

Forecast   **Available Hours Left**
(Work Order #)

		#12	#4	#6	#2	#5	#9	#7	#11	#3
Mech	35						33	13	3	
Welder	35	27				22	20	0		
Machinist										
Painter	20								10	0*
Tech	35		28	13	3	0				
Helper	40	32	25	10		0				

Totals __165__   __3__

*Only 10 hours of 40 hour paint job assigned. WO#003

Instructions: Subtract job work hours from available line total until balance reaches zero for each line or backlog runs out.

Figure 6.11 Allocating a partial work order.

available, The scheduler decides to use the painter as a helper for WO no. 011. Although there is painting work in the backlog, it has a lower priority than WO no. 011. The painter would still have some hours to begin the painting job and could finish it as carryover work the following week.

Finally, Fig. 6.11 shows the last work order that the scheduler places into the weekly schedule. The scheduler does not want to overschedule the crew. WO no. 003 requires 40 painter hours. The scheduler expressly states by writing a note on WO no. 003 that only 10 hours are being

---

ADVANCE SCHEDULE WORKSHEET

For week of: _5/10/99_ to _5/13/99_
For crew: _A Crew_   **By:**_C. Rodgers_   **Date:** _5/ 6/ 99_

Forecast   Available Hours Left

Mech	35	33 13 3
Welder	35	27 22 20 0
Machinist		
Painter	20	10 0*
Tech	35	28 13 3 0
Helper	40	32 25 10 0

Totals __165__   __3__

*Only 10 hours of 40 hour paint job assigned. WO#003

Instructions: Subtract job work hours from available line total until balance reaches zero for each line or backlog runs out.

Figure 6.12   Realistic completed Advance Schedule Worksheet for the A Crew.

scheduled for the next week. The scheduler expects the painter to begin the paint job on the painter's last day of the week. The scheduler stops the schedule allocation process taking up all but 3 hours of skilled mechanic work hours from the forecast. There are no jobs in the backlog that take as few as 3 hours that the mechanic could perform. The scheduler has allocated 162 hours of backlog work for the forecasted 165 hours the crew has available. The scheduler considers this to be a 100% allocation. The scheduler does not want to give the crew any occasion to suggest that too much work was allocated. The scheduler now has work orders arranged in several folders, Mechanic, Welder, Painter, and Technician. The scheduler did not place any work orders in a folder labeled Helper because in this case, no work required helpers alone. The scheduler takes the work order folders over to the A Crew supervisor. A Crew supervisor Jones will use the work orders to begin establishing a daily schedule for the first day of next week.

In actual practice, Fig. 6.12 shows the actual worksheet as the scheduler would have filled in the information. The scheduler would not have listed each work order at the top of columns. Rather, the scheduler would have written in available hours left as the scheduler selected jobs and placed them into folders.

Figures 6.13 through 6.17 illustrate using the scheduling worksheet to allocate the nonoutage backlog for B Crew.

The scheduler first copies over the craft levels and forecasted hours from the Crew Work Hours Availability Forecast worksheet for B Crew (Fig. 6.4). Figure 6.13 shows the Advance Schedule Worksheet after the scheduler has selected the first 12 jobs from the previously sorted backlog for B Crew. None of the these jobs requires the scheduler to consider using a craft skill level other than the ones planned on the work orders. For each work order, the scheduler merely subtracts the required work hours from each craft skill level line. Then the scheduler puts the selected work order into its folder for the pertinent lead craft.

After including the twelfth job, WO no. 042, the scheduler runs out of room on the worksheet. The scheduler simply labels the worksheet "Page 1" and begins a second page. Figure 6.14 shows the second page and the next five work orders selected. Similar to the first page, none of the jobs requires the scheduler to consider using a craft skill level other than the ones planned on the work orders. For each work order, the scheduler subtracts the required work hours from each craft skill level line. Then the scheduler puts the selected work order into its folder for the pertinent lead craft.

Beginning with WO no. 25, the scheduler must decide to work persons outside of their top skill levels. WO no. 25 requires 17 technician hours, but only 9 are available. The scheduler decides to use 8 mechanical hours to make up the difference leaving 88 mechanical hours and 0 technical hours available afterward. Likewise, WO no. 038 requires using 3 mechanical hours as technical hours. There are no such decisions to make for WO no. 044 and WO no. 031. Each requires helper or mechanical hours which are available. Figure 6.15 shows the worksheet that results from including these four work orders.

Finally, Fig. 6.16 shows the resulting worksheet after the scheduler includes the last three jobs in the backlog. The scheduler uses the remaining 13 helper hours for WO no. 32. Since this work order calls for 40 helper hours, the scheduler also uses 27 machinist hours. This decision leaves 11 machinist hours and 0 helper hours. WO no. 43 calls

ADVANCE SCHEDULE WORKSHEET

For week of: 5/11/99 to 5/14/99

For crew: B Crew   By: C. Rodgers   Date: 5/ 7/ 99

Page 1

Forecast   Available Hours Left

(Work Order #)

		#35	#36	#29	#28	#41	#37	#33	#34	#39	#26	#23	#42
Mech	107	101				97							
Welder	38									36	33		
Machinist	68			60			46	40					
Painter													
Tech	107	87			83		69	63				43	35
Helper	135	123	111	103	99	91			89		69		

Totals   455   _____

Instructions: Subtract job work hours from available line total until balance reaches zero for each line or backlog runs out.

Figure 6.13   Input of original labor forecast and first 12 work orders for the B Crew.

for 40 technical hours of which none are available. Therefore, the scheduler uses 40 mechanical hours. Likewise, the last work order in the backlog, WO no. 22 is planned for technical hours. The scheduler uses mechanical hours instead. Now all the backlog has been allocated.

In actual practice, Fig. 6.17 shows the actual worksheet as the scheduler would have filled in the information.

Without the allocation system, the crew may have not realized it had the ability to complete the entire backlog. Instead, the crew may have only concentrated on completing the higher priority work. In addition,

```
┌─────────────────────────────────────────────────────────────┐
```

ADVANCE SCHEDULE WORKSHEET

For week of: __5/11/99__ to __5/14/99__

For crew: __B Crew__   By:__C. Rodgers__   Date: __5/ 7/ 99__

Page 2

Forecast   Available Hours Left
                    (Work Order #)

Forecast	#40	#21	#30	#27	#24	Available Hours Left
Mech			96			
Welder	23					
Machinist			38			
Painter						
Tech		29		9		
Helper	59	58		38		

Totals _____   _____

Instructions: Subtract job work hours from available line total until balance reaches zero for each line or backlog runs out.

Figure 6.14 Using a second page and allocating the next five work orders.

the allocation process identifies how many extra craft hours are left. There are 45 craft hours left, 11 mechanical, 23 welder, and 11 machinist. The scheduler has a basis to suggest using B Crew labor to assist A Crew next week. The B Crew Mechanic and Machinist could not only complete WO no. 010 and no. 013, but replace 10 of the helper hours on WO no. 007. This would free up 10 hours of A Crew welder hours. Along with the 23 B Crew welder hours left over, this is almost enough hours to complete the priority-4 welding job, WO no. 001. There are two common practices to execute this assistance. One method would be for

ADVANCE SCHEDULE WORKSHEET

For week of: 5/11/99 to 5/14/99
For crew: B Crew   By: C. Rodgers   Date: 5/ 7/ 99

Page 2

Forecast   Available Hours Left
(Work Order #)

	#40	#21	#30	#27	#24	#25	#38	#44	#31
Mech			96		88	85		65	
Welder	23								
Machinist				38					
Painter									
Tech		29			9	0			
Helper	59		58		38	35	33	13	

Totals _____   _____

Instructions: Subtract job work hours from available line total until balance reaches zero for each line or backlog runs out.

Figure 6.15   Allocating the next four work orders. The first two work orders use other than the minimum labor skill that was planned.

a crew supervisor to give the other supervisor several work orders. The other method would be for a crew supervisor to plan to loan the other supervisor several persons for a day or two. For example, presume that on the scheduler's recommendation, the A Crew supervisor gives the B Crew supervisor work order nos. 010 and 013. In addition, the B Crew supervisor arranges to give the A Crew one day of machinist help and two days of welder help toward the end of the week, if necessary.

ADVANCE SCHEDULE WORKSHEET

For week of: 5/11/99 to 5/14/99
For crew: B Crew   **By:** C. Rodgers   **Date:** 5/7/99

Page 2

Forecast   Available Hours Left
(Work Order #)

	#40	#21	#30	#27	#24	#25	#38	#44	#31	#32	#43	#22
Mech		96			88	85			65		25	11
Welder	23											
Machinist			38							11		
Painter												
Tech		29		9	0							
Helper	59		58		38		35	33	13	0		

Totals _____   45

Instructions: Subtract job work hours from available line total until balance reaches zero for each line or backlog runs out.

Figure 6.16 Allocating the final three work orders in the entire backlog.

Formal Weekly Schedule Meeting

Although the planner can coordinate the schedule with the operations group and the crew supervisor, it is highly recommended that the maintenance manager implement a formal weekly schedule meeting. This meeting not only legitimizes the entire scheduling process, but also ensures that scheduling takes place week after week and year after year. The objective of the meeting is to finalize a list of work orders for which both operations and maintenance agree maintenance should complete next week.

```
ADVANCE SCHEDULE WORKSHEET

For week of:  5/11/99  to  5/14/99
For crew:  B Crew    By:C. Rodgers   Date:  5/ 7/ 99


Forecast                    Available Hours Left


Mech      107 | 101 97 96 88 85 65 25 11
Welder     38 | 36 33 23
Machinist  68 | 60  46 40 38 11
Painter       |
Tech      107 | 87 83 69 63  43 35 29 9 0
Helper    135 | 123 111 103 99  91 89  69 59 58 38 35 33 13 0



Totals   455                                          45


Instructions:  Subtract job work hours from available
line total until balance reaches zero for each line or
backlog runs out.
```

Figure 6.17   Realistic completed Advance Schedule Worksheet for the B Crew.

The attendees should include persons from both the maintenance and operations sides of the house. Persons required include the maintenance manager, the operations coordinator, or other operations leaders who can intelligently speak for production. Each planner and each crew supervisor should attend. A plant engineer might attend if he or she has an ongoing project that might affect maintenance or operations. It is important that all the required attendees have alternates who will take their places when they must be absent. The maintenance manager over the crew supervisors will lead the meeting.

The agenda should be as follows:

1. *10 minutes Review.* schedule compliance score from last week for each crew and key reasons if the schedule was not met. Itemize important jobs not completed.

2. *10 minutes.* Heads up on schedule progress this week if important jobs will not be completed.

3. *30 minutes.* Present proposed schedule for next week for each crew showing total available labor hours and total hours of jobs scheduled. (It is preferred that the planner previously had asked both the operations coordinator and the crew supervisor about any "must include" jobs before assembling the schedule to bring to the meeting.)

4. *10 minutes.* Agreement in principle if jobs need to be added or deleted from the schedule.

5. *1 minute.* Special notice of intercraft coordination among crew supervisors if needed.

This is a working meeting and the bulk of the time for the meeting should be the operations coordinator and crew supervisor combing through the listed work orders to see if they are acceptable. The operations coordinator, in particular, should mention if there are any jobs that the operators might have difficulty clearing up (tagging out) for the crews to work with a day's notice next week. The operations coordinator might also express a preference for which day maintenance should work certain jobs based on knowing the general operating plan for the plant next week.

Follow-up activities after the weekly schedule meeting include the planner for each crew revising the schedule if necessary and publishing the schedule as soon as possible. The schedule is a simple list of all the jobs maintenance intends to do next week. Publishing this list to operations, maintenance, engineering, and plant management helps implement Schedule Principle 2 which states that schedules are important. It helps encourage plant personnel to think ahead and lessen interruptions of a set schedule.

Staging Parts and Tools

Staging is not essential to maintenance planning. Planning and scheduling can achieve high productivity without staging parts and tools. Moreover, staging can be a time waster. Nevertheless, staging can significantly boost productivity for completing more work.

Staging means physically moving a part or a tool out of its regular storage place to where a technician can more easily obtain it before a job. Staging items reduces the time a technician would otherwise spend gathering parts and tools before a job. A net reduction of time for the company comes from a combination of increased specialization, planner expertise, and reduced opportunity for delay trips. First, the person staging the part normally stages more than one job at a time, perhaps as many as needed for the whole crew. The employment of staging reduces the overall number of trips to the storeroom when handling several jobs at the same time. The person staging the items also gains a better than usual familiarity with the storeroom, further reducing time to procure any single item. Second, if the person who plans the job also stages the items, there is another advantage. The planner who specified the item to begin with has the best idea of exactly what item the job requires. This familiarity may speed the process of obtaining the right part at the counter. Third, staging helps keep technicians on the job. Any trip away from the job to a storeroom or tool room can escalate beyond a simple delay to obtain the originally intended item.

Staging is similar to having the operations group clear a piece of equipment. The maintenance group prefers to have the equipment cleared before arriving on-site with three technicians. If the operations group did not have the equipment ready, the technicians would have to wait or lose overall efficiency finding other jobs to fill the wait time. Therefore, the operations group might normally clear the equipment during the night shift based on a request from the previous day's scheduling meeting or schedule. The technicians then arrive and go to work in the morning without delay. Likewise, the maintenance group prefers parts and tools to be ready for work. There might be overall inefficiency if two of the technicians have to wait for one technician to check out a special tool or wait in front of the storeroom counter.

Consider several tasks by a school maintenance person as an illustration of some of the benefits of staging. The school had identified a broken easel and a broken toddler table. The maintenance person serving several schools showed up at 11:00 AM. The school secretary handed over a bag containing a replacement bolt and plastic fitting for the easel and another bag with a washer and bolt for the table. The maintenance person took the staged items and fixed both pieces of furniture in 20 minutes. The school secretary had saved expensive maintenance service time by staging the parts needed. The secretary had determined what items were needed by simply examining the furniture and describing the problem to the appropriate school furniture supply company. This company had mailed the exact required parts the preceding week.

Consider another example from a modern electric power plant. The planner scoped the job to replace a flanged valve. The planner staged a sling, gasket material, and 48 bolts, washers, and nuts along with the replacement valve. The planner did not stage a necessary come-a-long as this device was standard issue in mechanic tool boxes. The planner placed all the items together on a pallet in the tool room. The planner identified the job as staged on the work order and attached a copy of the work order to the pallet to aid later identification. The two assigned technicians later transported the materials to the job site and expeditiously completed the work.

With staging in place, envision a technician arriving at work already knowing what job to start on from the previous day's schedule. The technician picks up his or her tool box along with a bag of parts staged the night before and heads to the job site.

The following sections discuss what might be practical to stage, where items could be staged, who should stage the items, and how to go about staging items.

What to stage

The maintenance group should consider staging all the items the planner included in the job plan. When planning a job, the planner identifies the items that the planner anticipates the job requires. The key word is anticipated. Just as the planner includes time for only anticipated delays, the planner plans for only anticipated parts or tools. The planner estimates the anticipated job cost using these anticipated times, parts, and tools. The planner may include an equipment parts breakdown with the job plan, but this is only a list. The work plan expressly identifies anticipated items for this job. Therefore, it follows that if the job plan calls for certain items, those items could be staged.

However, a number of questions remain. What if it is uncertain what parts a job will need? Perhaps a high chance exists that the anticipated parts will be unnecessary. In addition, perhaps a high chance exists that unanticipated parts will be necessary. If there is a high probability that the technician will have to go to the storeroom for unanticipated items, why bother to stage the anticipated items the job will certainly require? Moreover, what happens to unused staged items on a job? What about time expended to return those items to stock?

Consider a technician that has already started a job where the planner has anticipated the use of several parts. The technician soon finds out the job requires only a single specific part. The technician can obtain just that item from the storeroom. Thus, there are no extra items taken to the job site and no leftover items to return. Why not have the planner just reserve the anticipated parts rather than stage them as well? Reserving through advance notice to the storeroom rather than the additional step of staging may be all that is needed.

As one can see, these questions complicate decisions regarding whether to stage items.

Jobs vary just as do plant sites and plant processes such as receiving and returning storeroom items. Therefore, the following guidelines help the maintenance force make better decisions about staging.

Always stage anticipated items that are:

1. Nonstock and purchased especially for the job.
2. Certainly needed for the job and there is little likelihood any other items from the same place will be needed. Example: Job to replace air filter.

Favor staging anticipated items where:

1. There is high likelihood item will be needed.
2. There is low likelihood other items from the same place will be needed.
3. Technician time is valuable.
4. Technician time is limited.
5. Persons to stage items are readily available.
6. Equipment downtime is valuable.
7. Equipment downtime is limited.
8. Distance to the storeroom or tool room is excessive.
9. Availability or accessibility of storeroom or tool room is limited.
10. It is relatively difficult later to transport items to the site if they are not staged.
11. Item is easily returnable to storeroom or tool room.
12. Item is disposable if unused or lower in value than would be worth technicians' time to return.
13. There is some experience with planning and scheduling.
14. There is high maturity and sophistication of the planners to anticipate items correctly.
15. There is high confidence that the job will start the week or day scheduled.
16. The storeroom frequently has the wrong parts.
17. The storeroom is unstaffed or not very secure.
18. The storeroom carries very low stock levels.
19. The storeroom has a lot of stockouts.

Do not stage items that are:

1. For unscheduled jobs unless a nonstock item was exclusively obtained for a job.
2. Difficult or impractical to move repeatedly due to size or storage requirements.

3. Difficult or impractical to move repeatedly due to legal tracking requirements.

Scheduling is frequently the most critical factor influencing staging. Chapter 3 presented the case that if work could reasonably be expected to start in a given week, staging could be practiced. If there was no advance scheduling to tell when maintenance expected to start jobs, it might be impractical to stage parts or tools. One problem arises from taking parts or tools out of stock for a job that the maintenance crew may not start for quite a while. The staging makes the parts and tools unavailable for other jobs. The second problem arises from the physical storage of the staged parts and tools. Temporary holding places for staged items become overwhelmed and items become lost. On the other hand, if the maintenance group schedules a job for the next day or the next week, staging parts and tools can boost productivity. Staging a part or tool for a scheduled job makes sense because there is a commitment to starting the job. In that case, the staged items become properly unavailable for use elsewhere because they really do need to be reserved for jobs about to begin. In addition, because their jobs are about to begin, there is limited danger of overfilling a designated staging area.

Where to stage

There are various possibilities for where the maintenance force may stage items, each with advantages and disadvantages. These possibilities include central staging areas, scattered staging areas, job sites, crew ready areas, and technician benches. Moreover, combinations of any of these approaches may be the most practical for a particular plant situation.

A *central staging area* would be an area where any item could be staged. The area could be part of another operation such as the tool room where technicians come to a counter to request their items. The area could otherwise be one dedicated for staging without a counter where some or all technicians have open access. Using a central staging area gives fairly good security to keep parts from being lost. Persons have little doubt where an item is staged because there is only one possibility. A central staging area lends itself to uniform procedures, especially with a counter operation. Unfortunately, the central staging area may not be better than leaving the items in the storeroom if the technician still has to go to a counter and wait for the attendant. Despite this concern, staging storeroom items in the tool room may still be a good idea. The maintenance group may have better control over the staging area than the storeroom. Many companies place the storeroom under the control of a group other than maintenance. The storeroom management may be unwilling or unable to make its checkout procedure user friendly. In this case, having a few persons stage the anticipated items out of the

storeroom to a more readily accessible tool room for the bulk of technician activity makes sense. Geographic accessibility also makes a difference. The storeroom may be more remotely located than the staging
area for most jobs. There may even be several storerooms scattered
about the plant site for various types of goods. Having an efficient operation to stage items to a central location might increase overall efficiency.
Technicians would not have to be as familiar with the various storerooms
to obtain a part if they could go to a central staging area. A central staging area has several disadvantages versus other staging options. It may
be better to stage items closer to job sites to speed up work. Also, a central staging area may still require the technicians to make a side trip.
Each extra trip during the day invites technicians to add unnecessary
delays. Technicians might check with the supervisor or technician
friends "just to see what is going on." They might run by the machine
shop to use the telephone "for a minute." Then the technicians must refa-
miliarize themselves with the jobs when they return. Mostly due to
simple human nature, the delays add up when the staging area does not
support technicians staying on the job site.

 Scattered staging areas attempt to remedy the shortcomings of central staging. In this arrangement, there are several designated areas for
staging. They are scattered throughout the extensive plant area. The
locations are close enough to the general areas of work to avoid inviting any unnecessary side trips. They are also located out of regular traffic pathways, but easy to find. These staging areas are not necessarily
elaborate and may be simply formed with lines of yellow paint. One
aluminum rolling mill operation outlines squares of space on the plant
floor in this manner. A more complex area might involve a shed. This is
how one steam plant set up a staging area for specialized turbine tools
on the turbine deck. Such a secure area may have a counter with an
attendant and be open only during certain turbine work. At other times,
limited access is available to supervisors or certain technicians. Other
variations abound in between a secure shed with a counter and a painted
square space. A shed, room, or cage could exist with or without a door
lock or padlock. Large wooden boxes could be placed in strategic locations. Expensive or inexpensive shelving can be utilized. A closely associated issue is having access to tools at various locations throughout the
plant. A particular job on a burner might require specialized tools that
the plant uses nowhere else. It would make sense to have the burner
tools located in a tool box on the burner deck ready for use whether a
job is in the backlog or not. It could be argued whether this is scattered
staging or a scattered tool room. In either case, the objective is to have
items on hand to reduce travel and delays during the job. The disadvantages of a scattered staging strategy revolve around having less control and requiring more coordination. There is less security to prevent
missing parts. It takes increased effort to take items to more than a

single location. Nevertheless, having scattered sites close to the work areas can reduce delays to improve productivity.

Staging material directly at a job site provides the greatest advantage of not having to move parts repeatedly and limiting reasons to leave the job site. On the other hand, this arrangement provides the least security against missing parts. It could also contribute to a more hazardous plant site with parts located every which way. There is also danger for damaging items before they can be used if they get in the way of other plant maintenance or operations. The amount of coordination involved in finding each specific job site to receive items could be a further problem. Even with good coordination, there is a significant chance of putting an item at the wrong site and in effect losing it. On-site staging might make sense for a rotating spares program where there is a designated storage accommodation for the extra equipment. It is also practical to deliver large, nonstock equipment directly to the job site. It may also work out for items that are not readily transported later such as medium to large motors, valves, piping, and pumps. Note that the size of these items also makes it unlikely that they would be carried away from the job site. One plant makes the last duty of the day for tool room attendants to move heavy equipment such as cranes to the sites for the next day's jobs. Smaller items such as gaskets, small valves, and fittings are more practically staged in crew ready areas or on technician work benches.

The *crew ready area* is a likely place to stage relatively small items. Many crews have a designated area where they check in each morning. The crew may also use this area as a break room. Having a shelf or side area in this room for staged items reduces the need for an extra trip to a central staging location. This is especially true of the first job for the day and usually for a second and third job if breaks can be coordinated between jobs. The supervisor has some measure of control over such a location which may be an advantage over scattered locations or job sites. The supervisor also may be more interested in the staging operation with its visible presence there. A supervisor receives excellent feedback of how staging is working when the crew receives items in the supervisor's presence.

Technician benches cannot be overlooked as likely places to stage items. Many companies have a work area of sorts for every technician. This is especially true of many I&C technicians and electricians. It is also common for mechanics and other mechanical crafts to have work benches. These work places allow technicians a clean shop area to work on devices and equipment. Cleanliness gives a great boost to any maintenance task. Companies would rather a technician be free to concentrate on keeping gasket faces clean than to worry about where to obtain gasket material. Planning ahead to ensure parts availability and staging to deliver parts helps the technician's concentration. Technician benches are natural places to consider staging items. The technicians consider these areas their personal spaces and check in there just as they would their crew area each morning. In addition, technicians conduct many of the maintenance tasks in whole or in part on their bench. Therefore, staging items here might mean the technicians need not make any trips elsewhere. Consider a job to replace a gasket on a leaking flange. On this job, the technician needs to cut a gasket from a roll of gasket material. The planner has noted the size of the gasket needed as well as the gasket material. The day before the job, the supervisor has someone from the tool room drop off enough gasket material on the technician's bench for the job. The next day, the technician cuts the gasket, grabs rubber boots, and heads to the job site with the tools to fix the leaking flange. Drawbacks to staging on technician benches include a limitation on item size and possibly limited space available. The same coordination and security drawbacks of having multiple and scattered staging locations exist. An additional problem, unique to staging items at the technician benches, is that the person staging items must know the name of the assigned technician. The crew supervisor makes the assignments with the daily schedule. Because the weekly schedule does not identify the assigned technician, the person would not be able to use the weekly schedule to stage the items. Staging would have to be done daily from the daily schedule.

Who should stage

Normally the planner stages items, but there is equal opportunity for the scheduler, materials purchaser, tool room personnel, storeroom personnel, crew person, or even the supervisor to perform this function. Staging must not come ahead of other planning duties if management has excessively limited the number of planners. The materials purchaser is probably the best person to stage nonstock items received. In addition, if the staging is based more on the daily schedule than the weekly schedule, the supervisor or his or her designee would be more appropriate to do the staging. For simplification, the following section calls the person who does the staging, the staging person.

The process of staging

After the scheduler makes up the weekly schedule, the staging person reviews the scheduled work orders. This person might review them before delivery to the crew supervisors or later in the supervisors' offices. (Alternately the planner could review, and mark, the work orders and direct the staging person.) To avoid delaying the receipt of the weekly schedule by the crew supervisor, the staging person may find it more desirable to work from the crew supervisor's office. The person would also have better access to the work orders. The staging person marks the work orders for which items will be staged and makes a copy of each work order that will have staged items. Then, the staging person takes the work order copies to the storeroom and tool room and collects the items to stage. The staging person takes the collected items to the appropriate staging locations and attaches the work order copy to groups of items for each job. The person places items in appropriate containers such as bags for bolts and loose items, each marked for the appropriate work order. If there are several containers for the same work order, the staging person marks the work order copy as having four bundles and each bundle as 1 of 4, 2 of 4, 3 of 4, and 4 of 4, each with the work order number. These activities may take one or more trips.

The staging person must relocate the original work order for any items that were supposed to be staged, but could not be staged for some reason. The person marks these work orders accordingly. The staging process might discover an unexpected problem with item availability that will impact a job. In this case, the staging person informs the crew supervisor and planning so that the job will not be assigned until planning remedies the problem. In the case where the daily schedule is utilized, the same process is followed on a daily basis.

What happens to staged items for a job that is not started in its assigned day or allocated week? The items that were staged for this job are not collected and returned. It is presumed that this job will likely be assigned on a subsequent day or allocated in the next week's work. The job is simply considered already staged. This situation could get out of hand if jobs are routinely staged that never start. Staging areas could become overrun and inventory stocks could become depleted. This scenario is more likely for the plant that stages without any scheduling. Staging accomplished after some scheduling effort keeps these problems to an acceptable level. Daily staging for only the next day's work lowers the probability of these problems even further.

One thing to realize is that there is a limited time available to stage items. The difficulty comes in the timing because the advance schedule is only an allocation of work and does not denote which day activities might begin. If the staging person waits to use the advance schedule, there is not much work time after which the schedule becomes available before the work week begins. If the crew works Monday through Friday, 7:30 AM to 4:00 PM, the advance schedule would be available Friday afternoon for the staging person. The staging person could extend the work time available by working a weekend or evening shift to stage items. If the crew works another type shift, other times may work better. Consider a crew that works Monday through Thursday, 10-hour days. The staging person could work Monday through Friday 8-hour days and have Thursday afternoon as well as all day Friday to stage. In addition, if the supervisor prepares the first day's schedule soon after receiving the weekly allocation, the staging person could first concentrate on staging parts for that first day's work. Then the staging person could spend a small amount of time each day reviewing the next day's schedule from the supervisor and staging the appropriate parts.

In summary, staging can help improve maintenance productivity, but it is not essential to effective planning and scheduling. Furthermore, staging can become somewhat complicated to execute properly. Among a number of guidelines, scheduling control and experience dictate the successfulness of staging. Imagine a common household situation. One adult plans to hang several pictures and shock the swimming pool one night during the coming week. This person "stages" a hammer, several nails, and a bag of pool shock on a kitchen counter as the items are encountered during other weekend tasks. However, the items disappear from the counter as the week proceeds and wind up in various drawers or garage shelves. The person's spouse has put the items away to avoid clutter. The tasks of hanging the pictures and shocking the pool have become more complicated now as the location of the necessary items has become uncertain. Industry commonly experiences the same primary problems hindering these maintenance tasks, namely an improper staging area and an imprecise schedule for the work.

Outage Scheduling

Although routine maintenance provides the greatest opportunity for improvement and so is the focus of this book, this section gives the associated keys to understanding the concepts of outage scheduling.

The vast majority of plant work orders are tasks for the standing work force. The standing work force at the plant maintains the plant day in and day out continuously. Planning leverages that day-in and day-out maintenance work. Plants frequently overlook the opportunities within routine maintenance because outages receive so much attention. Managers view outages as extremely important. They see unscheduled outages as tragedies and extended outages as fiascoes. In fact, plants organize outage events so well that they efficiently accomplish large quantities of work leaving everyone impressed with how much work can be done. They often attribute this great amount of work to extra effort based on the obvious urgency of the situation. However, the success is also due to the organizing effort, primarily the advance allocation of a specific quantity of work to complete. Similarly, planning and scheduling for routine maintenance can help accomplish an amount of work that can equally impress plant management. Planning and allocating a week's goal of work to a crew not only creates the same sense of urgency as for an "important" outage, but provides tools to manage and improve upon past problems. Nonetheless, it is worth discussing two keys to outage scheduling. First, planning provides accurate time estimates for larger jobs because larger jobs consist of a multitude of small jobs. Second, the scope of the outage must be controlled by managing the identification and inclusion of the small jobs.

An outage is normally considered the taking of an entire unit out of service. An outage is not simply shutting down a redundant process line or piece of equipment. Technicians many times can perform maintenance without taking any equipment out of service. Sometimes, technicians require taking only certain equipment or areas of a process out of service for only a brief period in such a manner that allows the unit to continue producing the product. For example, operators might briefly take makeup water equipment out of service provided the plant has reserve tanks of water available. On the other hand, technicians cannot perform some maintenance tasks without causing the plant to make some or all of a unit unavailable for service. Maintenance may be able to complete some work while the plant runs at a reduced capability. For example, technicians may be able to work on one of two boiler feed pumps while a unit runs at half load. Maintenance situations may require shutting down an entire unit, require no shutdowns at all, or require varying in-between unit conditions. Even with a requirement to be off-line, many plants can shut down an entire unit with hardly any advance notice or appreciable problems, operational or economical. Perhaps the company has not sold out its product line leaving open time. Perhaps technicians can complete maintenance during one shift and the unit can operate another shift during the same day. Plants normally call the taking of an entire unit out of service an outage. Outages may be major outages scheduled every so many years to overhaul major pieces of equipment on a routine basis or they may be short outages, either scheduled or unscheduled.

Some persons also call major outages overhauls or turnarounds. Major outages are when a plant schedules one or more major process systems for extensive routine replacement, refurbishment, or other maintenance. The work requires shutting down the entire unit. These maintenance activities cannot be accomplished when the unit is operating and are often too large in scope to be done during shorter outages whether scheduled or unexpected. Many companies utilize set schedules for these events such as once every 5 years. However, the advent of predictive maintenance programs many times allows systems to run longer. Plants can then schedule major outages when sophisticated inspection methods predict the process system equipment needs attention.

On the other hand, a short outage is an event between major outages requiring taking the entire unit out of service. Short outages may be unscheduled, such as the sudden requirement to repair a burst boiler tube. The unit cannot run with the boiler tube losing process water, so the plant must take the unit out of service almost immediately or within the next few days. The plant strategy may also be to schedule short outages to take a unit out of service to perform maintenance tasks that do not require immediate attention. The plant performs these tasks in anticipation that they will lessen the likelihood of later unscheduled or surprise outages.

An evolution in the timing of short outages takes place as a plant increases its reliability through proper maintenance. A plant with fairly poor reliability usually has something break just when the plant otherwise needs a short outage. A number of work orders sit in the outage backlog. Yet rather than plan for a time to bring the unit off-line, the maintenance group knows the unit will "trip" or otherwise require shutting down for some unforeseen need on a regular basis. Take the example of a steam plant with poor water chemistry control. Boiler tubes burst periodically requiring shut down for repair. When a tube bursts with an estimate of, say, 12 hours to repair, the maintenance group springs into action also working the other outage jobs. The crew starts all jobs in the outage file with estimates of 12 hours or less. The crew may start longer jobs if the unit is not again needed immediately or management desires the unit to return in as good a shape as possible. Longer jobs that management decides not to start would wait for a future outage.

As a plant evolves from total reactive maintenance to more proactive maintenance, short notice outages begin to become less frequent. As fewer unscheduled, short notice outages appear, plant reliability improves. Yet the outage backlog first increases in size because fewer outages naturally occur. As plant reliability improves, the plant finds fewer opportunities to execute waiting outage work orders. Then the plant experiences some short notice outages occurring for new reasons. Previously, the maintenance group would have taken care of certain situations before they became too serious by doing some work during a short outage. However, when no outages come up, these situations worsen and themselves bring the unit off-line. For example, if technicians always repacked certain valves during outages, frequent unscheduled outages were available for this work. As reliability of the rest of the unit rises, the valves themselves begin to cause outages because the unit cannot wait. The evolution of the maintenance department's effectiveness continues until there is a predictability that the unit should be brought down for maintenance. At this point, management commitment to continuous reliability brings the operations and the maintenance groups together to schedule short, regular outages for routine maintenance in addition to the standard, infrequent major outages. The entire plant including operations and maintenance must adapt to a strategy of planned short outages to execute short notice outage work (SNOW). The plant must accept scheduling an outage in advance when there are a number of serious work orders on the SNOW list or there is a sufficient amount of SNOW work. The timely execution of the outage work prevents unscheduled outages. Overall plant reliability and availability increase through the strategy of short scheduled outages reducing the occurrence of infrequent, but serious, unscheduled ones. The evolution continues as maintenance and plant engineering perform defect elimination work to identify and replace equipment that requires excessive routine downtime. Moreover, maintenance, plant engineering, and the corporate project group perform defect elimination work to install plants or systems that incorporate lessons from the past. The evolution results in a superior performing plant capable of full capacity as needed and having minimal outage requirements for maintenance.

The evolution of the maintenance department's effectiveness changes with how it approaches short outages. As a plant rises from routinely having poor reliability to become a superior performing plant, fewer unscheduled outages become available for work. Therefore, regular, scheduled, short outages become more frequent.

Planning work orders for outages

Many plants have work orders that they can only execute during an outage. To help the scheduler quickly select the outage jobs, the plant keeps a SNOW list or SNOW grouping of work orders. The SNOW list or grouping identifies or even keeps together the work orders that must be done on an outage, but not necessarily the next major outage.

There are only a few differences between planning work orders for short outages and for routine maintenance. Because technicians have limited time during the actual outage to gather parts and information, the company puts more emphasis on planners identifying and reserving anticipated parts. The planner has time to do this for even reactive jobs when an outage has not yet started. On the other hand, planners place a high priority on quickly planning outage work orders. They never know when an unexpected outage may suddenly occur requiring finished plans.

Individual work orders may make up some of the work for major outages, but not necessarily all. Large tasks such as certain turbine work may instead involve special outage books of notes from previous outages. The planners should take advantage of requesting help from the specific supervisors and technicians that worked particular areas of previous major outages to determine estimated times and labor requirements. The outage books should also identify parts and tools from previous outages.

Key concepts in scheduling for outages

Many individual work orders of jobs make up outage work. Therefore, the scheduler can utilize the concepts of advance scheduling developed for routine, weekly scheduling. This allows the scheduler to make accurate enough assessments of time frames for the large quantities of work involved in outages. This concept provides the first key to understanding outage scheduling.

Because outages consist of many individual jobs, the scheduler can apply the concepts that make weekly work allocations an accurate tool. A scheduler can use planned work order estimates to determine accurately the duration and labor hours for major outages. In a routine week of maintenance, a scheduler can allocate the right amount of work even though the time estimates for small jobs have a tremendous amount of variance in individual accuracy. For example, the planner's estimate for replacing a single control valve may vary considerably. Yet a weekly schedule allocation of 100 jobs smooths out any variations of individual jobs. The work force might therefore have confidence that it could accomplish the overall amount of work in the scheduled week. The scheduler may similarly consider a single large job to consist of many small work orders and therefore accurately estimate labor requirements for the large job. The scheduler and planners together can therefore estimate the total duration and labor requirements of overhauling 20 control valves with a satisfactory degree of precision. A major outage consists of many large jobs and the handling of each large job can be approximated to the weekly scheduling process. Consider the major overhaul of a large steam turbine consisting of many individual tasks on many individual systems. The scheduler and planners together can reasonably estimate the total duration and labor requirements of restoring the turbine itself. The restoration involves many small tasks such as disassembly, inspection, lifting, transporting, machining, coating, polishing, transporting, lifting, assembly, fastening, and alignment. There are also a myriad of miscellaneous work orders that the plant identified over the past few years that could only be done during the overhaul. The planners have already planned these work orders. The scheduler can group them to determine their group labor requirements. Overall, the scheduler uses the concept of the grouping of small tasks allowing overall estimate accuracy.

Note that during routine maintenance, the crew forecast of labor available for a single week determines how much planned work the scheduler assigns. For a major outage, the amount of work is the independent variable. That means the amount of work determines the length of the outage considering basically a set amount of labor each week. After determining the initial estimate of the outage duration, management can evaluate options to increase the work force. Management may supplement the regular labor force using 24-hour, around-the-clock work shifts or contract labor. Many plants have labor sharing agreements to help each other during major outages. For outage maintenance, the scheduler adjusts the outage time to match a given amount of work, then considers special labor arrangements. For nonoutage, routine maintenance, the scheduler adjusts the amount of work to match a given amount of labor hours.

The overall outage can be managed through critical path methods (CPMs) and other special scheduling techniques. However, these techniques show the large groupings, not the minute, individual maintenance tasks. That is why these techniques can be successfully used with the outage.

Because a major outage consists of many individual jobs, the crew supervisor must create daily or shift schedules as the outage proceeds. A scheduler can set the major activities and overall times for the outage, but cannot control individual jobs. At the beginning of each shift, the crew supervisor must ensure persons understand their assignments based on the progress of the previous shift. The supervisor must provide this coordination during a major outage even though technicians usually stay on the same equipment and do not move around too much between different areas. An entirely different group of technicians may have worked the previous shift. In addition, the technicians may be executing work that the plant performs only once every 5 or 10 years. That means the average technicians may have drastically limited experience with the work over their entire employment at the plant. On the other hand, the often older crew supervisor may have critical personal experience.

The scheduler gathers appropriate SNOW work orders and sets the labor requirements and duration for a short, scheduled outage in a similar manner to a major outage. The scheduler and management determine the best crew and time arrangements considering labor availability and shift options.

A single event often drives a short outage. Consider first an unscheduled short outage. Consider again the boiler tube that erupts and causes a unit outage. Maintenance must repair the tube. This single task causes the unit to be on outage and unavailable. For such a task, the pertinent crew supervisor and management estimate the duration of the outage, say 18 hours. The scheduler then takes all of the work orders that have been waiting for an outage and selects the ones that can be done in 18 hours or less. Maintenance crews then complete as many of the jobs as possible. So for a short, unscheduled outage, the primary job sets the time frame. Labor availability determines how many jobs the crews can accomplish. The scheduling consists of the scheduler or supervisor taking all the outage backlog jobs with an estimated duration within that time frame and then considering the persons needed to work the jobs. If there are any backlog jobs that have not yet received planning, the supervisor guesses the time requirements and includes suitable ones in the outage scope. The supervisor writes down all the jobs on a daily schedule sheet with hours for each of the crew members. This is very similar to the regular weekly and daily scheduling routine except for one difference. The amount of hours the crew has for the week does not drive the schedule, the outage backlog selected from the outage file by duration does. The supervisor reassigns persons from their nonoutage tasks currently under way and plans overtime as needed to get the work done.

The second key in outage scheduling is that the scope of the outage must be controlled. If the scheduler has a specific amount of work, the scheduler can develop specific time schedules and labor requirements. If the company allows the amount of work to vary, the time schedules and labor requirements cannot help varying as well. Although the techniques of accurate scheduling are important, the overwhelming key ingredient in making an outage go well is agreement on the scope. The scope itself is less important than agreement on the scope. Many times an outage will start with one scope of work, but as soon as the plant takes the unit off-line, the scope doubles.

This is a simple concept to apply to an outage consisting of 100 work orders. The scheduler can reasonably set the labor time requirements from reviewing the job plans. Then the scheduler can establish the overall schedule based upon management preference of crew shifts. However, suppose the amount of outage work orders suddenly jumps to 150. The scope of the outage has changed. It has increased by the 50 new work orders. The scheduler must change the schedule.

This consideration is especially important for major outages where the initial part of the outage may consist of inspections of major machinery. If the inspections reveal more serious damage than anticipated, the sudden inclusion of more work may extend the entire outage. This is through no fault of the scheduler. After the inspections determine the extra work, the scheduler must then analyze options of labor or critical path changes for management review. This is why outages including major inspections are difficult to schedule precisely. The advent of predictive maintenance technologies has greatly assisted scheduling for major outages. Through sophisticated technology, predictive maintenance (PdM) allows more precise determination of maintenance needs before the plant shuts down major machinery to begin outages.

Knowledge of the scope of work provides only part of this second key to outage scheduling. The rest comes from understanding that the scope must be controlled. The plant must identify work as far as possible in advance. The plant must not include new work on the eve of or even during the outage whenever possible. Late announcements of work destroy schedules and labor arrangements. Late inclusion of work causes that work to be poorly planned. The ensuing confusion may also cause incomplete execution of new work hastily identified. It might also contribute to hasty decisions to delete other work from the schedule. The company may have already used the set schedule to make arrangements for production and sale of product which may become expensive to change. Additional labor necessary to maintain a set schedule with an increased scope may be more expensive or less qualified than could have otherwise been arranged. Finally, there may be insufficient advance lead time to procure material necessary to execute the work. The maintenance group would not be able to execute that work.

To reduce these changing scope problems, the scheduler first begins identification of future outages as far as possible in advance. Many companies have 1-, 5-, and 10-year outage plans. These plans give approximate periods of all anticipated outages. The 1-year plan might set specific days or weeks, whereas the 10-year plan may set specific months or seasons. As the outage time approaches, management allocates preparation time for the schedulers and key company personnel to begin defining the work scope and time frames. The particular scheduling horizons depend on the specific type of outage, equipment lead times, and labor resources. Most of the scope definition work for a larger outage is straightforward 6 months to 1 year before the scheduled outage starts. The initial project team makes use of outage files and books. Even work completed every 5 or more years is still repetitious work. Crews executing outage work must record feedback to make files most useful. The period of about 3 months before an outage starts is when scope additions multiply if left unattended. During these 3 months, it seems almost everyone knows about new work requirements. However, with less than 3 months before the outage, there may not be enough material lead time or other time to prepare adequately for the new work. Therefore, the scheduler must pay close attention to the control of the outage scope.

Schedulers commonly control the scope of outages through the use of lists. The scheduler may issue a list of known work with a stated main purpose for the outage 6 months to a year before the start date. The scheduler has an initial meeting with supervisors to identify maintenance needs. Then, with increasing frequency until the start of the outage, the scheduler continues to meet with the supervisors and issue lists of identified work. Planners or special outage planners finish planning the identified work tasks as necessary to have them ready to execute during the outage. An outage project manager ensures this cycle of publish, identify, and plan continues up to the start of the outage. Management support and organizational discipline help ensure giving serious attention to timely identification of work. Management should not appreciate late identification of outage tasks. If managers do not wish to have an unexpected outage extension, they must be willing to freeze the outage scope and not allow routine additions at some point before the outage begins. The scheduler might issue a statement at some point before the outage that "Any work identified after this week must be arranged and managed by the originator of the work." The early start of giving attention and listing the work greatly reduces the occurrence of later scope additions.

During the course of an outage, the scheduler continues to meet with the crew supervisors to identify completed work. The scheduler also continues to publish lists of work remaining. The scheduler compares actual completion of blocks of work against expected completion to measure the progress of the outage. In order to encourage supervisors to attend meetings during the execution of the outage, the scheduler should limit the meetings to only 15 or 20 minutes.

After the outage is over, the scheduler should still schedule one or two meetings with the crew supervisors, planners, and other personnel, possibly even including technicians. Just as a planner needs feedback to improve future plans, the scheduler and planners seek information to improve future outages. The final meetings should address what went right as well as identify where the team could make improvements. These meetings should occur soon after the outages before personnel forget information and ideas.

The plant handles scheduled short outages very similarly to major outages in regard to work scope. Smaller outages that the plant determines are necessary only months, weeks, or days before starting, and have similar but less extended preparations made. The operations and maintenance groups together arrange to have a short outage at a future scheduled time to maintain the plant's capability of operating at full capacity. This is not a recovery from a trip or loss of capacity. The key to how much work the maintenance group will accomplish for a short scheduled outage is the plant decision on the scope of work. The scheduler must continually encourage the routine identification of work even if the work can only be done during an outage. This ensures the backlog contains the necessary work orders for an outage before the last moment. The scheduler assesses what routine preventive maintenance to include. The scheduler reviews any predictive maintenance recommendations or other corrective maintenance in the backlog. With this information, the scheduler prepares a preliminary scope of all jobs to include in the outage for review by the maintenance crew supervisors and plant management. The scheduler takes the scope of work and then assesses how long the outage should last based on discussions with crew supervisors for crew availability and management preferences for off shift and overtime work. The crew supervisors create a daily scheduling sheet(s) for the outage to get all the work done.

As previously discussed for unscheduled short outages or trips, the primary failure is often the item that determines the length of the outage. The scope normally consists of all SNOW work orders that the planners estimate have a duration of equal length or shorter.

Management must be willing to set a freeze on the scope of even short outages. It must resist the addition of new work after an unscheduled short outage begins or after setting the scope of a scheduled outage. Otherwise, management must accept the responsibility for outage schedule changes.

A final necessary note includes the involvement of the operations group. Timing governs the success of outage work in such a way that one must not forget the time involved for operator tasks. Operators must not only clear equipment, but return equipment to service. Managers and schedulers must include the operations group in all outage meetings to discuss clearance and restoration to service of individual components as well as overall unit shakedown. Operators may have a preference for the timing of the return of specific components to levelize their activities. Their knowledge may also profoundly affect the overall scheduling of the outage. For instance, a major steam plant may sometimes require days or weeks to clean boiler water to allow return to full capacity. As with any maintenance work, the maintenance technicians are responsible for delivering equipment that is ready to run. However, the operators have the responsibility of testing and returning the equipment to service. Schedulers must include the operators when considering maintenance work.

Two key concepts help the understanding of the scheduling of outages. The first key is realizing that the scheduler can only have great accuracy when scheduling *blocks* of work orders. Grouping of many tasks into blocks of work for an outage tends to smooth out the inaccuracies inherent in the estimates of individual tasks. The second key is realizing that the overall ability of being able to schedule outage work accurately lies in the control of additional work to the original outage scope. Schedules or resources must change if the work scope changes. A scheduler can schedule the work scope of an outage. The plant must control the scope.

Quotas, Benchmarks, and Standards Addressed

The terms quota, benchmark, and standard see frequent usage in maintenance and should be addressed in light of the current scheduling context. The normal concepts implied by these words do not lend themselves very well to a planning and scheduling operation.

Quotas establish amounts of work that a crew must do. For example, management might dictate that a certain maintenance crew must always complete 60 work orders every week. At first glance, the development of a planning and scheduling system seems to have that very thing in mind. However, a quota is more of a mandatory figure that does not allow very well for consideration of current events, quality, or actual jobs ever being different from the plans. Quotas do not provide the answer to superior maintenance. Rather, proper planning and scheduling set schedule expectations and realistic goals based on current conditions as well as provide the means to improve upon past jobs. A technician must be able to point out why a job should be extended in the case of special circumstances. There may be valid schedule pressures and there may be times a technician cannot be the sole judge to delay a job. However, these factors must be maturely worked out for the optimum benefit for the plant as a whole. Rather than setting a strict quota of work orders to complete, management should consider how many technician work hours the crew has available and assist the crew to select that same amount of estimated work hours of backlogged jobs that would most benefit the plant. To emphasize that a planning and scheduling operation does not promote quotas, consider the following. If it is not a quota to assign two persons to a job, why would it be a quota to assign 2 hours to a job or two jobs to a person?

Compensation arrangements that take into account production amount without regard for quality form a type of quota. Management should avoid these situations or watch them very closely. For example, wages based on number of jobs or number of "book hours" might encourage a technician to pay insufficient attention to quality. It is true that the supervisor has less need to motivate such piece workers; but are they performing proper work? Instead, paying technicians by the hour, but giving them sufficient amounts of work to do requires more supervisor effort, but appears to keep quality higher in focus.

Benchmarks are self comparisons to how other facilities perform in certain areas in which one has an interest to improve. For example, management may compare how much it spends yearly on maintenance versus the most profitable companies in similar or dissimilar industries. The comparisons may let management know if its own maintenance department is relatively effective. However, a comparison itself may not tell management how its own company can improve. The other facilities may have certain factors that provide for their particular success. Simple benchmark comparisons do not readily reveal these factors. The other facilities must be carefully studied for benchmarking exercises to become most useful. How similar is the other facility? Perhaps the other plant has a greater abundance of skilled labor available. Perhaps the other plant has already made the transition from reactive to proactive maintenance. Perhaps the other company invests regularly to replace troublesome equipment. Perhaps the company has an active program for maintenance personnel input into new plants being built. The point is that simple benchmark numbers may not tell why plants perform differently. There are often key factors that must be understood beyond the benchmark number. So it is with planning and scheduling. A benchmark comparison that shows a plant has a planning department and schedules its maintenance does not tell if or why those programs are beneficial. A benchmark telling that wrench time is different does not tell why it is different. The planning and scheduling principles in

Chaps. 2 and 3 embody obvious factors that influence how well planning and scheduling work. In addition, the entirety of Chap. 1 points to other factors necessary for planning to function. A simple benchmark would be difficult to use among different plants in many of these regards. Consequently, benchmarking among plants to control a planning operation may not be as beneficial as closely adhering to the planning and scheduling principles and other guidelines presented in this book. Nevertheless, the principle underlying benchmarking—to visit other plants to pick up good ideas—may invaluably assist to improve any aspect of plant operation and maintenance. Benchmarkers should try to visit best performers. They should attempt to understand how, not just how much.

Another use of the word benchmark implies less of a cross-company comparison and more of a simple goal or idea of what one's performance should be. The planner's estimate of a job's labor hours do set somewhat of a benchmark in this regard for the technician's consideration.

Standards present a combination of quotas and benchmarks, but on individual jobs, weekly schedules, and daily schedules.

A real standard applied to a specific job would dictate beyond any doubt how long a job should take. One way an engineer or analyst would establish a standard would be to study a task as it is repeated many times by different technicians. The analyst or engineer together with some of the technicians might then decide the best way to perform the work. Then technicians trained to perform the work in this best manner would be timed. The resulting labor hours and duration would become the official job standard. This approach may be practical for an automobile repair shop where industrial engineers have studied specific tasks repeatable on identical vehicles maintained in similar shops. This approach is less applicable to an industrial plant where technicians may repeat specific jobs only once or twice over the course of a year. There is limited opportunity to observe and study such work. The manufacturers may have some ability to specify standard maintenance tasks for their equipment. On the other hand, manufacturers may be better qualified in the manufacture of the equipment rather than in its maintenance. In addition, the great diversity in the shop arrangement and personnel skills of the persons using the equipment may make generic standards useful for a starting point, if practical at all. PM tasks may lend themselves to locally developed standards because technicians may execute them on a more frequent basis than once or twice a year. Nonetheless, the great productivity improvement available through planning and scheduling comes from having some schedule control, not necessarily incredibly precise estimate accuracy. Typically, the introduction of proper planning and scheduling in a maintenance organization dramatically improves productivity. Addition of precise job standards contributes a relatively smaller productivity gain. Normal job planning simply does not study jobs thoroughly enough to set precise time estimates. The planned estimates are accurate enough for scheduling work, but not for establishing official job standards. Fortunately, the ability to schedule the work in a reasonable fashion is all that is necessary. Nonetheless, when used less formally, the term standard is somewhat applicable to the planner's estimate of a job's labor hours. The planner's estimates are said to be the job standard. One must keep in mind, however, that these are not engineered standards. They are simple, initial determinations of a skilled planner.

In addition, just as the planner's estimate is an informal, but useful, standard for a job, the weekly schedule is the scheduler's standard for the week of work and the daily schedule is the supervisor's standard for the day of work. The week of work represents the standard of what the crew should be able to accomplish in a week. The day of work represents the standard of what the crew should be able to accomplish in a day. Having standards helps the plant control the maintenance work. The plant assesses with job feedback, weekly schedule compliance, and supervisor inspection how well the standards are met. Comparing performance against the standard helps planners, schedulers, and supervisors adjust their practices to improve performance.

Setting estimates and assigning work through planning and scheduling does not necessarily involve engineered standards, production quotas, and cross-company benchmarks.

Summary

This chapter described the specific activities that accomplish weekly scheduling. For advance scheduling, a scheduler simply allocates an amount of work orders for a week. The scheduler does not set specific days or times to begin or complete each work order. The scheduler works with the crew supervisors to establish forecasted labor hours and then selects that quantity of work order hours for the allocation. Specific methods, routines, and forms help the scheduler select the best mix of work for the plant. For daily scheduling, the crew supervisor selects work orders mostly from the weekly allocation. However, the supervisor also maintains the flexibility to reassign the crew for emergency and other urgent work that may arise. The supervisor may follow different methods and procedures to select and assign the appropriate work each day. All of these methods and procedures have certain elements in common, primarily giving each crew member a full shift of work based on planner job estimates. As plants gain experience with planning and scheduling, staging certain job items may further improve labor productivity. These techniques of scheduling greatly assist maintenance in improving its labor productivity. This chapter described the exact steps of scheduling to clarify the concepts of scheduling. After understanding the concepts, readers may implement alternative steps than the ones prescribed. This chapter should allow companies to implement their own effective scheduling.

Routine day-to-day maintenance offers the greatest opportunity for planning and scheduling to make a difference. Companies typically execute maintenance outages with much success already. Consequently, this book does not dwell on the actual scheduling of outage maintenance. However, the scheduling of outage maintenance relies on particular concepts inherent in the practice of routine maintenance scheduling. One concept is the increased accuracy of time estimates achievable for blocks of work made up of smaller jobs with less precise time estimates. Another concept is the control of inclusion of the smaller jobs that make up the larger blocks of outage work. If the scope of work continually changes for an outage, the overall outage schedule must change as well. The chapter addresses these concepts as keys to outage scheduling in addition to describing the routine scheduling of maintenance work.

Daily Scheduling and Supervision

As discussed for Scheduling Principle 5 in Chap. 3, the crew supervisor schedules work orders on a daily basis. Formal daily scheduling assigns specific individuals to specific work orders in a manner that accounts for each labor hour available and works toward the goal of completing the work allocated by the weekly schedule. The supervisor also supervises the work in progress in the field. These matters of daily scheduling and supervision of current work execution are not the responsibility of the planning department, per se. However, they are of such importance to the planning and scheduling system that this handbook has an obligation to explore their specific tasks. That is, while these tasks do not belong to the planning department that creates job plans and issues the weekly schedule, they are part of the overall planning and scheduling system. Note also that the method presented is only one method. Others are acceptable provided they work toward the goal of weekly work.

The chapter presents an actual description of a supervisor working a normal day and then more specifically addresses assigning work, coordination with operations, handing out work orders, and duties each day.

A Day in the Life of a Maintenance Supervisor

Chapter 5, Basic Planning, begins by describing how a planner actually goes about planning jobs with the episode of *A Day in the Life of a Maintenance Planner*. (Chapter 9 depicts the same planner in a system that uses a CMMS.) The following account shows a crew supervisor at work with an emphasis on the planning and scheduling aspects of daily supervision. This episode helps the reader understand how the supervisor benefits and interacts within the planning and scheduling system.

So, what does a *Supervisor* do? Industry wisdom claims that the process of planning and scheduling dramatically improves maintenance productivity. We have seen what a planner does throughout the day, but what exactly does a supervisor do? The following narrative follows a supervisor through a normal day to see.

6:45 AM.   Maintenance Supervisor Terry Brooks arrived at work Tuesday morning looking forward to another day helping the maintenance department keep equipment working to support the operation of the plant. Fortunately, no one had called him in the middle of night for an emergency call out; that meant maintenance had achieved some measure of reliability for the equipment. Terry immediately went to his office and checked his voice mail and email. There were no urgent messages for work needed, again a good sign. In addition, operations had not notified him that they had been unable to clear any of the requested jobs from the previous day.

Terry took the daily schedule for today (Tuesday) that he had handed out yesterday and attached it to his clipboard along with a blank schedule for tomorrow (Wednesday). He headed down to the crew break area.

7:00 AM.   As crewmembers began to arrive, Terry checked them off as present, handed the lead technicians all of their work orders for the day, and watched everyone go to their first assigned tasks.

The technicians first had to go to the clearance office in the control room and "sign onto" the clearances for their jobs. At this plant, operators cleared equipment and required any technician who had to work on cleared equipment to sign the clearance form in the clearance office for that job. After the technicians finished jobs, they would go back to the clearance office and "sign off" on the clearance forms. After a lead technician (empowered by the supervisor) informed operations that work was complete, the operations group checked that all technicians had signed off the clearance form (to ensure they were no longer working on the equipment) and restored the equipment to service as necessary. This procedure caused a bit of traffic through the control room area, but the plant considered it time well spent for safety.

7:45 AM.   The daily schedule for today assigned everyone work for the entire 10-hour shift. No one had called in sick, so Terry did not have to rearrange anyone's schedule. As the last person left, Terry headed up to the morning meeting. On the way, he stopped by the planning department to drop off the stack of completed work orders from yesterday.

8:00 AM.   The morning meeting was short and sweet. Its purpose was to review all the latest work orders for proper priority and craft. Most of the plant supervisors attended as did the operations coordinator and planning supervisor (who also chaired the brief meeting). As a group, they raised the priority of one work order and lowered the priority of two others. One reason Terry attended was to see if operators had written any urgent work orders for his crew that he should work today with or without a plan. It looked like there would not be a schedule breaker for him today. The meeting also gave him a feel of the general condition of the plant.

8:30 AM.   Terry headed out into the field to help his crew. He did not necessarily visit every job underway, but started with jobs that seemed most likely to be of significance. As he approached each technician on a job he would ask, "Is there anything I can do for you?" Even if he had not approached their jobs, technicians could radio him for special help. Still, his presence helped move some jobs along. Adamson had been on a vacuum pump job for the second day without seeming to make much headway. Terry spent about an hour with him. Together they reviewed the progress so far, hunted down an O&M manual in the planning department, and settled on an action plan to finish the job.

10:00 AM.   It was now 10 o'clock and the day had been successfully launched. Everyone was here on task with no schedule breakers. Terry swung by the engineering office to meet with an engineer who wanted input on a valve actuator modification planned for a valve in Terry's area of the plant.

10:30 AM.   Terry went back to working with technicians in the field until lunchtime.

12:30 PM.   After lunch with a couple of the other supervisors, Terry checked his email and headed back out to check on-job progress. This time he began filling out the blank daily schedule form for tomorrow. The schedule form was easy to use because it had technician names written across the top and blanks down the side for entering work order numbers. Terry had a tablet of the blank schedules. As he encountered technicians, he questioned each for the estimated completion time. This afternoon reinforced the fact that job time estimates were only that, estimates. What had appeared to be a straightforward job on the dock gutters for Capper, Glade, and Young for 4 hours had exploded into a rebuild of the screens and curbs. On tomorrow's schedule, Terry wrote down the work order number. He gave each of the three technicians 5 hours for the next day as carryover work. Conversations with other technicians from his visits or by radio indicated two other jobs would run long and so Terry planned for some carryover work on tomorrow's schedule. On the other hand, it appeared Kingsley and Jensen would finish all of their work for the entire day very early, so Terry gave them several new jobs from the weekly schedule stack. He handed a small job to Kingsley that he ought to be able to finish today and a larger job to Jensen that would probably carryover a few hours into tomorrow. During this time, Terry received two radio summons from operations personnel. Terry returned the calls on his cell phone, both of which dealt with operations needing special work. Terry convinced the operators in each case to write work orders with urgent priorities so that they could be planned and he promised he would work them tomorrow. Terry also called the planner to give a "heads up" for planning the reactive work.

1:30 PM.   Terry continued to visit job sites and fill out the schedule form. At one point, Kingsley and Jensen gave him completed work orders for their completed jobs. Terry reviewed the completed work orders and saved them for the planning department. After a while, Terry had a handle on which jobs the technicians would complete today and which jobs needed time tomorrow. Several times Terry abandoned the scheduling effort to help technicians. He particularly wanted to spend some time with Sanchez reviewing the auxiliary feed pump job. One problem was not a job problem, but involved discussing whether Johns could take off several days during an upcoming outage.

2:30 PM.   Terry returned to his office after swinging by the planning department to pick up completed plans for the urgent work orders operators had brought up earlier. Although these reactive job plans had little beyond "peek at the history, peek at the job, and put on a plan," they were invaluable in giving him guidance on job scope, skills needed, and time durations. Many times, they also helped identify past parts used for the equipment. Above all, he could count on the planner using a quick planning approach that would not slow him down from going ahead and assigning the work for tomorrow. Terry finished the rest of the daily schedule for tomorrow. Based on an earlier discussion with the technician, he guessed how long another carryover job might take. He then added the urgent now-planned work orders he promised to start tomorrow. Next Terry looked through the stack of remaining work orders scheduled for the week. The weekly schedule did not specify which day each job should start. It was really more of a stack of work orders for which Terry's crew had the labor hours to accomplish. Balancing the more important work orders against the technicians he had available with different skills for tomorrow, Terry looked through the stack choosing work orders and doling them out on the daily schedule form. He assigned work until everyone had 10 hours of work since everyone on the crew worked a 10-hour shift. Terry used the planner estimate of hours in each case.

3:00 PM.   Terry took Wednesday's proposed schedule to the daily clearance meeting with operations. He also took a clearance request form for each new job on the schedule, a copy of the front of each proposed work order, and the entire stack of remaining work orders in the weekly allot-ment. In the meeting, led by the maintenance manager, each mainte-nance crew supervisor explained his or her list of proposed work for tomorrow and handed the operations supervisor the clearance forms with a copy of the work order front page. (This front page helped oper-ations understand the work better in order to assist with any clearance needs.) Terry then returned to his office and posted a copy of it both in the crew break area and on his office door.

3:30 PM.   Terry then returned to see the dock gutters job. Glade was the lead tech on the job and did not need any special help from him. Glade had already got help from the planning department expediter to order a special part that would arrive tomorrow.

4:00 PM.   Terry walked to his office and began to review the finished work orders received so far for the day. He had already received two from Kingsley and Jensen, but Johns had left one in his in-box. On Jensen's work order, he added comments to clarify the feedback for the planner.

5:15 PM.   Later, Terry began receiving signed timesheets from everyone as they returned from jobs and the control room after signing off clear-ance forms for completed work. Terry approved these timesheets and put them in the maintenance clerk's in-basket for entry into the plant accounting system. Sanchez handed him a completed work order on the small pump alignment job. In addition, Barber had finally wrapped up the boiler weld job and gave Terry a completed work order. He glanced at them for now, but would finish reviewing them in the morning. Both lead techs said everyone had signed off the clearances so operations could restore the equipment.

5:30 PM.   At the close of the day, Terry walked to the parking lot. He thought about the part he played in the high availability the plant enjoyed. Although initially suspicious, he had begun to enjoy receiving the weekly batch of work orders from planning that matched his crew's skills. This relieved him of having to dig continually through the entire plant backlog to find work. It also gave him a sense of mission that replaced his previous notion of being here simply to respond to urgent operation needs. With specific productivity goals in mind (the weekly batch of work), Terry felt that he knew where his crew was in terms of plant expectations. Terry was grateful that plant management also gave him full permission and even encouragement to break the weekly sched-ule for true plant emergencies and urgent work that could not wait until the next week. Terry also appreciated that he could devote his attention to the persons in his crew and their actual execution of work instead of lining up future work and administrative tasks as had been the case in the past. He knew where he fit in the plant maintenance system and could sense his impact on excellent plant availability and performance.

Assigning Names

Different exact methods exist to schedule work for a single crew day. The important elements are that the scheduling method attaches particular tasks to each crew member's name and that the method fills each crew member's available work hours. For example, the crew supervisor should assign 10 hours of work to each crew member available for 10 hours. The supervisor bases the assignments on the planned estimates for the work orders. Depending on the industry, the supervisor might also denote the exact time during the day when each job should start or end. This would be more the case for a product line taken out of service with exact scheduling from the operations group. It would be less the case where operations can clear up certain areas for a day of maintenance without significant production problems. This chapter's outage section later addresses the urgency and coordination of outage work.

The crew supervisor normally schedules the daily maintenance activities for the crew. The scheduler might be able to do the operation of daily scheduling, leaving the supervisor free to manage work in progress and people issues. However, the daily scheduling is normally too integrated into the management of the crew for the supervisor to transfer away this duty.

Figure 7.1 shows a typical form that supervisors might use to assign names to work orders. The supervisor uses the form for a specific day. The form consists of a grid with spaces for work orders on the left and crew member names on the top. A single line near the bottom takes care of nonwork order time such as vacation. Codes allow identification of specific types of nonwork order time. The bottom line provides a space for totaling each technician's time. These totals help the supervisor assign enough work to fill each hour of the shift for each person.

Figure 7.2 shows how the B Crew supervisor has added the name for each member of the crew. The supervisor can add the names one time and then copy enough forms for a supply of preprinted, daily scheduling forms. The company's graphics department might publish these daily schedule forms in pads.

Figures 7.3 through 7.7 illustrate the use of the daily schedule form for B Crew using the previously developed Crew Work Hours Availability Forecast (Fig. 6.4) and the allocated backlog for the week (Fig. 6.17 and Table 6.6). First, Supervisor J. Field adds the day and date for the day being scheduled. Field is scheduling for Tuesday, the first day of the work week for B Crew which works 10-hour days. Figure 7.3 also shows the craft skill level for each person: M, T, S, W, and H for mechanic, technician, machinist, welder, and helper, respectively. Such designation is normally unnecessary because supervisors are familiar with their technicians' capabilities.

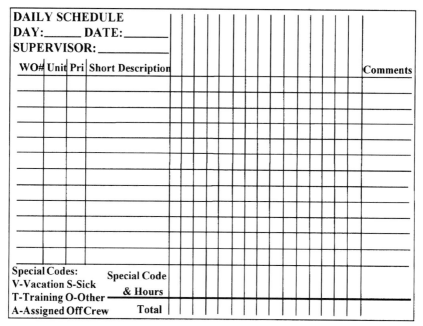

Figure 7.1 Form for daily scheduling.

Figure 7.2 Example preprinted form with crew member names.

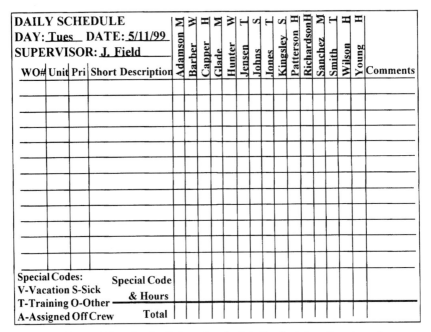

Figure 7.3   Input of date for the B Crew to schedule the first day of work week.

The supervisor first enters unavailable hours for different crew members. The supervisor marks the crew members who are unavailable for the next day for training, leave time, or other reason. The supervisor marks the form by placing the number of hours any employee is unavailable under the employee's name along with the proper reason code on the special code line. Field had approved 10 vacation hours for welder Hunter that day. Next, the supervisor adds any carryover hours from the preceding work day. The supervisor writes down the work order number and brief title of any carryover work that will run into this next day's schedule. The supervisor marks down the hours needed during the next day for carryover on the same horizontal line as the work order, but under each involved technician's name. Whenever the total hours for any individual reach ten, the supervisor writes 10 on the bottom line of the form. Figure 7.4 shows the results of these entries. Note that at this point, six persons of the 15 person crew have been assigned their entire 10 hours.

Next the supervisor must consider any urgent work that has come up since the time the weekly schedule was established. There may be several new urgent jobs that should not wait until the next week to begin. The supervisor adds to the daily schedule any urgent job that should begin that day. Such reactive work is planned, but planning on reactive work usually provides only an adequate scope and estimates of craft needs and work hours. This plan is all the crew supervisor needs to work

DAILY SCHEDULE DAY: Tues DATE: 5/11/99 SUPERVISOR: J. Field																			
WO#	Unit	Pri	Short Description	Adamson M	Barber W	Capper H	Glade M	Hunter W	Jensen T	Johns S	Jones T	Kingsley S	Patterson H	Richardson H	Sanchez M	Smith T	Wilson H	Young H	Comments
016	0	1	Boiler structure		10								10						Carryover
017	0	3	Fab pump shaft									10							"
019	2	2	Vacuum pump	10															"
015	2	2	Underdrains						10										"
Special Codes: V-Vacation S-Sick T-Training O-Other A-Assigned Off Crew		Special Code & Hours						V 10											
			Total	10	10			10	10			10	10						

Figure 7.4   Indication of craft skills. Input of unavailable crew hours and carryover work.

the job into the schedule. With even a rudimentary plan, the supervisor keeps control of the schedule because planning has identified craft and time. The supervisor makes the decision to work these jobs into the current work load knowing that they will hinder the crew's completing all the weekly scheduled work. The supervisor would prefer to wait until the following week to begin all newly identified work for several reasons. First, jobs already scheduled for the current week have a greater likelihood of being staged leading to higher crew productivity. Second, waiting until next week gives the new jobs a better chance of having their items staged. Third, the current week's schedule was put together with some thought toward including all PM work plus sufficient other proactive work. The new job is reactive and will presumably keep proactive work from being done. Fourth, there is less chance to coordinate other resources such as other crafts. Management measures schedule compliance to encourage crews to get more work done. Working on the weekly allocation of work orders is a high priority for the crew supervisor. Nonetheless, the crew supervisor must redirect crew resources to work on truly urgent work for the good of the plant. The overall scheduling process is geared toward doing more work and doing more proactive work. However, the overall goal of the plant is not just better future availability, but availability in the immediate present as well. Urgent reactive work cannot be ignored. The crew supervisor must address those areas needing immediate attention. This reasoning is why the supervisor is responsible for daily scheduling and the immediate direction of the crew. The crew supervisor deviates from the weekly schedule when necessary and later explains such variances to management.

The supervisor does not have to consider emergency work when planning the next day's schedule. Crews always begin true emergency work immediately. If it can wait until the next day or the start of the shift, it may be urgent work, but it is not emergency. The supervisor immediately reassigns technicians from their current assignments for any emergency work. The supervisor makes note of these reassignments by marking up the current day's schedule, but there is nothing to mark on the schedule for tomorrow unless the current work will carry over. Then it is treated on tomorrow's schedule as carryover just as any job still in progress. If an emergency job were to come up in the middle of the night before the start of the next shift, the schedule would have already been completed. The supervisor simply revises the schedule as necessary to reflect the already assigned persons.

The weekly schedule still helps improve weekly productivity even when crews do not start scheduled work as a result of emergency or other urgent jobs arising. When reporting variance from the weekly schedule, it is difficult to explain why only one emergency job and three urgent jobs kept ten scheduled jobs from starting.

In the current illustration for B Crew, no emergencies or urgent jobs have come up for consideration in the current day's work or for the following day. This plant places an emphasis on preventing urgent situations. Personnel report small problems that can be handled proactively before they grow into large problems. Management discourages anyone including engineers from requesting work at the last minute that they could have identified earlier.

Next, the supervisor inspects the week's worth of work consisting of the folders with work orders. The advance scheduler had prepared these folders for each lead craft. Generally the supervisor starts with the folder that represents the bulk of the crew. This folder has the most work since there were more persons for which the scheduler had to give work. The supervisor selects one job at a time and places the work order number on the daily schedule form along with a brief description of the work. Then the supervisor assigns the proper crew members to the job by allocating the work hour estimates under their names. The selection of jobs and assignment of persons is not an exact science. The supervisor prefers to consider higher skill work ahead of lower skill work to help keep assignment options open. Another preference the supervisor follows is to assign a large job to several persons, then keep those persons together for the day by assigning smaller jobs to fill their available hours. Selecting the larger, high priority jobs first leaves smaller jobs for fitting into schedule slots to make up a person's whole shift. The supervisor's skill in working with personalities comes also into play. This consideration is beyond the ability of the scheduler to make when creating a weekly schedule. Which crew members work best together? Which employees do not get along with each other? Which employees do not cooperate well with anyone? Which employees enjoy a certain type of work? Which employees need a challenge? Which employees possess the necessary skill to accomplish a unique, critical task? Which employees seem to need extra coaching and supervision? Which employees work best independently? Which employees need more experience working together with others? Which employees want opportunities to develop their leadership capabilities by leading larger jobs? Which employees rank high on the overtime list? These employees might be assigned to critical jobs that could run over requiring overtime. Which employees have the most familiarity with a particular system? Which employees need more experience in a certain system? Which employees need more experience developing a particular skill? Which employees most successfully knock out a series of small jobs? Which employees always seem to stretch their jobs to mid-day or to the end of the shift? The supervisor's knowledge of the crew allows taking these considerations into account when assigning the allocated work on a daily basis.

The supervisor should resist assigning more than the estimated hours on the work order. For example, if the work order has an estimated requirement of two mechanics for 6 hours each, the supervisor would put "6" under two crew member names, not "8" for each to fill up an 8-hour shift. The job may end up running over, but the supervisor does not want to begin by anticipating the job will consume the entire shift. The assignment of exact work order hours would be a good check area for concerned managers. Checking the daily schedule may indicate a problem with most jobs seeming to be always planned for 10 hours or scheduled for 10 hours just because that is the shift arrangement.

The supervisor continues selecting tasks from the folders and distributing their hours among the crew. A work order may be attached to other work orders for the same system. In this case, the supervisor prefers to assign all of the attached work orders to a larger group of persons for the same day rather than assigning them to a smaller group over several days. Working all the jobs for a single system at the same time shortens the time the system must be out of service. It also contributes to a sense of accomplishment for both the crew and the operations group when they complete all the work on a system.

The supervisor continues selecting tasks until all crew members have their available work hours assigned to work orders. After assigning a number of the work orders, the supervisor has to exercise some further judgment in selecting work orders. The best persons to which to assign a particular work order may have too few hours left in the shift to complete that work order. The supervisor is free to assign technicians to start a bigger job that they will not finish in the shift. If assigning a 10-hour job to a person with only 3 hours left available, the supervisor would place a "3" under the person's name across from the new job. The supervisor could also immediately begin the next day's schedule with the job at hand placing a "7" in the pertinent crew member box. That schedule may be adjusted later for different carryover hours when actual job progress is assessed. On the other hand, there may be smaller jobs with lower priorities, but fewer hours that could be completed on the same day. The supervisor is free to pick any of the jobs of lower priority in the week's allocation for assignment anytime during the week.

If the supervisor runs out of work from the weekly allocation, then he or she checks with planning to obtain planned work from the unscheduled backlog. This work is in the planning group's waiting-to-be-scheduled file. If there is no planned work waiting, the supervisor checks with the planner about working jobs that are not yet planned. The supervisor should resist assigning work without a work order.

The supervisor totals up each person's assigned hours under the person's name. The hours for each person should add to the person's paid work hours. For example, if 10-hour shifts are employed, then there should be 10 hours totaled up under every employee.

Figure 7.5 shows the first two jobs the supervisor has selected from the week of allocated work. The supervisor has selected the top job from

DAILY SCHEDULE DAY: Tues DATE: 5/11/99 SUPERVISOR: J. Field				Adamson M	Barber W	Capper H	Glade M	Hunter W	Jensen T	Johns S	Jones T	Kingsley S	Patterson H	Richardson H	Sanchez M	Smith T	Wilson H	Young H	
WO#	Unit	Pri	Short Description																Comments
016	0	1	Boiler structure	10									10						Carryover
017	0	3	Fab pump shaft						10										"
019	2	2	Vacuum pump	10															"
015	2	2	Underdrains					10											"
036	0	1	Unloading arms		6	6												6	
029	0	4	Dock gutters		4	4												4	
Special Codes: V-Vacation S-Sick T-Training O-Other A-Assigned Off Crew		Special Code & Hours						V 10											
			Total	10	10	10	10	10	10			10	10				10		

Figure 7.5   Input of the first two jobs from the allocated weekly backlog.

the Mechanic folder. The supervisor first selects WO no. 36 requiring a mechanic and two helpers for 6 hours each. This work order has another work order attached for the same system. This second work order, WO no. 29, requires only two helpers for 6 hours each. After briefly reviewing the work order, the supervisor decides to assign it to the team of three persons doing the first job. The supervisor assigns each of them 4 hours to complete this job and makes a note on the work order. The supervisor has changed the plan of 12 helper hours to assign 8 helper hours and 4 mechanic hours. (This is a little unusual to change a plan in this manner, but it is acceptable.) Now three more persons have 10 total hours for the day.

Figure 7.6 shows the result of the next three work orders selected by the supervisor. The supervisor decides to start none of the FO, Fuel Oil Storage and Transfer System, work on Tuesday because neither welder is unavailable. One is involved with carryover work and the other is taking vacation. The supervisor selects WO no. 28 requiring a machinist and a helper and then adds a smaller job, WO no. 27, to finish up the machinist's hours for the day. Because WO no. 030 is on the same system, the supervisor assigns it next.

Next, the supervisor has 20 technician hours, 9 mechanic hours, and 10 helper hours available. The supervisor selects the large job, WO no. 35, requiring 20 hours for one technician to begin 10 hours. The supervisor

DAILY SCHEDULE
DAY: Tues   DATE: 5/11/99
SUPERVISOR: J. Field

WO#	Unit	Pri	Short description	Adamson (M)	Barber (W)	Capper (H)	Glade (M)	Hunter (W)	Jensen (T)	Johns (S)	Jones (T)	Kingsley (S)	Patterson (H)	Richardson (H)	Sanchez (M)	Smith (T)	Wilson (H)	Young (H)	Comments
016	0	1	Boiler structure			10								10					Carryover
017	0	3	Fab pump shaft										10						"
019	0	2	Vacuum pump	10															"
015	2	2	Underdrains								10								"
036	0	1	Unloading arms			6	6											6	
029	0	4	Dock gutters			4	4											4	
028	0	1	Pump impeller							8				8					
027	0	2	Cat underdrain							2									
030	0	1	Trn pump align									1	1						

Special Codes:
V-Vacation S-Sick
T-Training O-Other
A-Assigned Off Crew

| Special Code & Hours | | | | | | | | V / 10 | | | | | | | | | | | |

| Total | | | | 10 | 10 | 10 | 10 | 10 | 10 | 10 | | 10 | 10 | | | | | 10 | |

Figure 7.6 Input of the next three jobs into daily schedule.

also assigns a 40-hour job, WO no. 31, to the mechanic and helper to begin a total of 19 hours. Finally, the supervisor assigns a 17-hour job, WO no. 25, to the last technician, Smith, to begin 10 hours. After this scheduling, the supervisor finds that a single person has less than 10 hours assigned. The helper Richardson has only 9 hours assigned. The supervisor sees no single-hour helper jobs and is reluctant to have a helper begin another job alone that will carry over. Therefore, the supervisor adds the helper's last hour to the cation underdrain fabrication job, WO no. 27. This will give Richardson, an apprentice, some time in the machine shop. Figure 7.7 shows the completed daily schedule form with names for all crew members assigned to work orders with all shift hours accounted for, adding up to 10 hours each.

Coordinating with the Operations Group

With the schedule filled out, the supervisor then coordinates with the operations group. The supervisor must communicate with operations for two reasons. The supervisor first requests that the operations group clear the pertinent equipment for the next day. The supervisor must also determine if any testing is advisable as the maintenance crew returns equipment to service. Many plants conduct a daily schedule meeting each afternoon to coordinate the next day's work between these two groups. The meeting consists of the crew supervisors and the operations

DAILY SCHEDULE
DAY: Tues   DATE: 5/11/99
SUPERVISOR: J. Field

WO#	Unit	Pri	Short Description	Adamson M	Barber W	Capper H	Glade M	Hunter W	Jensen T	Johns S	Jones T	Kingsley S	Patterson H	Richardson H	Sanchez M	Smith T	Wilson H	Young H	Comments
016	0	1	Boiler structure	10													10		Carryover
017	0	3	Fab pump shaft								10								"
019	2	2	Vacuum pump	10															"
015	2	2	Underdrains						10										"
036	0	1	Unloading arms			6	6											6	
029	0	4	Dock gutters			4	4											4	
028	0	1	Pump impeller							8			8						
027	0	2	Cat underdrain							2			1						
030	0	1	Trn Pump Align										1	1					
035	0	1	Piping									10							2 day job
031	2	3	Aux feed pumps											9	10				2 day job
025	0	2	Caustic leaks													10			10 of 17
Special Code & Hours								V / 10										
Total				10	10	10	10	10	10	10	10	10	10	10	10	10	10	10	

Special Codes: V-Vacation   S-Sick   T-Training   O-Other   A-Assigned Off Crew

Figure 7.7   Completed daily schedule for the B Crew.

supervisor. At this meeting, maintenance may also advise regarding the progress of on-going work in addition to coordinating the start and testing of new work.

Clearing means that the operations group is making the equipment available and safe for maintenance work. For example, the operations group may clear the equipment by draining process lines and closing necessary valves and electrical breakers. Then, depending on particular plant policy, the maintenance personnel may follow behind and lock out valves and breakers to their own satisfaction and security. For simplicity, this book presumes the operators do all clearing and maintenance personnel do not perform separate locking. Companies with additional requirements would simply add their required procedures to the illustrative process described here. The company should have a standard process for the maintenance group to notify the operations group of clearance needs. With a daily schedule meeting, maintenance can hand over a printed clearance request form. The maintenance crew could also deliver a clearance request form at any time to the operations group. However, maintenance should keep in mind that the operations group also schedules their activities and desires timely advance notice. Plant policy might allow less formal notice as phone calls or emails to the operations group. The plant might utilize a CMMS computer to flag the backlogged work orders needing clearance for the operators. In a daily scheduling meeting, each crew supervisor gives operations a copy of the crew's proposed daily schedule and a printed clearance request form for each work order needing clearance. The clear request form should describe the work proposed along with times needed for clearance. A copy of the work order form with special clearance instructions makes a good clear request format because the operators see the actual, maintenance work document. The supervisor has the original work order forms in hand at the meeting instead of just the schedule sheet to allow reference to any planner notes regarding special circumstances. The face-to-face daily meeting allows the operators to commit to clearing up the necessary work areas during the evening shift for an around-the-clock operating environment. Otherwise, the operators can commit to when they would clear and have ready equipment the next day if they would not accomplish it during the night. At this meeting the operations group would also explain any particular requirements if it could clear the equipment for only part of the day. The supervisors also record feedback on the pertinent work order forms for new information regarding special requirements. The crew supervisors also note restricted times for maintenance work on the daily schedule forms and the work order forms which they will give to the field technicians. The operations group can also give immediate feedback for any jobs that it would not be able to clear at all. The supervisors make changes to the daily schedule if needed. The supervisor also has brought the weekly backlog of work to the meeting in case a new work order or two needs to be selected as replacements for jobs that would not be cleared.

During the daily scheduling meeting, the operations group and the crew supervisor agree regarding any special testing after job completion. For example, a particular strainer that has shown a past tendency to leak after maintenance should be tested before the technician has completely demobilized from the site. The decision of which jobs to test really depends upon the type of work done and the confidence in the maintenance skills and so would vary from plant to plant. The operations group or the maintenance group may have definite preferences for particular work orders to test. The planning group may recommend by noting within the job plans if the jobs should receive special attention. Management should establish a clear policy when possible regarding testing. The crew supervisor makes appropriate notes on the work orders which require testing after completion.

Handing Out Work Orders

After the daily scheduling meeting with the operations group, the crew supervisor publishes the daily schedule with any necessary revisions. The supervisor might do this by posting the daily schedule form on a crew area bulletin board or outside the supervisor's office. This allows individuals to think about their assignments for the next day. Spreading the news keeps crew members feeling more a part of a team. If published in time, there may be productivity improvements as well because technicians finishing their current jobs could take their tools and somewhat set up for their next assignments. The technicians could also review the job plans or technical manuals to familiarize themselves with the equipment. Technicians could even start to gather parts and special tools if they were not going to be staged. In effect, the technicians can stage their own jobs because they have advance notice.

The handing out of the physical work orders may depend on the plant's policy with the operations group certifying that it has cleared the necessary equipment to allow safe maintenance work to begin. (Some jobs do not require clearance and the supervisor marks this information on the work order form if the planning group has not already done so.) There are different ways to accomplish permitting work areas, some less formal and some more formal and rigorous. The exact procedure would be dictated by company policy. The exact policy does not appreciably affect the enhancement to crew productivity of planning and scheduling. The process presented here is where operations maintains a master clearance form for each cleared job that technicians must "sign onto" before they work.

After the operations group clears the necessary areas, technicians sign onto the clearance form.

The supervisor may physically hand the work orders out in a number of ways. A common practice is through a morning check-in meeting. The supervisor meets with the whole crew at shift start. The supervisor physically hands out work orders to the lead technicians making sure everyone knows where they will be working for the entire day. The supervisor may hand out all the work orders of the day or just the first ones, planning to distribute the rest of the work order forms as individuals, pairs, or groups of technicians near completion of their first jobs. The advantage to handing out all the work orders for the day at the beginning is that technicians can plan their pace for the day better. They understand the goal for the day. This factor probably outweighs a slight disadvantage for them having to keep track of the paperwork. A lost paper work order form would be a significant problem in an entirely paper system with no other record of the work. Arrangements where a computer database has at least a record of each work order make the problem of lost forms less important. Fully utilized CMMS computer systems have the least worry over lost paper forms where the computer record of the work order is the master and any paper copy serves only as a field device for the technicians. Of course, the trade-off becomes the risk of having the computer system available and functioning. Paperwork would be that much harder to keep up with, if emergencies frequently seem to interrupt work. Another common practice is for supervisors to pin work orders under lead technician names on a crew area or supervisor office bulletin board. Cubbies, file folders, or special mail slots can also be used for this purpose. Alternately the supervisor could write work order numbers on a dry erase board with lined columns and rows. Some supervisors, whether they have daily crew meetings or not, prefer personally contacting individual technicians throughout the day making sure each technician has the right work order and knows where the next assignment is.

Figures 7.8 and 7.9 illustrate possible aids for handing out work orders by placing them in slot or box structures. Everyone should be aware of work permit policies when setting up a system for technicians to have access to work orders. The following arrangements might be duplicated by writing work order numbers on dry erase boards. Figure 7.8 shows a cardboard, wooden, or metal mailbox arrangement. A supervisor might create tentative daily schedules for the entire week and distribute the entire allocated backlog. The supervisor might also distribute work orders for only one day at a time. One problem with a mailbox system is that a paper work order could only be placed in one box under a single name, usually the lead technician. Each column might be headed by a group of names or only lead technicians for teams. However, this presumes the same groups of persons would remain together throughout the week or longer. A dry erase board avoids this problem because the number of a work order could be written under several names at one time.

	Adamsn	Barber	Capper	Glade	Hunter	Jensen	Johns	Jones
Fri								
Thu								
Wed								
Tue	19	16		29/36		15	27/28	35
Mon								

	Kingsly	Pattersn	Richdsn	Sanchez	Smith	Wilson	Young	
Fri								
Thu								
Wed								
Tue	17	16		31/30	25			
Mon								

Figure 7.8 Optional mailbox arrangement to distribute paper work orders to lead technicians.

The supervisor might keep the actual work order forms to hand out when teams are ready for them. Technicians themselves might print each work order form from a CMMS computer when ready. Figure 7.9 would be applicable to maintenance environments where technicians work as individuals (or on stable teams) and where specific start times are important. Each technician has a column of mailboxes or slots where the supervisor inserts work orders for different times of execution. At the beginning of the shift, the supervisor changes the heading of the previous shift to the following day of the work week. For example, at the beginning of Monday, there are headings for Monday and Tuesday. At the beginning of Tuesday, the supervisor changes the Monday heading to read Wednesday.

One method is not advised for assigning or handing out work orders. Some supervisors take the entire backlog of work from the computer whether or not it has been planned or scheduled and immediately assign technician names. In this manner, supervisors distribute their entire

Monday

	Dave	Jan	Max	Rex
7-8	119	120	121	
8-9	123			122
9-10		124		
10-11	125	126		
11-12				
12-1			127	
1-2				
2-3		128		
3-4	129			130
4-5				
5-6				

Tuesday

	Dave	Jan	Max	Rex
7-8				
8-9				
9-10		131		
10-11				
11-12				
12-1				132
1-2				
2-3				
3-4				
4-5				
5-6				

Figure 7.9 Optional slotted board arrangement to distribute paper work orders to individuals for specific start times. The second day was still being established.

backlogs within their crews giving everyone a share of the plant work orders. This procedure generates the problem of not having any schedule expectations. Everyone has an individual backlog to work. When they finish one task, they should move on to the next. However, a major point of planning and scheduling is setting schedule expectations. Supervisors that assign individuals unplanned or even unscheduled work orders lose these advantages. Moreover, a CMMS computer has worsened this problem. Assigning a technician a work order in parallel to the planning process many times may result in a technician printing out the work order before a plan is available. With a pure paper system, at least the work stays in the planning channel unavailable for the technician to work prematurely.

During Each Day

Daily scheduling is a continual cycle. During each shift as technicians work jobs, the supervisor manages from the current day's schedule and creates the next day's schedule. As technicians finish jobs, they return the work orders to the supervisor and sign off of the master clearance forms so operations can begin to put the equipment back into service.

Even if there are no special testing needs, the technicians and supervisors must not wait until the end of the shift to turn in work orders and return work permits. Informing the operations group promptly upon completion of individual work orders allows more time to return equipment to service and restore plant capacity. Operators can operate the equipment sooner after completion of maintenance. If initial operation reveals problems or concerns, technicians are then also still on shift to return to jobs and make corrections or advise. Were maintenance to wait until shift end to return all completed jobs, operators would also not be able to unclear all the jobs at once. Allowing operators to work throughout the shift restoring equipment to service levelizes the operations group's efforts.

In addition to coordinating with the operations group, many jobs naturally run over or under estimated times. The supervisor ensures that the technicians give feedback to help the planners estimate future jobs and the supervisor adjusts the daily schedules for today and tomorrow as needed. Other situations may arise and cause the supervisor to adjust the schedules, such as plant emergencies or unexpected meetings.

Daily scheduling is not simply a form filled out at the end of the day for the next day's activities. Even more important, it is not simply handing out work orders at the start of the shift. Throughout each day, crew supervisors use the daily schedules as a tool to control work.

Supervisors must also give technicians time to complete work order feedback as addressed by Chap. 5.

Summary

This chapter describes the daily work activities within their overall planning and scheduling context. For daily scheduling, the crew supervisor selects work orders mostly from the weekly allocation. However, the supervisor also maintains the flexibility to reassign the crew for emergency and other urgent work that may arise. The supervisor may follow different methods and procedures to select and assign the appropriate work each day. All of these methods and procedures have certain elements in common, primarily giving each crew member a full shift of work based on planner job estimates. These techniques of daily scheduling help maintenance improve its labor productivity. The chapter describes the exact steps of daily scheduling to clarify the concepts. After understanding the concepts, readers may implement alternate steps than the ones prescribed. This chapter should allow companies to implement their own effective methodologies.

Forms and Resources Overview

In the introduction to the April 1995 issue of *Maintenance Technology,* editor Bob Baldwin (1995) identifies a concern about the knowledge drain occurring in many companies as senior employees leave the workplace. From his participation in maintenance seminars and roundtables, it appears that many of these persons "are first-line supervisors who are among the few people that really understand the machinery in the plant." Baldwin states that "Maintenance planners are in the best position to stop the knowledge drain." Baldwin holds that "The smart planner taps into the collective knowledge of maintenance and reliability people wherever he finds them and institutionalizes that knowledge in the planning process."

Indeed, many companies initiate planning primarily to capture knowledge as supervisors near retirement. It is interesting that the knowledge in danger of being lost is not the 20 or 30 key lessons or experiences of the past 20 years. Rather, it is the thousands upon thousands of minute details of particular jobs on particular equipment. It is the knowledge that a certain fan must have its bearing removed from the inboard side because of an imperceptible taper not mentioned in the technical manual. It is the identification of the only gasket material found to hold up for the fuel oil control valve, no. 8 flange connection. It is the specific maintenance detail that makes the difference between a 4-hour job and a 10-hour job for countless maintenance tasks. It is the ability to capture this information and apply it to future work precisely when needed that makes the planner's job as a simple file clerk so valuable to these companies. This chapter covers the practical aspects of using files and other resources to apply information.

This chapter discusses the types of resources and forms planners use. The discussion includes how planners and other maintenance groups employ these areas of information assistance. Resources include areas such as the plant files and plant schematics. Forms help the maintenance group collect and use data and information.

The planner does not plan every job from scratch. Of a 20% increase in wrench time for the crafts, at least 5% might be attributed to avoiding the repetition of job delays encountered on past similar assignments and thereby moving up the learning curve. The planner must do research to create job plans. Other than just relying on personal experience, the planner has a wealth of information available. The planning department is the proper location to keep general equipment information as well as specific equipment information. Vendor catalogs and manuals are examples of general equipment information. History and data for individual pieces of equipment are examples of specific equipment information. Planners must not store information haphazardly, but in an organized fashion allowing easy retrieval. Planning Principle 3 addresses the importance of storing information at the individual component level. When technicians first work on a piece of equipment, the planner creates a minifile to record information solely for that equipment. This minifile holds all the recorded information from that job, both the initial plan and any feedback received from the field. The next time the equipment requires maintenance work, the minifile contains useful information. The planner adds information from subsequent jobs for that equipment and soon the file contains comprehensive maintenance data. This chapter describes how the various information resources in the planning department assist the planners and others. Chapter 8 covers the use of a computer management maintenance system (CMMS) computer. Understanding the need for such resources and forms precedes any consideration of computerization.

Forms

Forms supplement job plans in two ways. Forms that are procedural in nature help technicians remember task steps. Other forms are data gathering in nature to help collect information. Many forms possess elements of both natures. Figure 8.1 illustrates a largely procedural type form that planners might attach to alignment job plans. This company insists on certain steps being taken for alignments. Instead of including the standard alignment instructions in the job plan, the planner simply attaches the form to the job package. Forms also provide an organized method of collecting and recording information for current analysis or future reference. Forms keep information from being scattered and lost in an otherwise wide variety of media such as e-mails, scraps of paper, and the like. They also guide what should be recorded so a pertinent piece of information is not forgotten to be collected. Another example of a form would be a work order form.

ALIGNMENT CHECK SHEET

___1. Break coupling and remove old grease.
___2. Inspect coupling for damage.
___3. Secure coupling. Run motor if magnetic center of motor is unknown. Mark motor shaft.
___4. Shut down motor. Set motor shaft to electrical center. Set coupling gap per coupling instructions.
___5. Inspect hold down bolts and washers for damage.
___6. Install jacking bolts on all corners of machines to be shimmed.
___7. Correct machine to be shimmed for soft foot and defective sole plates.
___8. Correct stationary machine for soft foot and defective sole plates.
___9. Remove all shims. Clean or replace as needed.
___10. Pack proper amount of grease into coupling and reassemble.
___11. Align machine. Use PdM group instructions for considering thermal growth.
___12. Tighten all fasteners. Loosen jacking bolts. Turn shaft to ensure free rotation.
___13. Attach form to work order.

Technician_____   Date_____

Figure 8.1 Sample form for planners to attach to job plans.

Figure 8.2 illustrates another form that might be used in maintenance. The planner would attach the listing to all pertinent burner jobs. This type of listing is very similar to a "standard plan" discussed later with the exception that the form usually has a form of sign-off to ensure either certain data is collected or certain actions are executed. A standard plan also normally applies to only a single piece of equipment,

UNIT 2 BURNER CHECK SHEET FORM

Burner Number:_____

Date:_____

TECHNICIAN INITIALS EACH STEP
WHEN COMPLETE

___1. **Replace orifice, swirl, and spill plates**

___2. **Replace both gaskets.**

___3. **Verify correct lance setting.**
       Setting is: _____

___4. **Torque feed tube to 200 ft lb.**

___5. **Torque cap nut assembly to 77.5 ft lb.**

___6. **Adjust spider to provide 1/8 inch to 3/16 inch gap between outer tube and washer.**

Figure 8.2   Sample form for planners to attach to job plans.

whereas a form might be more broadly applied to many pieces of equipment. The characteristics of job plans, standard plans, forms, and datasheets certainly have much overlap. The important thing is to recognize where something can help the maintenance effort. The forms might help personnel adhere to a certain plant standard or procedures when necessary. The forms might help collect useful data. The forms might also help simplify job plans because supplementary information is included in accompanying forms.

The planner has a collection of blank forms that can be attached easily to job plans. When the job might encompass a certain procedure such as alignment, the planner attaches the pertinent form to the job plan. Planners should actively seek to have copies of all forms used by maintenance technicians. Planners should also create new forms when practical to aid collecting useful information. In addition, the crew supervisors also maintain sets of blank forms for use as needed to help ensure that technicians collect proper job feedback on all jobs. Planners should take the lead in helping the supervisors keep supplies of useful forms on hand.

The plant also uses other forms that the planning department might not directly use itself. Figure 8.3 shows a deficiency tag, a form that the originator attaches to deficient equipment before the planning department begins its work. The deficiency tag serves several useful functions. It allows other persons to see that the deficiency has already been written up on a work order. In addition, it cautions persons that extra care and attention might need to be exercised with regard to the equipment. It also helps planners and maintenance technicians find the correct equipment needing attention.

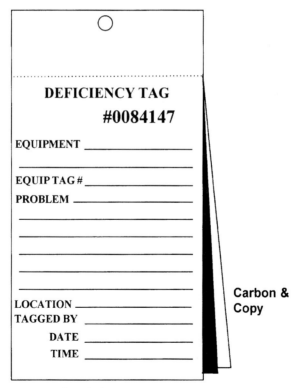

Figure 8.3   Sample form showing a typical type of deficiency tag that plant personnel use.

Appendix D includes a blank copy of all the forms presented throughout this book. All of the forms in this book may be copied and used by the purchaser of this book.

Resources

This section reviews the types of information that the plant possesses, how they relate to planning, and how the planner uses them. This section does not necessarily set guidelines, but rather gives an overall perspective of what exists.

Component level files—minifiles

Planning Principle 3 explains why the planner keeps a file at the component level for nearly every piece of equipment. This file is called a *minifile.* A *system file* would contain information on many pieces of equipment in a system.

When planning receives a work order to work on a piece of equipment, the minifile enables a planner to obtain the following information quickly:

1. A listing of what was done on the equipment previously including work scope, work hours, duration, and cost.

2. Schedules for normal checks by preventive maintenance, predictive maintenance, and operations.

3. Equipment technical data including specifications, standard plans (if available), safety data, vendor information, and nameplate data.

4. Part breakdowns and plant inventory information such as part numbers.

5. Special notes for peculiar information concerning this equipment such as a left-hand threaded assembly.

6. Actual copies of previous work orders with job feedback of problems and corrections on previous plans and information.

This above information would not be readily available if it were contained in a large system file.

The minifile allows the planner to link pertinent information directly to the piece of equipment as it is learned. As the maintenance force repeats jobs and the planner adds information to the file, it becomes easier to plan work orders because more information is readily available. In addition, the minifile permits the planner to recognize problems from previous jobs and minimize their impact on the job currently being planned. For example, if on a previous job the craft had to find an unexpected part or use an unanticipated special tool, the planner would ensure that item is available for the current job. In this manner, the entire maintenance organization reduces delays, job by job, more quickly through the planning function.

Making a minifile. This section shows a specific method to make a minifile. This is not the only way to make such a file, but it is important to review a step-by-step method to illustrate the idea rather than just read a general explanation of the concept. This section presumes the use of the sample equipment coding structure in App. J.

The minifile begins with the physical piece of equipment in the field. The equipment should have a physical tag number already hung on it using the plant's system of numbering. The file number must match the equipment number. The planner uses alphanumeric character labels to affix the equipment tag number on the end of a side-label, manila folder. Any dashes (–) are left off to allow more space on the folder. The planner hand writes the equipment name on the outside of both sides of the folder. This allows double-checking the equipment's identity later when locating it by equipment number. It also helps persons less familiar with the equipment coding system to find files. The planner also puts his or her name and the date of file creation on one side of the outside of the minifile. It is preferred to have a manila folder with a pocket to avoid spillage as persons handle files. A legal-size rather than letter-size folder allows easy insertion of any drawings or equipment manuals. Figure 8.4 shows how the manila folder's side label would appear for equipment number N02-FC-003.

A piece of equipment may not have a component tag number because the plant has not numbered the system. In this case the planner should still create a minifile using labels for only the known portion of the eventual number. The planner should know the first five digits of the equipment number (plant, unit, group, and system codes) even if the equipment has not been given a unique number to form the last three digits of the overall tag number. The plant's strategy of assigning equipment numbers may involve the planning department placing unique numbers as equipment is maintained. In this case the planner would place a physical tag on the equipment when scoping the job and use this same number on the minifile label.

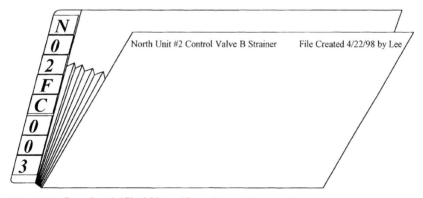

North Unit #2 Control Valve B Strainer   File Created 4/22/98 by Lee

Figure 8.4   Sample minifile folder with equipment tag number.

Planners should not underestimate the value of making a minifile for almost every piece of equipment on which maintenance performs work. A rule of thumb affirms that if certain equipment in the plant is important enough to have a separate identification number, it is important enough for the planners to make a minifile for it.

Planners make minifiles at the *File Creation Station*. This place in the planning office has all necessary supplies to make minifiles.

When making a minifile, the planner adds each of the standard minifile forms shown in Figs. 8.5 through 8.8. The planner completes the forms only for information known at the time, as found during subsequent planning research, and from job feedback. These forms may not all be necessary for all planning organizations, but are shown to illustrate the different types of information a minifile contains. The forms help organize information. The completeness of minifiles varies across different planning organizations. At one end, some groups have preprinted forms to collect all imaginable information and save every work order form along with copies of technical information from manuals gathered while planning jobs. At the other end, some groups merely save the work orders that they consider to have valuable information and throw away the rest. Many groups save and organize physical information in a manner that falls in between these extremes. The confidence a planning group has in the reliability of a CMMS computer system and the system's features may also affect what it saves in physical paper files.

EQUIPMENT HISTORY DATA SHEET   PAGE ___

SYSTEM _____

EQUIPMENT NAME _____

EQUIPMENT TAG # _____

HISTORY (Normally does not include PM's or other routine inspections.)

Work Order #	Job Description (What was done)	Complete Date	Print Revision Needed?	Work Type	Work Priority	Actual Hours Duration	Actual Labor Hours	Total Work Cost $
		/ /						
		/ /						
		/ /						
		/ /						
		/ /						
		/ /						
		/ /						
		/ /						
		/ /						
		/ /						
		/ /						
		/ /						
		/ /						
		/ /						
		/ /						
		/ /						
		/ /						

Figure 8.5 Sample form to include in the minifile to track history.

EQUIPMENT TECHNICAL DATA SHEET

EQUIPMENT NAME _____

EQUIPMENT TAG # _____

EQUIP LOCATION _____

Does this equipment have a "Standard Plan"?   _____

Manufacturer code and name _____

Vendor name, location, person, and phone number to contact:

_____

_____

NAMEPLATE INFORMATION

Model # _____ Serial # _____

_____

MSDS # AND ANY OTHER SAFETY INFORMATION: (Is a Confined Space
Permit needed? Describe any past accidents. What safety considerations
such as scaffolding, high temperature/pressure, chemicals, etc. are there?)

_____

_____

_____

OTHER INFORMATION AND SPECIAL NOTES FOR THIS EQUIPMENT

_____

_____

_____

_____

_____

_____

_____

_____

_____

_____

Figure 8.6   Sample form to include in the minifile to organize equipment technical data.

Equipment History Files (Including system files and minifiles)

The Equipment History Files consist of all the previously completed work orders however they are filed. The plant may have saved them sorted by system or simply chronologically. Simple chronological files would have all old work orders saved by date. There may be a file for each year or each month full of completed work orders. The plant may have

PARTS INFORMATION SHEET

PAGE ___

EQUIPMENT NAME _____

EQUIPMENT TAG # _____

	Stock Number	Total Quantity Used for this Equipment	Quantity Needed for this Job*	Description
1				
2				
3				
4				
5				
6				
7				
8				
9				
10				
11				
12				
13				
14				
15				
16				
17				
18				
19				
20				
21				
22				
23				
24				
25				
26				
27				
28				
29				
30				

*Use this column after copying this page to issue with a specific job.

Figure 8.7   Sample form to include in the minifile to organize spare parts information.

saved the work orders already by specific equipment and so already has component level files. On the other hand, the plant may never have saved old work orders. The plant also may have made extensive use of craft or blanket work orders and not accumulated much history information. Furthermore, the plant may not have had a work order system of any kind to authorize and complete work.

EQUIPMENT NORMAL PM's AND CHECKS

EQUIPMENT NAME_____

EQUIPMENT TAG #_____

By Maintenance Personnel

What	Frequency	Route Designation*

By Predictive Maintenance (PdM) Personnel

What	Frequency	Route Designation*

By Operations, Lab, or Other Personnel

What	Frequency	Route Designation*

*if any

Figure 8.8   Sample form to include in the minifile to identify pertinent PM routes and details.

The planning organization must begin to use component level or minifiles. The intent of the specific equipment file is to provide a one-stop file that contains not only the past work orders, but any other information specific to the equipment. Having each minifile contain only information for a single piece of equipment makes gathering information easy and practical enough to benefit the planning operation. The planner places specific technical information and work history into the minifiles as it is encountered on jobs. The planner must not place new information into any of the other old files that exist where supervisors or planners kept information on equipment before the institution of minifiles. It is frequently recommended to build the minifiles from the current point forward. That is, keep the old files from which to research information occasionally, but do not keep filing work orders and specific equipment information there. The importance of these old files cannot be overlooked, yet it is frequently difficult to easily assimilate all this information into a new minifile system. It simply may not be practical to go back and refile all the old work orders before minifiles started. Planners or technicians may discover information there while planning or working new jobs. That information is kept with the work order and filed by the planner in the minifile at the completion of the job. On the other hand, previous files kept by manufacturer or equipment name might be easily converted to numbered minifiles and be worth the effort to do so.

The planning department normally uses an open filing system. That means that the files can be seen without having to open cabinet drawers. A shelf arrangement might best facilitate this requirement. The folders have side labels for easy identification on the shelves. Planners arrange the files from left to right, alphanumerically by the equipment tag number codes on their folders. Minifiles are placed after any existing system file in the physical filing order.

For example, the following shows the correct order of several North Unit 2 files.

N02CP	System file for condensate polisher system
N02CPAR5	Minifile for AR5 control valve on polisher system (polisher had a preexisting number system for control valves that was incorporated into the plant-wide system)
N02CP006	Minifile for #6 manual valve on polisher system
N02CPCR4	Minifile for CR4 control valve on polisher system
N02DP	System file for boiler feedwater pump system
N02DP024	Minifile for B boiler feed pump on boiler feedwater pump system
N02FC003	Minifile for control valve B strainer on fuel oil service pump system

Technical Files

The Technical Files consist of technical information from manufacturers for only the equipment in use at the facility. For example, there may be a valve manual with specifications and procedures to maintain several valves produced by a manufacturer. This manual would not be kept in the Equipment History Files because it pertains to valves in more than one system. When an individual valve is worked on and information is used from the manual, the planner would copy over that specific information to that valve's minifile.

The maintenance department keeps the Technical Files in a section of the file area separate from the Equipment History Files. They are arranged from left to right, alphabetically by the manufacturer's name. To hold small technical bulletins, the files contain pocket folders identified with the first few letters of manufacturers' names. In addition, there are file sections for specific O&M manuals filed by unit and certain other technical manuals or documents identified in the following sections of this chapter. As specific information is used on a job, the planner copies that information over to the equipment's minifile.

A point of considerable controversy surrounds whether the planner should routinely send O&M manuals or similar source documents into the field as attachments to work order plans. Ideally, the planner would find the information needed within such manuals and might attach a copy of only certain pages at best. In the real world, the planner sometimes only gives the technicians a head start with the most likely information needed. Often the technician must supplement the planner's information through personal expertise or research. Does this case make it sensible to have the equipment manual normally handy? Consider persons working on their own cars. Would they like to have the vehicle manuals available? One would think so. This reasoning suggests that the planner should routinely attach such source documents to job plans. On the other hand, closer inspection gives another side to the issue. What are the most frequent maintenance tasks one performs on a car? These tasks include changing oil, filters, spark plugs, or coolant. Hardly anyone would think of referring to an owner's manual. Even in the case of changing water pumps and mufflers, one would more likely refer to the instructions in the box of the new water pump or muffler than to any vehicle owner's manual. The same thing occurs in the maintenance of an industrial plant. One finds that reference to an O&M manual does not occur routinely during maintenance. Therefore, the preference is against the planner routinely sending O&M manuals or similar source documents into the field as attachments to work order plans.

Attachment files

Sometimes planning departments create small booklets of important equipment information that they keep in the minifiles. These booklets contain especially critical information culled from the more massive O&M manuals. Planners bind these booklets with report covers so that they can be sent out on jobs.

Such larger attachments may be attached by paper clamps to work orders or they may be kept in files in the planning department. The planning department might keep the attachment in a special "Attachments" file in the planning department. This file keeps attachments in the planning department filed by crew and work order number (they have a copy of the work order attached) where technicians can come get them when ready.

Vendor Files

The Vendor Files consist of ordinary sales catalogs from vendors. These catalogs are arranged left to right, alphabetically by the vendor name on the catalog. Vendor files are kept in another section of the file area. The Vendor Files may contain a set of Thomas Register® books or similar vendor reference. The plant may be looking into having CD-ROM versions or Internet links for this purpose.

The planner may research these files to find information for new equipment under consideration for purchase. Information for equipment already in place might also be found. For instance, a planner might find details for a particular valve by examining a vendor's catalog of over 100 valves. The planner would then copy the particular valve's information sheet over to the minifile to avoid having to make similar searches in the future.

Equipment parts lists

Equipment parts lists or bills of materials as discussed in Chap. 5 are a very valuable resource for planners. Planners file this information in the minifiles. Planners must continually develop the listings of parts for equipment as it is maintained. Planners must insist on receiving parts information when new equipment is purchased.

Standard plans

Planners create standard plans for jobs where technicians might not be expected to remember particular sequences or job procedures as a normal part of their craft skill. These standard plans are not an attempt to dictate actions to technicians. Rather, they help the technician build upon past successful work. Planners store standard plans in their respective, equipment minifiles. The following two listings illustrate two styles of what might constitute a standard plan. Planners might create much more complex standard plans including pertinent manufacturer manual pages, exploded view diagrams, and vendor contact names, and other useful information. The planners might keep these standard plans available in special notebooks or binders to include as attachments to work orders.

Planning intends to create so-called standard plans for as many jobs as possible. The planner's intent is not to insult craft technicians, suppress unique skills available to trained craftpersons, or rob them of the opportunity to adjust to job conditions. The planners count on the technician skill level that the planner specifies in the job plan. The planner wants to capture the process of the job so that technicians can later concentrate on craft technique rather than unique equipment steps that may vary widely. Even specifying technique such as welding procedures should not insult technicians who help develop the procedures through giving feedback for future newer craftpersons. Finally, technicians "own" the jobs after receiving them. Crews and crafts are free to take the responsibility of deviating from the specified job instruction provided that they report actual actions they take instead.

The following are examples of standard plans for a somewhat complex job that might be developed initially or what might have developed after a few iterations of the maintenance job over time.

Example standard plan for B Fuel Oil Service Pump

Task and craft

1. Mechanic: Inventory parts and materials.
2. PdM: Pretest for vibration.
3. Operator: Clear and tag out.
4. Mechanic: Remove coupling guard.
5. Mechanic: Break coupling.
6. Mechanic: Remove drain plug no. 067.
7. Mechanic: Drain pump case.
8. **Disassemble pump.**
9. Mechanic: Remove seal pipe from cover no. 046.
10. Mechanic: Remove vent valve.
11. Mechanic: Remove cap screws no. 004 and remove head.
12. Mechanic: Remove gasket no. 009.
13. Mechanic: Break casing flange bolts.
14. Mechanic: Install $1/2$-in jack bolts on corners.
15. Mechanic: Remove casing half. Be careful of gasket.
16. Mechanic: Remove rotating element.
17. Mechanic: Using a puller, remove coupling half.
18. Mechanic: Using a bearing puller, remove bearings.
19. Mechanic: Remove casing rings and mike clearance.
20. Mechanic: Packing sleeves, remove and replace if needed.
21. Mechanic: Check rotating element for runout. max. 003/ft.
22. **Reassemble pump.**
23. Mechanic: Install new casing rings on impeller.

24. Mechanic: Heat bearings with induction heater to 200°F.

25. Mechanic: Place bearing on shaft.

26. Mechanic: Install new oil seals in bearing covers.

27. Mechanic: Install bearing covers on shaft.

28. Mechanic: Clean bottom half of casing.

29. Mechanic: Place rotating element.

30. Mechanic: Check antirotation pins for bottomed out in casing.

31. Mechanic: Replace upper half of casing.

32. Mechanic: Pull down bolts on corners (check rotation of binding after tightening each bolt).

33. Mechanic: Install all casing bolts torque to 200 ft lb.

34. Mechanic: Reassemble coupling.

35. Mechanic: Realign pump and motor.

36. Mechanic: Install coupling guard.

37. Mechanic: Contact operations to service bearings (replace oil).

38. Mechanic: Clean up area.

39. Maintenance crew supervisor: Sign off all clearances tags and permits.

40. PdM: Check vibration reading during test run.

41. Maintenance crew supervisor: Release for service.

Tools

Mechanic's box

Foot-pound torque wrench

Alignment kit

5-gal bucket of oil

Rags

Induction bearing heater

No. 2 bearing puller

No. 4 wheel puller

Quantity, part, and manufacturer's part number

Two bearings, SKF3303

One bearing, SKF3304

Two casing rings, W101543

Two packing sleeves, W101544

Four oil seals, W102343

One casing gasket, W100001

Craft, number of persons, labor hours, and durations

Two mechanics, 20 hours each, 20 hours total

One PdM, 2 hours, 2 hours total (step 2, 40)

One operator, 1 hour, 1 hour total (step 16)

Example standard plan for Fuel Oil Pressure Temperature Control Valve replacement

1. Clear and tag the equipment.
2. Prepare rigging.
3. Loosen all flange bolts.
4. Tighten up on chain-fall to support valve.
5. Remove all the fasteners.
6. Drop valve to floor.
7. Remove gaskets and clean flanges.
8. Rig new valve.
9. Position valve in line.
10. Insert three bottom bolts.
11. Drop in gasket on both sides.
12. Install all remaining bolts and nuts.
13. Tighten all bolts in a cross-tightening pattern.
14. Torque bolts to 200 ft lb.
15. Open and close valve to check for binding.
16. Release to operations and check for leaks (optr 1 hour).
17. Operate valve and adjust packing if needed.
18. Stroke valve (I&C 1 hour).
19. Return valve to service and clean up area.

Parts and materials

8-in 150-lb flanged gate valve

Two 8-in Selco® flange gaskets

12 $^3/_4$-in × 5-in grade 8 bolts

12 $^3/_4$-in nuts

Tools and equipment

1-ton chain-fall

1-in × 6-in nylon straps

Four-wheel cart to transport valve

Craft, numbered persons, labor hours, and durations

Two mechanics, 10 hours each, 10 hours total

One operator, 1 hour, 1 hour total (step 16)

One I&C, 1 hour, 1 hour total (step 18)

Lube oil manual

Many plants maintain a lube oil manual specifying lubricants for various machines. The manual might not specify all equipment by equipment tag number or equipment name. Instead, the manual might specify a single grease for universal use and particular oils for certain classes of equipment. Planners would keep a hard copy of this manual in the Technical Files section of the plant filing area. As minifiles are made and lubricants are selected for jobs, the planner should copy the designated lubricant choice to the minifiles. The creation of the lube oil manual and its maintenance would normally be accomplished by a plant engineering group rather than the planners.

MSDS

Planners keep a master set of the Material Data Safety Sheets discussed in Chap. 5 in the Technical Files. Planners copy individual MSDS sheets to the minifiles as they include them on jobs. Depending on plant policy, inclusion of MSDS sheets on all jobs may be unnecessary. It may be necessary only to include only certain sheets on jobs and to have the MSDS sheets available.

Plant schematics

These drawings identify equipment on process flow diagrams for each system. They help the planning department most when each system includes tag numbers. The plant might keep master schematics in the engineering or drafting areas of the plant rather than the planning area. If the planning department kept a set of schematics, it would be kept in the Technical Files. The plant may not have such drawings available and choose not to produce them. The plant may prefer to have equipment otherwise identified by maintenance records or computer listings and databases.

Figure 8.9 shows a plant schematic with equipment numbers matching the numbers labeled on the minifile folders and hung on tags on the

Figure 8.9 Sample equipment system schematic.

equipment physically in the field. The schematic shows only the unique portions of the tag number for each piece of equipment. The plant, unit, group, and system codes shown on the schematic title block would be added for each component. For instance, the bubble shown for "003" identifies equipment N02-FC-003, the Fuel Oil Service Pump Control Valve Strainer B.

Rotating or critical spares program

As discussed in App. A, a rotating spares program comprises entire replacement assemblies kept in stock that maintenance can quickly exchange for failed equipment. Later under controlled, nonemergency conditions, maintenance forces can rebuild the failed components and put them into stock.

After the plant has utilized a rotating spare, the maintenance force must follow through to ensure that the failed component is repaired for future readiness. The following procedure outlines the general process which may be modified to suit a particular plant:

1. The crew supervisor assumes the responsibility to write a work order to refurbish the rotating spare. The supervisor or designee tags the item with a deficiency tag. The crew returns small rotating spares to the tool room. The crew moves larger items to special staging areas at the tool room's direction.

2. The planner consults with the crew supervisor knowledgeable about the failure and equipment to determine if the repair should be made in-house or sent out to a vendor shop.

3. The planner plans in-house repairs as normal work orders and plans the return of the refurbished item to the original storeroom location.

4. The planner handles vendor repairs as normal procurement of services and has the refurbished item returned to the original storeroom location after suitable inspection.

Security of Files

As discussed in Chap. 2, the planners arrange the files so that supervisors and plant engineers do not require the planner to look in the files; they look in the files themselves. These persons must still work with files and information. This is why open paper files that are easy to see with side labels on individual folders are preferred. This is also why planners normally keep the files in a common area, not within individual planner cubicles. The files need to be accessible, especially in the middle of the night.

On the other hand, if other persons have easy access to the files, there is a valid concern for security. Having easy access to file information might mean that files can easily disappear. The only copy of a critical O&M manual for a particular piece of equipment may be irreplaceable, especially if the equipment is old and the manufacturer is no longer in business.

Generally, having the file area located so that persons must first pass through the planner area is adequate. This arrangement strikes a balance between making the files accessible and making the files less prone to wander off by knowing who is there. Supervisors may want to designate that only certain individual technicians may access the files depending on the competence of the technicians in this regard. Other security measures include the planners rarely, if ever, sending the only copy of any document into the field as a work order attachment. Avoiding having a single copy is a reason planners make copies of documents found in Technical Files for the minifiles. Planners may want to require projects and engineering groups to give them two copies of manuals so one can be marked as a field copy and the other as an original to remain in the office. The planning office should also have a copier and a research table to assist persons to make any copies in the planning office rather than borrow books and information sources. Microfilming certain manuals or documents may be an option if an original had to be replaced later. Similarly, computer scanning may be an option that might also enhance researching and planning jobs, especially from remote locations. Night time access is restricted because the planning offices are normally locked at night. Maintenance supervisors have the only other maintenance key rights to this area. Other persons requiring access during the night time would at least have to register their entry with operations supervisors that have keys. Some plants do not hesitate to place planners on call to respond to night time emergencies to help facilitate file access and security. Some plants have enabled planners to access computer files from home to assist supervisors or others who call with information needs making planner office entry unnecessary. Finally, management commitment and organizational discipline must consider the offense of being careless with taking or returning a file as a serious matter.

Summary

Rather than just rely on technical skills of planners and technicians, a planner also has many resources available to help with planning work. Most of these resources revolve around making the minifiles as useful as possible over time. Forms including datasheets help collect information and the planner makes frequent use of them. In developing the minifiles, planners perform another valuable plant service; the planners institutionalize plant knowledge. The planners gather the day-in and day-out information that the technicians use to execute jobs. The planners link this information to the equipment itself in the planning process to aid future work.

Perhaps the biggest resource in many plants is the CMMS or computerized maintenance management system. All in all, the computer can be thought of as a set of minifiles and forms. The CMMS electronically links forms to equipment to make data and other information available as needed regarding that equipment. The following chapter covers the use of a CMMS computer.

9

The Computer in Maintenance

More than a few persons describe their maintenance planning function by how well their computerized maintenance management system (CMMS) works. Many of these persons profess that "everything will be perfect" when they fully implement their CMMS. Similarly, many persons wonder why planning "does not quite seem to be working" even though they have installed an expensive CMMS. Unfortunately, planning is not simply "using a computer" and just because a company has a computer does not mean it even has a planning function. The CMMS can be a tremendous resource for planning, but it is not planning itself.

Perhaps unreasonable expectations contribute to disappointment for many purchasers of maintenance computer systems. Larry Beck (1996) relates survey results of manufacturing executives that indicate they typically require minimal cost justification before implementing such systems. Many of these executives expect productivity gains, but may not have specific reasons why the CMMS should deliver such improvements. David Berger (1997) says that the "failure rate" of CMMS package implementation is commonly thought to be about 50%, depending on one's definition of "failure." Christer Idhammar (1998) places the "success rate" for CMMS packages at only 18%.

Companies must look at computer systems as simply another tool. They are certainly not the ultimate remedy to one's maintenance troubles. Nevertheless, a CMMS is an important tool, an information tool. Moreover, a CMMS is not a tool for just the maintenance staff; it is a tool for the entire plant or company. Thus, many companies call their computer system an "asset" management system rather than a "maintenance" management system. The system helps with information for more than just the maintenance staff. In addition, the term "maintenance" has a bad connotation from the prevalent culture that thinks of maintenance as merely repair of broken equipment. With regard to maintenance, this information tool is one key to further improvement after the basic processes are correctly in place. The CMMS does not provide a *step* improvement as does implementing the basic process of planning and scheduling, but the improvement is significant.

Planning is also a tool, not the ultimate maintenance solution. The CMMS extends beyond planning, and even maintenance itself. However, the planning system also extends beyond the CMMS. Appendix L examines the six planning principles and six scheduling principles against the CMMS and concludes the CMMS contains information; it should not dictate planning strategy.

Regardless of planning not being synonymous with a CMMS, planners are the major users of such a system when in place. The CMMS soon becomes their most valuable tool. In this perspective, a chapter on the computer's role in maintenance is most justified. This chapter cannot be all inclusive of the role a CMMS would play in a maintenance organization because this book is about maintenance planning and the influence of a CMMS extends beyond planning. Nonetheless, the involvement of planning with a CMMS must be addressed.

The previous chapters of this book establish specific strategies and techniques of planning. In this context, the following sections address computerized maintenance management systems. This chapter addresses types of computerization, major benefits, specific cautions, selection of a system, specific helpful advice for planners, and some particular advanced features that while uncommon, would be helpful. These sections also refer to App. L which gives more detailed information on features and aspects of maintenance software projects including horror stories the reader might never imagine.

A Day in the Life of a Maintenance Planner (Using a CMMS)

Chapter 5, Basic Planning, began describing how a planner actually went about planning jobs with the episode of *A Day in the Life of a Maintenance Planner*. However, that episode portrayed a planner using only paper files. The following account provides a helpful illustration of the same planner in a system that uses a CMMS. It also identifies the planning principles and concepts along the way in parentheses.

Maintenance Planner David Clemons arrived at work Wednesday morning looking forward to another day helping the maintenance department boost its productivity (vision of planning and scheduling) with planned work orders (mission of planning).

After checking his e-mail, he opened the CMMS (other tool, computer) to find new AUTH work orders (other tool, work order system) for the mechanical crafts. These were the ones waiting for him to plan. There were several "reactive" as well as "proactive" type work orders (reactive versus proactive maintenance planning). David could quickly sort out the reactive work orders because another planner had come in earlier to "code" the new work orders (work order coding). David needed to go ahead and plan the three reactive work orders almost immediately, certainly before lunch. The marching orders for reactive work were "Peek at the history. Peek at the job. Put on a plan." Above all, the planners must not slow down a crew that wanted to go ahead and work a job. (Of course, emergency jobs were not planned at all, even though the planner might help a crew find special information during the job execution, if requested.)

David printed out a copy of each reactive job for note taking. The first job was a simple welding job and required only "minimum maintenance" attention (minimum versus extensive maintenance planning). Someone wanted a handrail welded where it had come loose. It was a reactive job apparently in the sense that it was in a high traffic area and should not be roped off for long. The other two jobs needed "extensive maintenance" consideration. One was for a clogged polisher underdrain and the other for noticeably high vibration on a potable water pump. He might be able to find useful history for these two jobs. Thankfully, the originators recorded pertinent information (other tool, organizational discipline) on all the reactive work orders. Perhaps most importantly, they had identified all of the component tag numbers (equipment numbers and history) for the extensive jobs. This allowed David to scan the CMMS quickly for previous work order history. At this point, David was mainly interested in scanning the history descriptions for past failures, if any. That information might help him know what to look for when conducting his field inspection. David noticed the plant had worked on the Unit 2 polisher underdrain once before for clogging and on the pump twice before, once for vibration. David also glanced at the cost for previous work for any excessive expenses that might affect a repair choice he might consider. Nothing looked out of the ordinary.

For the three reactive work orders, David then made a field inspection to "scope" them. He did this simply by taking the work order copies and viewing each job in the field. He mainly wanted to verify that the job was as described and then visualize how he would approach doing the work if he were the technician assigned. Also of interest would be any special circumstances, such as insulation removal or scaffolding needed. The handrail looked straightforward, but David noted the welder should have a safety harness and tie-off due to the elevation. David found the other jobs as well and felt that he had a perspective on what might be involved. This done, David returned to his desk in the planning office (separate planning department).

For each job, David needed a job plan. By definition, for the minimum maintenance job, there was no history and he would plan it from scratch.

David began creating the plan by specifying a single welder for the craft and one hour for the time. He then wrote a satisfactory procedure by merely stating that the job was to weld the handrail back in place. He remembered to add a requirement for the safety harness. Simple as it seemed, this was now a planned job. As in many cases, the craft technician did not need a detailed procedure as much as the crew supervisor needed craft and time specifications (weekly or daily scheduling). The supervisor needed this information to manage his crew productively. (But the welder would appreciate the safety "heads up.") David then looked at the existing job plans in the CMMS history for the other two jobs involved. From this information and his personal experience (planner skill), he made a suitable plan for each job in the CMMS and attached it to the work order. He did this by insuring the job plan adequately, explaining the work needed, and including *easily available* information. (Since this was reactive work, David did not go out of his way to find other technical data from O&M manuals.) In addition, as in the case of most plans, he was careful to plan the general strategy of the job and not spend undue time including "how to" details unnecessary for a competent technician (technician skill) (other tool, technician training). In the history, David noticed that the last time the underdrain needed cleaning, the technician had managed to clean it without entering the confined space of the vessel. That had saved 7 hours of technician 15 time plus what it would have cost in time for a "confined space" permit and an extra person to watch the entry. David now could use a plan improved after a past learning experience. On the other hand, the pump history showed the impeller coming loose had caused a similar problem although it had not happened in the last 2 years. David presumed this was the cause of the current problem. In this case, while David was not going to have an "improved" plan, the history pointed him toward the most likely problem and also saved him time in creating a plan. After attaching the plan used last time for the pump, each job now had a "planned package." David changed the status of each work order to PLANNED in the CMMS.

After finishing the reactive work orders, David began to close work orders previously completed by maintenance. The field technicians had been given printed paper work orders to carry in the field for execution of the jobs and to record feedback. These work orders had been handed back to crew supervisors as the technicians completed jobs. The crew supervisors had marked the jobs as COMPLETE on the CMMS, but the planner still needed to close them. David scanned the written information (getting feedback) on repairs made, delays encountered, and parts and tools used for each job. David also knew that a clerk would enter daily timesheets from each technician into the CMMS that would assign labor costs to each work order. Along with parts charged from the storeroom, the CMMS would total the cost for each work order. It was critical to have history cost information to help guide future repair or replace decisions. On one job, it was not clear what extra part the technician had used. David paged the technician to ask him so the plan for a future job would have the part available. After making changes to several of the existing job plans in the CMMS (using feedback), David gave the paper work orders to the planning clerk to type the written feedback onto each work order in the CMMS, mark them as CLOSE in the CMMS, and then file the actual paper copies.

Time for the morning break. Things were going at a good pace. Not only was all the reactive work planned before lunch, but David had been able to close out the finished work orders and would get a start on planning the other work orders (1:20-30 ratio of planners to technicians).

After break, David concentrated on the "proactive" AUTH work orders in the CMMS. Two jobs required extensive maintenance planning and two jobs required only minimum maintenance planning. David printed out a copy of each proactive job for note taking. On the first extensive job, a thermography route (other tool, predictive maintenance) had shown a slight leak for a valve. A check of the CMMS history showed that this valve had a history of leaking. The second extensive job, for a pump, had no identified component tag number because schematics and coded tags were still being developed and hung for that section of the plant. David still looked around in the CMMS to see if he could find any past work orders under the general system code, but wasn't able to find anything (system versus component level). Of course, there would not be any history for the minimum maintenance jobs.

David put on his hardhat and safety glasses and went out to scope the proactive work orders with a field check. He noted that although the valve was in somewhat of a high-pressure service, it had flange connections and would not require a certified welder (lowest skill required). David decided to include scaffolding in the plan. David then looked at the valve reported to be running hot. He was not sure what the solution to this would be. Since the pump job had no component tag number, David attached a temporary tag directly on the pump by the nameplate. One of the minimum maintenance jobs was as expected, but he had to clarify the other one with the originator. On this job, an operator had requested maintenance to install a screen over the wharf gutters to keep trash from clogging up the drains. After looking at the job, David thought that only the drains themselves should be screened so that the gutters could still collect trash and keep it from the river. David had time to try to hunt down the originator for proactive work because no one would be chomping at the bit for the plan. David then returned to the office.

David studied the current plan and history in the CMMS for the pressure valve. The valve's history showed that the seat and disc had been reconditioned as well as replaced, both within the last year. David decided that the present valve was marginal for the service and planned the job to replace the valve with an upgraded valve from the warehouse. In this case, David called the plant engineer to get a second opinion. The engineer said the upgrade was okay. David remembered to update the CMMS equipment module with the new valve type that would be used. David briefly wrote out the job procedure as replacing the valve and specified a mechanic and a helper (primarily because of the valve weight) for 3 hours each. David also didn't forget the scaffolding. One of David's jobs was to go ahead and call the scaffolding contractor so the platform would be ready before the supervisor (daily work) assigned technicians to the job. David marked the job as PLANNED.

It was now lunch time. David had planned the reactive jobs, closed the completed jobs, and planned one of the proactive jobs. Thankfully, no one had called him with any requests to help find information for any emergency jobs already started. And, none of the supervisors had called him to make a plan for a new reactive job they wanted to start after lunch.

After lunch, David took time to look in the O&M manual for the pump. Since the job was proactive, David had plenty of time to research the Technical and Vendor Files (resources). Fortunately, the manual had a troubleshooting section in the back. David decided that for this type of vacuum pump, the most likely cause for running hot might be a bad intake valve reed. David wrote the job plan to inspect all the intake valves for bad reeds and take appropriate action. David then included the inventory part numbers for likely parts: the channels, reeds, and backing plates. David specified that this would take a mechanic two days. Hopefully, good feedback would help improve a future work plan. David marked the job as PLANNED.

He then quickly wrote plans for both of the proactive, minimum maintenance jobs and marked them as PLANNED.

The proactive work planned, David now looked in the CMMS for any new AUTH jobs. Yes, there were eight new jobs. David was not the maintenance planner doing the morning coding that week, but he went ahead and coded them. He coded two of them as electrical, one reactive/minimum maintenance (ELEC-RM), and the other reactive/extensive maintenance (ELEC-RE). He coded one of the others as I&C and proactive/extensive (I&C-PE). David coded the last five for himself, three MECH-RE and two MECH-PE. David wished he had looked earlier in the CMMS for the reactive mechanical jobs, but he'd go ahead and plan them now.

The first reactive job stated that the Unit 2 Control Valve B Strainer had a high-pressure drop. As soon as David looked in the computer, he saw that this valve occasionally became clogged from rope or other debris. A quick field inspection did not reveal any unusual circumstance other than operations was not doing very good housekeeping in this building. One of David's first goals was to create a proper job description. He wanted the job description to be in field technician, not operator, language. The operator's job description had said, "The strainer had a high pressure drop." David noted that the plan he would use from the computer had a job description that said the technician was to "clean the strainer." In addition, David knew that this work order was for the strainer, not for an overall cleanup of the area. David then looked at the past planner estimate. He decided that even though the last job had taken 7 hours for two mechanics, he would still keep the current estimate calling for a mechanic and a helper for 5 hours each. David also was pleased special tools from the last job's feedback had already been added to this equipment's job plan. The technician should take a 2-inch combination, impact socket, and impact wrench to the job. David easily attached the upgraded job plan from the computer and he changed the computer status to PLANNED.

David also quickly planned the other two reactive jobs easily, having scoped them in the field earlier when he had scoped the first one. Taking a short break in the middle, he soon marked them PLANNED as well.

David then looked at his watch and called the reliability engineer to see if he had time to meet. The engineer had wanted David's advice on the specific craft requirements for several PM's he was setting up through the reliability centered maintenance (RCM) program (other tools, PM/RCM). David decided that since he had planned ten jobs today, he could put off the two new proactive jobs until tomorrow to help the engineer with PM without being bogged down there. Ten jobs was probably a good standard that would keep him ahead of the mechanics since the mechanics not only had his planned work to do, but had PM tasks that would keep them busy as well. Setting up PM's was engineering stuff to a large degree with the engineers setting up equipment requirements, but it made sense for the engineers to get field experience built into their job plans. David always helped, but knew that his primary task was to plan new work orders (future work) coming into the maintenance department.

At the close of the day, David walked to the parking lot. He thought about the part he played in the high availability the plant enjoyed. The backlog of planned work allowed the scheduling (advance schedule) of planned work to match (schedule 100% of hours) the forecasted available craft hours (forecast labor) for the next week. The weekly schedule set a work goal and made the advance coordination of other crafts and material staging possible. Simply providing a work goal through advance scheduling had already helped the maintenance force boost its productivity (wrench time) up to 45% from the 35% industry norm for good maintenance groups. Yet beyond that, keeping the plant on a constant learning curve by using information from previous jobs had boosted wrench time to 50%. Technical data was available and previous job delays were avoided. Now as the plant developed its tools, inventory, and other capabilities, including the CMMS, wrench time was slowly creeping up to 55%. At 55%, the productivity of the 30 people for which

David planned would be the same as for 47 people working at only a 35% wrench time. This was an incredible improvement, the same as if the maintenance force had added 17 free technicians (Wow!). The benefits of planning actually involved productivity and quality savings. The productivity savings came from reducing delays during and between assignments. The quality savings came from correctly identifying work scopes and providing for proper instructions, tools (other tool, tools), and parts (other tool, storeroom). The productivity improvement also freed up craft, supervision, and management time. This allowed them to focus on troublesome jobs requiring more attention and an opportunity to do more proactive work. This proactive work included root cause analyses (other tool, RCFA) on repair jobs, project work (other tool, project work) to improve less reliable equipment, and attention to preventive maintenance (other tool, PM) and predictive maintenance (other tool, PdM). David felt good that his work in planning contributed (other tools needed) to a cycle of continuous improvement.

What Type of Computerization

A company might already be computerizing much of its maintenance without a formal CMMS through basic software already in use. However, if a company does pursue a CMMS, the company has different levels of CMMS size and networking to consider. This consideration leads to more dependence on in-house IT support and vendor CMMS expertise. A company must also consider whether to develop one in-house or pursue a commercial system.

Software already in use

Before addressing what type of CMMS one might pursue, note that many maintenance planning groups benefit from computerization without buying a CMMS. All planners should have access to the Internet and email. One mechanical maintenance planner helped an electrician find an exact procedure for replacing a faulty device manufactured by a company that had been out of business for over 10 years. The planner searched the Internet and had a vendor email him an exact replacement procedure within a half hour of looking. Finding information and purchasing items on the Internet may boost a maintenance organization. Searching for MSDS information may be practical although legal requirements may dictate having on-site records. Other computer programs without a CMMS may help maintenance. A simple database program may adequately store nameplate equipment information. A clerk may easily be able to type and print work orders for PM's with a simple spreadsheet and word processor. Planners can easily develop plans and improve them over time with word processors and three ring binders.

A company should not purchase an entire CMMS with the sole benefits expected to be the automatic printing of work orders and the replacement of a single clerk. Nevertheless, there are benefits from using a formal CMMS that this chapter explores.

Single user or larger network

There exists a wide variety of types of computer systems. The first computers were mainframes and some company software applications today still operate on them. Mainframe computers themselves are expensive as are the in-house Information Technology (IT) groups staffed to develop and maintain the software for them. Some mainframe maintenance applications are fairly sophisticated and helpful. However, in-house IT groups usually prioritize basic company applications such as payroll, human resource systems, and general ledger accounting ahead of maintenance functions. Many organizations with only mainframe computer capability do not have much of a maintenance computer system. These maintenance groups may have certain codes tied into the payroll system and can use this limited information for certain analyses. With the advent of personal computers, many maintenance groups began to use maintenance applications on individual computers not tied into any company wide systems. A single individual who operates the CMMS for everyone keys individual computer applications. With networked personal computers, maintenance applications allow more than a single person to review or manipulate information. Many companies are transferring their software away from mainframes to networked computers using client server structures or web applications. More powerful personal computers and networks allow large databases and programs to interact from central computers with local computers. Small systems run from individual personal computers or simple networks may not need IT support. More complex networks and client server applications require IT support.

Creating versus purchasing
a commercial CMMS

The developers of maintenance software may be in-house personnel or vendors. The advantage of using in-house personnel is having control over the customization of the system. Unfortunately, as mentioned above, in-house IT personnel may give maintenance work less attention than desired. In addition, in-house IT and maintenance personnel even working together may not possess as much knowledge or skill as could be obtained commercially. Then, once IT writes a program, it may have limited resources to upgrade the software continually to take advantage of new computer technology or maintenance practices. CMMS vendors

are usually in a position to incorporate the most successful maintenance practices across a number of companies and even industries. The
market also drives CMMS vendors to keep up to date with computer
technology. Theoretically, instead of a small, in-house IT staff developing software for a single maintenance group, a CMMS vendor can maintain a large staff developing the best software and sharing the cost over
many maintenance groups. Moreover, the CMMS vendor might specialize in maintenance software whereas maintenance might not be a
central concern to an in-house staff. Finally, an in-house staff generally
appoints a single individual to the maintenance of maintenance software. If this individual was the developer and left the company, the
maintenance group may suffer due to few persons being able to pick up
the software issues. Contrary to this situation, if an in-house individual left a company that had vendor supplied CMMS, the CMMS vendor
would be able to provide assistance for continuity.

The conventional wisdom in the maintenance industry appears to be
working with the IT group and selecting a commercial CMMS that
closely mirrors one's current maintenance process. A company would pay
the vendor or a vendor-certified consultant to provide some initial customization. Then, ongoing maintenance of the software itself would be
through a license fee to the vendor with in-house personnel performing
some ongoing customization. Under no circumstances would the in-
house group receive the "source code" from the vendor and make modifications that might render the system unrecognizable to the vendor's
specialists. What a maintenance group desires is a system that automates how it currently conducts maintenance. The maintenance group
also desires some value added functionality that allows accomplishing
certain tasks not practical without a computer. It does not want to procure a CMMS in the hope that the computer will teach it how to conduct maintenance properly. If a company does not already have a good
maintenance process, it should wait to computerize.

The industry sometimes calls such minor customizing "configuring"
when it does not change the source code or otherwise hinder future
upgrades from the vendor. Companies must configure the software to
meet their own particular needs to achieve an adequate CMMS for effective maintenance.

Benefits with the CMMS

The CMMS helps maintenance in two ways. First, it automates and facilitates existing processes to improve efficiency. Second, the computer
can add value to produce benefits otherwise not practically achievable.
All of these benefits could be lumped together as better plant reliability and better cost control, mostly through improved information.
However, this section delves into these benefits to define them a little

more specifically. Arguably, the benefits of having a computer system, mostly from a maintenance planning perspective, are

- Standardizing Work Processes
- Inventory Control
- Information for Metrics and Reports
- Finding Work Orders
- Linking Information to Equipment
- Common Database
- Scheduling
- PM Generation

Standardizing work processes

Standardizing work processes is a two-edged sword, a benefit, and a caution. Consistency improves reliability and standardization improves consistency. This is one reason Chaps. 2 and 5 gradually move plans of general strategy to more procedures based plans. Similarly, because maintenance is still more of an art than science, any consistency helps the maintenance department. This is an enormous benefit to maintenance. Inefficiencies cannot creep as easily into a set process. In addition, maintenance management can observe consistent practices with an eye toward improvement and best practices. The CMMS helps standardize processes throughout the maintenance department from ordering parts to generating PM's, essentially because a CMMS often only allows a single rigid way to do something. Yet, it is this rigidity that may hinder maintenance. Who is to say that a CMMS has the best practices? What if a company has valid practices it wishes to automate and a CMMS thwarts these with alternate practices? This situation commonly leads to disaster with the CMMS. The "Cautions with a CMMS" section later describes this problem. Nevertheless, the benefit of standardizing maintenance practices is substantial. Arne Oas (2005), a CMMS consultant, says it well, "The main benefit of maintenance management software is to help and encourage the user to focus on good operating and maintenance practices. These are the areas where all the possible savings are derived." Notice the wording of "help and encourage." Management should very carefully approach standardization.

Inventory control

Computerizing the inventory system produces the overwhelming largest value-added benefit. Knowing part availability and allowing economic order quantities to maintain part availability give a major financial saving. It is not unheard of that computerizing inventory might reduce a $5 million inventory level to $1 million in a single year. The company savings from inventory alone justifies the purchase of a CMMS. On the other hand, many company storerooms have already computerized their operations with inventory only software without the purchase of a CMMS. The value of the benefit of converting an existing computer inventory system to one contained in a CMMS package is less significant. Nonetheless, many computerized storerooms utilize mainframe systems and must convert to nonmainframe systems for financial reasons. Early involvement of maintenance management with storeroom management may help the company make a strategic decision to procure a CMMS for everyone to utilize. Regardless of whether the inventory system is standalone or part of a CMMS, planners should be able to locate and reserve parts for specific work orders.

Information for metrics and reports

Simply providing easily obtainable reports gives the maintenance group its second largest benefit from a CMMS. This is another value-added opportunity and not merely an automation of an existing process. Management and others need information to work within and improve a maintenance organization. A CMMS provides valuable information many times not otherwise available. How much did a particular piece of equipment cost the company over the last 5 years? What are the most costly systems in terms of maintenance expenditures? What is the current backlog of work orders? What is the current backlog of work waiting on parts? What is the current backlog of work for electricians in terms of work hours? What work orders caused lost availability over the last week? Is this trend rising? What work orders could be worked tonight if there was a short notice outage? What is the trend in percentage of reactive versus proactive work orders over the last 2 years? How much have failure and breakdown work orders been reduced? Which crew has the greatest amount of failures occurring in the systems it maintains? Which work orders are waiting for planning? Which work orders have been completed, but not closed because of drawing revisions needed? Are there any planned work orders that can be worked along with the emergency job just started in this system? How many hours are spent for PM? How many hours are spent for corrective maintenance that is generated from PM work? Just knowing that there are 651 open reactive work orders may cause management to take more decisive action than a previous feeling that the backlog was around 50 work orders. As one can see, management can determine the degree of success of the maintenance program's efforts with information readily available from a CMMS. To make the CMMS information usable requires effort on the part of the planning organization. The planners normally "code" work orders to allow future reporting or analysis. Codes such as type of outage required, type of work, and equipment system are necessary. The planners are persons most familiar with the plant coding structure and provide consistency when they assign codes to new work orders each day. Whenever they handle them, planners should also scan work orders for "sense" and correct inaccurate information.

Finding work orders

The next most valuable benefit of the CMMS is the finding of work orders. This includes work order loss prevention and determination of current work order status. It also includes sorting work orders, a benefit similar to the report benefits. First, the manual method of assigning paper work orders to crews involves a risk of lost work orders. If the physical piece of paper becomes misplaced, the work request might be forgotten. This might not appear to be a problem for the plant management used to thinking of maintenance assignments as missions to correct obviously failed equipment. How could these work assignments be forgotten? However, the maintenance group accomplishes its critical objective of preventing problems from happening in the first place through regular identification and correction of minor deficiencies. If one of these type paper work orders becomes misplaced, its loss may go unnoticed until the minor deficiency becomes a glaring problem that maintenance could have avoided. A maintenance crew that normally focuses upon urgent work orders may tend to put off working less urgent work. Even a good filing system may occasionally lose paper work orders that maintenance does not immediately execute. The CMMS avoids this problem because the work order is not ever considered to be the physical document in the first place. A work order is literally what the term suggests, an order to do certain work. That identification of work resides in the computer as the actual work order. Anyone can run a report showing a listing of all work orders at any time. If a technician loses a printed out document describing the work, the computer still shows the work required. The technician or supervisor can print out another copy of the work description if desired at any time. This benefit does not require a CMMS. It could be gotten from a simple database of work orders recorded by a clerk each day. The management commitment required under such an arrangement would be the insistence that supervisors are responsible for the work on the database listing, not the physical documents they have in hand. In addition, if a paper work order were lost, someone would have to recreate the document from the database. These arrangements usually run into some difficulty with paper documents lost after work completion either in the field or the minifiles without the clerk registering their completion. Then, the database is obviously less current than the actual holding of paper work orders as an indication of work not completed. On the other hand, once the maintenance group adopts a more complete CMMS, supervisors seem to take better ownership of helping update work order status. The supervisors begin using the CMMS database to organize and manage their own work. Management's claim is that the CMMS has the real work order and its status also appears more credible. The second part of this benefit is realized for the entire plant as the status of each work order on a CMMS begins to have real value. Before the CMMS utilization, the originator of a work order had few means of determining if the work requested had been done or would be done. If written a month before, the originator might well presume the maintenance group had put off the work order for the indefinite future. The work order had gone into the "black hole." With most network CMMS packages, the originator can easily find the pertinent work order through a number of different search options. The originator might be able to hunt by work order number, originator name, date of approximate entry, equipment number, or any number of choice key words of problem description. After finding the right work order, the status of the work order is evident whether waiting for approval, planning, scheduling, or closing of the paperwork. One major power station bought a sophisticated CMMS package for the sole reason of allowing operators to find work orders and their status. The planners, schedulers, and supervisors also use the CMMS package's ability to find work orders or to sort out certain work orders. A planner might easily find all the new reactive work orders to plan for the day. A scheduler might sort all the planned work orders in order of priority, system, and job size to facilitate scheduling work for a crew. A supervisor might sort out all the work orders already assigned and in progress as a listing to help manage the day's activities. Reports and better visibility of equipment work orders help the plant optimize equipment maintenance and capital investments timing as well as make better repair versus replace decisions. The ability of a CMMS to give a view of "what is out there" in real time is an important benefit.

Linking information to equipment

A closely related benefit to making reports available is the linking nature of the CMMS. The CMMS is an automation of the minifiles linking history, parts, procedures, safety data, tools, and other information directly to specific equipment. Operators can keep track of standard clearances. Maintenance personnel can keep track of warranty and service agreements. Planners and engineers can keep track of cost. The CMMS connects all these different plant data to the equipment.

Common database

Beyond this joining benefit, the CMMS provides another plant advantage. In the past, different plant departments kept various independent records of important equipment information, many times duplicating the work of each other. For instance, three different groups might have a record of nameplate equipment data. One problem from this arrangement was when the maintenance or project group replaced the physical equipment. Not all of the plant departments might update their nameplate data. In time, the different departments ended up with conflicting information in their records. The advent of computerization made this problem of conflicting plant information worse. Personal computer spreadsheets and databases allowed easier organizing of data and departments began to collect more information. A hodgepodge of plant data resources exists at many plants. The network CMMS provides a solution to this situation when everyone can access a common database for different types of equipment information. Authorized persons in each department can update a common database a single time with the latest information. The database should represent the latest information any of the departments possesses. Work order and equipment information can also be remotely viewed through the CMMS. Instead of journeying to the planning department to review minifile contents, engineers and others can find what they want on the computer. Instead of journeying to the engineering department to review equipment technical data, planners can find what they want on the computer. A plant should consider implementing a CMMS ahead of the proliferation of multiple databases around the plant if possible.

Scheduling

The CMMS provides another benefit to the planning department with regard to manipulating scheduling information. An advance schedule should normally be a simple allocation of work and a daily schedule should involve the supervisors' personal knowledge of crew individuals for best work assignments. Nevertheless, a CMMS might facilitate some of these efforts. In addition, the CMMS allows easy "what if" reviews of different alternatives. The CMMS also allows easy "publication" of the schedule to anyone interested. This promotes better craft coordination as well as coordination with the operations group for equipment clearances.

PM generation

Next, the CMMS helps the maintenance department by automating the PM generation of work orders. A small plant may avoid having a clerk to generate PM work orders manually from a master spreadsheet. The cost of having a clerk generate PM work orders manually may not be significant for a bigger plant. However, the volume of PM work orders for a bigger plant may be large enough to cause some concern over having them correctly issued and assigned. The CMMS precisely sets the PM work orders each time they are issued. Although these PM logistical advantages may seem slight, the importance of correct PM to the maintenance group's mission is significant.

Problem diagnosis and root cause analysis support

David Berger (2005), contributing editor to *Plant Services Magazine*, specifically adds problem diagnosis and root-cause analysis to the key benefits a CMMS should deliver. Indeed, the CMMS is "emerging as an important problem-solving tool." The planning department certainly plays a role in problem diagnosis and somewhat of a role in more formal root-cause analysis. The planning department must accurately record work order information for the engineering department to analyze later for in-depth analysis.

Cautions with the CMMS

On the other hand, the computer may not affect certain maintenance functions at all or it may add complications to certain functions in the maintenance process. As discussed above, maintenance must approach standardization with caution. However, other cautions of computerizing worthy of mention include:

- Faulty processes
- System reliability and speed
- Data protection
- Improper costing
- Letting everyone see work orders and status
- Creating unnecessary metrics
- Unwisely eliminating all paper
- Expecting a CMMS to do everything
- Expecting a CMMS to think
- Overuse of templates
- User-friendliness
- Cost and logistics

Faulty processes

Of course, the most obvious downside of computerizing is automating a faulty maintenance process or philosophy. Computerizing a poor maintenance process will not help maintenance. Karl Kapp's (1996) "USA" admonition must be heeded. Understand, then simplify, then automate. The computer will help a company execute its faulty processes faster to the detriment of plant reliability. The computer may create the illusion of progress and maintenance advancement when equipment performance has not improved. One must guard against the computer becoming a distraction to real maintenance improvement.

Similar to automating a poor process is expecting a poor process to excel when confronted with a CMMS that demands a "perfect" process. David Berger (2005) laments that companies think a CMMS can straighten out processes, "You can't simply automate without first changing the attitude and behavior of the stakeholders, and the processes for which they are responsible."

In addition, why would a company think a CMMS has a perfected maintenance process? Many CMMS programs do not adequately meet many common maintenance expectations and needs. The vendors of these CMMS programs are primarily software companies at heart. They hire expert programmers to integrate many fast-changing technologies of computer hardware and software. Yet compared to the computing technology, these companies see maintenance as much less sophisticated. They may not even have a true maintenance professional on staff. Then, if they do have a maintenance professional on staff and let the person be a player, they vote on critical capabilities. In a vote among five computer group leaders and a single maintenance person, computer solutions win over maintenance needs every time. The resulting expertise from a CMMS is the same as from the IT group in-house, too little maintenance input.

Reliability and speed

Another significant concern is the reliability of the computer system. This may be a function of the particular CMMS package, the plant computer equipment, or both. The more a plant counts on the CMMS to assist with daily maintenance functions, the more the CMMS has to be available. It is not acceptable for the computer system to crash routinely as planners and others utilize the CMMS to perform their jobs. If the operators write work orders in the middle of the night, the CMMS should routinely work in the middle of the night. If the planners report to work at 6:30 AM to code work orders, the CMMS should routinely work at 6:30 AM. The IT department must accept the same philosophy as the maintenance department that its job is to provide an operating machine, not provide a repair service when called.

On the other hand, the planners and the maintenance department should understand their jobs well enough to be able to carry on in the sudden absence of the computer system. There should be a backup plan such as allowing operators to submit emails or phone calls in the absence of the CMMS. Planners should be able to scope jobs and prepare work orders with plans in the absence of the computer. Those paper minifiles also become very useful in the middle of the night when technicians with little inclination to access the CMMS need a simple piece of information on certain equipment. Do not be too hasty in discontinuing the collection of paper documents.

Closely associated with reliability is speed. Speed is everything. A paper system never requires the planner, supervisor, or technician to wait. The computer should respond similarly with commands returning information click, click, click; and not click and wait, click and wait, click and wait. Do not underestimate the importance of speed. Appendix L discusses reliability and speed in more detail.

Backup system

The above section discussed the plant having a backup plan to work on equipment in the event of CMMS downtime. As the plant becomes more reliant on the CMMS, the CMMS must reliably keep data secure. The CMMS must have a backup or parallel system that does not compromise plant maintenance information needs. The plant must protect the CMMS databases from corruption and data loss. The plant errs in becoming overdependent on electronic files if the CMMS cannot keep those files.

Cost assignment

Another area of caution is cost assignment. The planner's estimate, the job hours recorded on employee timesheets, and the hours recorded on the returned work order itself may all differ. In addition, none of those three amounts may accurately reflect the hours the technicians did spend on the job. Computerization only automates what data is entered. Users of reports must be familiar with the possible shortcomings of analysis made from computer data. In general, the planner estimate might be the most accurate cost to use in reliability analysis. This is because employees spending longer than what a smooth job would require might spend extra time due to poor scheduling and supervision rather than equipment problems per se.

Employee evaluations

On the other hand, the workforce should strive to collect the best data possible on what jobs did actually require. Planners might be able to review this data to fine-tune estimates or otherwise become aware of job delays. Technicians must feel free to enter real times and information on problems without fear of penalty. If management tries to evaluate employees on specific times versus job estimates, technicians will adjust reported times accordingly. Instead of reporting 7 hours on a 5-hour estimated job, a technician will find a way to report only 5 hours. This is similar to holding employees accountable for wrench time. Wrench time is a measure of the overall planning and scheduling effectiveness, not employee willingness to work. David Berger (2004) says, "In my view, it's always a mistake to use CMMS-based data when building an argument for giving someone a poor performance review."

Goldfish bowl

Another new problem potentially arises after computerizing, especially when a company fully implements a sophisticated CMMS. Before having the CMMS, paperwork orders moved sequentially from originator to planner to scheduler to supervisor and to technician. There was always the occasional urgent work order that went from originator straight to the crew supervisor, but normally technicians did not receive unplanned work orders for routine maintenance. The workflow system kept the technicians from even knowing about the work prematurely. However, this changed with the advent of the information sharing CMMS. Now technicians and supervisors can routinely view new work orders for equipment in which they have an interest. The new CMMS is a goldfish bowl because everyone can see the work orders. A common scenario as this presents itself. An operator writes a work order Thursday night. On Friday morning, a technician views the work order, prints it out, and completes the work, all without changing the computer status. However, also on Friday, a planner plans the work order and changes the status of the work order to "waiting to be scheduled." On Monday, the planning department clerk receives the completed work order form from the field crew and changes the work order status to "closed." The planner reviewing the work order paper is frustrated that the technician worked on the job in an unplanned manner. In addition, later reports indicate the job was planned and give the crew credit for completing planned work. This type of situation is not uncommon and requires management commitment and organizational discipline to manage. A successful planning effort that aids technicians with information from past jobs and that aids supervisors with scheduling control information encourages crews to seek job plans before beginning work. A proactive, professional maintenance organization also helps.

Unnecessary metrics

Computers also tempt managers to compile unnecessary metrics. Computers allow collection and arrangement of data much more easily in many cases, but certain metrics are unnecessary for proper management. The compilation of them wastes the time of analysts and managers. This is part of the distraction capability of a CMMS.

Eliminate paper?

There may be some areas that are impractical to computerize any time soon. For example, some maintenance practitioners profess wanting to eliminate paper documents altogether. Many companies have paper morgues bursting with old documents. Yet, is paper elimination a practical objective? It seems that there should be room for a 1- to 2-inch minifile for each of 10,000 pieces of equipment. Is it also practical to scan entire O&M manuals into computer or microfilm archives and expect easy retrieval? Would the additional step of searching discourage planners from utilizing this resource? It may be best to allow planners easy access to fast scanners to scan specific sheets as they find them in O&M manuals and attach them to specific work order plans for jobs. In this manner, the planner builds the computer files while building the paper minifile with particular O&M pages.

In addition, having paper work orders in files encourages better feedback in most cases. A plant should not go paperless for the sake of saving a wall of shelf files if it means compromising better maintenance and reliability.

The *Maintenance Planning and Scheduling Handbook* recommends creating minifiles and saving most work orders even when using a CMMS.

Jack of all trades, master of none

Another idea to consider is that a computer software that does everything may not do any one thing particularly well. CMMS programs tend to handle certain areas fairly well such as equipment and work order history databases. However, other portions of the overall CMMS program may simply be after-thoughts to provide a "complete product." Areas where the general CMMS vendor may have less expertise may be predictive maintenance and instrument calibration. Many CMMS packages acceptably deal with their limitations and form alliances with other software vendors. A CMMS vendor with less experience with document management and imaging may incorporate the successful abilities of an imaging software developer. In this manner, the CMMS package is able to bring a "best of class" performance together under a single umbrella program. In addition, some CMMS vendors are adept in allowing their programs to communicate and integrate with existing software and programs that a plant already utilizes.

PdM.   Predictive maintenance (PdM) might be one area best left separate from the global CMMS. The PdM software has evolved to a high degree of performing trend analysis and analyzing complex data. Many plants appropriately leave the PdM group running its own software and keeping its own database of trends. However, the plant should insist that the PdM group utilizes the plant CMMS for maintaining equipment nameplate data and writing work orders. The PdM group should also utilize the same equipment identification numbers in its software. In effect, the PdM group functions as an independent laboratory providing specialized service to the plant.

Calibration.   Instrument calibration systems have highly evolved software to record past calibration history and set points. Instrument technicians can download information from these systems into handheld calibrators that plug into field instruments. It is perfectly acceptable for the CMMS to initiate a work order for the instrument group to calibrate an instrument or instrument loop requiring them to access the instrument software. The instrument software is the I&C tool that the CMMS specifies. The CMMS dictates the plan and schedules the work in relation to all the other work going on in the plant. The IT group might also have the instrument software and the CMMS link and exchange data such as equipment name, physical location, and instrument specifications such as name, model number, and serial number to save double entry.

Keep a perspective of the overall advantage to the plant of having specialized software and a CMMS. Do not abandon unique advantages of specific software simply to have a single CMMS for everything.

Artificial intelligence

Closely related to instrument calibration is autogeneration of work orders. Many times, single events upset processes and cause "outliers" for process control. These events should create alarms that operators investigate. Did the process return to normal parameters or did the process remain out of control? Alarm management is a significant issue in modern plant operation. Operators respond to alarms and correct processes when appropriate. Operators write maintenance work orders when appropriate. Operators do not write maintenance work orders for every alarm.

If CMMS autogenerates work orders for alarms, at least set the alarm points outside the 3-sigma control limits of the process.

Templates

Templates warrant a special caution with regard to the planning principles set forth in this book. Templates provide "quick and easy" solutions to common equipment problems. The CMMS provides a master generic troubleshooting guide to boost planner efforts. For example, a planner has a pump problem. The planner searches the CMMS for pumps. The CMMS shows common problems a pump might experience. The planner selects which problem the pump has symptoms for and the CMMS lists common solutions for that problem. The planner then selects the desired solution and the CMMS delivers a procedure for the job plan. The problem with this approach is that the planner usually already has adequate information to help determine the equipment problem. For one thing, the planner is a superior technician with specific experience with plant equipment. For another thing, individual equipment usually fails in only one or two favored modes that the minifiles should have well documented after even only a short time.

Furthermore, the technicians are usually skilled enough to make many repairs without having a generic procedure. What the technicians do need is specific part numbers and other information recorded from previous jobs to avoid past delays. Generic repair procedures do not contain that information because that information is specific to individual pieces of equipment, not types of equipment. Specific troubleshooting guides included in specific equipment O&M manuals are more helpful than generic templates contained in some CMMS packages. Overall, most templates are too general in nature.

User friendly

Whether everyone will use the CMMS depends on how user friendly the system appears and the interest of the workforce. With the advent of email to disseminate company wide information, all levels of many companies are rapidly becoming computer literate. The rapid acceptance of using a CMMS may surprise many persons. One plant installed a CMMS at first only to track paper work orders and to provide equipment data for planners. Management at the plant decided that everyone should have the option of writing work orders directly into the CMMS. The person actually responsible for implementing the system expected only 10% to 20% of new work orders to be directly entered. It was a great surprise that over 95% of the subsequent month's work orders were entered directly. The plant soon made the direct entry mandatory and eliminated the need for a clerk to enter work orders each morning. Appendix L addresses the issue of user friendly computer systems in more detail.

Cost and logistics

Unfortunately, the decision to utilize a CMMS does not result in a computer being up and running the following morning with everyone fully familiar with how to use it. The selection, purchase, and implementation of a CMMS are very important issues. Implementing CMMS software can be very expensive and drawn out. In addition, the result of a helpful system may not be certain. Consider the following actual results of one released version from a major CMMS company before corrected with a patch:

- Server crashes after selecting "cancel" from a pop-up window.
- Cannot print multiple work orders if any has an attachment.
- Meters can cause countless PM's to generate and this action cannot be canceled once started. (This would be a case where it would be better for the server to crash.)

Rather than simply automate a perfected internal paper maintenance system, the company may have hundreds of similar "bugs" to resolve when implementing new software on such a scale as the CMMS. Appendix L further clarifies problems with major software projects and offers helpful advice to "survive" them.

One thing that surprises many companies is that the purchase price tends to represent only a fraction of the investment the company will make, perhaps only 20%. The company personnel to install and administrate the system consume an enormous amount of time. There is also the matter of computer equipment upgrades either initially or over time. The leading software vendors continuously upgrade their product. Usually a company pays a yearly license fee to have access to free upgrades and technical support. The CMMS vendor usually makes available an upgrade every other year or so and has a new major release every 3 to 5 years. These changes to the software come about because of the research being done by these vendors and because computer technology is evolving at breakneck speeds. Thus, a company has to spend time and effort simply staying current with the latest CMMS software from the vendor already selected and utilized. If a company does not choose to install the latest software, the company risks soon having a system that neither the vendor nor the IT department can well support. This is not all bad. As computer technology evolves, companies may be upgrading their equipment anyway. In addition, the expenses paid to coordinate CMMS upgrades are investments in proper maintenance. Poor maintenance is very expensive as well.

Implementing new software also requires other expenditures and efforts. The data will have to be initialized into the new system including inventory and equipment information.

Selection of a CMMS

Other major issues concern who and what is involved with selecting a system. A team usually selects a system following a predictable routine. The actual installation may proceed all at once or in predetermined phases over several years. Again, App. L gives further information concerning software projects.

Team

The persons involved in selecting a CMMS tend to be the planning supervisor and a representative from the IT group being led by a plant engineer. This group has frequent interaction with the maintenance manager. The maintenance manager may have to ensure that the IT group does not overwhelm the group with inadequate maintenance process knowledge. It is also important that the plant engineer understands the planning and maintenance processes. Another person on this team should be a person who will become the system administrator. This person will set passwords and grant other rights to users. This person may also create standard reports and do some screen customization. It is preferred that the system administrator be a plant engineer or maintenance representative rather than an IT person for several reasons. The plant needs a local representative available to answer questions who understands the plant maintenance organization and processes. The system administrator choice should provide a person who will fulfill the role for years to come rather than as a current IT assignment. Many persons feel that it is better to have as a system administrator a maintenance person who becomes interested in computers rather than a computer person who becomes interested in maintenance. The system administrator should coordinate larger tasks of modification with the IT department or the CMMS vendor. Planners play a major role in utilizing a CMMS, but being a system administrator consumes a lot of time. It may be practical to have a planner fulfill this role if there is an abundance of planners, but this is usually not the case. Of course, for standalone, single user versions, planners may be the only users that directly enter or extract information. A planner would naturally be the choice system administrator. With this team of persons, the company might begin to select a CMMS. Teams such as this also frequently employ the use of a consultant to do much of the legwork and supply more familiarity with CMMS systems during the selection process.

Process

The actual selection process tends to be somewhat the same for many companies. This is a very simplistic overview. Obtain a clear idea of the objectives of why maintenance (or management) wants to buy a CMMS. Assemble a team for the selection. Obtain internal views of what system features are desired. Determine initial selection criteria. For example, some companies desire a CMMS vendor to be an established company that will stay in business. Some companies want specific industry experience. Some companies want geographical closeness for support. Some companies require a specific operating platform. Survey technical magazines for summaries of systems. List likely systems. Gather literature from these companies. Call and ask them questions to see how they view maintenance. Expect to establish a long-term relationship of support and continuing improvement. The ability of the company and vendor to work together as a team in the future should be evaluated along with the other technical requirements. Develop ideas of prices to review with management because complete systems range in price from under $1000 to well over $250,000. Invite vendors to make presentations and demonstrate systems. Refine and weight selection criteria. Investigate selected systems. Check the claims of vendors where appropriate. Ask for unabridged user lists and call users. Visit other plants with the software. Call the CMMS support lines for responsiveness. Negotiate and sole source or bid final candidates. Do not expect any one system to be perfect.

Again, remember the initial price tends to represent only a fraction of the total cost, perhaps only 20%. The company personnel to install and administrate the system consume an enormous amount of time. There is also the matter of computer equipment upgrades either initially or over time.

Implementing new software also requires other expenditures and efforts. The data will have to be initialized into the new system. An inventory system already computerized might have to be downloaded into the new system. Who will make the links? Does the IT department have time to write these "scripts" and execute them? Many companies have a consultant who helped them in the CMMS selection or the CMMS vendor themselves help make the transfers. If this is planned, it should be part of the selection process. Computerizing of inventory for the first time will require data entry. Does the maintenance organization have the time to do this? Can temporary help be hired? Who will direct their efforts? Can the consultant take care of this? Similarly, does equipment already have unique equipment tag numbers? Is there an existing database to transfer? If not, who will decide how to number equipment? Who will hang the tags and enter them into the CMMS? Will the CMMS have to communicate with other company programs such as payroll or inventory? Who will program these relations? The IT department typically is cautious with new programs interfacing with existing ones. Who will enter all the PM tasks? Could the consultant supply clerical help to transfer an existing PM listing into the CMMS? In addition, as the new system is implemented, many specific details will have to be handled as they arise. Questions such as what to do if the new program mandates the use of a different term for "waiting for scheduling" will arise daily.

Also, do not presume that a company must install a CMMS fully all at once. It may never need full installation. If the company bought the CMMS for inventory and work order tracking, labor calendars and vendor registries might wait for use. The company might initialize the equipment module gradually as needed to reference work orders. Similarly, the planners might initialize the planning module gradually as they become familiar with the CMMS. On the other hand, the company would be wise to initiate automatic generation of PM's sooner. Be cautious that some CMMS systems use relational databases to such an extent that one cannot implement only a portion of a CMMS package. The decision on how to proceed with package implementation rests on many factors. These factors include mature judgments of the value of each feature of the CMMS, personnel and money available for implementation, and management's desire, willingness, or patience.

Management may be willing to allow full implementation, but may not have the patience to provide support over a schedule of several years.

The success of the system does not rest on full implementation. Success rests on being able to meet the original objectives for which it was purchased. If the company purchased the system to allow originators to find out what happened to their work orders, the CMMS must meet that objective. Implement other features of the system only as they benefit the company.

Specific Planning Advice to Go Along with a CMMS

The following briefly captures helpful advice for planners using a CMMS. Appendix L amplifies this list, but more from the perspective of what a planning group should request from a CMMS.

Use the planning module. Unbelievably, many companies buy a CMMS and do not plan in the planning module. Instead, they plan jobs directly on each work order. They bring up each new work order and add job steps directly one by one. These companies should be planning the jobs in the master plan area where they can save the job plan and then attach it to the work order. The advantage of this proper method is the planner can later bring up this saved master plan after receiving feedback and improve it to make a better starting point for the next plan. Otherwise, a planner is simply typing each job from scratch. A similar bad habit is the planner leaving each master plan in the planning module too generic as only a template. These planners mistakenly think they should attach the generic template to a new job and then make it special for the job. Instead, the planners should make the master plans as specific as possible for each job improving them continually with all lessons and information available.

Add reference and history sections as described at the end of Chap. 5. Computers lend themselves well to extended job steps that itemize references and changes made to job plans.

Continually improve physical descriptions to print out on work orders. (However, obviously planners can do this just as well with paper files.)

Always select that the inventory parts used on jobs automatically become listed against equipment spare parts. In other words, the CMMS adds the used part on the standard spare parts list for that equipment. Common advice against this practice suggests that improper parts or trivial parts may end up on the lists depending on what happens on work orders. This advice suggests not automatically adding anything or at least not adding certain parts (rags, bolts, etc.) automatically. The danger of this advice is that it ends up in nothing ever being added. It is much better to have everything ever used on a piece of equipment added to its spare parts list than having nothing added. The planner can sort through a list of 100 items used on the equipment in the past much more easily than an entire storeroom catalog of 10,000 items. The argument that the planner must sort through 100 listed items to find 30 valid items is meaningless when contrasted against the argument that that the planner otherwise must sort through 10,000 items to find 30 valid items. Do not make the mistake of thinking the list is perfect. Everything in planning is a starting point. The experienced planners still must use their knowledge. It is also helpful that the planner can improve the specific equipment spares list and delete inappropriate items used in the past.

Even with a CMMS, issue paper work orders into the field to capture the best feedback. (Again, see Chap. 5.) Also, continue to file the paper work orders into minifiles for research and reference. The major benefits of a CMMS do not include reducing paperwork. The major benefits most focus on promoting better information. The benefit of reduction of paper is so minimal in most cases, it is usually wise to continue capturing paper history

Try to incorporate equipment numbers into job plan numbers. This makes it easier to use job plans for similar equipment as a starting point. For instance, different job plans dealing with equipment N02-FC-003 might be numbered N02-FC-003-1, N02-FC-003-2, N02-FC-003-3REBUILD, and N02-FC-003STRAINER.

Have originators enter the equipment name at the beginning of the work order description even though there is a separate field for equipment name. For example, the problem description line would state "2A BFP leaking at seal." Reports that list the work orders will then make more sense.

Participate in CMMS user meetings. These meetings allow the CMMS vendor to grow and develop its product to be more helpful. Try to influence the product direction. In addition, the meeting allows the user to learn from not only the vendor presentations, but from other users.

Network with other users of the CMMS to exchange ideas and to gain contacts to call for advice. Some CMMS systems have on-line user forums not even sponsored by the vendor to exchange ideas.

Buy computer typing games to improve planner keyboard proficiency if needed.

When a company first implements a new CMMS, a major question always arises regarding old closed work orders. Should they be entered into the system? While it might seem relatively simple to have a clerk enter them, it is not always practical, especially if the old work orders have no explicit equipment numbers. Even if the equipment numbers are present, this effort may be unnecessary. The repetition of maintenance operations on specific equipment soon generates helpful information for maintenance planning purposes. Maintenance does not have a critical need to enter past work orders. A strategy of work order entry "from this point forward" is frequently satisfactory.

Advanced Helpful Features for Planning and Scheduling

From the standpoint of planning and scheduling as presented in this book, the following represents some helpful features that may not be common on some CMMS systems.

The CMMS would make it easy to enter actual job feedback on each step rather than a general feedback field at the end of the work order. It would also be easy to update actual parts and tools used.

Most entries would be done from a minimal number of screens, preferably a single one that resembles what is printed out as the work order form.

The planner could go to an O&M type module that contains most equipment and job plan information, just like an actual physical O&M notebook. The planner can cut and paste from this manual to form job plans for individual jobs. The manual would have a "master" procedure detailing everything that could be done on the pump complete with general steps, parts, and tools. This plan would actually be developed by the planner as new jobs are planned and feedback is received. With each new job, the planner would check the master to see if the details are present from which to cut and paste. Otherwise, the planner would add new steps to the master to represent this latest job.

The entire CMMS would be backed up daily automatically or continually, along with the network automatic routines.

The CMMS inventory function would automatically record the previous use of parts by equipment. This would provide an ongoing development of a parts list for each piece of equipment as maintenance works on jobs. Unfortunately, many CMMS inventory modules do not provide this automatic feature because they presume the plant has a complete set of parts lists for all its equipment. The modules primarily concern themselves with tracking quantities on hand and reservations for specific work order numbers.

The PM function could handle routes on multiple equipment and register the history with the individual equipment.

The company could *easily* add fields to link or track particular codes or information as needed.

One could *easily* find work orders and their status.

Supervisors could *easily* update job progress to help schedulers obtain backlog information.

Supervisors could *easily* keep on-line calendars to help schedulers obtain crew forecast information.

The scheduling routine would "work persons down" into lower skilled, but higher priority jobs as necessary to create the weekly allocation.

Supervisors could *easily* assign names into a weekly schedule for a single day.

Schedule compliance would be based on jobs started.

Timesheet information would form the actual cost collection, not feedback on work order forms.

Company would be able to implement the CMMS partially or in phases.

Planners could save email by equipment in the CMMS.

Summary

Obviously, planning encompasses more than utilizing a computer. Nevertheless, a modern CMMS can be an important information tool. The planners need an accurate filing system and the CMMS links a tremendous amount of information to individual pieces of equipment.

Many companies that implement CMMS packages are disappointed with the results. This disappointment appears to stem from having only vague expectations of expected results. Disappointment does not need to be the case. CMMS software contributes to the bottom line when purchased for specific information reasons. With appropriate software, management can control and reduce inventory. Ad hoc and regular reports can provide management with necessary information to control the efforts of maintenance. The maintenance group can better visualize, determine, and manage its backlog. The maintenance group should be wary of becoming distracted with computerizing instead of maintaining the plant, but generally, the computer should have a positive impact on maintenance. The CMMS cannot help a planning system floundering with the basics of planning, but can help in specific areas of the planning process.

This chapter has also briefly characterized some of the wealth of available information in the literature for guiding companies in selecting and using a CMMS. If a statement could succinctly sum up a company's proper CMMS guidance as it seeks improvement, it might be with a statement by Nicholas Phillippi (1997). He says, "The best investment protection is a thorough understanding of the existing maintenance processes and application of the maintenance system in concert with these processes."

Understand, simplify, and then automate.

Consideration of Preventive Maintenance, Predictive Maintenance, and Project Work

This chapter covers the specific interfaces of these important areas with planning for the overall success of maintenance. Appendix A describes the concepts and importance of preventive maintenance, predictive maintenance, and project work along with their general relationship to maintenance planning. This chapter, after the development of the planning principles and practices, considers in practice how a planning and scheduling system ties into PM, PdM, and project work. Companies strive to do more preventive maintenance, predictive maintenance, and project work to lessen the incidence of reactive maintenance work and increase plant reliability. Planning can facilitate the use and effectiveness of each of these preferred types of maintenance.

Preventive Maintenance and Planning

The planning department issues schedules and PMs even if an engineering group with a living RCM program creates and reviews them.

Starting with a basic system, visualize a planning clerk who each week types up work orders with preventive maintenance tasks due the following week. The clerk looks at a spreadsheet or notebook that tells which PMs to create each week. Some PMs have frequencies of every week, while others may have frequencies of only once every 2 years. Many other PMs fall due in between such extremes. The clerk's listing of PMs organizes the various PMs to allow easy determination of which PMs to issue at what time.

A PM listing probably gives only a very brief description of each PM task. A planner should plan each of the PMs with a plan that the clerk can reissue each time the PM is due. The plan should have a clear scope and craft requirements including numbers of persons, work hours, and duration. The plan should also list anticipated parts and special tools. If none of the existing PMs has a plan, the planners could begin to plan them as the clerk originates them.

The scope is of great importance in a PM work order. The scope of a PM job plan should encompass more than simply changing a filter or greasing a fitting. One of the greatest tools maintenance management can bring to bear to improve equipment reliability is promoting close involvement with the equipment. Even if the PM is an apparently simple task to grease a coupling, the plan should specify that the technician should note any unusual equipment conditions. The PM plan should empower the technician to make any minor equipment adjustments or minor repairs on the spot during execution of the PM if it does not require special clearances or coordination. For example, if the technician notices that the motor mount appears to be loose, the technician should tighten the fasteners while he is on the coupling PM job. On the other hand, the technician should write up as new work orders any equipment deficiencies that would take a significant amount of time or coordination to remedy. The technician must not become bogged down in adjustments or repairs that are not explicitly in the scope of the PM work order. If the technician has been assigned several PM work orders, the technician must work according to the assignment schedule to finish the assigned work. Otherwise, the technician might thoroughly restore all problems in the area of one PM, but not even start three other assigned PM work orders. If the technician suspects that the loose fasteners have affected alignment of the driven assembly, the technician should not then become involved in a lengthy alignment task, but write a work order for the situation. One of the primary tasks of any PM work order plan should be to identify any situations that require corrective maintenance.

Another aspect of the scope involves TLC. *TLC* is not "tender, loving care," but "tightness, lubrication, and cleanliness." Terry Wireman (1996) points out that not maintaining the basic conditions of cleanliness, lubrication, and tightness contributes to 50% of all breakdowns. That means that if a plant had a backlog of 400 serious equipment deficiencies, about half could have been eliminated through proper PM in this area.

A brief scan of a company's work history reveals the following situations as illustrations.

1. Generator foaming, tank level high alarm on constantly, linkage loose, retightened, switch caked with dust deposits, replaced missing cover.

2. Sump pump, no foundation bolts in pump base, aligned and bolted up piping.

3. Induced draft fan motor amps pegged meter then dropped to zero with sparks from junction box, root cause of improper torque of terminal block, power feed set screws.

4. Boiler feed pump tripped, inboard end throat bushing damaged, retaining nuts on both discharge, and suction pump shaft sleeves were loose.

5. Entire unit derated due to turbine generator, gear segment spider of the valve operating mechanism binding up with excessive wear on spider bushing, lack of lubrication noted throughout entire valve train.

6. PdM collected lube oil noting that the B induced draft fan inboard and outboard motor bearings were full of water.

7. Lab results of last four consecutive oil samples for boiler feed pumps indicate upward trend of rust contaminants caused by moisture.

8. Decrease in bearing lube oil level was not detected due to a clogged level gauge glass.

9. Instruments supervisor notes that nuisance alarms have gone down significantly now that they routinely vacuum out control cabinets. The vacuuming (with a special low static vacuum) operation initially found cabinets with over an inch of dust buildup and many loose terminal wire connections.

Cleanliness helps in a number of ways, not the least are reducing contamination sources and allowing clean surfaces to reveal the presence of new leaks. The very act of cleaning brings technicians close to the equipment for observation. Cleanliness avoids expectations that dirty equipment normally fails now and then. Dirt and grime also add undesirable insulation conditions that may affect equipment performance.

PM plans should include general cleaning and wiping down of equipment if possible under the existing maintenance culture. Filter replacement should not be neglected as candidate PM's for regular replacement. Rags for cleaning should be included in many plans as special tools.

Proper fastening merits serious attention as well. A group of senior technicians was asked if adding thread lubricant would increase or decrease fastener torque requirements. *Half* of them were uncertain of the answer. If a certain bolt requires 100 ft lb torque, adding a lubricant would decrease the requirements to about 75 ft lb depending on the lubricant. Lubricant reduces the friction necessary to set the fastener threads; all the remaining torque goes toward stretching the fastener to provide a tension that holds the assembly tight. If the technicians increase the torque, say to 125 ft lb, their efforts might damage the bolt by the application of 50 ft lb over the 75 required. The bolt would fail if stretched beyond its ability to recover. Such a failed bolt would later appear to be loose causing a vicious cycle of technicians believing they were not providing enough torque.

PM plans should call attention to the general tightness of equipment assemblies. They should specify torque requirements or attach torque charts where appropriate. Some plans should include torque wrenches as a special tool. Bolts have a limit to the number of times they can be tightened and loosened. Planning for parts should encourage the replacement of fasteners sometimes when technicians must otherwise dissemble a component for inspection or service. The cost of fasteners is much less than that of equipment failure.

Lubrication is the most commonly recognized portion of a PM program. Unfortunately, an improper lubrication PM can be more damaging to equipment than having no PM service at all. For instance, many times a technician introduces dirt into a bearing by not wiping a grease fitting beforehand. Although general technician skill should prevent such practice, PM plans can help by including rags for wiping. PM plans can also specify removing old grease where components can be dissembled. PM plans can also guard against using oil with any chance of moisture content by specifying new containers where practical.

The emphasis on all PMs to inspect equipment for abnormal situations makes the planner lean away from specifying only helpers that would be too inexperienced to detect them. Unfortunately, many plant cultures do value assigning experienced personnel to mundane tasks as PMs with a focus on cleaning equipment. Nonetheless, as planning and scheduling increase productivity and deplete existing backlogs, plants assign more persons to PM tasks because they identify more work for the crews. Planners should assign as experienced persons as allowed to perform PM work orders.

For time estimates, the planners should go somewhat against the previously established principle of not allotting any time for unanticipated delays. The planner should allow some extra time on all PM work for making unspecified excessive cleaning, small adjustments, or minor repairs. The time will also allow the technician to write work orders for future correction of more extensive deficiencies. The plan might specify that the technicians write any work orders or that enough information be given in the feedback to allow the planner to write work orders.

Finally, the plan must receive feedback to improve the PM work order itself. Were there any unusual delays? Are any other parts or tools desirable? Should the hours or frequencies be adjusted? Just as with any other work order, the planner should review the work order after job completion to determine if the PM plan requires improvement.

The PM frequency is one of the most subtle areas for the planner to manage. If a particular piece of equipment fails only once every 2 years, the appropriate frequency for cleaning or inspection may still be once every month. This is because PM frequencies must be set by age of installation, likely failure modes, and criticality to plant process instead of previous failure rate. Once the plant installs equipment, that equipment must work through an early period during which it has a higher than normal chance of failure. Newly installed equipment should be inspected more often than older equipment for signs of failure. After the proving period, PM frequencies might be lessened. Certain equipment also has certain favored failure modes. A valve may exhibit a sticking symptom several weeks before failure to operate. A flange may drip months before leaking bad enough to cause a problem. The sound of cavitation may indicate a future pump problem. The PM frequency should take into account the time between the appearance of a particular symptom and the time the equipment may be restored to proper operating time without experiencing a failure. In addition, certain failures may not interrupt plant operation because of being in nonessential services or having installed spare equipment the operators can utilize. If these failures do not cause more extensive repair operations than would be necessary to prevent failure, the plant may exercise a strategy of minimal PM effort or attention to set frequencies. On one hand, the planner wants to set PM frequencies to minimize failures and generate corrective work orders. On the other hand, the planner does not want to set excessively short PM frequencies that would overtax personnel resources. The planner must balance these plant needs.

One particular issue dealing with PM frequencies is that of sliding schedules. Some plants schedule the next PM work order a specified time after the first work order is complete. In other words, say a PM is due every 30 days. After the first time it is issued, the crew might not complete the work for 10 days. The planning clerk then adjusts the schedule to have the next PM come out in 30 more days, that is, 40 days after the first PM was issued. As this cycle continues, instead of the PM being done every 30 days, it is actually done on an average of 40 days or more. A particular danger is if the crew delays in performing the work for several months. With this process, another work order would not be issued anywhere near the original desired frequency. Another version of sliding schedules is having the work order come out more frequently than really needed. For example, a PM comes out every 2 weeks, but really needs to be done only once a month. The crew typically tries to complete one of the work orders each month and ignores the other. Both sliding schedule arrangements are unacceptable in most cases. If a PM should be done every month, the work order should come out every month.

The planners should arrange a special file with the planning clerk to handle the PM job plans. The PM job plans should be located next to the planning clerk. They should be in files labeled with the PM identification number on the planning clerk's list. The clerk can then write a new PM work order and attach a copy of the latest plan from the file cabinet folder. After reviewing a completed PM work order, the planners should file it in the same filing system in its respective PM folder. The planner cannot utilize the minifile because many PMs cover multiple pieces of equipment on routes or as systems. The planner should make a minifile link, however, by identifying in each pertinent equipment's minifile what PM job plan number(s) covers its preventive maintenance.

The planners should also actively add new PMs to the list with the intent of encouraging routine involvement with most of the plant's equipment. The planners should particularly cover critical plant processes and equipment and previously identified high failure rate areas. Planners should obtain and evaluate equipment manuals for new equipment manuals or PM advice. Planners should welcome plant engineer input for new PMs needed, particularly as the result of investigation team efforts. Planners should also review the histories and feedback from all work orders they plan to determine if additional PM work orders are needed.

As the number of PM work orders increases, the sheer quantity could overwhelm the planning department. A CMMS computer often becomes valuable for managing the magnitude of the effort. At this point the value of checklists or simple technician ownership becomes evident. A supervisor may assign a checklist directly to a technician without a work order. A commonly encountered PM card system works much the same way because the cards are portions of a master checklist.

With checklists or cards, each PM work order begins to handle more equipment to reduce the amount of paperwork. Instead of greasing one or two pumps, a checklist might specify handling filters, alignment, and greasing for all 20 pumps on the first floor of the shop. With ownership, the plant may avoid paperwork altogether and simply assign different equipment to different technicians for PM activities.

With either of these two plant preferences, the planning department should still encourage the use of work orders specifying at least the crafts and hours. Experience has shown some problems in PM systems without work orders. For one thing, without a work order, the tendency is to think of the work as nonessential or fill-in work which may not be completed. For another thing, right or wrong, management reacts to work order volumes, and the supervisor who manages a crew without work orders is a candidate for losing persons to crews that have a visible backlog. Of course, with the planning and scheduling system, work orders are necessary to allow efficient scheduling of resources. Even with the computer assistance, planners should still be involved with evaluating work order feedback.

Checklists can be compiled with or without the computer and the planners can help technicians by typesetting their checklists. Checklists may reside in binders left with the technicians who receive a work order directing them to use their binders. Checklists might reside on the computer printing out each time with the PM work order. Steve Stewart (1997) of Tenneco and others recommend that these checklists be living documents. Checklist items on pieces of equipment should evolve from general inspections to more specific checks of particular details as experience is gained. "Check condition of bearing" might evolve to "Ensure clearance is less than 5 mils."

If the plant has no PM program or one that is clearly inadequate, the establishment of one is beyond the scope of this book. However, the general creation of such a program might begin with the prioritization of equipment with regard to plant importance or criticality. Then the plant would establish PM plans and frequencies for equipment in the order of equipment criticality.

As a final note, the planning group should bring the PM work orders into the normal routine of how it plans, schedules, and improves work orders from feedback. The plant should not think of PMs as fill-in work to do when it has time. PMs are real work.

Predictive Maintenance and Planning

The planning department plans and schedules PdM-initiated work orders.

Predictive maintenance (PdM) uses technology not available to the regular maintenance work force. It is only natural that PdM personnel "make the call" regarding the creation of new work orders. The often more extensive experience of the maintenance planner may sometimes disagree with the proposed work. Nevertheless, the PdM technology has the potential of greatly moving the plant's reliability upward. Even when some PdM predictions do not prove valid, the predictions involved in predictive maintenance show an important capacity for growth in accuracy. However, the expertise of the PdM persons can only grow if guided by experience. In addition, many of the initial predictions of the PdM group are correct in identifying specific problems, but it is only the few errors that receive publicity. Planners must resist the urge to compete with the PdM group for identifying work scopes. Planners must accept PdM work orders for jobs and translate them into the appropriate scope for the maintenance crews. On the other hand, the planners should continue to be a resource for the PdM group and be open for questions, especially those regarding past similar circumstances. Planners should also facilitate PdM being present when equipment is worked so that PdM personnel may quickly climb the learning curve. On PdM requested work, the planner can easily place a direction in the job plan for crew technicians to notify PdM personnel in time to witness certain events in question.

Regarding files, PdM normally uses separate software and keeps its own files of trends and past reports. Planners should presume PdM knows which equipment to watch and will write work orders when necessary. It is only these work orders which enter the planning files and plant-wide computer system. Planners should insist that PdM uses the same equipment tag numbers as the rest of the plant to ease communication problems. Planners can help PdM if PdM informs them of certain machines on the PdM "watch" list. PdM may feel reluctant to write a work order yet, but would like to inspect certain machines if they come down otherwise for maintenance.

Planners should also seek to utilize standards set by PdM for certain jobs. These jobs may involve alignment criteria, bearing clearances, or other rebuild tolerances. As with any plant engineering group, the PdM group becomes involved in newer technology generally ahead of the maintenance group. The PdM group can help update the technology of the maintenance force, especially with the active assistance of the planning group.

Regarding the time accounting by the PdM personnel themselves, PdM personnel function as do plant engineers, separate from work orders in their daily duties. The planning department would issue work orders for PdM routes only in the unusual circumstance where PdM persons were technicians with little specialized training and were working directly for the maintenance supervisors to run relatively simple routes such as for thermography. On the other hand, the plant considers any work orders initiated by the PdM group as predictive maintenance. For example, the PdM work to take vibration readings is predictive maintenance, but not normally conducted under a work order. However, the request by the PdM analyst to have the maintenance group replace a suspected chipped impeller is also predictive maintenance and is done through a work order.

Project Work and Planning

Similar to PdM work, the planning group plans and schedules work orders to implement projects as regular jobs.

The difference is usually in the larger nature of projects coming from the project group and the project group's inclination toward using contractors. The plant normally has the personnel to implement much project work, especially with the productivity of a planning and scheduling system. However, the project group usually favors contractors because of a contractor's ability to commit to a schedule and estimate. Plant management should treat projects as outages and plan them into the plant long-range schedule by working closely with the project group. Planners should be able to estimate and commit to project schedules with plant engineering and scheduler assistance. Planners should not become too distracted away from planning day-to-day maintenance, but can assist project work.

Even when not involved in the installation of projects, plant personnel should track and become involved with any new work to be done at the plant. The plant personnel seek to influence the installation of reliable equipment before it is purchased. The plant personnel should be involved with setting plant standards for the project group. The plant personnel should spend time as directed with corporate teams and review project plans seriously. Although many of these duties may be shared among plant groups, planners typically have the edge over plant engineers for valuable field experience. Maintenance supervisors or senior technicians are preferred for these assignments, but frequently management sees planners as somewhat more expendable. If a company has only one or two planners, it may overwhelm the planners' time to become involved in all the standards or projects. On the other hand, if there are 30 planners, it would make sense to have one planner involved in motor standards, one involved in pump standards, one involved in this project, and so on and so forth giving each planner a small additional duty with respect to front end loading.

Planners can best accomplish certain tasks for large projects. Planners must vigorously pursue collecting documentation to establish files and PMs. Planners are also in an overview position where they can cancel certain work orders to patch systems that are due for project replacement or overhaul. This position also affords the planners a good viewpoint for proposing new projects based on records of troublesome equipment.

Other than for major projects, planners can propose new projects by writing work orders or passing ideas to the plant engineering group. Planners can also be as helpful as possible to plant engineers who select equipment without becoming neglectful of their primary planning duties.

11

Control

This chapter finally arrives at the all important issue: How does one ensure planning itself works? Surprisingly, it is not on the basis of indicators, although two of the 12 planning and scheduling principles describe indicators. It is on the basis of the selection and training of planners.

Organization Theory 101: The Restaurant Story

Dr. Stephen Paulson (1988) tells the story of John Smith who retired from the Navy. John was at loose ends for a while and began meeting daily for lunch with several of his friends. John naturally enjoyed cooking and the lunch group usually met at John's house where he made the sandwiches. Everyone would always chip in to pay for the lunch. The company and the sandwiches were good and soon more friends were coming around at noon time. Someone eventually suggested that John should lease a small shop where they could spread out and be more comfortable. There seemed to be enough income from everyone's contributions so John found a small place in Jacksonville Beach where the group could meet. Thus, John's Sandwich Shop was born.

At first, everything continued as before. John made the sandwiches and would join in the company and discussions around several tables. The "organization" of the establishment, so to speak, was simply John Smith. John took care of everything from opening the door in the morning, making the meals, and collecting the money to bussing the tables, sweeping the floor, and closing the door in the late afternoon.

As word got around, some of John's friends started bringing their other friends making the place busier than ever. Soon, John had less time to visit and was spending more time making sandwiches. That was okay with John since he enjoyed cooking. However, he really needed help and so his wife, Mary, began coming in every day to help. The organization of the sandwich empire consisted of two persons on equal footing doing whatever needed doing. John did most of the cooking, but still helped Mary clean tables and collect money. They even bought a cash register to help make change. Communication was no problem. Whenever John needed something, he called over to Mary. Whenever Mary needed something, she called over to John.

The happy atmosphere of John's Sandwich Shop delighted friends and other customers. Business thrived and soon John and Mary needed more help running things. They both decided to hire someone and they brought Joe on board for wages. Joe reported directly to John, who gave direct supervision to his activities. Normally, John had Joe bussing tables and washing dishes. Occasionally, John would direct Joe to perform other specific tasks. These tasks included activities such as sweeping the floor or running an errand to buy certain supplies that were running low.

With summer approaching and business booming, John and Mary decided to expand and take a lease on the vacant shop next door. Remodeling to remove part of the connecting wall almost tripled the sandwich shop's floor space. They also decided to hire three high school students out of school for the summer. These students had no restaurant shop experience so John organized them behind a counter. John planned for customers to enter the shop and then proceed to a counter to place their orders with his wife. Mary would handle the cash register and pour drinks. For the students, John wrote specific instructions how to make three standard sandwiches customers could order. The first student would set out and slice the type of bread requested applying the required dressing. The second student would add the required meat and tomato or other required ingredients. The last student would place the sandwich in a basket with chips and a cookie and hand it to the customer. John planned to hang around and make any special orders himself. John also continued to direct Joe in his normal assigned duties. This arrangement worked very well and the business was very successful. John did not look forward to the end of the summer when the three students would leave to return to school.

Near summer's end, John and Mary decided to hire three experienced sandwich makers who had mentioned they would not mind working at John's Sandwich Shop. John had discussed their qualifications with them at some length. These persons had quite a bit of restaurant experience which would allow John and Mary to change the organization. The three professionals were able to handle operations behind the counter almost entirely without instructions. They also were able to expand the types of sandwiches being offered. Any of the three could make nearly any specialty sandwich imaginable. They could each handle multiple complicated orders at the same time. For instance, each person could handle making a meatball sandwich (light on the sauce), which required microwaving while simultaneously shredding lettuce to place on a cold ham sandwich. The resulting success behind the counter allowed John to increase the amount of time he could visit with his friends at the tables.

As the years went by, John and Mary opened another, identical shop in neighboring St. Augustine. John and Mary later both stepped back from day-to-day operations, allowing their son to run the original store and their daughter to run the new store. Both stores remained very prosperous. John and Mary still maintained a corporate ownership of the business, but their management style was to have a family business meeting once each year after a special dinner gathering. At this meeting, John would look at his children carefully and ask two questions very seriously. The first question would always be: "What is the net profit after taxes for each store?" The second question would always be: "Does a meatball sandwich in the Jacksonville Beach shop taste exactly the same as a meatball sandwich in the St. Augustine shop?" With these two questions, John and Mary managed the multiple divisions of John's Sandwich Shop.

The restaurant story pointedly illustrates the different basic organization structures that exist, each doing best with a particular type of primary coordination method. An organization is a group of persons with a common objective, such as the maintenance of an industrial plant. Where different persons work together, they must coordinate their work. Coordination methods and practices help direct the efforts of the different persons. Planning itself is a coordination means. Planning coordinates many of the specialized areas of maintenance. Most organizations typically utilize many different coordination methods at the same time, but they usually emphasize a single primary or dominant coordination method. *Emphasizing the right primary coordination method with the right type of organization makes an organization stable and effective.* Using the wrong kind of primary coordination method with a particular organization structure will cause unnecessary problems and inefficiency.

Dr. Paulson's story pictures the basic organization forms with their preferred coordination methods described by Mintzberg (1983). The account first shows John Smith as an individual doing everything by himself. In the second situation, John and Mary organize and function as an "adhocracy." The *adhocracy* structure of organization consists of different persons brought informally together. The adhocracy coordinates its activities through frequent meetings and exchanges of information ("mutual adjustment") much as do Mary and John with their conversations as they go about doing their jobs. The next organization where John supervises Joe represents a *simple structure*. John provides the coordination needed with his direct supervision of Joe. Next, the organization of the summer students represents a *machine bureaucracy*. A machine bureaucracy achieves efficiency by coordinating with explicit rules and procedures. John wrote rules for the students to follow in their assignments. An assembly line typifies this organizational structure. Of course, if the business environment becomes more complex and varied or undergoes rapid change, rules themselves might become too complicated or subject to constant change, making this type of organization unsuitable. As Mary and John transform the store to allow significant independent judgment on the part of the sandwich makers, coordination by a set of rules becomes impractical. The concentration on obtaining skilled employees becomes their preferred coordination method. This organizational structure is known as a *professional bureaucracy*. This structure must be coordinated with attention to staffing, that is, hiring and training. When one thinks of a medical physician, one realizes that there could not be a sufficient set of rules to handle the doctor's behavior throughout each day. The doctor constantly sees different situations calling for independent judgment and skilled action. The expertise of the doctor is much more important than the standard handbook of medical procedures the doctor sometimes consults. Does the doctor have the skill not only to select, but to execute the correct procedure? A professional bureaucracy coordinates itself through the procurement of skills. Finally, Mary and John oversee the company as a *divisionalized form* of organizational structure. Mary and John see two divisions as entities to coordinate although there may be a different form or forms of organizational structure within each shop. From their level, Mary and John best coordinate the effort of either shop with indicators. The use of indicators comprises the preferred method of coordinating this type of structure. Obviously, Mary and John could not manage each shop with direct supervision or constant meetings and communication. Other methods of coordinating are also not appropriate for their level of management in the organization. Variants exist, but these same basic organizational structures are found in organizations throughout the world. The restaurant story shows which particular method of coordination is most appropriate for each of the different basic structures.

Selection and Training of Planners

With regard to maintenance planning, how do these lessons apply? The identification of preferred coordination methods has direct application to the planning department organization. To begin with, maintenance work orders come in all sizes and shapes, from straightforward to incredibly complex and from ordinary to unusual. Because of the extreme diversity of jobs, planners cannot be given simple instructions for how to plan them. For example, there is not a one-size-fits-all approach to scoping jobs or determining what job details are critical for a job plan. Planners may be directed to scope a job, yet the identification of the correct work scope is entirely a creation of the planner's skill. A planner may be directed to research a minifile, but the planner must recognize what made the difference in the last two filed jobs. Similarly, only the qualified planner can adequately anticipate likely job problems or spare part requirements on equipment that has not yet been dissembled. *The planners must function as a professional bureaucracy*, being allowed to exercise discretion and personal judgment in the effective planning organization. Thus, *the primary coordination method to manage the planning group must be an emphasis on staffing.* The purpose of Planning Principle 4 (skill of the planners) becomes more clear as one realizes that having qualified planners controls the successful planning organization.

Does not every job require having a qualified person? Perhaps, but this concept is not the sense of the professional bureaucracy. A person to operate under direct supervision must be willing to submit to direct supervision. A person willing to work on an assembly line must be willing and able to follow specific instructions. However, the skills required within the professional bureaucracy implies neither of these qualifications.

Without the obtaining of skilled planners as the primary coordination method, none of the other coordination methods matter. First, planners do not need to share information continuously with each other to plan jobs. Neither is there time for each planner to communicate constantly with other experienced personnel regarding job requirements. Second, no planning supervisor could adequately directly supervise the activities of planners planning hundreds of diverse work orders each week. Third, as previously discussed, no set of rules or guidelines could possibly take the place of the skilled planner on the majority of maintenance plans. This explains one of the cautions with the template approach to job planning. When all is said and done, the planner's skills must come into play to use the templates and to provide enough job specific expertise. Finally, indicators may show whether the planning group is making a difference in the work force's productivity, but indicators cannot coordinate the activity of the planners. Indicators only show whether the planners chosen have the skills necessary to make the difference.

After the selection of planners, the emphasis of supervision should be on training and other support. A school system provides an excellent model. The principal does not directly supervise any of the classrooms. The principal is not even present in the classrooms at all times. Instead, the principal performs a primary duty by procuring qualified teachers. Then the principal sustains or enhances those qualifications by making training opportunities available to teachers. For example, these might include seminars about new techniques or concepts of learning. Next, the principal supports the teachers by supplying everyday needs to allow the teachers to execute their teaching skills rather than spend time gathering supplies. For instance, teachers should not have to worry about obtaining copy supplies, having adequate student desks, or providing proper air conditioning of the classrooms. The principal organizes the front office to support the teachers. Rather than have the principal direct the teachers, the teachers should almost direct the principal and front office group in their support needs. In this type of role, the principal coordinates and controls the smooth functioning of the school organization, a professional bureaucracy. If there are only a few planners, they may adequately report to the maintenance manager or superintendent who is also over the crew supervisors. The manager or superintendent would ensure their proper selection and provide ongoing support. If there are more planners, a planning supervisor or lead planner may be desired. In either case, the ongoing objective would be to provide training and support to the planners. Training should consist of establishing the vision, principles, and techniques of planning and scheduling. Training might also include instruction in the use of computers and a CMMS computer system. Support might include copy machines, paper supplies, computer resources as needed, filing supplies, or other physical necessities that would allow the planners to focus on planning jobs. The supervision over the planners would ensure adequate office support exists. Finally, consider that the objective of support is to keep qualified planners in place adequately performing their planning duties. *Planners' wages should be competitive to ensure that qualified persons have the desire to accept and stay in the planner positions.*

One sees that Planning Principle 5 (skill of the crafts) indicates that crew technicians also function somewhat within a professional bureaucracy. In addition, Scheduling Principle 5 (crew leader handles current day's work) lends this same structure to crew supervisors. This is because of the diversity of work orders and the increasing technological sophistication needed to maintain modern machinery. This is why one commonly hears admonitions to train and upgrade one's workforce. However, do not job plans provide work rules, which is the preferred coordination method of the unchanging assembly line? Not at all. The work plans provide support both to the crew supervisors and to the craft technicians. The job plans provide information on job scope, crafts, and hours to allow the supervisor to assign and schedule the correct skills. The job plans provide filing support to avoid previous delays and a head start on other job information for the technicians. A heart problem would be assigned to a cardiologist and a foot problem to a podiatrist in a hospital. A cardiologist and podiatrist still refer to "tried and true" procedures as references. The office staff and nurses would provide previous medical histories to help each doctor treat each patient. Similarly, planners primarily perform triage and file services for crews.

They do not normally dictate mandatory procedures. Their job plans provide support and helpful procedures.

There is simply not enough repetition of identical jobs to establish the planners or the technicians into assembly lines coordinated with absolute work rules. There is enough repetition of jobs to allow a planning function to support technicians in learning from past jobs and creating evolving procedures.

Within such a framework, the question of "How do I control planning?" implies a fundamental misunderstanding of the situation. Once the planners have been hired, the majority of the control action has been completed.

Indicators

A wider perspective makes indicators or metrics also important. The restaurant story suggests the plant manager might oversee the general operation of the maintenance and operations departments as a "divisionalized form" of organization structure. Without complete attention to the inner workings of each department, the plant manager might place heavy emphasis on indicators to control these departments. The managers of these departments should be responsible for indicating their efforts through indicators. Even within each group, opportunities exist for indicators to help coordinate efforts, though perhaps not as a primary means of coordination.

Persons can relate to overall plant availability or overall plant capacity fairly well. Figure 11.1 shows a sample overall availability metric. However, these indicators may be so global that they do not provide much assistance in determining what to do to improve their score. What factors have specifically contributed to maintaining a high availability or capacity? What factors have specifically reduced the overall availability or

Figure 11.1   The simplest measure of overall maintenance effectiveness.

capacity of the plant? Other indicators should support these global indicators. Subindicators to availability or capacity might provide better information for coordinating or managing resources. The following sections present common indicators of maintenance performance.

Planned coverage

Figure 11.2 illustrates planned coverage, a standard measure for a planning and scheduling system. Management desires that technicians spend more hours on planned jobs than unplanned jobs. This indicator is based on the actual hours technicians spend on jobs. The measure represents the percentage of these hours that are on planned work orders. The actual hours are measured regardless of the originally estimated hours of the planners. The metric utilizes actual labor hours as the unit of measure rather than quantity of work orders because the size of work orders can vary considerably. For instance, typical project work might normally be larger work orders than breakdown work orders. In addition, preventive maintenance (PM) work orders might normally be smaller than breakdown work orders. Management desires for maintenance forces to spend adequate time on the appropriate type of work. Therefore, the metric should utilize a time-based unit. On the other hand, management could cautiously use work order quantities if actual time values are not initially available.

Proactive versus reactive

This metric measures the reactive nature of the plant maintenance work. Management desires reactive work to lessen in proportion to proactive work (Fig. 11.3). This indicator is based on the actual hours technicians spend on jobs. The actual hours are measured regardless of the originally estimated hours of the planners.

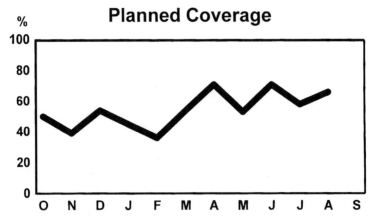

Figure 11.2   Management wants more labor hours spent on planned jobs.

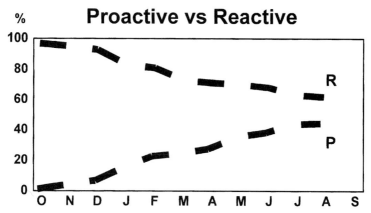

Figure 11.3 Management wants to spend more hours on proactive work than reactive work.

Reactive work hours

Figure 11.4 shows the absolute amount of reactive maintenance work. Management desires not only to perform more proactive than reactive work, it desires for the absolute amount of reactive work to decrease. The score of this indicator may be very erratic on a monthly basis and might be better measured on a yearly basis. The amount of reactive work may also initially increase as crews increase their productivity and perform more work of all types. This indicator is based on the actual hours technicians spend on jobs. The actual hours are measured regardless of the originally estimated hours of the planners.

Work type

Management needs information regarding the different types of maintenance work performed. Specific areas of interest are proportions of

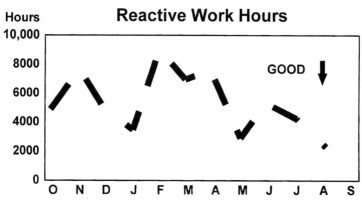

Figure 11.4 Management wants the overall amount of reactive work to decrease.

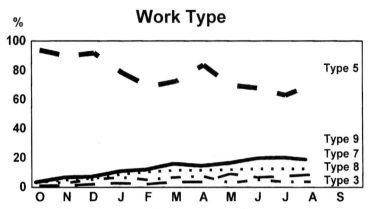

Figure 11.5   Another indicator of the proportions of reactive versus proactive work.

preventive maintenance, predictive maintenance (PdM), project work, and corrective maintenance versus actual failure and breakdown maintenance. This indicator is based on the actual hours technicians spend on jobs. The actual hours are measured regardless of the originally estimated hours of the planners. (See Fig. 11.5.)

Schedule forecast

Figure 11.6 shows an example of an indicator tracking forecasted hours. Note how the chart indicates carryover hours. A large proportion of these

Figure 11.6   Maintenance might track forecast hours to help coordinate the scheduling process.

hours could indicate a scheduling problem. This indicator uses hours taken directly off the form for Crew Work Hours Availability Forecast. The sample hours shown are for B Crew's forecast developed in Chap. 6.

Schedule compliance

As discussed in Chap. 3, weekly schedule compliance provides the ultimate measure of proactivity. Some plants prefer the term *schedule success* to clarify the objective to measure control over the equipment rather than over the supervisors. Figure 11.7 shows a sample chart with data illustrating B Crew's performance. This company measures PM compliance as well. Figure 11.8 illustrates a helpful worksheet to calculate the schedule compliance score. Figure 11.9 illustrates the use of the form with B Crew numbers. The scoring of compliance gives the crew credit for all jobs that started during the week regardless of whether they finished. Chapter 6 explains that this gives the crew every possible benefit of any doubt of compliance.

Howard Mathews, PE and Pamela Pifko, CMA of SaskPower developed a helpful report called the "Red Green Report." This type of report lists work orders from the previous week in three sections. The first section lists all the work orders that maintenance completed that were on the weekly schedule. The second section lists all the work orders that maintenance did not complete that were on the weekly schedule. The third section lists all the work orders that maintenance completed that were not on the weekly schedule. Each work order listed identifies its description, priority, estimated hours (if any), and actual hours. With these sections, management can easily see what scheduled work orders

Figure 11.7   Schedule compliance to measure schedule success.

ADVANCE SCHEDULE WORKSHEET #2

For week of: _____ **to** _____
For crew: _____   **By:**_____   **Date:**_____ .

TOTAL SCHEDULE
A. Total Hours Scheduled   _____
B. Any Available Hrs Left Unscheduled (_____)
    Why? (No backlog, etc.) _____

_____

C. Total Hours Returned   _____
    Any Hours That Were Unclearable (_____)
D. Sched Hours Worked (D = A - C)   _____
E. % Schedule Met (E = D/A x 100)   _____%

PREVENTIVE MAINTENANCE
F. PM Hours Scheduled   _____
G. PM Hours Returned   _____
    PM Hours That Were Unclearable (_____)
H. PM Sched Hours Worked (H = F - G) _____
I. % PM Schedule Met (I = H/F x 100)   _____%

Figure 11.8 Sample of a helpful form to calculate schedule compliance.

maintenance did not complete and what other work orders took their place. Management is interested to see that only higher priority, unscheduled work orders replaced lower priority scheduled work orders. Even then, management can question why the unscheduled work orders could not be planned and scheduled for the following week. The name "Red Green Report" derives from the practice many plants have of coloring the different sections to stand out. The completed scheduled work orders would be colored green. The unscheduled schedule breakers

---

ADVANCE SCHEDULE WORKSHEET #2

For week of: _5/11/99_ to _5/14/99_
For crew: _B Crew_   By:_C. Rodgers_   Date: _5/ 14/ 99_

TOTAL SCHEDULE
A. Total Hours Scheduled   _410_
B. Any Available Hrs Left Unscheduled (_ 45)
   Why? (No backlog, etc.) _Backlog_
   _____ _ran out_
C. Total Hours Returned   _84_
   Any Hours That Were Unclearable (_ 0)
D. Sched Hours Worked (D = A - C)   _326_
E. % Schedule Met (E = D/A x 100)   _78%_

PREVENTIVE MAINTENANCE
F. PM Hours Scheduled   _54_
G. PM Hours Returned   _0_
   PM Hours That Were Unclearable   (_____)
H. PM Sched Hours Worked (H = F - G)   _54_
I. % PM Schedule Met (I = H/F x 100)   _100%_

Figure 11.9   Sample of schedule compliance calculations.

would be colored red. This is a remarkable report that easily helps management understand the progress on the schedule.

Wrench time

Figure 11.10 shows a sample wrench time metric. This indicator utilized within maintenance measures the percentage of time technicians actually spend on the job. This would be time where otherwise available technicians are not involved in delays such as procuring parts, tools, or

Figure 11.10   Sample indicator illustrating wrench time performance.

instructions. Industry commonly refers to this time as *wrench time*. In-house analysts or consultants properly measure wrench time with a work sampling methodology. What is more significant than the time on the job is the analysis of the time and circumstances that delay techni-cians from being on the job. Appendices G and H provide sample work sampling studies.

One limitation of wrench time analysis is that it *makes no presump-tion of how productive a technician is while on the job.* On the other hand, one would presume that the on job productivity should stay the same so *increasing the amount of time on the job should increase the overall amount of work produced.* Planning Principle 6 in Chap. 2 explains that increasing the amount of time technicians are on the job is the purpose of planning. The measure of *wrench time indicates the effectiveness of the planning and scheduling process rather than the efforts of the tech-nicians themselves.*

Minifiles made

The creation of the minifiles described by Planning Principle 3 is of great importance. A planning supervisor may want to count the number of minifiles each month in the early months of a new planning organi-zation. See Fig. 11.11.

Backlog work orders

Backlog of work orders is a very ominous indicator. Experience shows that many management efforts to reduce the size of a backlog result in

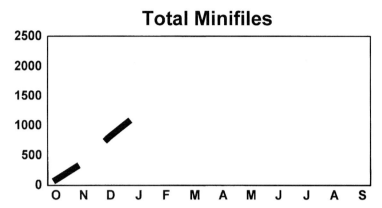

Figure 11.11   Ensuring planners understand the importance of the minifiles when starting a planning group.

a reduced amount of new work orders written rather than an increased number of work orders completed. The backlog is thus reduced by no longer identifying the work to which the plant should attend. *The generation of new proactive work orders especially suffers.* Other games played include writing larger work orders. Instead of writing a separate work order to take care of each fuel oil pump, technicians might write a single work order to take care of all three pumps. The backlog is thus further reduced by hindering the opportunity to keep good equipment records for each pump. If management intends to reduce equipment problems, it should track backlog by specific work type. The plant desires to reduce its reactive backlog, but increase its proactive backlog. The plant might define reactive work orders as failure or breakdown work orders plus other work orders as an urgent priority. The plant might define proactive work orders as project work, PM, PdM, or corrective maintenance except for ones that have become urgent. Through increasing the detection of proactive opportunities the plant can reduce its failures and reactive situations that hurt reliability. Management should *vocalize this vision with caution.* The simple command to reduce the backlog, but only the reactive backlog, can become confusing and counterproductive.

Work orders completed

Simply looking at backlogged work orders can be misleading. On the other hand, measuring the number of *work orders completed each month provides an excellent check and balance when used with the backlog number.* Management is interested in the maintenance group completing more work orders each month as one indication of productivity improvement. This indicator by itself might encourage the workforce to write smaller work orders. For instance, instead of writing one work order to repair a pump, the indicator would tempt personnel to write three separate work orders for disassemble, repair, and reassemble. Because management pressure to reduce work order backlog tempts personnel to write fewer work orders, these two indicators help balance each other when used together.

Backlog work hours

Plants making sudden improvement gains in productivity often quickly run out of backlogged work. Plants in highly reactive work environments should then take advantage of the opportunity to create proactive work orders or generate reactive work orders that would have been ignored in the past. The plant's objective with a stable work force should be to maintain at least 2 weeks of craft backlog. Such identification of work promotes the smooth operation of a productive maintenance department with planning and scheduling. Unfortunately, simple quantities of work orders do not indicate labor requirements. Prompt planning should establish a backlog in terms of hours. Dividing the normal paid hours of each craft into the backlog hours produces a number of backlog weeks. Because the paid hours are normally higher than the available hours for each craft, a goal of at least 2 weeks provides ample work. Management may desire to be aware of how many weeks of backlog are available for each craft. Further analysis by crew or specific skill level is unnecessary.

Summary

The management of the planners themselves is best conducted as a professional bureaucracy. That is, management emphasizes selecting personnel and training them. Management does not emphasize direct supervision, procedures, indicators, or frequent meetings for coordination. A great deal of importance rests on the qualifications of each individual planner. Organizations should select planners with an aptitude for planning. Organizations should train them in the principles and techniques of planning. The organization may obtain qualified planners either through hiring or developing persons with the necessary potential for success. Appendix M, Setting up a Planning Group, discusses how to accomplish the selection and training of maintenance planners.

While selection of planners handles the majority of planning control, management of overall maintenance does make use of several common indicators. The chief of these is overall availability. Other indicators include ones for measuring the proportion of work hours that are planned and the proportions of different types of proactive work versus reactive work. Management should use simple indicators of backlog with caution because the plant must generate a backlog to take care of maintenance. Schedule compliance helps determine if the maintenance force is controlling the equipment or if the equipment is controlling the maintenance force. Management measures crew forecasts and carryover work to help understand the functioning of the schedule process. Management uses these indicators to coordinate the efforts of the divisions or groups together.

12

Conclusion: Start Planning

What is maintenance planning?

Maintenance planning as envisioned by the *Maintenance Planning and Scheduling Handbook* IS NOT preventive maintenance (PM).

Maintenance planning IS NOT planning how to establish and organize a maintenance department.

Maintenance planning IS NOT using a computer.

Maintenance planning IS NOT providing a detailed procedure every time describing how to perform every maintenance task.

Maintenance planning IS NOT simply identifying spare parts and special tools before a job starts.

Maintenance planning IS providing file information to technicians to allow them to learn from past jobs and avoid delays. Planning also helps ensure the availability of anticipated spare parts and special tools.

Maintenance planning IS providing crew supervisors with job scopes plus craft and work hour estimates to allow them better to assign daily work.

Maintenance planning IS advance scheduling to allow managers to allocate work for crews based on forecasted labor availability.

Why do companies need maintenance planning? They need maintenance planning because it helps increase the amount of time technicians spend on direct work, actual work without delays. Maintenance planning reduces the time technicians spend in gathering parts, in finding tools, in receiving instructions, or in many other delay situations. In industry today, delays commonly cause maintenance crews to spend only 25 to 35% of their time on job sites making job progress. These delays do not even include time lost to vacation, training, or other type of administrative absences. Maintenance planning helps boost the direct work or wrench time of technicians to as much as 55%. A good company at 35% with 30 maintenance technicians would enjoy the effect of having 47 technicians if it had 55% direct work time. The company would add 17 extra technicians without cost to its maintenance efforts. The company with 90 maintenance technicians would see the effort of 141 technicians, a 51-person improvement. And lest one should think maintenance planning helps only the large corporation, the company with 10 maintenance technicians would see the effort of 15 technicians, a 5-person improvement.

Why have companies not taken advantage of such an opportunity? Companies have not exploited maintenance planning for several reasons. The biggest reason has to do with a belief that direct work time could not possibly be as low as 35%. Yet, study after study reveals that companies have a typical direct work time of 35% *at best* without planning and scheduling. Other reasons primarily include fundamental misunderstandings of exactly how a planning and scheduling system properly works. This explains the great frustration of many companies that have unsuccessfully attempted improvements through maintenance planning. The inner workings of a proper system have seen limited study because of its position in the organization. The plant manager often sees maintenance planning as too low in the organization to give it direct attention. The plant engineer often sees planning as "low tech" and so not of much interest. The maintenance manager often sees planning as requiring too much of a change to the existing process of maintenance and not clearly worth the effort. Resulting efforts at planning see implementation without clear guidelines, practices, or even vision of its purpose. These companies do planning only because they are "supposed" to do planning.

Planning does not just happen. Experience has shown that planning is a system with many subtleties requiring attention. The preceding chapters have established 12 basic principles that resolve the issues involved with a maintenance planning system. Specifics of actual practice help explain the principles. Companies might easily implement each principle to establish a maintenance planning system that allows attainment of dramatic maintenance improvement. The first principle requires keeping planners separate from the supervision of the individual crews. Separation best allows planners to concentrate on planning future work. The second principle is to avoid delays rather than to merely help deal with delays of a job that is already in progress. This principle takes advantage of the repetitious nature of most maintenance work and moves jobs up a learning curve. The third principle recognizes that planners can only practically retrieve prior job information for learning if linked to specific equipment. Specific equipment minifiles establish this link for paper documents. It is primarily within this linking principle that the computerized maintenance management system (CMMS) has its place. Besides helping with inventory tracking, the CMMS allows management to leverage key information regarding work orders and equipment. On the other hand, a CMMS can significantly automate and otherwise facilitate the overall efforts of maintenance planning. The fourth principle provides for easy, practical estimation of job requirements through the expertise of an experienced technician as planner. The fifth principle has this planner utilize the skills of the field technicians and avoid giving more procedural information than necessary on initial job plans. The last planning principle, wrench time, embodies the purpose of planning to reduce delays to help technicians spend more time on jobs. Planning also involves scheduling because reducing delays during individual jobs allows supervisors to assign more jobs. Six more principles address scheduling. The first principle specifically obligates each job plan to estimate time requirements and lowest craft skill levels. To reduce interruption delays, the second principle advocates the practice of not interrupting jobs already in progress through proper prioritization of work. The third principle commits crew leaders to forecast labor hours for craft skills for the next week. The fourth principle combines the forecasted labor hours with estimated job hours, sometimes utilizing persons beneath their maximum skill levels for the good of the plant. Although a scheduler in the planning department allocates the week's goal of work, crew leaders best handle the daily work assignments as established by the fifth principle. The sixth and final principle establishes the importance of schedule compliance. Not an end unto itself, this indicator measures the success of the crew taking control of the equipment. One overall consideration makes these principles practical, making the difference in their successful application. Planning must not constrain crews from immediately beginning work on urgent jobs. On reactive work planning can abbreviate its efforts while still providing helpful information. Together, these principles make the difference and make planning work.

In learning about maintenance planning, one also sees that there exists no magic answer by itself. Rather than being the latest management fad, one sees that planning provides a benefit by helping coordinate the rest of the maintenance group's resources. This brings up a valid concern. If maintenance planning assists in coordinating the rest of maintenance, how does one coordinate planning itself? The key to the coordination of activity within the planning group lies principally with the proper selection of the planners. Many times companies want to organize a new group by quickly hiring personnel and then drafting numerous procedures to govern their activities. Direct supervision or indicators then track adherence to these activities. This approach will not work with maintenance planning. Management must first, above all else, carefully select qualified planners. After selecting planners, management must then imbue them with the vision and general principles of proper planning. Then management must support them in their knowing how best to conduct maintenance planning. This book explains what maintenance planning is all about and why it works. Seek the advice in App. M on how to establish a new planning organization or transform an existing one.

WIIFM means "What's In It For Me?"

For the technicians, they have a file clerk that faithfully pledges to help them avoid the painful lessons and delays of the past. They have a head start from an experienced technician that anticipates the problems of the job about to start. The file clerk helps them codify these lessons into helpful procedures over time.

For the supervisors, they have the means of knowing how many jobs they can assign to which skilled technicians. They have a subset of appropriate work orders called from a much larger plant blocklog. They also have a means to leverage the knowledge of an experienced technician across all the jobs of the crew. Ironically, supervisor fear of losing "control" presents one of the chief impediments to planning and scheduling. Yet, the planning and scheduling system gives supervisors more control. They receive job information and a backlog service allowing them to assign jobs better and freeing them to spend more time with their crews.

For the managers, they have the means of improving productivity through knowing how much work that crews should execute each week and how to allocate it. Managers have the tool to assist 30 technicians achieve the effort of 47 technicians.

For the companies, they have a means of practically coordinating the expensive resources they have acquired for maintenance to improve and maintain superior plant reliability.

In conclusion, start planning.

Epilogue: An Alternative Day in the Life—May 10, 2010

This section has four short narratives typical of maintenance with proper planning help. These accounts are revised versions of the misadventures set forth in the Prologue, just before the Preface of the book. The Prologue recounts these situations as they might occur without the assistance of a planning group.

Bill, Mechanic at Delta Ray, Inc.

Bill reported to work on time and went straight to the crew break area. There the supervisor gave out the assignments for the day. Bill received the four jobs he was expecting because he had seen the daily schedule sheet posted the previous day. He was there for a 10-hour shift and the job estimates totaled 10 hours. There were three jobs for 3 hours each and a simple 1-hour job. The big jobs were fixing a leaking valve on the northeast corner of the mezzanine floor, replacing a gasket on a leaking flange for the demineralizer, and regreasing the coupling and laser aligning a boiler fill pump. The last job was to lubricate and change the packing on a valve that was becoming difficult to operate.

Bill had worn his old boots since he knew about the valve jobs. The first thing Bill did was review the job plans written on the work orders. All four jobs were staged in the tool room. Bill went to the tool room and presented the work order numbers to the attendant. The attendant shortly brought out a box with the parts for all four jobs. Copies of the work orders he had in his hand were attached to groupings of the parts. There was a 4-in valve body with two gaskets and eight bolts with matching nuts. There was a sheet of Teflon gasket material and 12 bolts with nuts. There was a package of grease for the pump. And there was an applicator for grease to be used on the valve. Bill wanted to go ahead and check out the laser alignment kit. Since there were only two of them, the tool room attendant told him that his supervisor had reserved one for him for the afternoon.

Normally, since Bill knew what size the bolts were for the mezzanine valve and de-mineralizer flange, he would just take a couple of wrenches with him. But since he had the pump job which was near the other jobs, he might need an array of other tools, so he decided to take his whole tool box along. In addition, the planner had not noted what size packing was needed, so any of the different size pieces of packing he normally carried in his tool box might be needed.

The first job was cleared as expected and Bill started unbolting the old valve. Bill remembered the old days when the work orders were not planned to specify whether a mechanic would be needed for a bolted valve or a welder might be needed for a socket welded valve. Bill noted that the work plan stated that he should replace the valve because the valve had failed after being repaired twice this year so far. "So that's why I'm not just repairing it," Bill thought. Bill replaced the valve, gaskets, and bolts and was soon cleaning up. Bill radioed his supervisor to turn in the work permit so the operations group could begin unclearing the valve instead of waiting until the end of the day. Bill was a little ahead of schedule and decided to go ahead and take a short break. This would be a good time to call the dentist from the break room.

After break at the demineralizer, the area was also cleared and Bill had the right work permit. The deficiency tag was hung near a pipe flange. The planner had noted that this was a water line and not an acid line, so Bill took the flanged connection apart at the demineralizer. Just as the planner had anticipated, in order to obtain access to the leaking flange Bill had to disassemble two other connections as well. With the gasket material in hand, Bill went to his work bench and cut out three gaskets using one of the old gaskets as a template. The planner had requested that Bill draw a copy of the gasket template on the back of the work order for keeping in the file for next time. Bill realized that with these gaskets, he could finish up this job in no time. That would allow him plenty of time for the other two jobs. Bill gathered up his gaskets and started toward the job. On the way he passed Gino cutting out some gaskets at his work bench. After stopping to compare notes for a few minutes, they both noticed it was almost lunch time, so they decided just to stay in the shop and talk.

After lunch Bill started reassembling the flanges. Most of the bolts looked in good shape, but bolts were inexpensive and the policy was to not take a chance on worn out bolts. Bill replaced all the bolts and nuts using the supplied items from the tool room. Soon Bill finished the job and he wiped down and cleaned up the area. He then radioed to his supervisor so this work permit could be signed off and taken to the control room. By then, there was about an hour left before break time so Bill took his stuff over to the boiler fill pump.

The pump was cleared and Bill sat down for a moment to look over the pages of the pump manual the planner had attached to the job plan. The coupling came apart easily and Bill wiped out all of the old grease. He then applied new grease and fastened it back together. Then Bill headed for the tool room to pick up the laser kit before break time. After break, he set up the laser device and aligned the pump. This time, however, the job ran into problems. No matter what he tried, the pump would not go into alignment. Bill finally called his supervisor, who came over to help. After a while, Bill told his supervisor that it did not look like he was going to finish his last job, repacking the valve. The supervisor said that was okay and it could be done tomorrow. The supervisor ended up by calling a mechanic with more expertise to look at the pump problem and the three of them finally got the pump aligned. Bill really enjoyed working with pumps and was glad they had enough time to do it right. The supervisor's main concern was writing down on the work order the specific alignment problem and resolution for the history file.

It was customary that the crew could use the last 20 or 30 minutes of the day filling out time sheets and showering. Bill had about 40 minutes left after the pump job and really did not want to start the repacking since sometimes a simple repacking job would run into trouble. But he did take his tool box, the lubricant, and the packing to the valve site so he could knock it out the next day. Then he filled out a time sheet and headed to his car at the end of the day.

On the way to his car, Bill felt good about his performance against today's schedule.

Sue, Supervisor at Zebra, Inc.

Sue considered herself a capable supervisor. She knew in order to keep the operations group satisfied, the maintenance crew had to keep the equipment from failing. She worked the crew steady and kept on top of all of the maintenance work that would keep problems from developing. Whenever a priority-one work order came in, she assigned it when someone was finished with what he or she was doing. She never interrupted jobs already in progress unless it was a priority-zero emergency. The crew knew the importance she placed on completing high priority work and was always willing to work overtime when required. Normally, however, the crew worked steadily on a batch of work orders that the planning group would assign for the week and did not require much overtime. The planning group would take her estimate of how many labor hours she would have for the next week and give her that much work to assign. Sue recognized that plant management supported preventive maintenance because the weekly backlog contained a sufficient portion of it. She was sure that this mix of work was in the best interest of the plant in keeping production at high capacity.

Her normal method of job assignment was to assign to each technician a full day's worth of work. She knew how long each job should take and what type craftperson she should assign because the planning group planned each job. Sometimes this required the art of deciding who would receive which jobs. She simply could not hold back the critical jobs for certain persons or she would not be able to complete all the work. She would simply have to provide coaching for persons whose skills were not yet up to par. She selected the day's work one day ahead of schedule from the week's backlog from planning. Thankfully, she did not have to dig through the entire plant backlog each time she wanted to assign a job. She knew the crew worked productively because they generally completed much of the work assigned each week and made few extra trips to the tool room or storeroom.

Lately it seemed that fewer higher priority work orders were being written with urgent problems. This enabled the crew to work without interruption to maintain the plant's capacity. She used to feel that sometimes the operations crews would exaggerate the priority of minor jobs just to make sure they were done. However, from looking at the recent work orders, there seemed to be a team atmosphere between the maintenance and operations groups with both sides writing low priority work orders to head off problems. She knew the crew was enjoying receiving planned work orders where they had a head start on what to do and could suggest future improvements. Sue's supervisory approach was to assign the work from the weekly backlog and monitor how much work the crew and individuals completed. She was usually in the field seeing where she could provide help to do work in a quality fashion. Since there was some schedule pressure, she sometimes had to remind technicians that doing the job right was more important than meeting the precise time estimate. She just needed feedback to know why they needed changes. Things were fairly calm. The crew kept moving along with the visible goal of daily work. Break times never seemed to be a problem, and neither were starting or quitting times. Things seemed to be fine.

Juan, Welder at Alpha X, Inc.

Juan received three planned jobs for the day. Two were 4-hour jobs for valves that were leaking through. Eli in the predictive maintenance group had used thermography to find the problems. The third was a 2-hour job to pull up and tighten the packing on a number of boiler valves to prevent problems. It looked like a full day of work.

Juan and his helper got to the site of the first job and Juan looked at the work order again. The west economizer drain root valve was leaking through. The work plan called for replacing the valve with one he had checked out of the tool room per the job plan tool list. The plan gave the following steps.

Cut the seat using kit instructions.

Cut new packing.

Cut out and replace valve.

Record title and page # of procedure from company welding manual.

Total time: 4 work hours for certifield welder.

Duration: 6 hours (including heat treatment).

Juan noticed the plan called for recording information to keep in the valve file for the future. Juan also noticed that the planner had identified the cost of the valve at $1200. The planner had also written special instructions for the operations group to drain this particular root valve during the night. The history file showed that a job done last year had been delayed 3 hours waiting for the leg to drain. Juan remembered that drawn out job.

Juan began cutting out the valve seat. It made sense that planning would take care of all the little details so Juan could go to work.

Jack, Planner at Johnson Industries, Inc.

Jack came in ready to go. As a planner for 20 technicians, he knew that each day he needed to plan about 150 hours worth of work orders. Standard preventive maintenance work orders that needed no planning would add about 50 hours. That would keep 20 technicians busy for a 10-hour shift. He could not afford to become bogged down.

Reviewing the work requests from the previous day, Jack got to work. He decided first to check the minifiles and then make a field inspection for eight of the most pressing work orders. He hoped to have them planned before lunch and start on another group. Jack found and took the minifiles to his desk for six of the jobs. The files had plans and part numbers used previously. Four had small equipment manuals. Jack hoped that knowledge of the previous work would help him scope the current jobs. He gathered the eight work orders on his clipboard and headed for the door. At the door he met George and Phil. They had just started a pump job and wanted the pump manual. Jack pointed them toward the technical file, OEM section. He cautioned them to copy what they needed if they found a manual. They should leave it on the table when finished so he could put it on the right shelf later. Then he went out the door.

Jack had no problem finding and scoping most of the jobs, but one job was hard to find. Jack radioed the operations coordinator assigned to help planning. He agreed to come to where Jack was to help. While he waited a moment, Jack hoped the other jobs he had planned last week were going smoothly. He knew the technicians or supervisors would take care of any problems and give him feedback on the completed work orders to help future jobs.

Jack and the operations coordinator found the elusive job site. Jack made a mental scope of the job and headed for the planner office. Once there, he noticed Jim searching the files for bearing size information on an unplanned job. Eventually, Jim called the manufacturer who was glad to help. Jack asked Jim to make sure to write down the tolerances on the work order so Jack could file it for future use. Even though Jack was very adept at finding information, there were 20 technicians to support. From the very first days when they began establishing the files, it made better sense for the supervisor and all the technicians to search for information the first time they performed a job. Jack's value added was filing the information once found and then retrieving it to avoid repeating a search. Over time, maintenance technicians usually repeated jobs requiring the same information. The technicians understood if they did not receive certain information on a job, that it was not readily available. Then they would have to hunt for it themselves and count on Jack to file it for the next time. Jack's job existed to support the field technicians and supervisors. He was able to give technicians a head start on jobs from past information. He was able to give supervisors time estimates and craft information to allow better job assignment.

After break, Jack sat down to write work plans for the jobs he had scoped. He wrote the craft, time estimate, and objective of each job along with parts information available from the file. He only had to consult the storeroom catalog for two of them. He called the storeroom to reserve parts for six jobs. He asked the purchaser to procure parts for the other two jobs. Jack had completed plans for 62 hours of technician work. He then chose another batch of eight work orders. Jack planned to check the minifiles for these jobs and complete some of the field inspections before lunch. That should allow him to finish the plans well before break. Then he could plan a few more work orders and still have time for reviewing and filing information from completed work orders before the day ended. Maintenance seemed to be in a cycle where the information from previous job plans and feedback made it easier to plan new work orders more helpfully. It just seemed that everything was fine.

Planning Is Just One Tool; What Are the Other Tools Needed?

This appendix explains that planning is only a "tool" and where it fits into the maintenance picture. Figure A.1 shows that while planning does not solve everything by itself, planning has a special relationship with the other areas of maintenance. Planning brings together, integrates, and even drives many of the other aspects of maintenance. This appendix identifies other necessary maintenance tools and their relationship to planning. Other tools needed include a work order system; equipment data and history; leadership, management, communication, and teamwork; qualified personnel; shops, tool rooms, and tools; storeroom support; process improvements; and maintenance measurement. In addition, the chapter considers essential reliability maintenance consisting of preventive maintenance, predictive maintenance, and project maintenance. The intent of this chapter is not to be a complete handbook on its own for maintenance management. However, planning is deeply integrated into most of the other aspects of maintenance. Therefore, this appendix finds it worthwhile not only to identify these other areas and their relationship to planning, but to illustrate common opportunities for improvement within them as well.

A tool is a device or instrument that helps one accomplish a task more easily. The tool is not an end unto itself. A mechanic uses a tool called a wrench to fasten bolts. Just by purchasing a wrench and putting it into a toolbox, a mechanic has not provided the employer any benefit from the wrench. The employer benefits only when the mechanic uses the wrench to fasten bolts during the alignment of a pump and the pump operates without interruption to produce a product for sale. The tool (the wrench) helps. On the other hand, what if the mechanic really prefers another tool, the pipe bender? Even if the mechanic prefers the pipe bender, it is the wrong tool for fastening bolts and the mechanic cannot use it for the

Planning is
not
"The silver bullet"

Figure A.1   Planning may not solve everything by
itself, but it certainly helps.

fastening part of the alignment procedure. If the wrench remains in the
toolbox during the alignment procedure, then the wrench itself is not helpful. On the other hand, if the mechanic pulls the wrench out of the toolbox, this correct tool still may not be helpful. The mechanic may incorrectly
use the wrench to apply so much excess torque that the bolts fail. Then
the pump soon fails and interrupts production, and the employer does not
benefit. Another aspect of using a tool is that frequently a tool is not useful
by itself. A wrench may not be the only tool needed to align a pump. A laser
device or reverse dial indicator device helps the alignment job. Training
for the mechanic to know a lubricated bolt requires less torque to tighten
than a dry bolt also helps the alignment job. The mechanic must use each
tool correctly for its intended purposes to provide a benefit.

Similarly, planning is just a tool. One that the company can use correctly, use incorrectly, or simply leave in the toolbox and not use at all.
Neither does planning work in a vacuum once the maintenance manager takes it from the toolbox. A maintenance organization does not "do
planning" simply because it is "committed to planning." A company does
planning because it understands the intended purpose of planning. A
company then still must use the planning tool correctly to receive a
benefit. Using the planning tool incorrectly or for the wrong reason does
not contribute a tangible benefit to the maintenance organization.

Planning as a tool greatly depends on the other maintenance
resources. Planning exists as a methodology to use the rest of the maintenance department's resources. Planning coordinates them together.
Consider a particular job to be done. The workforce consists of varied
crafts and skills. Planning connects the right craft and skill level to
this particular job. The storeroom has many parts. Planning connects
the right part to this particular job. The tool room has many special tools,
machines, and equipment. Planning connects the right special tool to
this particular job. The plant backlog has many jobs waiting. Planning
job scopes and time estimates allow scheduling this particular job at the
optimum time according to the other jobs in the priority order and
according to craft time available. Other jobs competing for assignment
may include preventive maintenance to reduce failures, projects to
lessen failures, or jobs recommended by predictive maintenance personnel to avoid failures. Management involvement ensures the use of
the planning process to integrate the other aspects of maintenance

rather than acceptance of the lower productivity and overall lower effectiveness that would result from individual efforts not working together. Planning works to coordinate all the other maintenance resources.

Because planning integrates the other plant resources, if the rest of those resources exist either marginally or not at all, planning is not going to be much help. Planning cannot coordinate things that do not exist. For example, what good is a job plan identifying a certain part if the storeroom does not carry that part and purchasing will not order it? What good is a plan identifying job time estimates if crew supervisors merely assign one job at a time expecting no timely completion? What good is a plan identifying only minimal mechanical skill if the supervisor assigns a highly skilled, certified welder? What good is a plan identifying certain gasket material if no equipment history file exists to tell that this material has not held up well in the past? The various maintenance resources do not help maintenance if they are underutilized. The planning methodology coordinates their utilization. The optimum utilization of the maintenance resources yields superior maintenance performance. However, first of all, these resources must exist.

The planning function can also help identify areas in the resources that do exist, but are weak and need improvement. Planning can help improve the weak areas because the planning function interacts with them so much. For example, the planning personnel can help improve a problematic purchasing process. Planning highlights the delays caused by poor purchasing. Planning actually drives the improvement and utilization of the other resources. This driving is not a by-product of the planning function, it is one of the main values of planning. Coordination is not merely information flow. Planning is not merely "Information Central." Planning is "Control Central."

Why does planning have to drive these resources and processes? Could not a manager exercise the same control? Indeed, a manager could perform these tasks of coordination. Even so, the coordination of the maintenance process to this extent demands a significant involvement in the routine maintenance processes themselves. How many managers would this take? Too many. In addition, the manager's best position is usually outside the routine processes. The manager ensures that the processes are working. The manager typically should not be a part of any process itself. On the other hand, planning deeply involves itself in the maintenance processes. Planning is one of the manager's best tools because of its routine involvement to keep processes working. One of a manager's responsibilities is to provide control. The manager uses the maintenance planning tool being "Control Central" to assist in this area of responsibility.

Some maintenance improvement programs fail because they only address certain areas and are not comprehensive enough to have a bottom line impact. If many of the other aspects of maintenance exist only marginally or not at all, the company must improve them. Simply starting a planning program will not do by itself. Moreover, simply improving one or two other areas by themselves may not help. For example, it does no good to give technicians better hand tools if there are still insufficient spare parts to utilize in the work. Consequently, there is limited improvement, but not great advance. That is why this chapter illustrates the existence of many common improvement opportunities. Understanding that all the areas depend on each other also helps. The essence of maintenance management manages continual improvement in all the maintenance areas. Management can mount an aggressive effort to improve system reliability simply by making a list of areas or "tools" to improve. The tools are not complex; what management needs is the desire to improve. Formally, maintenance could manage a program through regular meetings of supervisors who were responsible for the various aspects of the improvement program. A multifaceted maintenance management improvement program ensures continued maintenance effectiveness.

Several factors may hinder the improvement of the overall maintenance program. First, any normal company not in obvious danger of going out of business has probably developed a culture around the status quo. Resistance to change is natural. Second, many companies have witnessed trial-and-error approaches to maintenance improvement. Management by fad has tried all the latest one-size-fits-all maintenance programs. Third, be aware of the blind squirrel theory. The *blind squirrel theory* is that even a blind squirrel finds a nut sometimes. Applied to maintenance this means just because the maintenance program is doing something right, that does not mean that the overall program is going in the right direction. Fourth, early successes may hamper future improvement. As maintenance starts heading in the right direction, there may be a feeling that the program "has arrived." Employees may resist further changes. Fifth, there is a cycle in maintenance maturity. Mature programs have learned what needs to be done and have implemented programs to accomplish those objectives. However, as dependable equipment reliability becomes routine, there is a tendency toward complacency. The company forgets the reasons behind the current methods of doing maintenance. The company makes ill-advised changes and a slide begins back toward the time when equipment was not so reliable. "Tampering" makes changes to a program without factual basis and could just as well be detrimental as helpful. All of these factors may hinder the improvement of a maintenance organization. Consequently, maintenance management must be an intelligent force behind a comprehensive and continual development of the maintenance program. Even in mature maintenance organizations, management exerts effort to prevent drifting away from success.

In improving an overall maintenance program, realize first that all the areas must at least exist. It is more important to be doing the right thing somewhat inefficiently than to be doing the wrong thing very well. Maintenance organizations with great success do the right things. Yet in spite of the progress and results so far, they discover the majority of the areas within their programs have not yet achieved their potential. If effectiveness (doing the right thing) comes first, efficiency (doing it the right way) can then be improved. There continues an upward trend in availability of production capacity. The continued emphasis on all of the elements of the program furthers the success as key areas become fully implemented and mature. Make sure first that the right key areas exist, then strive for efficiency.

How much does an area need to be improved? The potential of an area may not necessarily be everything possible. Consider obtaining 80% of the improvement possible out of any single area and then moving on to the next area for attention. The last 20% may be beyond the point of diminishing returns and not ever worth the effort to maximize its full potential. Obtaining up to 80% in the next area to be addressed is overall worth more than going beyond 80% in the current area. The time to address any of the remaining 20% may be after most of the areas are near 80%. For example, first start using a work order form; later worry about having the perfect work order form. First worry about having the right parts in the storeroom; later worry about having too many parts in the storeroom.

The basic planning tool has been described along with the general importance of the other tools of maintenance. The following sections describe other necessary maintenance tools and their relationship and relative importance to planning. Other tools needed include a work order system; equipment and history; leadership, management, communication, and teamwork; qualified personnel; shops, tool rooms, and tools; storeroom support; focus on improved work processes; and reliability maintenance. Reliability maintenance consists of preventive maintenance, predictive maintenance, and project maintenance. Finally, the necessary tool of maintenance metrics shows where the maintenance program currently stands to guide improvements and to avoid losing ground.

Work Order System

A work order system provides perhaps the most highly leveraged tool a maintenance force can possess. This tool allows control of the maintenance work.

A work order system is merely a formal method of requesting and recording work done in the plant. In its simplest form, someone who wants some specific maintenance work performed fills out a specific document. The person turns that document over to the maintenance group who then uses the document to keep track of the work through its execution. The document is known as the *work order form* and the process of how the document is used is the *work order system*. As a system becomes more developed, the document itself may exist only on the computer and the work order itself is the *identification of the work*.

This book generally uses the term *work order* to refer to the identification of the work whether or not an actual physical paper document is involved and whether or not the work has been approved for execution. Some plants call the first identification of the work desired a *work request*. These plants use the term work order to describe work that the proper authority has approved or authorized for execution. In these plants a work request becomes a work order after some management or supervisory level agrees to perform the work. The work is first "requested," then the work is "ordered." Normally anyone may request work be done, but only certain persons can agree to perform the work. Usually a plant uses the same document for the work request as for the work order, the difference being the application of the required approval.

Why is the work order system so important? Think about a typical maintenance supervisor with eight maintenance technicians. Every day each of these technicians might complete two or three maintenance tasks. The entire crew might complete 40 to 60 work assignments in a 5-day week. The supervisor's crew handles a lot of work items. First, without a prescribed method to receive the work requests, the supervisor receives requests for work through a myriad of communications. They might include phone calls, emails, verbal discussions, yellow sticky notes, and notes jotted on envelopes at best. They might include mental imaginings at worst. Many persons needing work just presume or imagine that the crew supervisor knows everything that the plant needs. Second, after receiving all manner of requests, the supervisor must consider what work the plant has actually authorized and which work the crew should complete ahead of other work. Third, the supervisor must take the work selected for that week or that day and assign it to crew members. Finally, the supervisor must track the work for proper completion. A standardized work order system provides the documentation of all identified work to allow control in managing maintenance work. The work order system reduces confusion by standardizing how persons in the plant should request work. The work order form (or computer screen) also facilitates receiving desirable information by setting forth questions to be answered by the work requester. In addition to a description of the work requested, the work order form may request identification of equipment, location, and a statement of importance or suggested priority. With a consistent manner of identifying plant work, management can also set up a method of authorizing individual work requests instead of just having general guidelines. If desired, management or the appropriate level of authority can review work orders before giving authorization. A paper work order form also allows the supervisor to have a tangible handle on the workload and physically to hand out assignments consistently. (The supervisor can also similarly manage a computer listing of the work. The important consideration is that the work order system consistently documents all the work.) Each crew member becomes responsible for an assigned number of work orders. The supervisor can conceptualize and control the work. After completion of a work order, maintenance can send a copy to notify the requester of work completion. Maintenance can file the original paper document if not kept as a computer record. The primary need for a formal work order system is simply to allow better control of the process of accomplishing work.

If maintenance initially does not persuade the plant to use a work order system, the maintenance department should record each instance of work requested onto a work order document of its own and use it within the maintenance department.

In addition to helping the plant identify, keep track of, and manage the execution of work, the work order system has many other important values. The work order system facilitates keeping equipment histories, tracking cost, analyzing, and planning and scheduling maintenance work.

The work order documents (whether on paper or in the computer) assist keeping useful equipment histories and tracking cost. Maintenance can simply keep the paper forms in different files for different pieces of plant equipment. Equipment history is important for a number of reasons. History tells how reliable individual equipment is. History tells how individual equipment has failed in the past. History tells what parts maintenance used in the past. History tells who worked on the equipment. The work order system also allows tracking the cost of individual maintenance tasks. The recording of parts and labor as well as plant down time allows telling how much a piece of equipment cost the plant. Maintenance must include monetary cost in discussions with management for that is the language of higher management, bottom line cost. How many times have engineers or maintenance supervisors been turned down in their quests to replace older equipment with newer equipment simply because there is no cost data? These persons use statements such as "Pump XYZ really hurts us and we need to spend $5000 on a new pump" or "Trust us" to request newer equipment. Management frequently must turn them down when allocating scarce renewal funds. Instead, tracking maintenance and cost allows personnel to state that "Pump XYZ has cost $11,000 over the past 2 years in maintenance as well as $20,000 in lost production revenue. The fluid drive causes the primary problems. We could invest $5000 in a new transmission and eliminate this recurring expense." Maintenance personnel must factor cost into equipment decisions. If one does not know how much it cost to work on a piece of equipment in the past, one cannot determine the best economic decision of whether maintenance should repair or replace it in the future. One cannot arrive at the cost without work orders. Maintenance is an investment in equipment reliability. Tracking equipment maintenance and cost with a work order system allows determination and communication of wise investments of maintenance funds.

Analyzing maintenance work involves a variety of actions and benefits for the plant. A formal work order system makes coding work orders easy for a number of purposes. The plant can determine how much work it spends on reactive tasks rather than on preventive-type maintenance work. The plant can determine how many work orders are in a backlog, perhaps divided by craft or crew. The plant can determine how many work orders operating crews write versus maintenance crews. The plant can determine how long work averages staying in the backlog before completion. The plant can also use codes to facilitate completing the right work. Status codes assist in sorting out work ready for assignment or work already completed. Status codes can also tell what completed work still needs as-builts. Priority codes assist sorting work to keep less serious work from interrupting more important maintenance. Outage codes allow the selection of jobs to work during periods when the plant is in an outage condition. Equipment codes allow the sorting of work by plant or plant process. These codes help ensure assigning all the available work during the maintenance of a particular system. They also allow analysis of what type of work maintenance has already performed on a particular system. A formal work order system lends itself easily to many types of analysis. The consistency of the process of handling maintenance work lends itself to coding the work orders.

Beyond facilitating the management of the day-to-day work, tracking history and cost, and allowing other analysis of maintenance work, a work order system is the vehicle that allows maintenance planning and scheduling. Planning and scheduling are almost entirely dependent on having a controlled work order process. In a simple paper system, planners write maintenance plans onto (or attached to) the written work requests. Maintenance schedules consist of grouping the planned work orders into blocks of assigned work. In this respect, maintenance planning and scheduling are subsets of a formal work order process. In another respect, planning controls the overall work order system. Planning processes the work orders. Planning coordinates work order coding and authorization. With a paper system, the planning group files work order history and data into equipment files and then checks file information on new jobs. Planning would also be responsible for ensuring the supply of the work order forms themselves. With any computerization of maintenance, planning would also guide the development and be the major users of the system. The statement that planning is second only to a work order system in its leverage of the maintenance force revolves around the fact that planning could not exist without a work order system.

The intent of this book is to develop the concept and details of maintenance planning and scheduling. Nevertheless, the maintenance work order system is so intimately related that this section presents several pertinent details of a desirable work order system. In addition, App. J presents a sample plant work order manual with example forms and codes.

First, "No work order, no work." If the plant performs the preponderance or even much of maintenance work without work orders or on blanket work orders, there is inadequate control of maintenance. For all of the reasons above, the plant must document work. Make it known that maintenance must do all work on work orders, but be flexible. Try not to discourage work from being requested. Accept only 80% of the work being on work orders depending on the initial resistance or effort involved in ensuring compliance. Organizational discipline varies from plant to plant. Document emergency work on work orders after the fact. Discourage informal requests, but help people to write work orders.

Second, limit each work order to a single piece of equipment where possible because of the intent to file the work order to provide history and data for a single piece of equipment. If someone writes one work order to rebuild all the control valves in the plant, one has to weigh the convenience of a single work order against not having specific valve history. Planners may have to split up multiple equipment work orders after receiving them.

Third, work orders should be written to request work, not just document work already completed. (It may seem needless to mention this.) The work order documents the proper execution of work, but the work order also identifies work at the outset of the process of requesting work. The work order system via the work order facilitates the effective and efficient use of the plant maintenance resources.

Fourth, there are two schools of thought on recording action taken during the execution of a job. One school holds that the job should be executed as planned so that no comments are needed. Deviations should be approved by planning in advance and the plan reworked where needed. This book holds to a second school of thought where the technicians are greatly empowered to use their skill and training to accomplish a job as necessary. Plans only offer general job scopes counting on the skill of the technicians. Therefore, job comments on action taken are mandatory for adequate file information. These comments are part of the work order form. This methodology may not be suited for certain industries such as those with extreme regulation, but it is further discussed as one of the planning principles in Chap. 2.

Fifth, deficiency tags greatly improve the efficiency of the work order system and planning. The requester hangs a deficiency tag on the equipment needing service. The deficiency tags help for several reasons. Once equipment has a deficiency tag, other persons know someone else has already written a work order. The presence of the tag avoids someone writing a duplicate work order. The absence of a tag would indicate that no one had yet written a work order. Another reason is that the deficiency tag cautions potential users of the equipment of a certain condition. The most useful reason of a tag to planning is that it helps the planner and the maintenance technician find the equipment. Many plants have instances where the maintenance crew serviced the wrong equipment due to faulty identification. The deficiency tag helps reduce the potential for this problem as well as speeding up service by reducing search or question-the-operator time. The hanging deficiency tag may also help find the specific work order number if there is a cross reference from the deficiency tag number being entered into a computer work order data field.

Although a formal work order system appears to be merely a method of documenting maintenance work, it is one of the most important tools because it allows precise control of the maintenance process. A maintenance work order system is also the vehicle that allows maintenance planning.

Equipment Data and History

Having equipment information is a necessary tool for effective maintenance. Equipment information as a tool is basically the existence of plant equipment files. This tool helps the plant base the proper maintenance required on knowledge rather than on memory or trial and error. This knowledge comes from recording and referencing equipment data or specifications and the result of previous maintenance actions.

The amount of equipment information available to maintenance personnel varies greatly from plant to plant. The absolute minimum that a plant would have in the way of information would be the equipment nameplate data attached by the original manufacturers. At the other end of the spectrum would be plants with comprehensive libraries or document control sections. One would go to counter and check out necessary reference material through a librarian or clerk.

Beyond the minimum of equipment nameplates, many plants have equipment manuals from the manufacturers sitting on shelves or in file cabinets in supervisors' offices. The supervisors may have the original bid specifications and purchase orders detailing the original equipment specifications as well. The next step higher in available information would be the storing of work orders showing past maintenance performed on equipment. Many plants have computer databases with technical equipment data and work order history. Whether the plant relies on computer databases or paper hard copy, the plant should make use of both equipment technical data and maintenance work order history.

Some plants have either equipment files with no work order history or work order history with no equipment files. In the first situation, supervisors or technicians consult equipment manuals or equipment specifications as needed. Yet without a work order system, there is insufficient record of past maintenance performed on the equipment. When was the last time maintenance changed the oil? Does the technician realize that engineering had a new style impeller installed last year? Maintenance needs this information to plan properly for the correct maintenance actions for the present and future. These plants may use a work order system to manage ongoing work, but they do not file work orders after job completion. Every job is a new job, not referencing lessons learned from the past. On the other hand, in the second situation where the files collect only work order history, maintenance slows down continually to consult the field nameplates or call vendors to check on equipment details. Many vendors and sales representatives keep their own databases and records of equipment their customers own. In many cases the vendors maintain better records than does the plant, and the vendors greatly assist in decision making regarding plant maintenance.

A plant that has maintenance planning uses the planning department to operate the plant filing system. Maintenance planning uses equipment files to review equipment technical details and to review past maintenance performed to guide new strategies. If a valve keeps failing, perhaps planning should consult engineering for a new material. If unexpected safety concerns delayed a previous job, those concerns can be anticipated this time. After job completion, the maintenance planner files feedback from the field. The maintenance planner essentially becomes a file clerk for the crew. Because of the importance of the relationship between the plant files and planning, one of the principles of planning (Principle 3) in the following chapter addresses files. Chapter 7 and App. M also address the files.

Plant files whether electronic or paper should collect work order history, nameplate data, technical specifications, parts lists, manufacturer and vendor contacts, and correspondence. Work orders should include all work done on the equipment. Field technicians must include feedback stating exactly what they did and what parts they used. Technical specifications usually come from engineering and project groups when they purchase new equipment. In addition, engineering and project groups must understand the importance to procure and deliver as-builts and equipment manuals. Where the plant has previously kept poor records, the plant can begin collecting equipment specifications and data from manufacturers or when the specifications and data are otherwise discovered or uncovered during the maintenance of the equipment.

Proper maintenance uses the important tool of equipment information. Equipment data and history do not necessarily require a computer database, but the maintenance group needs records. Records become helpful only when they are created, maintained, and consulted. Planning makes the most of this vital area.

Leadership, Management, Communication, Teamwork (Incentive Programs)

Rather than distinguish the differences between leadership and management or dwell on whether or not they are actually tools, the point must be made that leadership and management greatly affect the success of maintenance. These intangible tools are of great importance in achieving the goal of superior plant reliability.

Leadership and management go hand in hand. Leadership must introduce the proper strategies of maintenance. Management intelligently provides for ongoing maintenance programs. Leadership is concerned with whether management or key leaders in the organization will understand and make the correct strategic decisions directing the entire maintenance focus. Leadership then motivates the organization to support the decisions. Leadership addresses these questions: Will planning be initiated as a group? Will the company implement predictive maintenance techniques? Management is more the ongoing stewardship of the selected processes.

Plant management must be interested in maintenance as a central part of company operations rather than as a necessary evil. Beyond having simple intelligence, does the maintenance manager really understand maintenance? Management of maintenance must extend beyond fixing things as soon as they break and be interested in advancing the way in which maintenance operates. This effort requires intelligent, dedicated, and active maintenance managers. The maintenance process must be managed. At stake is the capacity that the plant operates. Providing capacity is the maintenance mission; simple repair effort does not result in plant reliability. Can the maintenance manager speak the language of company profit in terms of finance and maintenance investment? Can the maintenance manager explain the job of maintenance? "Work harder" does not cut it.

With leadership and management in mind, maintenance success needs management commitment and organizational discipline. Management must sincerely and actively desire particular aspects of maintenance to succeed. Management must recognize and deal with elements in the organization not supporting those aspects.

Eleanor Kelley (1993) makes an excellent point about what makes a computerized maintenance management system (CMMS) project work that applies to almost any project or program. The company simply needs *two white knights*. One white knight is a corporate manager. This white knight comes to the plant asking questions and expecting answers. How many hours do crews spend on planned jobs? How many planners were put in place? The second white knight is a person in a key plant position. This white knight honestly believes in the new program enough to do the leg work to make it happen.

In the case of planning, someone high enough must insist planning happens. A corporate manager might drive the program and ensure that the plant places a capable person as planning supervisor. Alternately, for a plant not needing a planning supervisor, a plant manager makes sure a competent planner is in place. A maintenance manager could also make sure a competent planner was in place. However, planning requires a commitment to change the momentum and culture of the current way of doing business. Planning involves organizational changes including new positions. Sometimes plant level managers resist initiating change because they feel corporate management above them expects a certain status quo. Plant level management may be reluctant to expend the energy to create a planning function. Will upper management support or even understand the changes? Upper management will definitely understand if upper management drives the changes. The higher position the top person pushing occupies, the easier becomes the start of planning. On the other hand, while corporate management support is desirable, it is more common to have superior plants than to have superior corporations in terms of effective maintenance. This means that the practice of maintenance involves crucial attention to details difficult to establish corporation-wide.

The importance of details makes the second white knight that much more important for any maintenance program. There must be at the plant level someone with the energy, time, and freedom to make a program work. This could be the enlightened maintenance manager, but more often it is a person doing the work who really believes. This person must also have the proper understanding of the program's intent. If the program in question is predictive maintenance, will the assigned person be free to procure the right tools and have the authority to interact with the equipment and crews? Perhaps the first white knight will give actual authority. In the case of planning, the planners themselves or the key person directing the planners must be empowered to make it happen.

Management commitment also calls for organizational discipline. Along with the responsibility for making maintenance progress, management has the authority. In exercising authority, management and leadership must capably deal with issues such as active resistance to change, indifference, and vicious compliance. Organizational discipline comprises the willingness of the maintenance personnel to comply with the direction set forth by management. Management must lead the changes and not accept contrary behavior undermining the integrity of any maintenance program. This is especially important for maintenance planning where the workforce may not clearly perceive the mission. For example, the planning function sometimes becomes the scapegoat for nonperforming maintenance crews. The crews might complain that the planners do not deliver good technical information from the files, but in reality the crews do not give good job feedback to help the planners accumulate good file information. Not only must management understand how the process works, management must take corrective action at the proper points of leverage. Management must insist on proper job feedback to develop files. Management must insist on high crew productivity and use maintenance planning as a tool to go there. Management must insist on everyone being "on the same page," which is organizational discipline.

A final note on management commitment: Management must be able to support a concept through to implementation. Even with pilot programs, management usually does not have the luxury of waiting to see something first work in practice. Management must look ahead and determine what processes or changes they must bring into existence. If management does not wholeheartedly support a program or change until it is working well, it will never work well.

Leadership and management vitally affect maintenance overall, as well as any individual aspect of maintenance. Management commitment and organizational discipline are elements of leadership and management.

Communication is another aspect of maintenance closely related to leadership and management. A maintenance organization takes a big step forward when it realizes that it has room for improvement and that everyone can help.

Peter Senge (1990) states that "when people in an organization come collectively to recognize that nobody has the answers, it liberates the organization in a remarkable way." Managers must let their employees participate.

Plants need to reduce the problem of communication being restricted to the chain of command. Supervisors generally tend to resolve problems based on their own personal experience, which limits the opportunity for information sharing and innovative solutions. While decisions should be made through the chain of command, "everyone should be talking to everyone" to share information and ideas. The company vision should be to produce and sell a product to make money for everyone in the company. The maintenance vision should be to keep high availability for production capacity to make the product. The maintenance vision should not be: "Managers are smarter than supervisors who are smarter than technicians. I hope I can be promoted up the chain." If this is not the stated vision, is it what everyone is practicing or thinking? Plant and corporate management contribute to the success of open communication by frequently touring the work areas and holding open exchanges of ideas and information with the maintenance crews. Managers should frankly tie discussions to the corporate vision: "That's a valid idea, but it might not apply here because what we are trying to do is this... ." One of the main aspects of these exchanges should be a focus on quality. The overall indicator of maintenance quality is availability to run at capacity and produce product. Management contributes to the effectiveness of open communication to improve the maintenance organization.

Another aspect of communication is the reflection of teamwork in the plant's terminology. Is there really teamwork or does everyone use the terms "we" versus "them"?

Teamwork helps the plant accomplish effective maintenance. Management must bring resources together even as it is empowering and giving ownership to the workforce. The true coordination by management has specialized groups working effectively together.

One of the greatest tools to encourage teamwork and communication is the use of common incentive programs. Incentive programs can pull a team together.

Shared incentive programs give a common reward to everyone whether it is a common actual dollar value or a common percentage of salaries and wages. For example, one company gives an annual wage bonus of up to 5% to everyone. The company bases the bonus on achieving certain levels of plant availability, unit efficiency, safety, and dollars spent on operation and maintenance. If the targets achieved totaled up to, say, 4%, everyone would receive an additional check equaling 4% of their last year's wages. The incentive program made a significant effect on the communication within the workforce. Operations became less satisfied operating whatever was available and actively worked with maintenance to increase unit availability. Maintenance became less prone to suggest that operations did not care about the equipment. The groups came much closer together through focusing on a common language, the incentive program. The terminology became "How can we earn more bonus?" rather than "When will you fix it for us?"

Some companies feel that bonus plans reflect an improper attitude on the part of the employees or an ineptitude on the part of the managers. After all, are not these employees being paid already to do the best they can? Well, yes and no. Yes, the employees are being paid to do the best they can. No, they may not be doing the best they can regardless of the best of cultures. Shared incentive programs have improved performance even in companies already working well and having capable employees and managers. Profit sharing and employee stock ownership are similar aids.

Two reasons that incentive programs work is they take attention off the clock and direct it toward the goals of the company. Many maintenance employees actually think they are paid for the hours worked. In reality, the only reason they are paid at all is they perform maintenance so the plant may produce a product for sale. It is impractical in many companies to tie pay to piece work directly, that is, to pay for how many pieces or items of work an employee performs. There is also a serious quality issue if employees become excessively fixated on production or quotas. Nevertheless, a company can tie some compensation directly to overall goals. Employees not only generate some additional interest in the well being of the company as a whole, they may also come to realize how their jobs fit into the overall company vision. For example, one manager learned that a project engineer honestly did not know his job was related to availability, efficiency, safety, and cost at the plants. The manager discovered this when the manager first explained the overall, shared incentive program.

Some considerations to make in an incentive program are who receives a bonus, how tough are the goals, how constant are the goals, how much is the potential reward, how pertinent are the goals, and how easy are they to understand.

As previously mentioned, incentive rewards should be shared. It seems that in pushing for an incentive program, if the ones pushing for the bonus system exclude their own level and higher from reward, the program may be more acceptable to implement. If the very top, professional, corporate management pushed for installing the incentive program, everyone under them would benefit, but not they themselves. If the plant manager pushed for the program, everyone under the plant manager would benefit. The question to ask is: Which levels need a program to encourage them to work better and work better together? The persons above the highest of these levels should push for the program. The CEO of one company implemented a program and is the sole member of the company not to receive a bonus from it. Under the level of the person who set the program in place, everyone should both receive a portion and should feel there is an equitable division of the rewards. If the program only rewarded the managers, the employees would feel slighted. The result of implementing a program that rewards only managers might be to make rank and file employees even less productive. Some companies may have a steady stream of promotion into management, where the benefits of becoming a manager encourage rank and file employees to work more productively and cooperatively. However, many companies with a stable workforce and infrequent promotion opportunities would do well to consider a bonus program with more immediate rewards. In some companies, only the rank and file employees receive a share in a specific incentive program. The managers may be considered already well enough compensated, more professional in seeking improvement, or part of another bonus plan. These arrangements are counterproductive. The goal of the incentive plan itself is to provide a common focus or language. The managers need to be able to say "we" and be included just as much as any operations or maintenance group. Both the work and the coordination of the work are equally important. Managers and rank and file should share in the company's success under a common incentive program.

The toughness of the goals may influence whether behavior really changes. If the goals are too easy, they will encourage entitlement. Employees will expect the bonus every year and behavior will not change. Only in an unusual year does the plant not achieve the goal. Then everyone is upset. The results are no behavior change, extra cost for the program, and the potential of upsetting employees who count on the money. If the goals are too difficult, there will also be no effect on behavior as employees consider the goals unreachable. There may be frustration and mistrust of management ineptitude for such goal setting from ivory towers. On the other hand, stretch goals may encourage the workforce and provide a return for the company. Communication and education may also inform everyone what is or is not achievable. Some of the best of these type goals provide a target range where some improvement will result in some reward and great improvement will result in great reward. For example, if current annual plant availability has averaged about 85%, the program might set an initial level of reward at achieving 88% and additional levels of rewards at 90%, 92%, and 94%. Rewarding at initial levels, but allowing the reward to increase as behavior stretches, encourages employees to keep going.

Goals that always change cause much frustration with an incentive program. Every time employees achieve a goal, the company changes and sets the goal higher. The company feels that the correct behavior has been taught, employees have been shown the way, and higher challenge is needed. Consider instead a challenging set of goals that may take about 5 years to achieve. By rewarding initial levels to these goals as suggested above, employees have a continued interest in chasing a fixed target that does not appear to waver. The company may decide the real goal is, say, 94% plant availability and that reaching this goal in 5 years would be a significant achievement. The plant would set in place a constant 5-year goal of 94%, but allow some reward at levels of 88%, 90%, and 92%. If the plant stretches and achieves 94% the very first year of the program, that is great! Keep the goals in place to demonstrate management commitment to promises, sharing the success of the company, and assurance of 94% or better each year. Then after the 5 years, study the situation and set new 5-year targets. Keep in mind that setting a single year target of 94% may convince employees the goal is too hard. Setting a single target of 88% may cause frustration from later endless goal changing. Keep the same goals and pay-out formula (percentage of salary, etc.) for several years at a time.

Several guidelines for the value of the reward may be given. First, consider how much any improvement is worth. How much is a 5% improvement in availability worth to the company? How much would the value of the company increase as a result of achieving specific levels of the goals? How much would the company's profit increase? Try to base the reward to the employees on a percentage of this amount. It is not unreasonable to think of the bonus paid to the employees as a finder's fee for the big company profit. Also consider how realistic the profit or savings for the company would be. Are the numbers certain or very indefinite? It is not unreasonable to suppose if the company were sure to realize a $10 million windfall every year of the improved maintenance, that the company would want to hand out $1 million in bonuses. The company should be pleased at the prospect of being able to hand out $1 million every year after earning an extra $10 million. Second, the amount of money that motivates any particular individual varies. Rewards do not even have to be monetary. However, money allows an individual to purchase something that truly suits that individual. No company gift catalog program comes close to that ability. Consider beyond the $100 to $200 range. Consider a maximum bonus of 5% of one's salary or more. Some companies routinely give annual bonuses equaling 1 or 2 months' salary. On the other hand, these latter companies may base the bonuses upon the overall profitability of the company without specific goals or formulas. Also, if the pay out is routine, there may be less effect on behavior modification. Third, the goals should be specific and the goal formulas should be set so the employees feel the security of knowing better how much their additional efforts will pay back. The company should decide how much they are willing to pay before the 5-year cycle starts, not after the employees meet the goal. Fourth, the arrangement of the goals into levels of pay off raises interest even when the employees have not fully met the goal. A small bonus gives a taste of extra money without working extra hours and makes one think about achieving the higher levels in the coming year. Fifth, do not structure a new program to replace existing wages or take the place of raises that the company would obviously have otherwise given. Clearly relate the incentive program to extra effort. The company is willing to pay extra for levels of performance not commonly achieved.

Consider next the pertinence and clarity of the goals. First, a pertinent goal would be a goal related to one's work and also one that makes a big difference in the company. Yet remembering that shared goals are desired, the goal should not be one that only a few persons can affect.

For instance, the annual purchase cost of fuel may greatly affect the bottom line of the company, but only a few persons might handle the fuel buying process. Another problem with such a goal is that the cost of an item such as fuel may even greatly depend on events outside the control of anyone in the company. In contrast, if the goal involves throughput or labor hours per unit product, most of the plant employees realize they have an effect. Overall operations and maintenance nonfuel cost, overall plant capacity, plant availability, plant efficiency, and safe hours worked give examples of possible goal areas that might be both pertinent and shared for most of the workforce. Second, clear goals would be easy to keep up with and understand. At most a company should conceive of only having four or five goals in an incentive program. Having too many goals might simply make everyone just want to work harder in general rather than think about the specific areas of concern. The program ought to give regular feedback on how the effort is coming along and if things kept up as they are going, how much would be the reward. What areas could be better and why? What areas are fine and why? What specific things could we do or do better? Clarity also might have an opposing element to pertinence. Relating a goal to a general overall indicator may be more understandable to most plant persons than some specific, precisely measured, maintenance indicator that is technically superior. For example, tracking the overall number of unit trips may be more clear than tracking trips of large units over 100 MW during peak seasons when backup generators are unavailable. True, the latter measure may have a larger effect on the company bottom line. Nevertheless, working toward the former measure with a bit of common sense on the part of management may provide the same overall results and be much easier to keep up with and understand.

Remember that a properly structured incentive program may greatly assist plant management in bringing everyone together as a team to accept change and improvements in a maintenance program.

Planning most decidedly causes major changes in the maintenance process. Because of natural resistance to change, the implementation of planning requires organizational discipline coming from management commitment. Good leadership and communication help make changes more acceptable. Working together as a team toward a new challenge set forth by an incentive program may also make change more acceptable.

Qualified Personnel

Another critical aspect of maintenance is having qualified persons to perform the maintenance. All the organizing and managing in the world will not help if the plant utilizes mechanics not possessing the necessary skills to accomplish the work. Maintenance planning will not help if there is no one to plan for or if the workforce lacks proficient craft skills. How many persons of each craft are necessary? How should they be hired? How much training should they need? How about multicraft jobs? How about ongoing training for experienced technicians and supervisors? A superior maintenance program addresses these areas.

Management allocates human resources as one of its most important jobs. Management must provide enough qualified personnel resources to perform maintenance. This statement applies to company managers if the company performs its own maintenance or to the managers of a maintenance management company contracted to perform maintenance. Management needs to know how many persons are necessary and hire or develop the necessary skills in the persons utilized.

One key value of a work order system earlier discussed is in identifying work needs. Planning then transforms that work backlog into work hour estimates for particular craft skills. Having a handle on how much work exists in a demonstrable form, the work orders and plans, allows managers to determine precise labor needs. If the work load normally consists of 80 hours of electrical work, the work load could support two electricians. However, plants that do not have even a work order system would have to wonder if they had enough work to keep an electrician or two busy. If their two electricians stay busy but are not working on work orders, it is fair for higher management to consider if only one electrician might be able to handle the work load. Many plants have their personnel resources outright stripped away by corporate management that sees no visible backlog of work or record of work completed. These plants use a handful of blanket work orders if they use any work orders at all to maintain the entire plants. After losing a workforce through being unable to prove a work backlog, these plants suffer. Lack of work documentation cripples management's ability to manage the quantity of craftpersons. Right or wrong, management sees labor needs through a visible work order backlog, especially when planned with labor hour estimates. Plant management can thereby intelligently determine which work the maintenance group should perform and use that as a basis to determine labor needs.

Management invests in plant capacity by providing sufficient, qualified maintenance persons. Management can make a poor investment by not providing enough persons to maintain the plant and capacity suffers. Management can also make a poor investment by providing more persons than are needed to maintain the plant. The plant may preserve capacity at excessive cost because personnel are very expensive. Management's first responsibility is to provide sufficient personnel to run a maintenance operation, but these persons must also be qualified.

Management must have the proper persons in place. Managers must spend time putting the right persons into place. Putting the right persons in the right places sounds like obvious good practice, but many times managers attempt to legislate their way to success. The real world sees this problem frequently. Managers quickly hire or shuffle persons from crew to crew, reorganizing overnight; then they will spend the entire next year overusing rules and procedures to try and make the wrong persons perform the way the managers want. Managers want to get the personnel placement quickly out of the way so they can go back to their "real" job of making it work. These managers should realize that their real job is the allocation of resources, not just the direction of a set workforce. Different persons have different talents. Managers must spend time before hiring persons and before reorganizing persons. Managers must spend time training and assisting persons become the best persons they can become in any particular position. Qualified persons in the places they belong can accept empowerment and help the organization succeed.

Classification

Different companies train and classify their personnel in very diverse ways. For the sake of illustration, this book considers a workforce consisting of mechanics, welders, machinists, painters, electricians, and instrument and controls (I&C) technicians. This book calls all of the preceding craftpersons as "technicians" or "techs" for the purposes of general discussion. Furthermore, the workforce includes apprentices and trainees for each of the above classifications. Apprentices are generally less skilled than the technicians themselves and trainees even less so. Apprentices and trainees generally do not become assigned to jobs by themselves such as jobs requiring only a nominal skill level. Apprentices and trainees go as helpers with technicians so they can learn their crafts. This conceptualization of the crafts in the workforce helps the discussion to address the pertinent aspects of maintenance and the principles of planning. Maintenance planning activities would use the established classifications in a specific company to communicate the skills necessary to execute work.

This book uses also the terms workforce, labor, craftpersons, maintenance personnel, field technicians, and technicians interchangeably. In general, the use of one of these terms refers to the actual persons doing the hands-on work of maintenance. It is these persons who would receive and execute planned and scheduled job assignments. The scheduling sections of this book, especially Chaps. 3 and 6, make a greater distinction of actual classifications.

Crew supervisors are also giving way in industry to crew leaders. This signifies the adoption of an approach away from looking over shoulders to an emphasis on coaching and motivating. So this book uses the terms crew supervisor and crew leader interchangeably as well.

Hiring

Hiring employees with high potential for achievement begins the process of having qualified employees. Management must put considerable thought behind the overall process of hiring. Management must understand what skills the maintenance group needs in order to assess the potential of job applicants. For example, do the maintenance tasks require reading and comprehension skills to allow job plans to be used? Then management must have a sufficient number of applicants from which to choose. Are the wages and benefits sufficient to attract applicants with sufficient potential to succeed? If a company needs to hire for 14 positions and only has 14 applicants, this could be a sign of trouble. One company advertised for a class of new apprentices and had over 1000 applicants for a handful of positions. This company was able to be quite selective in the process of hiring. Because of the importance of not rushing the hiring practice, a company might anticipate ahead of plant expansions and gradually staff up as applicants with enough potential become known. Sometimes a company might run lean and not hire when applicants simply are unqualified. A company might make use of some contract or temporary personnel to assess abilities before hiring someone. Hiring the wrong persons costs a company dearly from the standpoint of wasted effort in hiring and training only to end up with an unqualified technician to meet company needs. Management must carefully handle hiring to acquire only candidates that have the potential and ability to succeed.

Training

Keith Mobley (1997) says it is not unusual, but typical, that employees do not have sufficient experience, skills, or tools to do their jobs. The actual training most employers provide is primarily dedicated to mandatory training required by government agencies such as OSHA.

Training time takes time away from maintaining equipment. Nevertheless, state-of-the-art science and engineering continually places new equipment in the production loop. Training programs qualify employees for maintaining them. Just as management keeps equipment from becoming out of date and obsolete, management must update the skills of human assets correspondingly. Mechanics, electricians, and I&C persons must keep abreast of new technology as new materials are developed and applied. For example, manufacturers are integrating electronic sensors into even the most basic of mechanical machines. Attrition also replaces experienced persons with new hires that may not yet have the necessary skills. Management must make continuing investments in training time to maintain the workforce itself. A workforce without the skills necessary to maintain equipment properly does not keep plant capacity very reliable. The goal of maintenance is to provide reliable plant capacity, not simply to provide an available workforce.

Not only does training improve the skills, it improves the coordination of the specialized workforce. Training goes hand in hand with communication. Joe Spielman (1997) of General Motors says that often a surly, uncooperative technician is someone whose skills the company has allowed to become outdated. Persons must have the skills if the company expects them to go out and use the skills. A company organizes to bring certain skills to bear in certain areas. Do those skills really exist? Just because a plant has always had a mechanical maintenance crew does not mean the mechanics can work on the latest mechanical equipment in the plant. Most organizing schemes recognize that employees have skills, but the actual practice of maintenance must ensure keeping those skills up to date.

Management makes an important investment in ensuring that they have qualified employees through training of maintenance personnel. Maintenance training encompasses new apprentices, existing technicians, and maintenance supervisors. Training yields ever increasing benefits in future years.

Apprenticeship. One step in having qualified employees could be the establishment of an apprenticeship program. The program could be informal with apprentices simply assigned to assist experienced field technicians. The program could be more formal and include a structured classroom with the field training. The company should address two issues up front with any apprenticeship or trainee program. These issues are the expected contribution of the apprentices to the current work load and the eventual promotion of the apprentices.

First, an apprenticeship program provides skilled employees in the years to come. The company hires apprentices to learn, not necessarily to be an extra set of hands. Management hired them to be the future. Most of their work should be with skilled persons on jobs the apprentices could not do themselves. In this manner, the apprentices learn to become skilled persons. The apprentices may be an extra set of hands, but only to assist experienced technicians. The time spent on a job requiring two persons may be shorter if the supervisor assigns two experienced technicians. However, assigning one experienced technician and an apprentice creates a worthwhile training period while accomplishing the plant's work. Field work does not take as much time away from maintenance as does classroom training.

Second, the company should address promotion. A primary concern is any promotion system in place that requires competition among apprentices for promotion to technician. This type system inherently discourages employees from sharing techniques and building team-work. Apprentices will not help each other and some technicians may be reluctant to help certain apprentices. Therefore, have a policy that allows promoting all apprentices when they become qualified. Keeping from having too many technicians requires management planning to keep a steady low number of apprentices coming up through the ranks to replace technicians who leave the workforce. Management should be flexible to withstand occasional periods of having a slight abundance or scarcity of technicians.

Technician.   Apprenticeship seems to be primarily a method of bringing newer, green employees up to speed. Apprenticeship brings up their skills to the level of existing technicians. Existing technicians also have a continual need to stay up to speed or current with technology as the plant evolves. After promotion from apprentice, additional programs keep skills current for existing technicians.

Some plants have recognized an overwhelming shortfall in skill level and suddenly decide to do something about it. These plants may have to establish formal classroom training just to bring apprentices and technicians up to the current state of the art. Perhaps the plant may suffer briefly or use contract labor to supplement the workforce while management provides intensive training. Other plants make a com-mitment to gradual upgrading of the workforce being conscientious to assign the few skilled technicians in more of a teaching role out with the other employees on jobs.

Many plants merely want to stay current with workforce competency. To continually possess a competent workforce takes management com-mitment in the area of skill improvement. Skill improvement can be con-ducted during or apart from normal work hours. The company has a number of options. During work hours, plants can bring in specialists or training companies on a routine basis and either require or encour-age employee participation. Plants can send employees to schools or classes. Apart from normal work hours, plants may establish tuition reimbursement programs for desired course work, conduct night time classes themselves, or encourage employees through pay program inducements to seek out training. Pay programs may be as simple as having pay grades that compensate employees better that have higher skills or certificates. Other employees seeking better wages must better their skills. One problem in this area is in deciding how the employees should demonstrate mastery of new skills. Finally, companies might stay attentive toward hiring new employees who already have the req-uisite skills as natural attrition causes other employees to leave the com-pany. Since all new skills have to come from somewhere, essentially this latter philosophy is having someone else do the training.

Plants need existing technicians to have up-to-date skills. After promotion from apprentice, additional programs keep skills current for existing technicians.

Multicraft programs.   Multicraft programs make sense when they allow jobs that require more than one skill to be completed with fewer persons assigned. This lowers coordination problems of putting the right persons in place or of overstaffing a job. Both individual job productivity and overall workforce productivity improve.

A multicraft program or system gives craftpersons additional job responsibilities they normally would not have at the plant. For example, there is a job to rebuild both a pump and its motor. The mechanic that must rebuild a pump might also have to perform the field work for the pump motor. The mechanic might have to disconnect the motor for delivery to the electric shop and then reconnect the motor later. Before multicrafting, the mechanics and electricians would have had to coordinate their field work. This coordination may have caused some waiting for one craft or the other. In addition, the field work may have required only a few hours while the most important work was to be in the mechanic and electrical shops. By having only one person make the field trip, maintenance might save labor or time overall. It is true that an electrician probably would be more efficient or productive in actually disconnecting the motor. On the other hand, overall productivity would be worse because the electrician spends time to set up and travel to the job site. Another example is that some plants require all welding be done by trained welders instead of mechanics. Teaching a mechanic to be able to use a torch to cut off rusted bolts rather than having the mechanic always wait on a welder may save job and company time and expense. The company not only requires fewer persons to complete some maintenance tasks, the company has more choices of whom to assign work. A company that has taught its mechanics to do light structural welding would not have to tie up its scarce certified welders for some simple welding tasks. So multicraft programs basically train employees to take on more job responsibility to give the company more flexibility in assigning work and reduce coordination problems for better overall productivity.

Multicraft programs can run into problems for several reasons. These problems include employees resisting acquiring new skills, insufficient work to keep skills proficient, and mistrust or fear of losing work by other crafts.

One might think that individual job satisfaction would encourage employees to desire more varied work experiences or more control over individual jobs. Nevertheless, employees previously kept in fairly rigid classifications may resent having to learn new skills. On one hand, a multicraft program obviously has a company benefit; but what is the benefit to the employee? One reason to accept multicrafting is job security through company competitiveness and survival. Management shares the vision down through the line on the importance of multicraft helping the company. Management's consistent, previous fair dealings also help encourage employees to do what it takes. Another reason to accept multicrafting is ultimate job security. The employees become more valuable to themselves if they can do more. The employees thereby gain real job security beyond even the immediate company. Nevertheless, the company might avoid needless hard feelings by sharing the monetary value of the multiskill arrangement. One company created a new classification for multicraft mechanics where mechanics learned light structural welding and certain light machining skills. The company made the new classification significantly higher paid because the company could define the financial benefit. Promotion into the new classification was voluntary and provisional. The company immediately boosted the persons accepting the promotion to the new wage level and allowed a given time period to acquire the new skills through an on-company-time training program. Anyone who desired could stay in the old mechanic classification and anyone not able to learn the new skills had to revert to the old classification. Only a few persons close to retirement elected to stay in the old mechanic classification. Eventually only another very few persons that did accept the promotion were unable to learn the new skills and had to revert to the old classification. Employees should be open to multicraft opportunities to help the company and themselves. Management should consider paying more for multicraft classifications beginning even before the technicians acquire the necessary skills.

Sometimes the maintenance group trains persons to acquire new skills where insufficient work exists to go around to keep the skills proficient. This can lead to fewer persons than before having a valid proficiency in a vital skill area. It may be reasonable that all mechanics should be able to use torches to burn off rusted bolts, but is there enough structural welding work for all the mechanics to practice? Is there much high pressure welding work? If there is not enough high pressure welding work available, it does not make sense to train or pay everyone to become certified in high pressure welding. There may be enough flange facing and bolt cutting opportunities to justify training all the mechanics in some light machining. There may not be enough heavy machine work opportunities to justify making everyone a certified machinist. Management must be attentive to the exact identification of actual opportunities frequent enough to justify the extra training and perhaps wages involved in multicraft arrangements. Companies should restrict multicraft arrangements to these opportunities.

In some organizations because of heavy tradition or union restrictions, persons may be unable to work outside narrow classifications. A real win–win opportunity exists for the company and the employees if the company is willing to pay a premium for the new skills before the employees acquire the skills. The company could propose a new classification that persons from both of the previous, narrow areas could enter and immediately make higher wages. Mechanics to do light torch work and welders to do light mechanical work could enter a new class of "mechanical technician" paying higher than either former position. The company might alleviate the craft fears of losing work by allowing everyone affected the opportunity to learn and practice some new skill. The key to this whole area is an intelligent management assessment of the situation. Management must have a conviction of the overall profitability for taking advantage of specific multicraft opportunities.

Another problem concerns the previous exclusive owners of the skill. Experienced practitioners of certain skills may be concerned about the abilities of newcomers to that skill area. For example, electricians may have valid concerns about the competency of mechanics trying to disconnect motors. This concern may be alleviated in two ways. The first is the initial thoroughness of management including receiving craft feedback when selecting candidates for what work to multicraft. Management might ask the electricians what electrical work a mechanic might be able to do that would help either the electricians or the company overall. The second is including the primary skill holders in leading the training. If the electricians themselves train the mechanics, the electricians have more confidence in the capabilities of the mechanics.

Planning utilizes the multicraft classifications when figuring which skills individual jobs need. For example, the planner indicates that a mechanical technician could do the light welding portion of a mechanical job rather than adding a certified, high pressure welder to the work plan. Another example: the planner indicates that a mechanic could disconnect a motor rather than adding a requirement for an electrician. The crew supervisors subsequently have the right to assign welders or electricians to the respective jobs. The planner has provided helpful information that increases the supervisor's flexibility when assigning work and given the opportunity for overall productivity increases.

Multicraft programs benefit the company when they allow jobs that require more than one skill to be completed with fewer persons assigned. Management must be careful when establishing such a program to avoid common problems. Planning helps promote the use of the multicraft classifications.

Certification. A plant may wish to add a certification program to its maintenance effort. A certification program would recognize and reward special expertise or effort developed in a person's primary skill area. For instance, everyone in a certain classification would receive the established wages, but certified employees would receive additional payments.

Such certification pay depends on having passed prescribed certification requirements.

Certification arrangements have advantages and disadvantages over creating new classifications. One advantage includes more acknowledgment that skills must be maintained. A disadvantage might be the administration of continually checking skill levels. Another advantage may be the company's relative flexibility in establishing or changing certification programs to meet needs. A disadvantage might be an inconsistency in how the management acts. Perhaps the primary advantage of the concept is that management can demand high enough standards for a really necessary skill. The company can tailor the pay benefits to entice the desired number of persons to develop or keep the pertinent skills. Establishing an entire classification on the other hand might tempt management to water down the eligibility requirements or by practice establish an errant tradition of promotion into the classification without sufficient merit.

In many plants, employees follow different career paths according to their interests in specializing in different areas or skills that the plant needs. The plant may hire persons with different skills to match plant needs. The plant should be flexible and try to match persons with jobs and training opportunities that match their interests. There might be skilled areas that the plant needs covered where either there are no employees that have those skills or the plant cannot keep employees with the necessary skills.

A case in point: the plant might have many mechanical technicians that can do some light welding or torch work, but the plant needs at least a few highly skilled, pressure vessel welders. Management does not desire to add a new classification, but is willing to pay extra for extra skills. A certification program would encourage and train those employees with the aptitude and desire to become such expert welders. The amount of special certification pay arrangements would be just enough to provide the plant with the numbers of high performers the plant needed. The plant would use the certification program to guide existing employees to develop necessary skills.

In another case, the plant realizes that among its technicians, certain performers far exceed the average technician in their capabilities. Management wishes to keep these employees from seeking employment elsewhere. These employees may be able to distinguish themselves by passing certification programs.

A plant may also be unable to keep persons with necessary skills in certain areas. The company might lose highly skilled persons or other high performers to the market. The plant underpays those persons according to the market with respect to the skills in question. They develop the skills and other companies hire them away. A certification program may raise pay to help this situation.

Planners use certification designations on plans to help identify special skills certain jobs require. The designations facilitate communication among the planners and supervisors regarding job needs and assignments.

Certification programs recognize and reward skills that the company values without creating new classifications. This may allow the company to tailor a program to adequately compensate high performers or attract only the required numbers of persons to a particular high skill area.

Supervisor. Supervisors who run crews make a tremendous contribution to maintenance success. These persons essentially control the culture in many plants. Craftpersons respect these capable men and women who have worked their way from the jobs of the crafts themselves to positions of leadership. These persons have daily contact with almost all of the craft technicians. The technicians solicit and listen to the opinions of the crew supervisors. The supervisors have a profound effect on the attitude of the plant overall when it comes to accepting or resisting change. In addition, the supervisors play an actual part in many maintenance processes, whereas managers typically remain outside of the plant processes. For instance, supervisors make job assignments. Therefore, the execution of many processes as prescribed by management lies directly under the control of the supervisors. For these reasons, management must give careful regard to the development or training of supervisors.

Maintenance supervisors and assistant supervisors who run maintenance crews should have their job descriptions and training programs oriented more toward being coaches and encouragers and away from their traditional roles as schedulers and overseers. Such crew leaders need to be in the field with their crew members. All too often, the "system" saddles these key players with endless administrative tasks such as checking time sheets, purchasing parts, attending a myriad of meetings, or even writing up employee evaluations. Supervisors should be coaches in the field, not clerks, planners, or secretaries. A guideline for supervisor time should be 6 hours in the field and 2 hours in the office for an 8-hour day.

Why should not the supervisors plan work instead of having special planners? Many plants utilize their supervisors to fulfill planning duties. The problem with this arrangement is that the very word "planning" implies something done before the work begins. The supervisor needs to be with the crew during execution of the work, something that cannot be done if planning chores preempt supervisor time. Planning works as a service for the supervisor to make work ready to go before job assignment and execution. Planning also works as a service for the supervisor after job completion to file feedback and manage equipment information. The planning department accomplishes activities both before and after job execution and frees the supervisors to work with their crews.

A superior maintenance program has qualified persons to perform the maintenance. Management must establish competent hiring and training practices for employees at all levels including supervisors. The mechanic must be at the plant with the necessary skills to accomplish the work. Supervisors focus on coaching crews during job execution. Planning interacts with the issues of craft skill and supervision. Planning uses the established designations for skill levels in the job plans. Planning helps free the supervisor from activities before and after job execution.

Shops, Tool Rooms, and Tools

Shops, tool rooms, and hand tools contribute to a superior maintenance program. Shops are special areas within the plant where technicians may work on equipment taken from the field. These shop areas may contain special work aids such as work benches for laying out parts and special tools such as lathes or other machining equipment. Tool rooms help manage keeping and sharing large or expensive tools occasionally required by technicians. Such tools would not be practical to keep in standard tool boxes because of their large size or infrequent use. Tool rooms might also issue individual consumable items taken from the storeroom in bulk such as ear plugs or fasteners. Hand tools are items that a technician would normally carry in the tool box. The frequency of use and size of the items might dictate what a technician would prefer to carry routinely. Shops, tool rooms, and hand tools have a significant role in the management of maintenance and planning. A few words regarding them are appropriate.

Maintenance shops aid in doing the work right. Additional logistics may be involved in moving field equipment to shops for maintenance. Yet, working in a clean, equipped shop rather than in a field environment (hanging upside down off a boiler comes to mind) favors quality. One of the ideas behind a shop is to allow cleanliness. Contamination with dirt or grime affects the success of many maintenance jobs. A technician should be able to concentrate better doing the job in a clean environment without having to worry about debris on flange faces. In addition, stores and special tools normally close to shops reduce travel delays when tasks require additional resources.

Shops should be properly outfitted to be most helpful to maintenance. A properly outfitted machine shop contributes to quicker and higher quality machine jobs. Consider upgrading all of the plant shops including mechanic shops, welding shops, paint shops, electric shops, and instrument shops. Upgrades should be done based upon recommendations from the persons who must use the new facilities.

The use of the shops to accomplish individual maintenance tasks may be planned by the planning department or left to the discretion of the field technician. The planner generally considers the technician to be competent in using plant resources to accomplish maintenance assignments. The planner may call for a valve rebuild and the trained technician would know to perform the job on the technician's shop work bench.

A tool room consists of a special area where technicians can borrow tools that would not ordinarily be found in their tool boxes. From a crane to a come-a-long, the right special tools make technicians more effective. Certain tools should be managed. For example, a tool room could properly store ladders and have them ready for use. Depending on the size of the maintenance operation and the complexity of the tools, the tool room increases in complexity. For small groups, having one or two certain technicians responsible for the tool room in addition to their other duties might be fine. The tool room might simply be a storage room where all technicians have free access. Larger maintenance forces might benefit from having a restricted access or a check out counter arrangement. Some tool rooms might benefit from having a full staff with a dedicated manager. The staff issues the tools and manages the maintenance of the tools themselves.

The plant maintenance force benefits when a knowledgeable, experienced maintenance technician works in the tool room. Many times a technician will come to the counter and describe a problem. The tool room attendant can suggest the proper tool. One oil well service company places the job of tool room attendant above the field technician in the company's line of promotion. The tool room attendant knows what tools technicians need, why they are important, how they work, and how to keep them working.

In addition to a central tool room, certain areas of the plant may have specialized tool needs. The plant might benefit from having certain tools remotely kept near those areas. A secure shed could contain specialized turbine tools on the turbine deck of a steam plant. Such a secure area may have a counter with an attendant and open only during certain turbine work. Otherwise, limited access would be available to supervisors or certain technicians. A particular job on a burner deck might require specialized tools not used anywhere else in the plant. It would make sense to have the burner deck tools located in a tool box on the burner deck ready for use.

Even when the tool room has a check out counter, free access to certain items facilitates productivity. Open cribs contain bolts, washers, cleaning rags, and common consumable items in the waiting area of the tool room. Candidates for the free access cribs would be consumable items that are relatively inexpensive and frequently used. They may also be consumable items that technicians may need to inspect during selection. A technician may want to inspect the different washers available to select one with a desired thickness. These open cribs keep technicians from wasting expensive labor time waiting in check out lines for inexpensive items. The tool room attendants would replenish the cribs as needed.

Maintenance planning identifies special tools on job plans. Rather than make several trips to a tool room as they identify needs, technicians can gather previously identified special tools before beginning a job. The technicians' feedback on special tools actually used on particular jobs aids the planner plan work better in the future.

Hand tool management also contributes to maintenance success. Many plants suffer from technicians having inadequate hand tools available. Efforts to identify and equip individuals with proper hand tools improve job quality and productivity. The objectives are to reduce problems caused by using incorrect tools and reduce delays in finding correct tools. Maintenance planning presumes technicians are adequately equipped with the common tools of their crafts.

Maintenance programs should consider the effectiveness of current shops, tool rooms, and hand tools. Maintenance planning coordinates the use of the special tools of the tool rooms, whereas the shops and hand tools are more of an extension of the qualifications of the individual crafts or technicians.

Storeroom and Rotating Spares

Maintenance must also consider inventory management when preparing for success. Five areas merit special comment. All of these areas concern interfaces between storeroom and maintenance groups and justify special attention. These areas are the control of the storeroom, identification of usage, standardization, use of critical spares, and use of open cribs.

First, many maintenance departments should but do not have authority over their storerooms. The companies position the management of the storerooms separate from even the plants in some cases under special storeroom departments or under the purchasing departments. The companies feel this arrangement provides a check and balance. The processes of operating a storeroom may be more similar to those of a retail store or purchasing group than to the running of large machinery in an industrial plant. The company is trying to protect against the storeroom operation not being managed properly as a storeroom. Storeroom operations can demand sophisticated controls and procedures. In other cases, the company feels that the large monetary value of the storeroom justifies a separate management group. The value of storeroom facilities and inventories certainly is large enough in many plants to be treated as significant and managed with qualified personnel. On the other hand, the less visible but real value of proper maintenance and reliable plant capacity nearly always favors the maintenance group having ultimate control over the storeroom.

In situations where the maintenance department does not control the storeroom, maintenance management must give constant vigilance to the interfaces between the storeroom and maintenance group. Improvements within the group that a manager controls receive most of the attention and processes hopefully become smoothed out and efficient over time. Interfaces between groups receive much less attention and frequently offer major areas of improvement. Managers of storerooms and maintenance must communicate and constantly watch over areas to avoid suboptimizing their individual systems to the overall higher expense to the company. Suboptimization might occur when the storeroom's major goal is to reduce inventory regardless of the cost to maintenance who might order parts frequently on an emergency basis or lose plant capacity.

Second, storerooms carry such a large value of stock that a natural tendency to reduce inventory exists. Stock is expensive, but less so than poor availability of plant capacity. A plant must have a handle on stockouts before it can consider reducing stock. What is the incidence of not having requested items? If a plant has frequent stockouts, it cannot yet consider reducing stock. Effectiveness must come before efficiency. A storeroom must adequately supply maintenance with parts before the storeroom can identify an overstocking problem. If a storeroom is not adequately supporting maintenance with the parts it needs, management might want to consider an aggressive program of identification and purchasing. The company priority order should be first an increase in plant reliability, second an increase in maintenance productivity, and third reduction in excessive inventory.

Maintenance may consider other options rather than keeping inventory on site. With a successful maintenance program that has few breakdowns, a planning function should know the need for certain parts ahead of time. Under those circumstances a plant may decide not to stock items that are readily available on a 24-hour notice. A plant may also set up blanket purchase orders with suppliers for common materials to reduce purchasing efforts at the times maintenance needs the items. Management should be cautious not to lose control over having critical items available for circumstances that could suddenly restrict plant capacity.

Third, inventory store quantities may be excessive because the plant has so many different types of equipment performing the same service. Companies can easily determine the lowest purchase price for an individual pump and perhaps make an effective judgment of the operating cost in terms of energy used over several years of projected use. Companies less often consider the maintenance cost or the cost of keeping spares on hand. These latter costs are very difficult to determine, but that does not mean they are insignificant. If the plant values reliability and the cost of inventory stores, it should consider some equipment standardization guidelines. Technicians enhance plant reliability when they can work enough on the same type equipment to develop a close familiarity and when they know that the storeroom has ready spare parts. A management convinced of the overall company benefit should allow maintenance to dictate certain equipment standards for purchasing and engineering to follow. Maintenance management should take on such a course of action as an important responsibility, not one simply to be recognized, but to be continually managed.

Many plants have decided that standardization is the most important key to reliability, high capacity, and profits. These companies have standardized and reduced entire assembly lines or processes to standardize as much as possible. They have even standardized or restricted allowable suppliers and vendors to a select few with whom they develop close relations and high expectations for consistency.

Another aspect of inventory standardization within the sole control of the storeroom concerns duplication. Many different suppliers deliver virtually identical parts, sometimes even from the same manufacturers. Although identical, these parts arrive with different vendor identification numbers and the storeroom stores them as different items. The storeroom may have eight categories of a certain type gasket with each bin containing ten gaskets. In reality the gaskets are the same and the plant keeps an excess of 80 gaskets on hand. The storeroom may have opportunities to reduce its inventory by intelligent identification of parts for plant use. Standardization of equipment and suppliers also reduces these problems of multiple part numbers.

Fourth, another success in the inventory area can be a rotating spares program especially for a plant designed with few equipment redundancies and limited physical room for adding them later. A rotating spares program identifies critical equipment whereby the plant purchases entire replacement assemblies to keep in stock. In the event of failure, maintenance can quickly exchange complete assemblies for failed equipment. Then under controlled, nonemergency conditions, the maintenance group can rebuild the failed components and put them into stock. The term *rotating spares* comes from the plants who usually find most of their critical spares have rotating elements such as pumps. The term also could apply to the fact that the spares rotate to exchange places. One spare "rotates" into service as the failed assembly "rotates" into the shop for rebuilding and then into the storeroom to take the original stock item's place. Some ships use an arrangement known as *bulkhead mounted spares.* The crew mounts the spare assembly right alongside the service equipment for easy use. In addition, a storeroom for a ship may not be readily accessible when needed. Most industrial plants, however, prefer spares be kept in the cleaner and better controlled environment of a storeroom rather than allow them to clutter an operating area.

The use of a rotating or critical spares program directly benefits plant capacity during emergency maintenance. If troubleshooting cannot quickly find an easy fix, the plant may suffer as extended diagnostic work or repair time drags on excessively. Yet with a rotating spare available, the supervisor may call for its use after only a few moments of unsuccessful troubleshooting to reduce the potential of an extended down period. The supervisor or manager makes an educated call. The supervisor might realize uncertainty of the in-place repair time that could last either 2 or 24 hours. On the other hand, exchanging the entire assembly would require a guaranteed 5 hours. The use of the rotating spare reduces plant uncertainty in emergencies. Moreover, the use of the rotating spare reduces uncertainty for scheduled maintenance. Consider a particular deficiency with the in-place equipment that has not yet caused a problem, but that maintenance needs to correct. The plant may decide to have the equipment exchanged with the rotating spare at the plant's convenience. Then maintenance may carefully examine and maintain the removed equipment in a shop environment. Management should consider the use of a rotating spares program to improve maintenance success.

Fifth, open cribs for certain inventory items as discussed for the tool room might be advantageous. The company must choose between specifically accounting for every minute inventory transaction and reducing delays encountered by maintenance technicians having to wait for service. Maintenance management might weigh the cost of having an extra counter clerk against the value of not only possible missing stock, but possible lower productivity and plant capacity. The plant might also want to consider if it sends contradicting statements to employees. The plant trusts and empowers the technicians in word, but in action does the company trust the technicians with easy access to any inventory items to increase productivity?

Planning drives the best use of the inventory stock. Rather than maintenance technicians determining what parts they need in the midst of a maintenance task, planning reserves likely parts before the crew begins the work. This manner of operation reduces stockouts at the time of work execution because of advance notice of part requirements. Parts identification that occurs days in advance of execution also allows inventory reduction. Storeroom or tool room personnel directed by planning can make use of blankets to pick up certain items from local vendors in time before job execution begins. In addition, planning acts on technician feedback. Often a technician that had a difficult time obtaining the correct parts for a task may simply be relieved when the task is completed. But planning has the time to evaluate the technician's job feedback and work with the storeroom or other processes to determine if stock levels need adjustment. The planner fulfills a clerical or administrative role for the technician to keep the inventory system working well. Many times management implements a process of some sort without providing an ongoing administration of the process. With regard to the interface of the storeroom and the maintenance needs for specific parts, planning fills this role. Planning also has a broad enough view of maintenance to play a large role in the determination of likely equipment candidates for standardization. It also aids the planning effort when standardization helps technicians to be familiar with equipment rather than technicians frequently requiring equipment information about unusual repair procedures.

Inventory control is an important tool of maintenance management. Certain areas of inventory management regarding interfaces warrant special mention. Management must continually be vigilant with respect to the effectiveness of the storeroom in supporting maintenance and plant reliability. Management should first ensure the availability of needed parts and materials before considering efficiency of storeroom operations. Standardization concepts help maintenance by reducing the incidence of unfamiliar, unique maintenance operations and allow storerooms to concentrate on proper stocking of fewer parts. Finally, use of open cribs may help reduce delays for technicians. Planning plays a key role in guiding and administrating the usage of parts from its vantage point in having a broad view of maintenance and time to pursue inventory issues.

Reliability Maintenance

While correcting equipment failures efficiently and effectively is important, anticipating and heading off failures is also a major part of the maintenance management tool box. This effort is made through concepts known collectively as reliability maintenance. Reliability maintenance concerns itself with keeping equipment from failing in the first place. Really, this should be the principal focus of any maintenance force. The maintenance group does not want to be in the business of fixing broken equipment while plant capacity suffers. Rather, the maintenance group wants to take necessary steps to keep equipment in proper operation and maximize plant capacity. Nevertheless, the term reliability maintenance in industry usually refers to specific programs maintenance management undertakes with regard to keeping equipment from failing.

Many maintenance improvement programs in the area of reliability maintenance have various names such as reliability centered maintenance (RCM) or preventive maintenance (PM) Optimization. Even with varying terminology, most of these efforts seem to revolve around three principal activities. These activities or tools are preventive maintenance, predictive maintenance, and project work.

This section of the book briefly defines the basic concepts of these three tools and their general relationship to planning. Guidelines for specific interfaces with these programs warrant the inclusion of Chap. 10 Consideration of PM, Predictive Maintenance (PdM), and project work. Planning is intimately involved in the execution of this work. Chapter 10 defines specific tasks of planning with regard to how these programs operate.

Preventive maintenance

Preventive maintenance is maintenance activity repeated at a predetermined frequency. The frequency may be based on calendar time or other occurrences such as service hours or number of starts. For example, someone may change the oil in a car every 3000 miles or every 3 months. What is significant is that the company does not schedule the activity based on a particular noticed problem, but upon an expectation that the regular maintenance will reduce or prevent the appearance of problems.

Because existing problems do not direct PM, one of the key elements of a PM program is deciding what activities to include. Consider a modern steam plant with over 10,000 pieces of equipment. Identification of what items need repair is relatively easy compared to identifying what items need PM attention regardless of condition. Most PM activities are set up based on experience from past failures, equipment characteristics, and vendor recommendations. Preventing past failures from recurring decides many PM procedures. If a pump has failed a number of times from alignment problems due to fasteners coming loose, a PM procedure might be set up to check fastener tightness for the pump each month. As the plant ages, the company has identified many common failure patterns that are being headed off by prudent PM action. Equipment characteristics such as age, type, or criticality may also cause the plant to set up PM procedures. Newer equipment is more prone to infant mortality and may justify routine checks of operating conditions after initial installation. A new pipeline may receive frequent walkdowns after construction to verify leak-free service. The type of equipment may dictate inclusion in a PM program. Equipment with rotating elements may be checked for tightness. Engines may have oil replaced. Devices with filters may have filters replaced. Criticality to the plant process may determine inclusion in a PM program. A single, stand-alone pump that is vital to production may receive more attention than an identical pump used elsewhere with an installed spare. Finally, equipment manuals from the manufacturer or vendor usually recommend routine maintenance procedures to keep the equipment in good operating order. However, because vendors are not involved in the specific application of their equipment, technicians and operators intimately involved with the specific installations should usually be the final judge on which recommendations are pertinent. In addition, it is possible that vendors may have a bias toward recommending excessive maintenance on their equipment.

This concern of excessive PM is valid. How much PM is too much PM? One hears percentages suggesting that 10 to 20% of maintenance should be preventive maintenance. This raises two important questions. What is the other 80% of maintenance supposed to be? Should not all maintenance be preventive maintenance? Consider that if the plant spends a lot of time on breakdown maintenance, then the plant does not spend enough time on preventive maintenance. Consider also another type of maintenance, corrective maintenance. *Corrective maintenance* resolves deficiencies after they are found so that operations does not have a problem with an actual breakdown. One aim of preventive maintenance is to find deficiencies that need corrective maintenance.

A less commonly perceived benefit of preventive maintenance is active involvement with the equipment on a routine basis. Technicians servicing equipment should notice some equipment deficiencies through their proximity to the equipment alone. The technicians become used to seeing the equipment in proper working order and notice deviations from their expectations. A plant should prefer that maintenance technicians notice and report the majority of maintenance problems. Why should not this be the operators' duty? Are not the operators at the plant using the equipment around the clock whereas the maintenance technician may come by the equipment only once each month? The answer goes back to the concept of reliability maintenance. The maintenance department desires to provide the operators with a plant to run at capacity. The maintenance department wants to avoid a reactive mode where maintenance only quickly fixes problems operators report. Many equipment problems derive from gradual failure modes. The equipment fails over a period of time where symptoms of impending failure become more evident. The operator who knows the equipment is working correctly may be content to report only the end failure if it occurs. The maintenance technician has time in the frequent PM service schedule to notice symptoms of equipment deterioration. Therefore, the technician can report the condition in time for maintenance intervention on an efficient, scheduled basis before the operator ever has time for alarm. The operator could be more aware of failure symptoms, but the preferred direction of maintenance evolution is for the maintenance force to provide an operating plant. This movement is being accelerated as plant operation is becoming more automated. Plant operators used to make frequent rounds to check on equipment. Now with downsizing and automation, operators spend more time in the control room and less time out on the plant floor. Therefore preventive and corrective maintenance go hand in hand to reduce breakdowns and increase plant reliability. It may be that the routine inspection aspect of preventive maintenance to find deficiencies is more important than the servicing activities themselves. This reasoning is inherent in the strategy of some companies to give "ownership" of equipment to certain technicians because it provides a responsibility to keep in tune with the equipment.

After it decides what items to include in a PM program, the plant must actively manage the program. A review system needs to be in place to reduce or otherwise adjust PM frequencies as the plant gains experience with failure patterns. If an item receives PM attention continually and never develops a problem, should the program lessen the PM attention? If equipment keeps failing, should the program increase the PM attention? Is the PM attention causing problems? PMs must also be created for new equipment. Many times a company pours resources into developing an extensive PM program, but does not give the ongoing administration of the program much thought. If a plant wants to avoid problems and have most of a plant's maintenance as preventive in nature, it should pay attention to the ongoing management of the program.

Preventive maintenance is often confused with *planned maintenance*. The PM jobs are planned in the sense that the plant plans or expects in advance to do this work usually on a scheduled basis. While this is a valid assessment, under this book's conception of maintenance planning, planning is done for preventive, corrective, and breakdown maintenance. Planning is the act of making the job ready to go by the advance, preparatory work of a maintenance planner. Therefore, all PM work orders are considered to be planned work every time they are issued. This is because they possess a planned scope of work, time estimates, material estimates, and the other elements of a planned job. This is true even though the planner may not have to do any further work after the initial issue of a PM planned work order. On the other hand, corrective and breakdown maintenance may be done on a planned or unplanned basis depending on whether the work orders go through a planning process to make the work ready.

Because preventive maintenance is not only a vital tool of maintenance, but maintenance activity itself, planning has a large degree of involvement. The planning department usually sets up and standardizes the PM jobs. That is, once the PM jobs are planned for the first time, they continue to be issued with the same job plan at their predetermined frequency. The company also usually charges the planning department with managing changes to the PM program. As the plant maintains equipment, planners review equipment history and suggest changes to established PM frequencies. Planners coordinate PM recommendations from investigation teams that may analyze equipment failures. Planners also review equipment manuals when new equipment is received to determine PM requirements. Preventive maintenance and planning go hand in hand.

Preventive maintenance reduces or prevents plant breakdown situations. Including and adjusting the correct preventive maintenance activities involves a balance between excessive labor and the appearance of breakdowns. Maintenance planning is in a position to administrate the PM program.

Predictive maintenance

A predictive maintenance program (PdM) goes far beyond normal, frequency-based preventive maintenance. Techniques such as vibration monitoring, oil analysis, and thermography detect early warnings of serious equipment problems. Plants commonly attribute instances of preventing catastrophic failure of major equipment each year to PdM. In addition, the knowledge learned from analyzing equipment facilitates the use of new alignment, rebuild, or other techniques to extend the trouble-free running times of equipment dramatically. PdM analysis also delays much routine servicing, thus completely avoiding the significant potential of reassembling error inherent in rebuilding equipment.

The idea behind predictive maintenance is that modern technology can detect some equipment problems much earlier than other previous means. Say a certain pump develops a vibration problem that would lead to failure in about 2 months if left to become worse. PdM diagnosis might detect the problem within a week as PdM personnel monitor the pump on a route taking vibration readings. The vibration signature indicates a cracked impeller. PdM personnel write a work order to repair the impeller sometime within the month at the plant's convenience. The maintenance work can be planned and scheduled for efficient execution. On the other hand, if the vibration had been left to previous modes of detection, an operator may not have noticed a decline in pump performance until the impeller actually broke apart. At that time, the failure would have shut down the plant process and damaged the casing of the pump. Another scenario might have had the maintenance technician or operator notice marked vibration only several days away from actual failure. In this case, an already set maintenance schedule would have had to been interrupted. Both latter cases interfere with maintenance efficiency and increase the potential of plant capacity loss. The early PdM detection gives the plant ample time to prepare for efficient maintenance. The advance notice allowed by PdM also allows the plant to alter a routine overhaul strategy as well. Past plant practice might have been to replace certain fans or pumps based on a calculated life expectancy. The plant would rather replace the equipment while it still ran well than to take a chance of it suddenly wearing out at an inopportune moment. This practice not only meant the plant might waste plant availability and labor discarding perfectly good equipment, but the practice introduced new or rebuilt equipment subject to infant mortality.

The use of PdM might avoid all of these situations. Near the end of an equipment's expected life PdM monitoring might pronounce the equipment in good shape. Continued monitoring might mean years of additional service beyond any traditional replacement period. So PdM techniques can be extremely valuable in monitoring equipment condition to give additional life and then timely notice for maintenance needs.

PdM uses many techniques. Among these techniques are vibration analysis, tribology or oil analysis, infrared thermography, boroscopic examinations, and ultrasonic testing. Some of the techniques involve trend analysis where the single observation is not as important as the direction of the trend of readings. Is the vibration becoming worse rapidly or is it staying about the same? PdM groups also provide special services to the plant in connection with their technologies. These services include balancing of rotating machinery, laser alignment training, preparation of standards for equipment repair, and quality assurance checks of repairs made. PdM also implements continuous monitoring systems that reduce the need for traveling PdM routes. Some of the issues involved in using PdM techniques are placement of sensors and proper interpretation of data by PdM persons. New technologies are being developed continuously, and one of the responsibilities of a PdM group would be to evaluate new ideas.

Some of the techniques first utilized by predictive maintenance become taken over by maintenance technicians. Laser alignment provides a good example. After demonstration that a laser-aligned machine outperforms otherwise aligned equipment, maintenance usually becomes interested in having all its equipment laser aligned. Soon PdM personnel train maintenance technicians to use laser alignment gear checked out from the tool room.

Due to the different technical nature of the work, a company typically does not organize a PdM group under the maintenance department, but positions it similarly to the plant engineering group if one exists. PdM might be a subset of plant engineering or the plant technical services group. The PdM group might even report to a corporate department away from the plant organization. Nevertheless, the PdM department works closely with the maintenance group because of their direct involvement with individual pieces of installed equipment. Many maintenance technicians have even been placed in PdM groups working alongside engineers doing the diagnostic work. To initiate work based on PdM recommendations, PdM personnel write work orders just as do any other persons requesting plant work. PdM submits work orders to the planning department for prioritization along with the rest of the plant's backlog. This statement necessitates distinction of types of predictive maintenance work. Generally, the PdM group does diagnostic work monitoring and testing plant equipment with special devices. The work orders they submit are also considered predictive maintenance work in the sense the work was initiated by PdM personnel. Plant maintenance technicians execute these work orders.

In any case, planners do not do any PdM diagnostic work. The PdM personnel generate work requests. Planners plan work requests. PdM identifies work needing to be performed. Planning takes this work identification and prepares proper job scopes, time estimates, material and tool requirements, and other information necessary for scheduling and assignment of the work to plant technicians. It is impractical for planners to use PdM techniques to check on suspicious areas. PdM persons constantly check all candidate areas of the plant to find suspicious areas. Proper PdM inspection should already have caught areas operating deficient enough to be noticed. PdM inspection is not considered to be a part-time duty for other personnel.

The PdM tool particularly requires management commitment and organizational discipline, especially for a new PdM group or one beginning to use a new technique. The PdM group must be allowed to learn. Frequently PdM makes a wrong call. The PdM group might determine that a pump has a cracked impeller as a case in point. When the plant takes the unit off line and maintenance opens up the pump, the impeller might be in perfect condition. After a few of these errant work orders, animosity develops between the maintenance and PdM groups. The maintenance force feels the PdM group wastes plant time and resources. Plant management must promote the use of PdM and communicate the learning curve problem. PdM must be allowed to be on site when maintenance opens up equipment to check results to help guide future calls. PdM and the maintenance personnel do well to keep a sense of perspective and humor during these early experiences. Plant management does well neither to override PdM calls nor to become involved in individual plant circumstances too frequently. Rather, plant management should set a policy that the company has set up PdM to have the expertise to make the call.

On the other hand, PdM persons should respect the years of experience of the crafts and planners. Many engineers pressed into predictive maintenance service do not have years of witnessing specific, repeated failure modes. Plant technicians have personally witnessed specific problems in this specific plant for years upon years. Planners have particular expertise in personal experience as the plant typically selects top craftpersons for planners.

Plants prefer advance warning of problems to improve maintenance. Predictive maintenance techniques reduce instances of surprise catastrophic failure and extend the trouble-free running times of equipment. Routine servicing can rationally be delayed to reduce the potential of infant mortality after servicing. PdM knowledge also introduces the use of new routine maintenance techniques such as improved alignment techniques. Planning does not perform the duties of predictive maintenance, but plans their requested work orders for execution. Because of the learning curve involved in PdM utilization, management commitment becomes almost a requirement. Plants want to do predictive maintenance rather than reactive maintenance and recovery from breakdowns.

Project maintenance

Reliability maintenance also includes project work. Project work makes up an important part of maintenance strategy. A project results in an addition or modification performed on a one-time basis. Project work intends to make the plant or involved equipment better. While normal maintenance aims to preserve the function of equipment by keeping equipment in its present condition, project maintenance aims to preserve the function of equipment by improving upon equipment. If a particular piece of equipment has inherent features that lessen its reliability, perhaps the equipment can be modified to make it more reliable. The plant desires to do reliability maintenance, not breakdown maintenance. The plant desires to do more preventive, predictive, and project work, and do less reactive work.

Many companies have separate project, engineering, and construction departments for major projects. Their completed projects to install equipment and systems will have to be maintained in the future. The outstanding opportunity for the current maintenance department to control its future is in "front end loading." Maintenance departments in general have accepted a role that whatever is installed, they maintain. This is a true statement, but harmful if it inhibits the maintenance department from working with the corporate groups. The overwhelming best time to make equipment more reliable is before the company purchases the equipment. If maintenance acknowledges that certain equipment and system designs are inherently more reliable than others, then maintenance must spend adequate time with equipment issues before procurement and installation. Typical corporate engineers or project managers do not have the maintenance expertise existing in the maintenance department.

Maintenance might accomplish front end loading through a number of activities. Maintenance might establish equipment and vendor standards, assign certain maintenance supervisors or planners to spend time with corporate groups on specific projects, and give a review of proposed project specifications serious attention. Planning performs an important activity with respect to corporate project work. Plants often assign planners to work with corporate teams or to review project specifications and drawings. Planners must also collect project documentation and establish new files for equipment. The best time to establish files is before the manuals have been delivered and lost through a lack of designated document control. Equipment operation and maintenance manuals, as-builts, and project specifications should all be collected for future reference. Maintenance and planning groups must not solely concern themselves with maintenance of existing equipment. They must take advantage of the great opportunity to increase the reliability of future equipment. They must become involved with corporate projects that will place new equipment and systems at the plant requiring future maintenance.

In addition to corporate projects, the plant also carries out project work at the plant level. Any work order that modifies equipment or restores equipment to perform at a superior level may be considered a project. Changing impellers from bronze to stainless steel would be an example. The plant should continually be evaluating project ideas for making the plant more reliable.

Theoretically, maintenance, operations, and plant engineering distinguish themselves around these issues. Operations sees itself as responsible for operating the current equipment. Maintenance sees itself as responsible for maintaining the current equipment. Plant engineering sees itself as responsible for the current design. So plant engineering might take the lead in evaluating the idea to switch materials of construction for an impeller. Actual practice greatly blurs these areas of responsibility and rightly so. Operations and maintenance should consider and propose new ideas, although the plant engineers ordinarily have the time and technical expertise to evaluate certain changes to plant design. Just as in PdM practice, plant engineers do well to include the experience of other personnel in their deliberations. Plant engineering frequently consults maintenance planning for two reasons. Generally planners are top technicians in their crafts and the planners also have the best access to plant records and files.

Planners also take a lead in changing the plant design with limits set by plant practice. After noticing that a certain type of valve commonly fails in a specific situation, a planner may have the expertise and experience to plan for the craft to install a new type valve. Depending on plant policy, the process application, and specific persons involved, the planner may or may not consult with engineering or operations. The planners continually seek to improve plant reliability as the plant gains experience through years of maintenance practice.

What types of maintenance situations call for projects and how does the plant decide which projects to implement? Anyone who writes a work order for routine maintenance to correct a discovered deficiency might consider if a modification might make the equipment avoid future deficiencies. The company specifically gives the plant engineers the tasks of considering the current design and thinking about trouble areas. Planners should be vigilant to notice repeated maintenance situations that could be avoided. Perhaps the most important times to seek to identify a project opportunity is when the plant loses capacity. Any event that causes the plant to lose capacity should be scrutinized for a project. Many times maintenance reacts to a capacity loss with a temporary repair that becomes permanent. Maintenance never upgrades the repair after resolving the initial emergency. However, some plants assign a specific cross-functional team appropriate to the failure to address and find the root causes of any specified events. These specified events might include any capacity loss and certain types of regulatory or safety concerns. The team not only searches for the root cause, but recommends any future work needed to provide a permanent repair beyond the initial one already in place. Some plants also provide root cause analysis training to craft technicians. These plants have them fill out a root cause analysis work sheet after certain failures even if they are not directly tied to capacity losses. Much project maintenance is essentially the repair of a failed mechanism beyond its original capability such as using a better material or different type of component for the service. On the other hand, some project work calls for significant changes to plant design and may also be costly. Many plants have specific processes to approve and prioritize such work. The plant may have a limited capital improvement budget and cannot afford all the projects that otherwise make economic sense to implement.

Planning performs a key role in project work whether at the corporate or plant level. The planning department plans the specific work orders that the plant craftpersons will execute. The planning department manages receiving and filing new information received to assist future maintenance work. Planning also helps the plant execute a front end loading strategy utilizing its position of plant knowledge to assist corporate or any group to procure equipment that has reliable performance and ease of maintenance in mind. Upon receiving any work order to plan, the planning department fulfills some of its most vital roles. The planning department's experienced personnel see all the proposed work and the history of the affected equipment. Therefore, the planning department not only proposes or initiates project upgrades, they might also detect and head off unwarranted changes to equipment proposed by persons with less familiarity or expertise. Planning may resolve the concerns with the originators or coordinate concerns with supervisors, manager, engineers, or operations as appropriate. In addition, the planning group might hold off current work orders to repair certain systems if they know a current project in the works will make the work orders unnecessary. Planning plays an important role in making project maintenance more effective.

Project work comprises an important part of the reliability maintenance strategy. By modifying the design, project work intends to improve plant reliability. Project work consists of modifications at the plant level as well as large, corporate projects. In either case, planning is well suited to initiate and coordinate plant needs to improve maintenance involving projects.

To summarize, reliability maintenance focuses on specific strategies to keep equipment from failing and involves the overall areas of preventive maintenance, predictive maintenance, and project maintenance. Preventive maintenance attempts to prevent problems from ever occurring by such means as keeping equipment clean and keeping equipment from coming loose. Cleanliness keeps contamination away and includes activities such as changing lubrication and filters. Keeping equipment from coming loose includes attention to tightening fasteners. Preventive maintenance also provides a close association with the equipment to allow early detection of the symptoms that would indicate equipment deficiencies or impending failure. These situations can be resolved through corrective maintenance before problems develop. Predictive maintenance uses technically sophisticated diagnostic equipment or trending analysis to provide earlier warning of developing equipment problems. Project maintenance takes aim at correcting the root causes of failure by modification to equipment or systems. Maintenance wisely invests resources in spending time and expertise in front end loading with project groups before the plant procures new equipment.

The planning function helps maintenance management utilize the tool of reliability maintenance. The planning group leads and administrates the PM program. This includes planning the initial PM work orders and issuing subsequent repetitions. The planning group changes the frequencies and modifies job steps as necessary. Specialized PdM personnel lead the PdM program. These persons issue work order requests for the planning group to plan as any other work order. Planning plays a larger role in project maintenance, often recommending new projects for plant execution. The planning group coordinates project input from plant engineering and operations. For projects coordinated by a corporate project group, the planning group plays a key interface role. Planning also receives operation and maintenance manuals after the installation of new equipment and develops new PMs. Planning also receives other project information such as specifications and as-builts to update plant maintenance files. Planning plays a significant role in reliability maintenance.

Improved Work Processes

There are always opportunities to improve. Not only by creating new tools, but also by refining existing tools. Interfaces among the tools usually are also prime candidates. An environment should exist where the maintenance supervisors and technicians themselves initiate a host of miscellaneous new methods of doing business. Examples are standby crews, alignment techniques, and shutdown checklists. Rotating standby crews with electronic pagers reduce time to respond to off-hour emergencies. New techniques for alignment might come out of predictive maintenance experience. Checklists and advance plans facilitate work for short notice shutdowns.

Some managers tend to see areas such as management commitment, communication, and development of craft personnel as areas to take action to address the overall environment of the plant. Other managers tend to establish formal teams or programs to give employees time to meet and propose improvements to plant equipment or work processes. Still other managers consider the culture itself. The culture stems from tradition, peer pressure, perceptions of management bias, previous labor problems, home family situations, individual personalities, status symbols, work hours, and a host of other general and plant-specific circumstances. These managers address individual circumstances their management intuition leads them to feel will make a difference. In any case, management must actively manage to encourage attention to current practices and awareness of potential areas of improvement.

Maintenance Metrics

The term *maintenance metrics* simply means the measures and scores of particular maintenance activities or results. Maintenance metrics involve selecting, collecting, analyzing, and presenting maintenance data. Measuring the total number of work orders in a backlog is an example of a maintenance metric.

Ron Reimer (1997) of Eli Lilly declares that metrics enlist management support to change specific company behavior. Create a metric for the behavior and show it to management. Let management react to the metric. Behavior will change. There is a lot going on in most companies. Management commitment is real, but management needs assistance seeing exactly what is happening. Maintenance metrics help clarify particular situations for everyone.

The use of maintenance metrics dictates their value. Many companies collect a myriad of metrics without properly using them to manage. Management can make many errors regarding maintenance metrics. First, management errs when it does not create metrics that are necessary or would greatly assist making proper decisions. Second, management errs when it has metrics created that are unnecessary to consult when managing. Third, management errs when it has metrics created, but makes decisions without consulting them. Fourth, management errs when it consults the wrong metrics when making decisions. Finally, management errs when it creates and consults the right metrics, but misinterprets their meaning.

Management can make decisions without formal indicators, but they enhance many decisions by the use of them. Some early automobiles were made without a gas gauge. Motorists had to refuel regularly to avoid running out of gas. Later models included a reserve gas tank to be engaged when the primary tank ran dry. Modern automobiles with gas gauges allow the operators to manage their time and refill at their convenience. The gas gauge is a metric. Simple measurements of industrial plant fuel inventory allow managing purchasing decisions that affect the company's very ability to stay in business. On the other hand, creating and maintaining metrics can be very expensive itself. While a manager might dictate the establishment of a new metric in the 10 seconds it takes to voice the command, the company might expend 40 hours per month in analyst time carrying out the mandate. In 40 seconds the manager can establish four new metrics possibly requiring a full time analyst. The actual labor time to create and maintain the metric might be minimal, but it might be significant. One plant figured that a subcontractor might be taking unfair advantage of being under the minimal oversight of a company supervisor. The company assigned a full time engineer to manage the subcontractor. One has to wonder if the extra monitoring of the subcontractor was worth the extra cost to the company of a degreed engineer. Consider also the opportunity cost for the analyst or engineer. This person presumably performed some profitable duty for the company before being assigned to compile the new metrics. Is the task of providing the metrics to management a better investment of time than the previous duty? Is there another more profitable opportunity in which to employ a person currently compiling metrics? Management must allocate scarce resources toward the activities most profitable for the company. Management should not be compiling metrics just "because everyone else does."

Management's insistence to create and consult unnecessary metrics frustrates analysts who waste effort compile the information. It also leads to faulty decisions and inappropriate cries of "wolf" that lend skepticism to the use of any metrics. The company should also be leery of collecting data that "might be useful one day." Computers add to the temptation to compile unnecessary metrics. Computers allow collection of data much more easily in many cases, but computer equipment, computer operators, data input clerks, and analysts to manage new metrics also add substantial cost.

Nonetheless, just because management establishes the proper metrics does not mean management will consult them when making decisions. Management may have the appropriate metrics available, but choose to consult nothing or the wrong metrics when making certain decisions. Even when consulting the proper metrics, management may misinterpret the results. To illustrate these cases, consider a simple metric showing a backlog of work measured by quantity of work orders. Based on a large backlog, management determines there is work to be performed at the station and keeps maintenance personnel at the station for its completion, a proper management decision based on an appropriate metric. However, even without consulting this available metric, management might decide to keep personnel at the station. Management might even consult another metric instead. The other metric might be previous work hours at the station. If the previous work hours indicate a certain quantity of persons normally work maintenance at the station, management may decide to maintain the same assignments. This is faulty management reasoning. Past work hours may be an excellent indicator of plant cost, but does nothing to show how much work the plant needs in the present. The wrong indicator was consulted. However, even consulting the appropriate metric does not guarantee appropriate interpretations of meaning. A large backlog of work orders may mean personnel are needed, but a small backlog by itself does not mean maintenance personnel are not needed. Consider a management interpretation that a large backlog of work orders means the plant has too many breakdowns. The large backlog of work may mean the plant has a lot of broken equipment or it may mean that many corrective and preventive work orders have been identified to keep the equipment from breaking. Many times management actually prevents the plant from generating enough work orders to head off problems by applying an undue pressure to reduce backlogs. A small backlog of work may indicate either that maintenance is overstaffed or that maintenance is not generating enough work to maintain plant reliability. The value of maintenance metrics comes from their proper use. The proper use of metrics leads to better maintenance decisions to improve equipment performance and reliability.

The maintenance manager should work with the person who would be doing the compilation when deciding what to measure and how. Managers have the experience and know what decisions they could improve with metric data. The analyst may know what is practical to collect. For example, a manager may desire a mean time between failure (MTBF) metric to help the plant focus on its most troublesome equipment. Theoretically, equipment that has a lower MTBF has more problems because its frequency of failure events is higher than other equipment. From a managerial standpoint, MTBF seems to be a relatively easy metric to develop. Simply track the number of days between failures for plant equipment and report the average number of days for each. However, consider these issues that the analyst has to resolve. Suppose certain equipment has never yet failed? The analyst could somewhat easily resolve this by not including equipment until it does break or selecting an arbitrary past or present date to begin for all equipment. Equipment with a known last failure date could be started with that known date. Next, the analyst must decide what is a "failure." Suppose the PdM group requests balancing a fan based on vibration trending. Maintenance subsequently takes the fan out of service, balances the fan, and avoids a "failure." Suppose maintenance performs a monthly PM to lubricate the fan and tighten its fasteners. During this PM work, a technician notices that the fan casing has considerable corrosion and writes a corrective maintenance work order to refurbish the fan casing. Maintenance subsequently takes the fan out of service to refurbish the casing and avoids a "failure." Would the analyst classify the PdM balancing, the PM servicing, or the casing refurbishment as equipment failures? The fan failed to stay in balance and the casing failed in the sense it corroded. However, in none of the cases did the fan fail to provide the plant with its intended gas moving service. Yet both maintenance operations caused the fan to be taken out of service. What if instead of needing balancing, PdM discovered extensive fan blade damage requiring the fan to be taken out of service immediately for two weeks? What if the plant could wait a week before taking the fan out of service? What if instead of PdM, the operator noticed unusual noises from the fan that upon inspection dictated a rebuild even though the fan was performing at its design level? Did the fan fail? The analyst also realizes that certain machines are rebuilt after a specified service life. Consider a pump that is on a 5-year rebuild schedule. Would the rebuild be counted as a "failure" for the purpose of MTBF? Probably not. However, the plant sometimes experiences infant mortality of equipment after rebuilds. Would the days between the rebuild and the new failure or between the new failure and the last real failure be utilized for the metric? The analyst would have to define what "failure" means for the metric. Next the analyst must consider whether to take different components on a single piece of equipment into account. One time the fan impeller fails and the next time a bearing on the fan fails. Is MTBF being tracked for the fan overall or for type of failure? Probably for the fan overall. Even so, the MTBF from the last impeller failure is hardly a measure of repair quality on the impeller if the bearing caused the next failure. Would the analyst be able to extend this reasoning and track MTBF for a whole gas path system if one time the fan fails and the next time a damper fails? Next, the analyst might have to decide how to interpret data. If a particular valve expected to last only 3 months in severe service has a MTBF of 6 months, how does that compare to a pump expected to last 5 years that has a MTBF of 2 years? Moreover, is the MTBF of two years statistically significant if only measured over two or three failures? Finally consider from a management perspective that failures should not happen. Maintenance's goal is to preclude failure, not simply lessen its frequency. Each failure might dictate changes in maintenance practice, perhaps a better quality rebuild, perhaps more PM attention, and perhaps more frequent PdM observation. The point of all this discussion is not to resolve the proper application of a MTBF metric. Rather, it is to suggest that management might dictate a metric without fully considering its practicality of being compiled or interpreted and its actual benefit in use. The manager with the analyst together makes a better team to decide what to measure.

Carrying this reasoning further, many managers involve supervisors in not only selecting what to measure, but in compiling the metric. A supervisor may be judged by the proportion of reactive work orders compared to proactive work orders for the equipment under his or her control. That supervisor might be more interested and better motivated when having a hand in collecting the data personally. Of course, administrative work can easily mount up and keep a supervisor from the prime duty of a field presence with the crew if the manager is not careful in this regard.

The purpose of this section is more to caution that the value of metrics comes from their proper use than to enumerate necessary metrics. However, it is appropriate to point out a framework of metrics and mention several pertinent ones regarding planning. Chapter 11, Control, further discusses the use of indicators for planning.

Considering a framework for metrics, the ultimate measure of quality is plant availability of capacity. The plant must be able to run to produce a product to sell for profit. It does not matter how hard everyone is working if the plant has poor availability. Plant availability is rather easy to measure, but difficult to attribute to specific maintenance factors or break down into availability for specific systems. For instance, condenser availability may be poor, but masked by even worse boiler availability. Every time the unit comes off line for boiler repairs, maintenance personnel also patch the condenser. Breaking down the metric for plant availability is not as simple as creating subindicators. That is why this chapter explains that all of the aspects of maintenance are interdependent and a conscious and motivated management must properly integrate them to achieve plant reliability. Metrics may be difficult to tie directly to plant availability, but must assist the manager to make better decisions related to improving plant availability. Then, only after the plant achieves high availability, the plant may consider efficiency. Managers can establish efficiency indicators much easier normally than availability indicators. Management may examine overall cost, cost per unit, overall labor hours, labor hours per unit, overall fuel use, or fuel use per unit of product produced. A concern here is that the manager should not dictate an improvement in the metric by simple math. For example, the manager knows fuel use cannot be reduced by simply dictating a 20% reduction. Why would the manager dictate the use of fewer persons as a means of achieving fewer persons? In other words, if it takes 10 craftpersons to maintain a plant, then simply reducing the labor force to 8 persons without changing some process or at least considering the existing work situation may not be intelligent management. Availability or reliability comes before efficiency and these areas are the ends of all metrics.

Chapter 11, Control, reviews pertinent metrics associated with planning. These indictors include planned coverage, work type, schedule compliance, and backlog of work hours among others. Chapter 2 covers a wrench time metric or indicator as a specific principle of planning. Planning helps improve the amount of work time where a craft is directly working on a piece of equipment rather than wasting time such as to gather parts, tools, or instructions, or even travel on the plant site. Planning does not improve administrative time such as time spent for training, meetings, or supervision, or vacation, the latter obviously. Administrative time can be quite large. Of the nonadministrative time, the plant desires that most of the work done in the plant be planned to maximize its efficient use of resources. The plant terms the planned portion of total work hours spent as planned coverage. The plant includes PM hours as planned hours. An associated metric is work type. The plant desires most of its work be on work to preclude or reduce failures rather than work on breakdowns. Schedule compliance measures the success of the plant in guiding its workforce productively toward work to increase plant reliability. Backlog kept by planned work hour assists the scheduling process and staffing process of the plant. Management considers these indicators associated with planning.

Notice that the above indicators primarily make use of hours, not numbers of work orders. Some studies indicate that, on the whole, work order counts can approximately show where time is spent. This suggests that in certain circumstances the instances of small work orders and large work orders balance out without skewing metric meaning. For example, a metric showing that the plant completes more preventive maintenance work orders than reactive work orders would suggest the plant spends most of its time heading off breakdowns. On the other hand, the plant might actually spend more work hours on reactive work. The metric might not reveal the true situation if the reactive work typically requires large work orders, whereas the preponderance of PM work orders are small tasks. The reactive work orders might frequently require two or more persons for several days. The PM work orders might normally require only several hours to make inspections or minor adjustments to equipment. Simple quantities of work order metrics are easier to establish than metrics using work hours, but they must be used with more caution. Work hours usually reveal proportional metrics better than simple quantity of work orders because productivity usually is the same across different types of work orders.

Even using work hours requires some judgment. Work hours are not only more difficult to collect, but can be difficult to define. Consider a work order that planning has estimated to take a single person a single hour to complete. However, the assigned craftperson took 5 hours to complete the task. Then the person reported that the job took 2 hours on the job feedback of the work order. Finally, the person reported 8 hours on the daily time sheet. The technician had no other jobs assigned that day so he took his time. When writing the work order feedback, he knew he had taken more than the estimated 1 hour, but was reluctant to claim it took 5 hours so he wrote down 2 on the work order. The timesheet was more problematic for the technician because it had to total to 8 hours for the entire day and he only had one job. The technician therefore used the only work order assigned to allocate time on the required daily timesheet. Payroll required a timesheet from each technician every day. Did the job cost the company 1, 5, 2, or 8 hours? The company certainly paid the technician for 8 hours, but the experienced planner had determined that the job warranted only a single hour. Did the low productivity cost any time? Did the supervisor's method of assigning only a single job at a time cost any time? Would a methodology of scheduling enough work to fill up a technician's day help reduce the cost of individual jobs? These are some of the considerations one must make when dealing with actual work hours. The job actually cost the company 1 hour. The maintenance process cost the company 7 hours. The timesheet is the desirable reporting device for hours because it accounts for all paid hours. Maintenance planning aims at remedying this problem. Management desires to have more hours on planned and scheduled jobs as a means to reduce this discrepancy.

Maintenance metrics are another tool of the maintenance manager. The use of the maintenance metrics dictates their value. Is the quality of management decisions proportionately increased? The maintenance manager should work with the person who would be doing the compilation when deciding what to measure and how. This is because there are so many judgments and interpretations involved with even some of the most apparent straightforward measures. Maintenance planning makes use of several metrics in particular dealing with work hours.

Summary

Planning is only a tool, but planning is a key coordinating tool that assembles, integrates, and helps manage many of the other tools of maintenance. These other tools include a work order system; equipment data and history; leadership, management, communication, and teamwork; qualified personnel; shops, tool rooms, and tools; storerooms; continual process improvements; and maintenance metrics. In addition, management considers essential reliability maintenance composed of PM, PdM, and project maintenance. On one hand, next to a work order system, maintenance planning is one of the most valuable tools maintenance management has. Yet on the other hand, maintenance planning is useless without the other tools. Maintenance management seeking to maintain high plant reliability places due emphasis on all the various aspects of maintenance. Nearly all of these areas present opportunities for management to improve their contribution to maintenance success.

The People Side of Planning

This brief appendix contains an important word about the people side of planning. Management must be cognizant of and deal with this issue in order to manage the planning program. An old adage says that engineers who get Master of Business Administration (MBA) degrees think of people as numbers. But, people are not numbers. People are humans. In fact, companies humanize machines ahead of persons. They think of "taking care of" and making a machine "happy" so it will not break down. Even the newer phrase "people are our greatest assets" could be construed to reduce persons to things.

The American Heritage Dictionary (1980) defines culture as "The totality of socially transmitted behavior patterns, arts, beliefs, institutions, and all other products of human work and thought characteristic of a community or population." The maintenance workforce is a community and the beliefs of the maintenance community affect its behavior and the work it accomplishes. Many companies wrongly think culture is simply a matter of "organizational discipline" which shows evidence of these companies' machine mentalities. Companies that do not recognize culture as a vital aspect of maintenance cannot fully engage the power of the workforce. Companies that do recognize culture as being important take into consideration employee beliefs and thought characteristics. This is a not a manipulative effort on their part, but a true effort to engage the minds of their employees as well as their hands.

Furthermore, frontline leaders, the crew supervisors, are the keepers of the plant culture. What these persons believe is the embodiment of the plant culture and affects how the workforce behaves. Supervisors do "take care of their people." The front line employees interact daily with the crew supervisors and express the frustration, problems, ideas, beliefs, and circumstances of daily work. To each expression, the crew supervisors respond with their vision of what is going on. This may or may not be the management vision. Why would the supervisors have a vision apart from the management? The supervisors form their vision based on management's actions more than management's stated vision. For instance, management expresses a vision of having a first class maintenance program. The supervisor sees that management has not allowed replacement of three retired mechanics and won't make the storeroom carry a routinely used item. The supervisors form a vision that management doesn't care. In spite of this, the supervisors devise ways to obtain the nonstocked item anyway and make provisions not to work the remaining mechanics "to death." Thus the employees feel that supervisors care even if management does not. The wise management recognizes the impact of the supervisors on plant culture and effectiveness.

The People Rules of Planning

In working specifically with planning and scheduling over the past two decades, the following themes seem to be recurring topics that greatly affect the emotions of maintenance organizations. Management that honestly and openly addresses these concerns and emotional needs are better able to implement the significant organizational changes a planning and scheduling program requires. The following themes have been expressed as rules to guide program thinking.

Rule 1: The planning program is not trying to give away the plant's work to contractors

Many technicians resist planning programs because they feel the plant is trying to write up formal job plans in order to obtain contractor quotes. If sufficiently scoped and planned, they think contractors can easily do the work instead of the plant technicians. In addition, the technicians fear giving feedback to improve the information and expertise the planners can provide to the contractors.

In general, plants should prefer to use their own regular workforce which is already being paid to be on hand, than engage additional resources. In addition, some maintenance forces find they have to be on hand to guide persons unfamiliar with the plant or go behind them to ensure proper maintenance. Only in the case of unusual specialty work or periods of unusually heavy work would many plants consider contractors.

The planning program is also oriented more toward in-house labor. Principle 5 requires taking into account the skill of the crafts and not waste time including unnecessary details that a skilled technician should know. This thought pattern extends to not wasting time including information that persons at the plant should normally know.

Planners cannot waste time on the multitude of job plans making them complete enough to give contractors firm specifications for work. Moreover, planning tries to fulfill a mission to be a technician library service so that technicians can use lessons of the past without having their individual lockers full of poorly organized technical details for various pieces of equipment. Planning is trying to give the technicians a head start, not give contractors a complete job.

Tell supervisors and technicians that the intent of planning is not to give away the plant's work. Tell them that planning is part of the maintenance group and not part of a projects group to engage outside contractors. Tell them that the planning program is part of an effort to increase the efficiency of the in-house group, so it can do as much of the maintenance as possible without contractors.

Appendix Q further analyzes contracting.

Rule 2: Planners cannot plan the perfect job

Management may have implemented planning in the past under an incorrect premise claiming that technicians would never have to hunt for parts because the planner would identify them all ahead of time. Technicians soon received job plans that did not have perfect parts information. The planner simply was not able to identify all the parts correctly for many jobs. This really did not bother the technicians too much as long as the planner was willing to help straighten out the job plan by searching for the extra parts the technicians discovered were necessary once the job began. However, the technicians also received a lot of unplanned jobs without any parts identified and would ask for planner help to search for parts for them as well. As a result, management was frustrated that the crews did not work much planned work.

The problem is that no one can possibly anticipate all the parts needed for maintenance work ahead of time. But because equipment needs repeated maintenance service over time, proper planning sets up a system to improve maintenance plans continually over time.

Tell supervisors and technicians not to expect perfect plans but head starts. Explain the cycle of maintenance and position the planning department as a library service to use technician feedback on imperfect jobs to improve future plans. Confide that it's expected that plans will not be perfect and they are only a starting point based on the information available. Tell them that if they run into problems on jobs they should resolve them as best they can consulting other technicians, supervisors, file information, or even vendors. Whatever problems they run into must be reported as feedback and they should expect planning to fulfill a solemn duty to use feedback information on future work so technicians will not have to "reinvent the wheel." Proper frustration is aimed at planning only if they fail to use the feedback information. Technicians should only have to hunt for the correct part a single time.

Rule 3: Planning is not designed to take the brains out of the technicians

Planning is not a plant strategy to allow hiring less skilled labor. Planning counts on skilled technicians receiving the job plans to execute the work. Planning does not waste time including information on job plans that skilled technicians should know. Including enough information on every job so that anyone could execute the work would slow the planners to a point where they could not plan all the work. In addition, the plant hires and trains skilled technicians because even with excruciatingly detailed plans, not just anyone could execute work requiring competent maintenance journeymen. Management should not implement planning expecting that it can now hire less competent technicians or reduce training needs for the current workforce.

Tell supervisors and technicians that the planning system depends on their skills and competence. The planners add value to the current practices by viewing and using historical efforts by these competent technicians to improve job plans. The planners then preview jobs to give the technicians a head start, but expect the technicians to respond to the actual circumstances of the job using all of their skills and abilities to perform the correct work needed.

Rule 4: The technicians own the job after the supervisor assigns it to them

In the past before the plant had a planning organization, when a supervisor assigned work to maintenance technicians, the technicians "owned" that work and responsibly made any decisions necessary to produce the proper maintenance results. Now with planning, many planners have a tendency to feel that they "own" the jobs even into the execution phase of the work. Sometimes the planners complain "The technician did not do the job as it was planned." Planners do not own the work after they finish planning it. The planning function gives the technicians a head start and then the technicians own the work to execute however they deem necessary. The proper planner attitude should be that they simply want the feedback about what the technicians actually did. Competent planners work with competent technicians jointly to develop better and better job plans to meet true equipment needs.

Explain to planners that they do not own the jobs and that it's okay for the technician to deviate from job plans. On the other hand, instruct planners that it is quite proper to complain on completed jobs about not getting enough feedback to understand what work the technicians actually performed.

Rule 5: Planners cannot make the perfect time estimate

Anyone that has ever worked on their own automobile or plumbing can testify that a job planned for a few hours one morning can easily take all day or a job thought to take a whole day might suddenly be finished in only an hour or two. Such is the nature of maintenance. The vast majority of maintenance work consists of jobs that technicians can accomplish within a single work day whether the jobs are estimated for a few hours or for an entire shift. On the other hand, maintenance work is not comparable to assembly line operations where industrial engineers can map out and set exact time standards. Maintenance work might mean technicians responding to equipment needs on any one of thousands of pieces of equipment in the plant. And each time the equipment is serviced, it might be by a different technician. Plus, the equipment might require attention only once or twice in an entire year. The best the planners can do is use their own extensive experience to *guess* how long a plan might take to execute if a good technician experiences a smooth job with nothing unexpected happening. As discussed above simply with parts, it is also impossible for a planner to anticipate correct parts on every job. The planners can consult history files for previous job actual times, but must remember that each of these jobs had particular technicians and particular circumstances of its own that affected the time. Thus, management cannot expect the planned jobs to have perfect job estimates for labor hours.

Of course, as explained by Chaps. 5 and 6, the accuracy of the estimates for individual jobs are only "plus or minus 100%." But these estimates are normally distributed and as many jobs run over their estimates as run under. This distribution means that given enough jobs, the estimates are very useful for scheduling a week's worth of work for an entire crew. They are also helpful for attempting to assign the proper amount of work to any technician for a single entire day.

Tell planners not to worry that they cannot magically estimate exact times for jobs. Tell them that management selected them in part based on their craft expertise and to take their best shot at the time estimate and forget it.

Rule 6: Management cannot hold technicians accountable to time estimates for single jobs

Because of the wide variation of the accuracy of individual job time estimates, technicians cannot be held accountable for hours expended on individual jobs. The planning system allows "best practice" assignment of work. Planners can allocate a reasonable amount of work for a crew for a week and supervisors can assign a reasonable amount of work for a technician for a day. Maintenance productivity improves because the plant assigns a reasonable amount of work.

Tell supervisors and technicians that it's okay if the hours are not met for a single work order. The technicians should aim their efforts at completing all the work assigned for a day. Even then, it's okay if all the technicians cannot complete all the work assigned in a single day. Tell them that time estimates are only estimates, but over the long run it is expected that on many days technicians will probably be able to complete all their work.

Rule 7: Showing what is not correct is often as important as showing what is correct

Planning has such a great potential for adopting bad practices that it is essential management discuss these practices. It is not proper for planners to help out on crews doing either technician work or helping them "chase parts." Planners cannot adequately retrieve records if they are stored in large system files rather than by their individual component equipment numbers. And planners need not put extensive job steps into job plans when those steps are unnecessary for the skilled technician. Talking about these false steps helps the planner better understand the correct method of achieving planner objectives. Many planners feel they are fulfilling the principles when in fact they are doing the exact opposite.

On the technician side, feedback on job plans gives the technicians the opportunity to say what is not right about an individual job plan. Incorrect steps can be crossed out and correct steps written in by technicians.

Explain to planners not only the correct principles of planning, but specifically mention incorrect actions as a way of clarifying expectations. Tell supervisors and technicians to feel free to mark up job plans and correct them.

Rule 8: Planners do not add value if they help jobs-in-progress

Planners might help jobs-in-progress for several reasons. Planners sincerely want to help technicians with jobs-in-progress. In the case of a planned job, it seems only "right" that a planner help straighten out a job not planned perfectly. Shouldn't the planners "redeem" themselves by helping correct their plans? In fact, the technicians feel very strongly that planners have an obligation to correct their errors. And for unplanned work, the planner often has the superior data and computer skills in the maintenance group for finding information. Why wouldn't the technicians seek out planners for help on troubled jobs-in-progress? It is very natural for planners to help jobs-in-progress.

However, if helping current jobs, planners merely replace activities the technicians used to do by themselves, with supervisors, with peers, or with vendors. Planners add value when they look at jobs and history before the job is assigned to give the technicians a head start. The problem with helping jobs-in-progress is that since planners cannot plan perfect jobs there are too many jobs that would require their help.

Helping nearly every job that requires more information keeps planners from focusing much attention on future jobs and the planning cycle of improving work through feedback collapses. And the scheduling program cannot advance because the backlogged work does not have planned time estimates. The future work being improved from planning and the scheduling process constitute new activities to the maintenance effort. Helping jobs-in-progress merely replaces what other persons were doing and does not add value.

Explain to planners why they do not add value when they do technician work for the technicians. Tell supervisors and technicians not to expect perfect jobs, but do their best to execute the work and resolve problems they can report as feedback. Explain to them the concept of added value and the importance of allowing the planners freedom to plan future work.

Rule 9: Everyone is an adult

This is a variation of Theory X and Theory Y. Theory X and Theory Y state that if management treats employees as if the employees are lazy, the employees will act as if they do not want to work, but if management treats employees as if they want to work and do a good job, they will. It is basically a self-fulfilling prophecy. In the case of maintenance, management should share information with technicians.

The most widely abused aspect of this concept occurs when plant policy dictates not putting time estimates on the technician copies of the job plans. In other words, if the planner plans a job for say 3 hours, the technician will not have the 3 hour estimate on the printed out work order. The thought behind this practice is "If the technicians can do the job in less than 3 hours, they will drag it out for the whole 3 hours!" One problem with this thinking is that without the time estimate, what keeps the technicians from dragging out the job for the whole day? Another problem is without the time estimate, how can the technicians let the supervisor know they are running into unexpected problems? If problems arise and the technicians know the estimate is 3 hours, they would notify the supervisor that it looks like the job will run over a few hours. Without the estimate, why would they notify the supervisor? If the estimate was a 16 hour one and the technicians don't understand why the job should take more than an hour or two, the discrepancy also brings up communication. They call the supervisor or reread the job plan to make sure they understand what is needed. Why would the technicians speak up if they don't have a time estimate that might indicate a problem with understanding the job plan?

Proper job assignment through daily scheduling greatly relieves any concerns that technicians might "milk" job plans. Simply by scheduling technicians a full day's worth of work minimizes the opportunities to slow down on any particular job. For example, don't simply assign one 3-hour job with the instruction for the technicians to come back as soon as they are finished. If they are working an 8-hour shift, assign the 3-hour job plus a 4-hour job and a 1-hour job. If they finish the 3 hour job in 1 hour, they need to start the 4-hour job because it might take 6 hours. And the 1-hour job may take an entire hour or more. Well, what if the technicians finish all three jobs (8 hours of work) in only say 5 hours? The quick management response might be that they were not given enough work. But isn't finishing three jobs better than finishing just the one job which might have been all that was completed without assigning a full day of work? At most maybe only two jobs might have been completed if the technicians came back for more work. Management's case for not placing time estimates on the work orders seems to be a case for limiting the amount of work completed.

Don't play games by not including time estimates on the work orders the technicians receive. Give the technicians as much information as possible on job plans.

Rule 10: Everyone should enjoy their work

Two areas of planning particularly concern enjoying work. One is a technician fear of downsizing and the other is a management drive for productivity that doesn't recognize human social needs.

Companies that engage planning with the idea of creating room for downsizing may not find planning successful. Planning benefits from maintenance-wide support. Planners need technician feedback. Planners need supervisor acceptance of a weekly work allocation. The fear of downsizing greatly diminishes enthusiastic support of new approaches such as planning.

Don't push for planning so that downsizing can follow. If the plant must downsize, do so and then implement planning to pick up productivity so that everyone can keep their jobs.

Management also sometimes has such a push for productivity that it cannot stand basic human social interactions on the shop floor. John Naisbitt in his acclaimed book *Megatrends* discusses High Tech, High Touch. This concept predicts that technology will never progress to the point where persons do not have to interact with each other on jobs. The more technology advances where persons can work apart, the more they create reasons for getting together. This means that occasionally technicians do chat with each other at work, but that does not mean productivity will suffer. Management best handles productivity by assigning enough work through planning and scheduling and allows employees to talk as much as they want to if they are getting all their work done.

Don't worry about supervisors and technicians spending too much time talking. Ensure that planners allocate a week's worth of work each week to each crew and that supervisors assign a full day's work each day to each technician. Then monitor Schedule Success.

Rule 11: Everyone should go home at the end of each day knowing if they have won or lost

Assigning a full day of work to each technician at the beginning of each day greatly *reduces* stress among technicians. In plants with the bad habit of assigning one job at a time, self starters and technicians who need more supervision each feel a certain amount of anxiety. The self starters finish jobs as soon as possible and go back for more work. Yet they do wonder what the next job will be and if it will be a worse job than the one they just finished. In addition, they also become an object of peer pressure to slow down to the average pace of the workforce. It also aggravates them that they push and push each day to accomplish as much as possible, yet the supervisor accepts the slower pace of the others. And pay is equal among the same rank crafts. Along the same lines and even worse, less motivated technicians may even dread the prospect of the unknown, of the next job. They may be currently working on repairing a ground level water line. What if the next job involves a down-in-the-mud sewer line? Maybe it is best not to hurry along the current task. They feel the faster technicians make them look bad. They wonder how fast should they go and how much should the other technicians slow down? What is a reasonable pace that the supervisor will accept? All of this can be avoided by simply assigning all the work at the beginning of the day. But how much work is enough work? Using the planner hour estimates allows the crew supervisor to assign the right amount of work. Yes, the estimates are plus or minus 100%, but given several jobs helps dampen out the variation, and having a respected planner make the estimates makes the daily assignment reasonable. Technicians then have a goal of how much work to do each day. They can speed through the day and ask for more if they finish everything. Or they can pace themselves to finish at the end of the day. And stress drops because the technicians know what the next assignment is. They know what to look forward to. Even better is the supervisor making the assignments the day before to reduce the stress of "I wonder what is going on tomorrow?"

Help the supervisors develop a daily scheduling routine not only as a tool for working down the weekly schedule to increase productivity, but because technicians appreciate seeing all of what is expected of them each day. Then they can go about accomplishing that goal.

Rule 12: Wrench time is not strictly under the control of the technicians

A low wrench time does not indicate lazy, slow technicians. Wrench time merely records the amount of time on jobs. It does not even measure how fast they work once they are on jobs. Wrench time identifies weaknesses in support areas. If the plant has a terrible storeroom without necessary parts, a situation where special tools are hard to find, technicians having to share hand tools, poorly written work requests, minimal explanations of work desired, and a system of assigning a single job at a time, how can technicians be responsible for low wrench time? Are technicians spending a great deal of time trying to find parts? Is trying to get the proper tools an area of common delay? Do technicians frequently have to leave the job in order to get more instructions? All of the preceding delay areas occur *during* specific jobs. In most cases, wrench time is low because of delays *between* jobs. This may be evidenced by high break time, late arrivals, early departures, or high unaccountable times. Are these signs of slow technicians? Not necessarily. Experience shows that they usually indicate supervisors are not assigning enough work. Many plants charge the technicians with doing as much as they can and assign them one job at a time. Technicians should not be responsible for finding their own jobs. Holding them responsible to find their own work leads to a general sense of apprehension in the workforce and average productivity finding a middle ground of mediocrity. Wrench time is a measure of the overall productivity of the work system, including planning and scheduling. It is not strictly under the control of the technicians.

Tell supervisors and technicians that wrench time measures how well management is supporting maintenance in terms of adequate warehousing, tooling, job plans, scheduling, and work assignment. Tell them management evaluates the delays identified by wrench time studies to see how it can better support technicians doing their work. Management does not want to pay technicians and then have a system that prevents them from actively working on equipment.

Rule 13: Schedule compliance is not strictly under the control of the crew supervisors

Just as wrench time is not under the control of technicians, schedule compliance is not under the control of crew supervisors. It's "okay" for the supervisors to break the schedule and work on an unscheduled job or even an unplanned job. In fact, the supervisor had better abandon the daily schedule even if it means reassigning the entire crew to work on an emergency breakdown. And, the supervisor had better deviate from the weekly schedule allocation to attack an urgent breakdown that can wait until tomorrow but not next week. Thus, the schedule compliance is more a measure of how reactive is the plant rather than how compliant is the supervisor. Many plants entering into a formal schedule program tie supervisor pay to schedule compliance. This is the definition of stress: Should I meet the schedule to help my pay or help the plant with emergency work and hurt my pay? In reality management does not want perfect schedule compliance, it wants perfect plant performance. Perfect plant performance is best achieved by maintenance crews quickly reacting to emergency work and productively pursuing regular maintenance work in the absence of emergencies. Productivity improves when supervisors begin each week with a weekly schedule, supervisors receive help setting up a daily schedule routine, and management examines weekly schedule compliance scores. Management does not have to tie supervisor pay to schedule compliance.

Tell supervisors that management wants them to start each week with a goal of work to accomplish and management is interested in knowing each week how much work got done. Tell them that management wants them to use some sort of daily scheduling. Tell them that the intent of scheduling is to line up and accomplish as much work as possible to reduce potential emergencies from happening. The measure of success of this strategy is the schedule compliance score because it measures the reactive nature of the plant. In fact, call schedule compliance schedule success because everyone wants the schedule to succeed. But also very clearly tell them that it's "okay" to break a schedule because it was just a starting point. Don't tie pay to schedule compliance.

Rule 14: It is better to train employees and lose them than to not train them and keep them

So much new equipment is technologically superior to items of the past that technicians, planners, and supervisors need ongoing training. In addition, techniques of modern maintenance and management improve over the years. Companies should take time to invest training in employees because these employees have careers spanning decades. One problem with training employees is losing them to other companies, especially the most valuable employees. Often these employees have improved their value through acquiring advanced skills and abilities. However, a problem with not training is ending up with more and more employees that are not contributing as much as they could with better skills. Instead of training, some companies hire more and more outside contractors and consultants to provide a technical expertise lacking in the plant. The medium range result of not training would be paying an unskilled workforce on hand and also paying a skilled consultant. The longer range result of not training would be losing the workforce through retirement and having an entire contractor and consultant base doing the maintenance. Or it would mean having a newly hired young workforce on hand and still having to hire contractors and consultants. Sitting back and taking a proper perspective seems to indicate that a company that wants an effective in-house workforce should train its employees continually. Extra costs for hiring and training new employees for ones lost and extra costs for raising wages to keep skilled employees would be offset by savings of not hiring outside forces and having skilled persons on hand familiar with the plant and its equipment.

Train employees and pay them what they are worth. Be conscious of the false savings of not spending money training.

Rule 15: Modern maintenance needs to do less with less

In Joel Leonard (2003) said that "Companies will continue trying to do more with less until they can do everything with nothing!" This statement points out the unreasonableness of simply cutting maintenance resources in an attempt to improve the bottom line. Nevertheless, this philosophy appears to guide many companies when they reduce maintenance budgets while at the same time add activities and duties to maintenance staffs. In reality, companies should invest more resources where they create a sufficient return or a profit and reduce investments in nonprofitable activities. Companies calculate net present values or returns on investments for judging new investments in capital equipment. Companies should also consider the returns or lack thereof when cutting investments in maintenance, whether it be labor, spare parts, or tools. Unfortunately, maintenance has a hard time quantifying financial figures concerning what might break if the plant does not replace a retired electrician. In this environment, Larry Bradley (2005), Technical Superintendent of St. John's River Power Park, suggests "doing less with less." He says that maintenance must recognize and stop those activities which do not produce sufficient return. For example, Bradley's plant stopped the routine calibration of pressure gauges that were not tied to transmitters. Instead, operators should report obviously faulty gauges and report process problems that might lead to performing special calibrations. This saves a great quantity of instrument technician hours each year.

Planning enters this discussion in two ways. One, companies that add new programs to maintenance each year while reducing its budget have a hard time implementing each new program. Maintenance may view planning as another program it will not have the resources to operate. Two, planning does have the potential to help identify areas of maintenance excess. Planners can see histories of individual equipment needed excessive attention and suggest projects to engineering. Planners can help supervisors not waste time assigning incorrect craft skills for uncertain time periods. Planners can help technicians not waste time repeating searches for parts, tools, and instructions that were sought out the last time the equipment was serviced. Less time spent on unproductive activities is less work for maintenance.

Carefully introduce planning as not another burden on maintenance, but as a means to help reduce wasted activities to make technicians more productive such as better utilizing lessons of the past through a library service.

Summary

The successful implementation of planning requires attention to the people side of planning and maintenance. The planning program profoundly impacts the maintenance organization because it completely changes the business process of maintenance. Because organizations consist of persons, management must consider the culture and mind-set of these persons. Carefully considering a number of issues that affect the beliefs and thoughts of maintenance employees greatly improves the likelihood of success for a maintenance planning program.

What to Buy and Where

The intent of this appendix for suppliers is to:

1. Identify the materials mentioned in this book helpful to a maintenance planning operation.
2. Identify the type of supplier for the material.
3. Give a starting place for procurement. The intent of this section is not to endorse any particular supplier.

Minifile Folders

These folders for creating equipment files are pockets to keep loose documents from spilling. The reinforced side gussets reduce tearing as persons repeatedly tug on the pockets to remove the file folders from the shelf. Three sizes listed below allow having small folders for most equipment, but larger sizes for equipment files expected to hold more documentation and manuals. Other styles are available besides one with side tabs designed for an open filing system. These products are available through many local office supply stores.

Smead or Pendaflex expanding end tab pockets
Legal size   Straight cut $1^3/_4$-in expansion

Smead or Pendaflex expanding end tab pockets
Legal size   Straight cut $3^1/_2$-in expansion

Smead or Pendaflex expanding end tab pockets
Legal size   Straight cut $5^1/_4$-in expansion

Company: The Smead Manufacturing Company
Phone 651-437-4111, fax no. 1-800-959-9134.
Attention: Smead Marketing Department.
www.smead.com

Minifile Labels

Color labels greatly simplify filing and retrieving information. The color combinations allow easy identification of incorrectly filed folders. The planner places the correct combinations of color labels on minifile folders to identify equipment. The folders listed above will hold eight large character labels on the side tabs. Plants with more than eight characters in the equipment number arrangement (not including hyphens or dashes) should use smaller size character labels. The plant should order a complete set of alphabetic (A–Z) and numeric (0–9) character labels depending on what characters could be used to identify equipment. Plants should order labels in rolls rather than sheets. The rolls make creating the minifiles much more easy. These products are available through many local office supply stores.

Tabbies Color Coded Labels
1-in high letters and numbers   No. 90100 (or 71000) series   (Specify rolls)
500 labels per roll

Company: Tabbies Division of Xertrex International, Inc.
Itasca, Illinois USA 60143-1171
www.tabbies.com   This site has a listing of local stores that carry Tabbies.

Miscellaneous Office Supplies

A new planner may benefit from office assistance that other office personnel take for granted. This is especially true as a company moves a mechanic or other type technician from the field into the planning office. The new planner may benefit from receiving a desk calculator, dictionary, and thesaurus as well as a calendar type organizer. Many planners benefit from a small hand held electronic spell checker. Computer typing tutor games help planners increase keyboard skills. These products are available through many local office supply stores.

Franklin Webster's Spelling Corrector Plus
Office Depot No. 350-959
About $20

Texas Instruments Display Desktop Calculator
Office Depot No. 222-059
About $9

Saunders Snapak Aluminum forms holder.   Helps planner take multiple work orders around in the field.
Office Depot No. 106-476
About $35

Typing Instructer Deluxe
Office Depot No. 835-273 (requires Microsoft Windows and CD-ROM)
About $40

Company: Office Depot
1-800-60-DEPOT(463-3768)   Fax no. 1-800-685-5010
www.officedepot.com

Equipment Tags

Many companies exist that can deliver generic or customized equipment tag numbers on many different types of tag material. Many plants find customized tags consisting of engraving on plastic tags suitable. These tags can be obtained from local trophy shops. A plant would give the trophy shop a list of the tag numbers and equipment names along with the size tag desired for each. The plant might specify the need for having a $3/16$-in hole drilled in the top right corner or both top corners to allow hanging with wire. The plant probably would elect to drill holes themselves later when attaching the tags. The plant might want to attach the tags without holes and wires by means of common silicone adhesive caulk. The plant should specify color tags. Appendix K, Equipment Schematics and Tagging, further discusses selection and utilization.

Recommended tag sizes and colors.

Normal size custom plastic equipment tags 2×3 in.
Two-color laminate material, $1/16$ in thick.
Letters: Gothic style, combined letter height should generally equal 40 to 60% of tag height for readability. This depends on number of lines, typically equipment number $5/16$ in high, equipment description $1/4$ in high.

Large size custom plastic equipment tags 6×8 in for large equipment.
Two-color laminate material, $1/16$ in thick.
Letters: Gothic style, combined letter height should generally equal 40 to 60% of tag height for readability. This depends on number of lines.

Super-large-size custom plastic equipment tags 12×16 in for large equipment such as tanks and whose tag numbers should be readable at some distance.

Two-color laminate material, $1/16$ in thick.
Letters: Gothic style, combined letter height should generally equal 40 to 60% of tag height for readability. This depends on number of lines.

(Optional minisize custom plastic equipment tags 1 × 2 in.
Limited use for short equipment numbers and only one or two word
equipment descriptions.
Two-color laminate material, $^1/_{16}$ in thick.
Letters: Gothic style, equipment number $^1/_4$ in high, equipment
description $^3/_{16}$ in high.)

Standard stock, high contrast colors:

Black over white (white letters)

Blue over white (white letters)

Green over white (white letters)

Red over white (white letters) (may have limited use due to possible
safety or warning color confusion)

White over black (black letters) (Preferred in low light areas)

Special order material available for special circumstances as

Decorator colors such as wood grains may be preferred in office areas

Ultraviolet safe for extreme sunlight in outdoor usage

More sturdy material for unusually high ambient temperature areas

The Trophyman (this local trophy shop does local, national, and international business).
1225 North 4th Street
Jacksonville Beach, Florida USA 32250
1-904-246-4919
Email: trophymn@fdn.com
The Trophyman can receive Microsoft text files to facilitate engraving tags.

Wire to Hang Tags on Equipment

This wire will not break after repeated bending and twisting. It comes
in convenient one pound, spool containers.

Aircraft Stainless Steel Lockwire   1 lb
Diameter 0.032-in wire, steel corrosion-resisting
Form 1, cond. A, comb. 302/304
Spec. MS20995-C (chem. and tensile *only*)
ASTM A580 nonelectrical

Malin Company
Brookpark, Ohio USA 44142
(800) 0167-9697

Deficiency Tags

Deficiency tags mark equipment that has been written up on work orders. A local print shop can make simple tags such as that shown by Fig. B.8 in App. B. LEM Products, Inc. makes a special tag with a self-laminating feature that allows the tag to be weatherproof after hanging. It also allows the originator to have a carbon copy to take back to the office with information to write the work order. LEM Products, Inc. inserts a brass eyelet in the top hole eyelet. Plant may specify whether it wants tag company to preattach a string or wire to the eyelet. Plant may specify a consecutive number sequence on the tags.

Lama-Tag LTC hanging style deficiency tag, blue with carbon, $3^{1}/_{4} \times 5^{1}/_{2}$-in, perforate with brass eyelet, attach wire, consecutively numbered starting with 00001.

LEM Products, Inc.
P.O. Box 190
Doylestown, PA USA 18901
1-800-220-2400   1-800-355-1414 (fax)

Shop Ticket Holders

Crews may wish to have reusable, plastic pockets help technicians keep work order forms clean and handy in the field. They come in boxes of 25.

Shop ticket holders No. 46912
C-Line Products, Inc.
1100 Business Center Drive
Mt. Prospect, IL 60056
1-800-323-6084
1-847-827-6661
1-847-827-3329 Fax
www.c-lineproducts.com

Open Shelf Files

Open shelves reduce filing space. They also encourage the use of files because of their visibility. Legal size files are preferred for ease of placing manuals and drawings in minifiles. Open shelves for filing systems can be procured from a number of companies.

Company: The Smead Manufacturing Company
Additional information available through fax no. 1-800-959-9134
Attention: Smead Marketing Department
651-437-4111

CMMS

This edition will not list any CMMS suppliers. To be honest, this is for two reasons. First, if nothing else, this entire book has set forth that maintenance planning is not a software project. Proposing a supplier herein might suggest that one needs to purchase a CMMS. A computer system is helpful, but not essential to implement effective maintenance planning. Chapter 9, The Computer in Maintenance, describes the computer's proper place in maintenance. Second, so many good systems are available in all price ranges that attempting to identify several would surely leave out others worthy of note. Chapter 9 and App. L suggest how one might go about selecting a system adequate for one's individual circumstances. Many journal articles and books are also available to assist in a CMMS evaluation.

Sample Forms and Work Orders

This appendix gives a complete set of forms (Figs. D.1 to D.14) used in this handbook. Purchasers of this book may use them as they are or modify them for maintenance in their organizations.

This appendix also shows sample work orders, all completed with technician feedback. These helpful examples illustrate proper information included at various stages of the maintenance process: requested work, coded work, planned work, and completed work. These samples may be used for a variety of purposes including determining what would be included in a work plan, what level of detail is needed for a work plan, and what is good feedback. It is one thing to explain what is needed, but samples showing what is meant prove helpful to the reader.

Work orders are typeset for ease of legibility to the reader of the *Maintenance Planning and Scheduling Handbook*. In actual practice, the work order request, plans, or completed work feedback might be handwritten or typed depending on the exact forms used or any computer system employed.

The following sample work orders (Figs. D.15 through D.26) are taken from the daily schedule illustrated in Chap. 7, Daily Scheduling and Supervision, for B Crew.

WORK ORDER

REQUESTER SECTION   Priority __ __

Equipment_____ Tag #_____

Problem or Work Requested: Def Tag #_____

Outage Req? Y/N   Clearance Req? Y/N   Confined Space? Y/N

By: Date & Time: **APPROVAL:**

PLANNING SECTION Assigned Crew:_____ Attachment? Y/N

Description of work to be performed:

Labor requirements:

Parts requirements:

Special tools requirements:

By: Date & Time: **Job Estimate:**   **Actual:**

CRAFT FEEDBACK (Modify plan sections above: actual labor, parts, & tools)

Work performed including equipment changes & any problems or delays:

Date & Time Started:_____ Date & Time Completed:_____

By: Date :   **APPROVAL:**

CODING

Figure D.1 Sample work order form.

CREW WORK HOURS AVAILABILITY FORECAST

For week of: ___/___/___ to ___/___/___

For crew: _____   By: _____   Date: ___/___/___

Craft	# Persons	Paid Hrs	Leave	Train	Misc	Carry-over*	Avail Hrs
_____	____ x 40 =	____ -	____ -	____ -	____ -	____ =	____
_____	____ x 40 =	____ -	____ -	____ -	____ -	____ =	____
_____	____ x 40 =	____ -	____ -	____ -	____ -	____ =	____
_____	____ x 40 =	____ -	____ -	____ -	____ -	____ =	____
_____	____ x 40 =	____ -	____ -	____ -	____ -	____ =	____
_____	____ x 40 =	____ -	____ -	____ -	____ -	____ =	____
_____	____ x 40 =	____ -	____ -	____ -	____ -	____ =	____
_____	____ x 40 =	____ -	____ -	____ -	____ -	____ =	____
_____	____ x 40 =	____ -	____ -	____ -	____ -	____ =	____
_____	____ x 40 =	____ -	____ -	____ -	____ -	____ =	____

Totals ____ x 40 = ____ - ____ - ____ - ____ - ____ = ____

*Carryover work is any work which has been physically started in the current period, but will not be finished and will run over into the forecast period.

Figure D.2 Crew work hours availability forecast form.

ADVANCE SCHEDULE WORKSHEET

For week of: _____ **to** _____
For crew: _____ **By:** _____ **Date:** _____

Forecast   **Available Hours Left**

Totals _____   _____

Instructions: Subtract job work hours from available line total until balance reaches zero for each line or backlog runs out.

Figure D.3 Advance schedule worksheet.

ADVANCE SCHEDULE WORKSHEET #2

For week of: _____ to _____
For crew: _____ By:_____ Date:_____.

TOTAL SCHEDULE
A. Total Hours Scheduled   _____
B. Any Available Hrs Left Unscheduled (_____)
   Why? (No backlog, etc.) _____

   _____

C. Total Hours Returned   _____
   Any Hours That Were Unclearable (_____)
D. Sched Hours Worked (D = A - C)   _____
E. % Schedule Met (E = D/A x 100)   _____%

PREVENTIVE MAINTENANCE
F. PM Hours Scheduled   _____
G. PM Hours Returned   _____
   PM Hours That Were Unclearable (_____)
H. PM Sched Hours Worked (H = F - G)   _____
I. % PM Schedule Met (I = H/F x 100)   _____%

Figure D.4 Advance schedule worksheet 2 to measure schedule compliance.

DAILY SCHEDULE

DAY: _____ DATE: _____

SUPERVISOR: _____

WO#	Unit	Pri	Short Description												Comments

Special Codes: Special Code

V-Vacation S-Sick   & Hours

T-Training O-Other _____

A-Assigned Off Crew   Total

Figure D.5   Daily schedule form.

ALIGNMENT CHECKSHEET

___1. Break coupling and remove old grease.

___2. Inspect coupling for damage.

___3. Secure coupling. Run motor if magnetic center of motor is unknown. Mark motor shaft.

___4. Shut down motor. Set motor shaft to electrical center. Set coupling gap per coupling instructions.

___5. Inspect hold down bolts and washers for damage.

___6. Install jacking bolts on all corners of machines to be shimmed.

___7. Correct machine to be shimmed for soft foot and defective sole plates.

___8. Correct stationary machine for soft foot and defective sole plates.

___9. Remove all shims. Clean or replace as needed.

___10. Pack proper amount of grease into coupling and reassemble.

___11. Align machine. Use PdM group instructions for considering thermal growth.

___12. Tighten all fasteners. Loosen jacking bolts. Turn shaft to ensure free rotation.

___13. Attach form to work order.

Technician_____   Date_____

Figure D.6 Sample alignment readiness form.

UNIT 2 BURNER CHECKSHEET FORM

Burner Number:_____

Date:_____

TECHNICIAN INITIALS EACH STEP
WHEN COMPLETE

___1. Replace orifice, swirl, and spill plates

___2. Replace both gaskets.

___3. Verify correct lance setting.
   Setting is: _____

___4. Torque feed tube to 200 ft lb.

___5. Torque cap nut assembly to 77.5 ft lb.

___6. Adjust spider to provide 1/8 inch to 3/16 inch
   gap between outer tube and washer.

Figure D.7   Sample burner checksheet form.

DEFICIENCY TAG
#0084147

EQUIPMENT _____

EQUIP TAG # _____
PROBLEM _____
_____
_____
_____
_____
_____

LOCATION _____
TAGGED BY _____
DATE _____
TIME _____

Carbon & Copy

Figure D.8   Sample deficiency tag form.

EQUIPMENT HISTORY DATA SHEET

PAGE ___

SYSTEM _____

EQUIPMENT NAME _____

EQUIPMENT TAG # _____

HISTORY (Normally does not include PM's or other routine inspections.)

Work Order #	Job Description (What was done)	Complete Date	Print Revision Needed?	Work Type	Work Priority	Actual Hours Duration	Actual Labor Hours	Total Work Cost $
		/ /						
		/ /						
		/ /						
		/ /						
		/ /						
		/ /						
		/ /						
		/ /						

Figure D.9   Sample minifile form for history.

EQUIPMENT TECHNICAL DATA SHEET

EQUIPMENT NAME _____

EQUIPMENT TAG # _____

EQUIP LOCATION _____

Does this equipment have a "Standard Plan"? _____

Manufacturer code and name _____

Vendor name, location, person, and phone number to contact:

_____

_____

NAMEPLATE INFORMATION

Model # _____ Serial # _____

_____

MSDS # AND ANY OTHER SAFETY INFORMATION: (Is a Confined Space Permit needed? Describe any past accidents. What safety considerations such as scaffolding, high temperature/pressure, chemicals, etc. are there?)

_____

_____

_____

OTHER INFORMATION AND SPECIAL NOTES FOR THIS EQUIPMENT

_____

_____

_____

_____

_____

_____

_____

_____

_____

_____

Figure D.10   Sample minifile form for equipment data. The "standard plan" indicates a booklet kept in the paper file. It might be updated on a word processor.

PARTS INFORMATION SHEET

PAGE ___

EQUIPMENT NAME _____

EQUIPMENT TAG # _____

	Stock Number	Total Quantity Used for this Equipment	Quantity Needed for this Job*	Description
1				
2				
3				
4				
5				
6				
7				
8				
9				
10				
11				
12				
13				
14				
15				
16				
17				
18				
19				
20				
21				
22				
23				
24				
25				
26				
27				
28				
29				
30				

*Use this column after copying this page to issue with a specific job.

Figure D.11   Sample minifile form for spare parts.

EQUIPMENT NORMAL PM's AND CHECKS

EQUIPMENT NAME_____

EQUIPMENT TAG #_____

By Maintenance Personnel

What	Frequency	*Route Designation

By Predictive Maintenance (PdM) Personnel

What	Frequency	*Route Designation

By Operations, Lab, or Other Personnel

What	Frequency	*Route Designation
		*if any

Figure D.12 Sample minifile form for PM's.

INFORMAL CHECKSHEET TO
BALANCE OBSERVATIONS

Mark the time started to locate persons within beginning half hour (B) or end half hour (E): Hand write a number beside each half hour to designate the week during which the observation occurred.

Weeks:

Time Period	Time	Mon	Tues	Wed	Thur	Fri
8	7:30-8:30	B E	B E	B E	B E	B E
9	8:30-9:30	B E	B E	B E	B E	B E
10	9:30-10:30	B E	B E	B E	B E	B E
11	10:30-11:30	B E	B E	B E	B E	B E
12	11:30-12,12:30 -1	B E	B E	B E	B E	B E
13	1 - 2	B E	B E	B E	B E	B E
14	2 - 3	B E	B E	B E	B E	B E
15	3 - 4	B E	B E	B E	B E	B E
16	4 - 5	B E	B E	B E	B E	B E
17	5 - 6	B E	B E	B E	B E	B E

Special comments:

Figure D.13   Sample form to ensure an even balance of work sampling study observations.

Figure D.14 Sample form to record work sampling observations.

Work order	**#015**
Requester section	Priority R2

Unit 2 polisher tank A (N02-CP-010) underdrain has high differential. Please clean.   Def. Tag #010897.
No outage required. Clearance req. Yes confined space.
J. Jones 5/4/99 1:15 pm.   **Approval:** S. Brenn 5/4/99

Planning section   Attached blank permit form.
B Crew. Obtain confined space permit. Vacuum out old resin. Blow out drain screen.
Labor: 1 Tech   20 hr   Total labor 40 hr   Actual 13
             1 Tech*   20 hr   Job duration 20 hr   Actual 13
                        *(certified confined space lookout)
Parts: None
Tools: 6' step ladder
             10 gal shop vac
             25' air hose with regulator

Planner: Rodgers 5/5/99 Job estimate: $1000 Actual: $325

Craft feedback
Cleaned underdrain. Used 2' extension on shop vac and did not need to enter polisher.   1 mech-13 hours
Job started 5/7 7:30 am. Finished 5/11 10:30 am.

S. Jensen 5/11/99   **Approval:** J. Field 5/11/99

Coding	Plan type RE	Group/syst CP	Crew B
	Work type 5	Equip type 06	Outage 0

Figure D.15 Sample completed work order to illustrate information added at each work process step.

WORK ORDER #016

REQUESTER SECTION   Priority S1

Stair steps on north side of Unit 2 boiler, several steps almost rusted through. Area taped off. Def. Tag #010627 No outage required. No Clearance Req. No Conf Space. M. Johns 4/23/99 7:30am   **APPROVAL:** J. Field 4/23/99

PLANNING SECTION

B Crew. Remove steps, repair, clean stair well, and paint. If needed, replace steps. 105' elevation.

Labor: 1 Welder 35hr   Total labor 70hr   Actual 90
       1 Helper 35hr   Job duration 35hr   Actual 45

Parts: Grade 8 bolts
      1 1/4 x 1 1/4 x 3/16" A106 angle iron $2.15/ft 20'
      1 can of spray cold galvanize

Tools: Needle scaler
      4" grinder with wire brush

Planner:Rodgers 4/26/99 Job estimate:$1800 Actual:$2300

CRAFT FEEDBACK

Removed 8 steps, repaired 5. Installed new bolts on all 8 steps. Used 10 ft angle iron and 2 cans spray. Time estimate was too short. Barber and Patterson-45 hours ea. Job Started 5/6 1:00pm. Finished 5/13 6:00pm.
M. Barber 5/13/99   **APPROVAL:** J. Field 5/14/99

CODING Plan Type RE   Group/Syst BJ   Crew   B
             Work Type 5   Equip Type 30   Outage 0

Figure D.16 Sample completed work order to illustrate information added at each work process step.

WORK ORDER #017

REQUESTER SECTION   Priority G3

Raw water booster pump "A"(N00-HR-001). Need new
pump shaft for inventory.   Def. Tag #NA.
No outage required. No Clearance Req. No Conf Space.
C. Rodgers 4/15/99   **APPROVAL:**C. Rodgers 4/15/99

PLANNING SECTION

B Crew. Fabricate new shaft per attached engineering
drawing #103. Spec is +/- 0.002"
Return shaft to inventory PMP-HB-001.
Labor: 1 Machinist 20hr   Total labor 20hr   Actual 15
                            Job duration 20hr   Actual 15
Parts: 2"x 48" 304SS material, non-stock Qty 1 Cost $145
                            Staged in Tool Room bin A-23-1.
Tools: 14" lathe
       Bridgeport Mill

Planner: Rodgers 4/15/99 Job estimate: $645 Actual: $520

CRAFT FEEDBACK
Shaft made and returned to inventory. 1 Mach-15 hrs.
Job Started 5/7 7:30am. Finished 5/11 at noon.

D. Kingsley 5/11/99   **APPROVAL:** J. Field 5/11/99

CODING Plan Type PE   Group/Syst HR   Crew   B
            Work Type 1   Equip Type 01   Outage 0

Figure D.17 Sample completed work order to illustrate information added at each
work process step.

Work order	**#019**
Requester section	Priority R2

Unit 2 "C" Vacuum Pump (N02-CV-022) running hot.
Def. Tag #010562.
No outage required. Clearance req. No confined space.
Hernandez 4/28/99 3:10 pm **Approval:** Stanwick 4/28/99

Planning section
B Crew. Inspect all intake valves for "C" Vacuum pump.
Take appropriate corrective action.

Labor: 1 Mech 20 hr   Total labor 20 hr   Actual 25
                                Job duration 20 hr   Actual 25
Parts: Channels   VAL-PU-015* Qty 10 Cost $7.18ea
        Reeds   VAL-PU-015* Qty 10 Cost $0.85ea
        Backing plate VAL-PU-015* Qty 10 Cost $24.60ea
                                *Reserved
Tools: Mech's box.
Planner: Rodgers 4/29/99 Job estimate: $550 Actual: $630

Craft feedback   Inspected all intake valves. Found
#1& #2 inlet valves with bad reeds and replaced. Ran
pump. Checked okay.
1 mech-25 hours. Job Start 4/7 10 am. Finished 4/12 4 pm.

S. Adamson 5/12/99   **Approval:** J. Field 5/12/99

Coding	Plan type PE	Group/syst FC	Crew B		
	Work type 9	Equip type 02	Outage 0		

Figure D.18 Sample completed work order to illustrate information added at each work process step.

WORK ORDER #025

REQUESTER SECTION   Priority R2

Caustic suction piping (N00-HC-036) has several small
leaks.   Def. Tag #010304.
No outage required. Clearance Req. No Confined Space.
A. Glade 4/28/99 10:00am **APPROVAL:** J. Field 4/28/99

PLANNING SECTION   Attached MSDS for caustic.
B Crew. Remove all 3" suction pipe. Cut and thread new
pipe. Install one 3" pipe hanger. (Planner to have
insulation contractor to remove and replace insulation.)
Labor: 1 Tech   17hr   Total labor 17hr   Actual 10
                        Job duration 17hr   Actual 10
Parts: 3" pipe   PIP-CD-200 Qty 15' Cost $1.80/ft
       Couplings   PFN-CD-200 Qty 3   Cost $2.30ea
       Pipe hanger HBW-HG-101 Qty 1   Cost $21.50ea
Tools: Mech box, 2- 24" pipe wrenches, safety gear,
       portable pipe machine.   (Insulation work $150)
Planner: Rodgers 4/28/99 Job estimate: $630 Actual: $500

CRAFT FEEDBACK
Installed new suction pipe and installed drain valve.
Valve VAL-GL-003, Nipple CUP-TC-010.
1 mech-10 hours Started 5/11 7:30am Finished 5/11 6pm

T. Smith 5/11/99
                    APPROVAL: J. Field 5/11/99

CODING Plan Type RE   Group/Syst IIC   Crew   B
            Work Type 5   Equip Type 09   Outage 0

Figure D.19 Sample completed work order to illustrate information added at each
work process step.

WORK ORDER #027

REQUESTER SECTION   Priority R2

Demineralizer underdrain monthly PM (N00-HD).
Frequency - Monthly. Mech crew.
No outage required. No Clearance Req. No Conf Space.
PM 5/1/99   **APPROVAL:** NA

PLANNING SECTION

B Crew. Check pressure differential on underdrains for all
tanks. Check sheer pins on baskets. Check machine shop
supply of sheer pins. Make tapered #7 sheer pins to
replace quantity used and to keep supply at 10 pins.
Labor: 1 Machinist   2hr   Total labor 2hr   Actual 4
                          Job duration 2hr   Actual 3
Parts: A-106 1/4" cold rolled
Tools: Machine shop

                    Job estimate: $50 Actual: $100

CRAFT FEEDBACK
Replaced and made two pins for "A" tank.
1 machinist-3 hours
1 apprentice (Richardson)-1 hour
Job Started 5/11 11:30am. Finished 5/11 3pm.
M. Johns 5/11/99   **APPROVAL:** J. Field 5/12/99

CODING   Plan Type PM   Group/Syst HD   Crew   B
            Work Type 7   Equip Type 06   Outage 0

Figure D.20   Sample completed work order to illustrate information added at each work
process step.

WORK ORDER #028

REQUESTER SECTION   Priority R1

"A" Potable Water Pump (N00-HP-010) very high
vibration. Need to check impeller.   Def. Tag #010573.
No outage required. Clearance Req. No Confined Space.
S. Chang 5/6/99 4:00pm.   **APPROVAL:** Simpson 5/6/99

PLANNING SECTION

B Crew. This pump has a history of impeller nut coming
loose. Open inspection door.   Check torque on impeller
nut is 200 ft lb. May need threads on impeller remachined.
Labor: 1 Mach   8hr   Total labor 16hr   Actual 8
            1 Helper 8hr   Job duration 8hr   Actual 4
Parts: Impeller nut   WAT-PU-002* Qty 1 Cost $5ea
Tools: Ft lb torque wrench.

                              *RESERVED

Planner: Rodgers 5/7/99 Job estimate: $405 Actual: $200

CRAFT FEEDBACK

Nut was loose. Retorqued to 200 ft lb. Should we make
checking this nut a quarterly PM?
1 machinist-4 hours   1 apprentice (Richardson)-4 hours
Job Started 5/11 7:30am. Finished 5/11 11:30am.
M. Johns 5/11/99   **APPROVAL:** J. Field 5/11/99

CODING Plan Type RE   Group/Syst HP   Crew   B
                  Work Type 8   Equip Type 01   Outage 0

Figure D.21 Sample completed work order to illustrate information added at each
work process step.

WORK ORDER #029

REQUESTER SECTION   Priority E4

Unloading wharf gutters have trash build up. Add screen
to keep out trash.   Def. Tag #NA.

No outage required. No Clearance Req. No Conf Space.

A. Glade 4/23/99 10:15am **APPROVAL:** J. Field 4/23/99

PLANNING SECTION

B Crew. Please clean out gutters and unstop both drains.
Cannot put screen over entire gutters because they help
keep trash from river. Cut hardware cloth and place over
drains to keep them from clogging.

Labor: 2 Helper 6hr   Total labor 12hr   Actual 15

                        Job duration 6hr   Actual   10

Parts: Hardware cloth from Tool Room.

Tools: Rotating wire snake with cutting bit.

Planner: Rodgers 4/26/99 Job estimate: $300 Actual: $375

CRAFT FEEDBACK

Completed as planned.   1 mech-5 hours. 2 trainees
(Young and Capper) 5 hr ea. Worked at same time as
unloading arms WO #036.

Job Started 5/11 7:30am. Finished 5/11 6pm.

A. Glade 5/11/99   **APPROVAL:** J. Field 5/11/99

| **CODING** | Plan Type PE | Group/Syst FW | Crew   B |
| | Work Type 9 | Equip Type 50 | Outage 0 |

Figure D.22   Sample completed work order to illustrate information added at each
work process step.

WORK ORDER #030

REQUESTER SECTION   Priority R1

Demineralizer "A" Transfer Pump (N00-HD-052) making
noise.   Def. Tag #010977.
No outage required. Clearance Req. No Confined Space.
K. Frank 5/6/99 2:30pm.   **APPROVAL:** S. Brenn 5/6/99

PLANNING SECTION   Attached alignment form.
B Crew. Check alignment with laser.

Labor: 1 Mech   1hr   Total labor 2hr   Actual   3
       1 Helper 1hr   Job duration 1hr   Actual   2

Parts: none

Tools: Laser alignment kit in Tool Room.

Planner: Rodgers 5/7/99 Job estimate: $50 Actual: $75

CRAFT FEEDBACK
Found pump out of alignment. Realigned and ran pump.
No noise. Vibration normal. Alignment sheet attached.
1 mech-1 hour guided apprentice (Richardson)-2 hours.
Job Started 5/11 1:30pm. Finished 5/11 3:30pm.
R. Sanchez 5/11/99   **APPROVAL:** J. Field 5/11/99

CODING Plan Type RM   Group/Syst HD   Crew   B
            Work Type 5   Equip Type 01   Outage 0

Figure D.23 Sample completed work order to illustrate information added at each work process step.

WORK ORDER #031

REQUESTER SECTION   Priority R3

Upgrade Unit 2 Auxiliary Feedwater Pumps.

(N02-DA-007 & -012)   Def. Tag #NA.

No outage required. Clearance Req. No Confined Space.

J. Field 3/18/99 11:00am.   APPROVAL: J. Field 3/18/99

PLANNING SECTION   Attached alignment form.

B Crew. Install 2 new pumps in place of old ones. Break
all pipes and disconnect motor. Unbolt pumps and install
new ones. (Planner to place copy of this work order in
each pump file for history.)

Labor: 1 Mech 20hr   Total labor 40hr   Actual 39

       1 Helper 20hr   Job duration 20hr   Actual 20

Parts: Pumps non-stock staged in Bin A-23-2 Cost $5300

       3/4" bolts and nuts for flanges from Tool Room

Tools: 1 ton chainfall, beam trolley, mech box, Laser
alignment box, 4 wheel cart.

Planner: Rodgers 5/3/99 Job estimate:$6300 Actual:$6275

CRAFT FEEDBACK

Installed new pumps. Suction pipe had strain, had to cut
and reweld. Alignment sheet attached.

1 mech-19 hour. Apprentice (Wilson)-20 hours.

Job Started 5/11 7:30am. Finished 5/12 6:00pm.

R. Sanchez 5/12/99   APPROVAL: J. Field 5/12/99

| **CODING** | Plan Type PE | Group/Syst DA | Crew   B |
| | Work Type 3 | Equip Type 01 | Outage 0 |

Figure D.24   Sample completed work order to illustrate information added at each work process step.

WORK ORDER #035

REQUESTER SECTION   Priority S1

#2 control room fire system has rusty looking union.

Def. Tag #010124.

No outage required. Clearance Req. No Confined Space.

Sotolongo 4/21/99 9:00pm. APPROVAL: Stanwick 4/21/99

PLANNING SECTION

B Crew. (Note: Operations crew must follow standard procedures when clearing fire protection system and notify proper authorities.) Replace union on supply header and all rusted pipe.

Labor: 1 Tech   20hr   Total labor 20hr   Actual 25
                        Job duration 20hr   Actual 25

Parts: 2" union Qty 1   Cost $18.50ea
       2" pipe   Qty 20 feet   Cost $165

Tools: 18" pipe wrench, 6' step ladder, portable pipe machine

Planner: Rodgers 4/22/99 Job estimate: $684 Actual: $820

CRAFT FEEDBACK

Replaced parts as requested. 1 mech-25 hours.

Job Started 5/11 7:30am. Finished 5/13 1pm.

O. Jones 5/13/99
                    APPROVAL: J. Field 5/13/99

CODING	Plan Type RE	Group/Syst KD	Crew B
	Work Type 5	Equip Type 09	Outage 0

Figure D.25 Sample completed work order to illustrate information added at each work process step.

WORK ORDER #036

REQUESTER SECTION
Priority E1

Oil Dock "B" Unloading Arm (N00-FW-014) dripping at flange onto wharf during last unloading.Def. Tag #010130 No outage required. Clearance Req. No Confined Space. F. Solz 5/5/99 3:00pm.   **APPROVAL:** K. Mulan 5/5/99

PLANNING SECTION
B Crew. Break flange, clean, and make new gasket. Replace gasket. Clean area.

Labor: 1 Mech   6hr   Total labor 18hr   Actual 15
      2 Helper 6hr   Job duration 6hr   Actual 10

Parts: Gasket material in mechanic's shop.
      3/4"x 3" bolts, 3/4" nuts, 3/4" flat washers.

Tools: Highlift, Rags, Can of degreaser,
      Plastic garbage bags.

Planner: Rodgers 5/6/99 Job estimate: $450 Actual: $375

CRAFT FEEDBACK
Completed as planned. Found gasket split due to over-torque. 1 mech-5 hrs. 2 trainees (Young & Capper) 5 hr ea Worked at same time as gutters WO #029.

Job Started 5/11 7:30am. Finished 5/11 6pm.

A. Glade 5/11/99   **APPROVAL:** J. Field 5/11/99

CODING	Plan Type RE	Group/Syst FW	Crew B
	Work Type 5	Equip Type 09	Outage 0

Figure D.26 Sample completed work order to illustrate information added at each work process step.

Step-by-Step Overview of Planner Duties

This appendix runs through the planner's duties step by step without commentary. This accounting of the duties provides a starting point for the required actions of specific planners in specific companies. Chapter 5 establishes the reasoning. Chapter 11 describes the importance of the planner selection. Appendix M describes the qualifications of the planner and issues related to planner hiring. Appendix N gives a sample formal job description of a maintenance planner.

I. New Work Orders
   A. Collect and review new work orders prior to the managers and supervisors meeting each morning. Code work orders and list work order numbers with a brief description on the Morning Meeting List. Copy each work order for the Planning Clerk to begin computer entry. Send a copy with the list to the morning meeting.
   B. Open any authorized emergency work orders for immediate work on computer systems to inventory and time sheet transactions.
II. Before Job Scheduling
   C. Select authorized work orders to plan per priority guidelines from waiting-to-be-planned file. Make a working copy of the work order and file the original work order form in planner active file.
   D. Determine the type of work order to determine the type of planning required per guidelines (proactive or reactive work order).
   E. For a proactive work order, determine the level of detail required for planning the job per guidelines (extensive or minimum maintenance).

1. **For a proactive, extensive work order:**
   a. Record the work order number and description on the Equipment History Data Sheet in the Equipment History Files for the component level file (minifile). If there is no minifile, create one. If a minifile cannot be created because there is no equipment number reported, create the minifile after the field inspection.
   b. Scope the job to determine the exact problem or situation using minifile and computer information, field inspection, and personal experience. Perform a root cause analysis if required per guidelines. When necessary, research Equipment Technical Files and Vendor Files for information; copy useful information to the minifile when found. When necessary, get knowledgeable maintenance or operator assistance including talking to Originator, reviewing Operating Crew Log Reports, and operating the equipment. Refer work orders requiring predictive testing, performance testing, or engineering to the Planning Supervisor for consultation and possible reassignment. During the field inspection, verify whether a unit outage is required and what type of outage. Verify whether insulation work is required and whether asbestos is suspected. Determine if clearance or confined space permits are required. Note if there is no component tag. Update work order form for all information discovered on the field inspection.
   c. Plan the general strategy of the job such as repair or replace to fix the probable cause permanently and outline overall steps when necessary such as "(1) erect scaffolding, (2) repair valve, (3) remove scaffolding." If necessary, attach a copy of useful information such as helpful procedures, sketches, and equipment specifications, drawings, or data using resources such as the History, Technical and Vendor Files. (Gather and develop these items as needed and always put a copy in the minifile.)
   d. Attach a Standard Plan if available.
   e. Plan the classification and number of personnel required including all crafts.
   f. Plan the number of work hours and duration hours to complete the job.
   g. Provide a listing of anticipated as well as other possible material required including complete equipment part breakdowns.
      (1) Coordinate with the Storeroom or Maintenance Purchaser for unavailable needed parts and put the work order in the waiting-for-parts file.
      (2) Reserve anticipated parts.

    h. Provide a listing of any special tools, equipment, or contractors (including insulation work) needed.

    i. Designate whether there are any job safety requirements and explain.

    j. Complete a cost estimate on work order form for all anticipated resources (labor, material, and special tools/contractor). Total all information on work order form. Confirm authorization with the planning supervisor if total cost estimate exceeds $5000.

    k. Attach any necessary data sheets identifying them on work order form.

    l. Update computer for planning information and status.

    m. File the completed planned work package in the waiting-to-be-scheduled file.

2. For proactive, minimum maintenance work order:

    a. Record the work order number and description on the Equipment History Data Sheet in the Equipment History Files for the component level file (minifile). If there is no minifile, create one. If a minifile cannot be created because there is no equipment number reported, create the minifile after the field inspection.

    b. Scope the job to determine the exact problem or situation using the minifile (if available) and computer information, field inspection, and personal experience. Determine if clearance or confined space permits are required. During the field inspection, verify whether a unit outage is required and what type of outage. Verify whether insulation work is required and whether asbestos is suspected. Determine if clearance or confined space permits are required. Note if there is no component tag or silver tag on the piece of equipment. Update the work order form for all information discovered on the field inspection.

    c. Plan the general strategy of the job such as repair or replace to fix the probable cause permanently.

    d. (No Standard Plan is necessary.)

    e. Plan the classification and number of personnel required including all crafts.

    f. Plan the number of work hours and duration hours to complete the job.

    g. Provide a listing of anticipated material. Include other possible material required including complete equipment part breakdowns only if already available in the minifile.

    (No parts ordering or reserving is necessary.)

    h. Provide a listing of any special tools, equipment, or contractors (including insulation work) needed.

    i. Designate whether there are any job safety requirements and explain.

    j. Complete the work order form for all anticipated resources (labor, and special tools/contractor). Total all information on the work order form.

    k. Attach any necessary data sheets identifying them on the work order form.

    l. Update the computer for planning information and status.

    m. File completed planned work package in the waiting-to-be-scheduled file.

  F. For a reactive work order, determine the level of detail required for planning the job per guidelines (extensive, or minimum maintenance).

   1. For a reactive, extensive work order:

    a. Record the work order number and description on the Equipment History Data Sheet in the Equipment History Files for the component level file (minifile). If there is no minifile, create one. If a minifile cannot be created because there is no equipment number reported, create the minifile after the field inspection.

    b. Scope the job to determine the exact problem or situation using the minifile and computer information, field inspection, and personal experience. During the field inspection, verify whether a unit outage is required and what type of outage. Verify whether insulation work is required and whether asbestos is suspected. Determine if clearance or confined space permits are required. Note if there is no component tag on the piece of equipment. Update the work order form for all information discovered on the field inspection.

    c. Plan the general strategy of the job such as repair or replace to fix the problem and outline overall steps when necessary such as "(1) erect scaffolding, (2) repair valve, (3) remove scaffolding." If necessary, attach a copy of useful information such as helpful procedures, sketches, and equipment specifications, drawings, or data only if already available in the minifile.

    d. Attach a Standard Plan if available.

    e. Plan the classification and number of personnel required including all crafts.

    f. Plan the number of work hours and duration hours to complete the job.

    g. Only if already available in the minifile, provide a listing of anticipated as well as other possible material required including complete equipment part breakdowns.

       (1) Coordinate with the Storeroom or Maintenance Purchaser for unavailable needed parts and put the work order in the waiting-for-parts file.

       (2) Reserve anticipated parts.

h. Only if already known or available in the minifile, provide a listing of any special tools, equipment, or contractors (including insulation work) needed.

i. Designate whether there are any job safety requirements and explain.

j. Complete a cost estimate on the work order form for all anticipated resources (labor, material, and special tools/contractor). Total all information on the work order form. Confirm authorization with the planning supervisor if the total cost estimate exceeds $5000.

k. Attach any necessary data sheets identifying them on work order form.

l. Update the computer for planning information and status.

m. File completed planned work package in the waiting-to-be-scheduled file.

2. For a reactive, minimum maintenance work order:

a. Check to see if there is a minifile with useful information. (No minifile creation or recording of information is necessary if there is no minifile existent. Record information if there already is a minifile.)

b. Scope the job to determine the exact problem or situation using personal experience only if the stated problem is unclear. A field inspection is highly recommended, but not mandatory. Determine if clearance or confined space permits are required.

c. Plan the general strategy of the job such as repair or replace to fix the problem.

d. (No Standard Plan is necessary.)

e. Plan the classification and number of personnel required including all crafts.

f. Plan the number of work hours and duration hours to complete the job.

g. Only if already available in the minifile, provide a listing of anticipated as well as other possible material required including complete equipment part breakdowns. (No parts ordering or reserving is necessary.)

h. Only if already known, provide a listing of any special tools, equipment, or contractors (including insulation work) needed.

i. Designate whether there are any job safety requirements and explain.

    j. Complete a cost estimate on the work order form for all anticipated resources (labor and special tools/contractor). Total all information on the work order form.

    k. Attach any necessary data sheets identifying them on the work order form.

    l. Update the computer for planning information and status.

    m. File the completed planned work package in the waiting-to-be-scheduled file.

III. After Job Scheduling

    G. Stage parts and tools for scheduled jobs per guidelines. Identify items as staged on the work order form.

IV. After Job Execution

    H. Receive the work order after job execution and review for feedback information completeness (including any attachments). Clarify feedback as necessary to allow improvements to planned work packages for future work. Complete the work order form for actual cost and totals.

    I. Discard the Minimum Maintenance work orders with no historical value.

    J. File other work orders in their respective minifiles. Update the Equipment History Data Sheet and other minifile sheets and information as necessary to allow improvements to planned work packages for future work. Make a work order copy for the Planning Clerk for final close out and distribution to the Originator. Indicate if a copy needs to go to engineering to show plant modifications.

    K. Initiate necessary *proactive* work orders for additional work based on feedback from completed work orders. Create new standing preventive maintenance work orders, if necessary. Also, adjust the existing standing preventive maintenance work orders, if necessary.

V. Other Duties

    L. Develop and maintain work packages for PM routes (and on individual equipment) to facilitate initiation of proactive work orders by maintenance forces performing PM work.

    M. Maintain History, Technical, and Vendor Files.

    N. Develop and maintain Standard Plan packages.

    O. Minimize unplanned work order backlog.

    P. Coordinate process problems with crew supervisors on a peer level.

    Q. Perform related duties as directed by the supervisor of the planning department.

Step-by-Step Overviews
of Others' Duties

This appendix follows in the steps of the previous appendix and runs through the relevant step-by-step duties of persons other than the planner without commentary. This accounting of the duties provides a starting point for the required actions of those specific persons in specific companies.

The following positions are included.

Maintenance Scheduler
Maintenance Planning Clerk
Operations Coordinator
Maintenance Purchaser or Expediter
Crew Supervisor
Planning Supervisor
Maintenance Manager
Maintenance Planning Project Manager
Maintenance Analyst

Maintenance Scheduler

These duties would belong to the planners if the company does not utilize a special scheduler.

A. Work with the Operations Coordinator to prioritize work orders for planning and scheduling.

B. Receive craft hours availability forecasts for the next week from the crew supervisors.

C. Develop a 1-week schedule for each crew of work orders to be worked from the waiting-to-be-scheduled file.

D. Begin coordination of work involving more than one craft technician.

E. Begin coordination of work involving advance notice to a contractor of more than a day or two.

F. Work with planners to stage parts for scheduled jobs.

G. Work with crew supervisors to coordinate work through planning that is approved to disrupt the current week's schedule of work.

H. Assist with measuring Schedule Compliance.

I. Work with crew supervisors to develop schedules for short notice outage work (SNOW).

Maintenance Planning Clerk

The clerk's duties may be classified onto ones dealing with new work orders, ones dealing with work orders after execution, and other duties.

New work orders

A. Open work orders to computer systems using work order copies. After decisions are made at the morning meeting, update entries using a work order worksheet. File past work order copies and worksheets in a chronological file.

B. Open authorized emergency work orders to computer systems for immediate work.

After job execution

C. Receive the work order after job execution and distribute to a responsible planner for review. After planner review, update the computer systems for job completion and closure.

D. Distribute a copy of work order form to the originator.

E. If the planner indicates the work order form contains any changes to equipment drawings or technical data, send a copy to the plant engineer.

F. File specified closed work orders in their respective files if requested by planner.

Other duties

G. Assist planners in maintaining History, Technical, and Vendor Files.

H. Maintain a database for established Preventive Maintenance (PM) work orders. Print and distribute PM work orders as scheduled.

I. Perform related duties as directed by the supervisor of the planning department, such as maintaining reports, creating files, coding work orders, and ordering office supplies.

J. Perform Purchaser duties assigned by planning supervisor.

Operations Coordinator

A. Review new work orders for appropriate priorities; adjust as needed.

B. Review work orders in the waiting-to-be-planned file to verify priorities for planning attention.

C. Review work orders in the waiting-to-be-scheduled file and verify priorities for scheduling attention.

D. Assist maintenance planners with troubleshooting and planning work orders from an operation's perspective.

E. Assist crew supervisors to coordinate clearances and other operations needs.

Maintenance Purchaser or Expediter

These duties would belong to the planners if the company does not utilize a special purchaser. Some of the duties could be handled by the plant administration group.

A. Order, track, and expedite nonstock equipment and materials for maintenance planners and for maintenance craft crews. Order from blanket purchase orders where applicable and track blanket totals. Order items not on blankets. Obtain quotations and availability as requested. (The research to select specific items is normally accomplished by the requesting planner or craftperson.)

B. Walk emergency purchase orders through to expedite orders.

C. Update the computer system information for orders.

D. Assist planning with creating or updating bills of material or illustrated parts breakdowns for equipment.

E. Establish blankets purchase orders for maintenance support.

F. Evaluate items for stock versus nonstock status. Process requests for making nonstock items into stock items.

G. Identify and resolve recurrent stockout problems.

H. Provide administrative support for off-site work. This includes coordination of shipment, processing of purchase orders, tracking progress, and receiving of equipment upon return. Others such as the planner, craft supervisor, or plant engineer develop the work scope and usually obtain quotations.

I. Meet with vendors, manufacturers, storeroom, planners, and craft-persons to avoid or resolve material problems.

J. Assist in determining acceptable "equals" or alternatives.

K. Assist in receiving purchases and coordinating quality assurance.

L. Participate on standardization committees to standardize stock materials and equipment or parts purchases.

Crew Supervisor

This is an overview of the duties of the crew supervisor only in regard to planning and scheduling.

Before job execution

A. Provide the Maintenance Scheduler with a craft work hours available forecast for the next week identifying hours by skill level.

B. Schedule work orders to be worked by crew individuals for the next day in accordance with the 1-week schedule, carryover work, new high priority reactive work orders, and other scheduling guidelines.

C. Coordinate any necessary clearances and any work involving other crafts for the next day.

D. Distribute the work order form to assigned lead technician.

E. Decide which material is appropriate to stage 1 day in advance and work with the planner to stage it.

During job execution

F. Reschedule work to handle emergencies and SNOW work.

G. Assist assigned crew personnel to resolve any problems with work packages utilizing planning files and information as necessary. Coordinate obtaining of unanticipated parts that are not in inventory through the storeroom (stockouts) or maintenance purchaser (noninventory items).

After job execution

H. Coordinate returning of equipment back to Operations.

I. Receive work order forms after job execution from technicians. Clarify feedback as necessary to allow improvements to planned work packages for future work.

J. Initiate necessary work orders for additional work based on feedback from completed work orders including "permanent fixes" and refurbishing parts.

K. Return the completed work order form to planning with all attachments and planned work package material.

Other duties

L. Coordinate process problems with planners on a peer level.

M. Perform related duties as assigned by the Maintenance Manager.

Planning Supervisor

A. Supervise the planning group, especially the selection and training of new planners. Support the needs of planners.

B. Coordinate insulation contractor work.

C. Coordinate engineering assistance.

D. Review plans where the planner estimate exceeds $5000. Confirm authorization with management where planner estimate exceeds $15,000.

E. Assist in the coordination of planning interfaces and concerns with crew supervisors and affected departments.

F. Maintain awareness and advise management of accuracy of computerized maintenance management system (CMMS).

G. Coordinate the receipt, distribution, and filing of operation and maintenance manuals. Coordinate their review for new PM recommendations.

Maintenance Manager

Read this book. Ask the right questions. Appendix P suggests some questions.

Maintenance Planning Project Manager

The company may assign a person the responsibility for implementing a new planning system or altering an existing one.

A. Develop and maintain maintenance planning vision, strategies, principles, and techniques. Advise management on planning performance.

B. Coordinate planning vision and strategies with supervisors and affected departments.

C. Help select a Planning Supervisor. Advise and coach the Planning Supervisor.

D. Help select planners. Train and coach planners and crew supervisors on planning techniques.

E. Manage implementation and utilization of any computer system
   1. Establish and manage a milestone schedule for utilization.
   2. Establish and manage guidelines for customization and provide overall direction to computer system administrator.
   3. Establishes signature authorities for system access and utilization.
   4. Coordinate with the computer system vendor on nontechnical details.
   5. Coordinate necessary initial training for managers, supervisors, engineers, operators, clerks, planners, maintenance personnel, and computer support.
   6. Coordinate with other groups in company using computer.

Maintenance Analyst

A. Work with management to determine appropriate metrics or indicators to establish and track.

B. Establish, track, and report appropriate metrics.

C. Perform in-house studies or coordinate outside help to conduct work sampling studies.

G

Sample Work Sampling (Wrench Time) Study: "Ministudy"

Two appendices on work sampling, Apps. G and H, contain the reports of actual studies. This type of productivity study is the primary measure of planning and scheduling effectiveness. Consultants typically conduct work sampling studies. Both example studies contain complete procedures for conducting an in-house study. Appendix G shows a streamlined, simple study requiring a minimum of effort. Appendix H contains a more traditional study with many more measurement observations. The traditional study contains a section validating the accuracy of streamlined studies.

This appendix includes the details, results, and analysis of an actual wrench time study, and a procedure for doing an in-house study without the need for a consultant. This appendix is a shortened version of a full-blown study that is shown in the next appendix, App. H. Such a ministudy adequately establishes a measure of the so-called wrench or direct work time of a craft for the purpose of improving a planning and scheduling operation.

(*Note:* Although this is a copy of an actual work sampling study, the names have been changed. Minor areas have also been altered to clarify peculiar company jargon and further author's notes have been inserted to call attention to particular details.)

WORK SAMPLING STUDY OF I&C MAINTENANCE, OCTOBER–DECEMBER 1993. FINAL REPORT, MARCH 25, 1994.

Table of Contents

Category Definitions

Study Results
  Collection of Observation Data
  Analysis

  Overall
  Personnel
  Unit Status
  Time

Conclusions

Recommendations

Attachment A: Procedure for Measuring Workforce Productivity by Work Sampling

Attachment B: Work Sampling Calculations

Executive Summary

A work sampling study was performed in-house by North Station I&C Maintenance. This study measured current performance and provided a baseline against which to compare future studies. It also helped station personnel become accustomed to being measured. The study used the same observation and analysis criteria as previous studies had done for the mechanical maintenance craft.

Current wrench time is 38.54%, almost 4 hours per 10-hour day, with a margin of error of 6%, as shown in Fig. G.1. The major adverse impact on wrench time is the combination of work assignment and waiting for

Figure G.1 Instrument and Controls group without planning department assistance.

instructions, both of which are information sharing processes. Together these two areas consume over 2 hours each day.

Hour by hour, wrench time changes during the day affected mainly by break, lunch, and meeting times. Other than that, work remains at a good steady pace all day extending even into the last hour.

Future studies should help distinguish wrench time for special groupings such as unit event, crew, or day of week. More observations being added to the database will allow making more statistically valid analysis.

Introduction

In October 1993, the Maintenance Management Improvement (MMI) program initiated a work measurement study of I&C maintenance at North Station. The objectives of the study were to (1) analyze delays for improvement opportunities, (2) establish a baseline against which to compare future studies, and (3) help station personnel to become accustomed to being measured as normal practice for a first-rate maintenance organization.

This study lasted a period of 7 weeks during October through December 1993. Observations were taken evenly over 10-hour, first-shift work periods, excluding the lunch period from 12:00 noon to 12:30 PM. Observations were only made of personnel within the plant grounds and did not include personnel going outside the main gate to the fuel oil dock nor to other company facilities. Care was taken to duplicate as much as practical the study practices of earlier studies for other crafts. This study utilized Attachment A, Procedure for Measuring WorkForce Productivity by Work Sampling.

The measurement focus of this study is on quantifying the amount of time spent by the work force in various types of activities when the employees are present to work for an entire 10-hour shift. The time spent performing work at a job site is direct work or wrench time. Time spent otherwise, such as in traveling or waiting, is indirect work. From quantifying and analyzing indirect work activities, action can be taken to reduce their amounts and increase the amount of time spent productively performing work at the job site. By conducting follow-up studies, the effect of such action taken may be determined.

The MMI Job Planning Coordinator performed the duties of observer and study analyst for this study.

It should be mentioned that as indicated by the margin-of-error percentage, results are more statistically significant for larger numbers of observations. This statement means that the accuracy of the results changes from being highly accurate for the overall I&C maintenance work force to being less accurate for an individual crew or craft skill level. Likewise, results for an individual employee are not statistically significant at all due to the low number of observations and are therefore not reported.

Category Definitions

The following are definitions for each of the work sampling study categories. Note that "other" categories are left available to capture special situations that clearly would not fall into set categories.

Note: As studies progressed, occasional activities were observed which did not fall neatly into one of the predefined categories. When necessary, after consulting with either the maintenance supervisor or maintenance manager, a category was selected and an appropriate category clarification was made in these instances. The criteria for altering a classification are usually twofold. First, preference is given to the benefit of the craft technicians, i.e., it gives them credit for more wrench time. It is very important to "bend over backwards" to avoid giving unnecessary criticisms of the study by the crafts themselves. Second, the work sampling categories are mindful not to include in direct work time, wrench time, any activity in which the planner could leverage planning time and avoid a larger delay later in the field.

In addition, as electrical crafts and I&C crafts were included in subsequent studies to the initial mechanical studies, the categories also evolved somewhat. For example, frequently I&C work requires careful study of plant controls schematics. While a planner finding technical drawing data could probably help a field mechanic save time later, an I&C planner doing too much beyond drawing identification would only be duplicating what the I&C field tech would later have to do. Electrician and I&C research of drawings is considered wrench time, but not for mechanical crafts. To avoid confusion, a consistent set of classifications is used as presented below and represents the most current evolution of the classifications as of 1997.

Working

1. *Working.* This category was used when persons were performing work (including troubleshooting) physically at the work site or shop. For the I&C and electrical crafts only, this category includes all troubleshooting, whether it is at work site or not. Also considered as this category were completing any job related paperwork or conferring with a planner to help the planner plan future work. If one or more persons of a work team were working and other members of the team were idle, then all members of the team were to be judged as working. An example would be the "hole person" who must be outside the boiler when others are working inside. An exception is that an I&C or electrical person away from the work site that is merely waiting for another person to troubleshoot is not in this category. (*Troubleshooting* is defined as determining the cause of a problem rather than finding information such as clearances, etc.)

(*Note:* It is Category 1 that is considered direct work or wrench time. The objective of planning and scheduling is to help keep the craft labor hours in this category. Examining how much time is spent in the other categories facilitates this objective.)

2. *Work travel.* This category includes walking to the work site; walking to the shop, supply room, tool room, or operations room, etc.; or traveling by vehicle in connection with work. Traveling within the work task location (e.g., as from one side of the boiler to the other) was not included in this category, but rather was considered as "working," Category 1 above.

(*Note:* It is invaluable to ask and tabulate why someone is traveling because travel is not done unto itself. For example, if getting parts causes most of the travel, then that valuable information guides the improvement of planning or other maintenance processes.)

3. *Set-up and take down.* This category has been infrequently observed and so is not used. This type of work is included in Category 1, Working.

4. *Work assignment.* Work assignment included any regular crew meeting for sharing information and any occasion involving reassignment to another task, upon completion of a job during the day. This category included being idle between jobs unless in the break room or other break area.

5. *Wrap-up.* Wrap-up was only used at the end of the day when persons were filling out time sheets, cleaning up after working, and meeting at the end of the shift.

Waiting

6. *Waiting for materials.* This category was used when a person was waiting at the storeroom for materials, or in the case of a team, when persons were waiting for another person to return with the supplies or materials needed for them to continue their job.

7. *Waiting for tools.* This category was used when a person was waiting at the tool room (or any other tool location away from the work site), or in the case of a team, when persons were waiting for another person to return with a tool(s) needed by them to continue their job.

8. *Waiting for instruction.* This category was used when persons performing the job were delayed by the necessity to acquire resolution of questions raised concerning some aspect of the job. It was also utilized as a category for capturing time expended answering telephone calls when a person was paged by another company employee. It was also utilized whenever a person was discussing any work related question with a co-worker (unless in the case of troubleshooting by I&C or electrical craft) or supervisor.

9. *Clearance delay.* Clearance delay was employed as a category when a team was delayed from working by the necessity to acquire clearance for a piece of equipment prior to continuing work on their project.

10. *Interference.* This category was used when a person or team was not able to perform their job until another person or team completed theirs.

11. *Other work waiting.* This category was never utilized.

12. *Waiting for operator.* When an operator or engineer was required to inspect or assist in the work and was not available, and therefore created a delay, this category was utilized.

13. *Weather delay.*

14. *Other.* This category was never utilized.

Other

15. *Meetings.* Though most meetings are administrative and not included in study time (for company safety meetings or other gatherings), there are meetings such as with operators, supervisors, or vendors that are brief and usually unscheduled, and therefore are included in the study.

(*Note:* We are only interested in studying persons who are available for work. That is, for persons available to the supervisor to work an entire 10-hour shift, how is their time spent? The planner and supervisor cannot do anything to leverage required administrative time and so that is of no interest here. However, administrative time away (vacations, illness, training) can be a significant management problem worthy of study in itself.)

16. *Training.* This category was never utilized.

(*Note:* See above discussion for Category 15.)

17. *Idle.* Idle was used usually at the job site when work, tools, equipment, assignment, etc., appeared to be available, the person did not appear to be performing work, and there did not appear to be any obvious interference or delay preventing the employee from performing the job.

(*Note:* This is an area where it helps if the study observer has some familiarity with the craft work being done. It is sometimes difficult to tell if a team is idle or being delayed. Questioning without craft familiarity usually does not lead to a confession of idleness.)

18. *Rest room.* This category was used when persons were in the rest room at times other than traditional break time, lunch or afternoon meeting.

(*Note:* An example of the considerations one has to make in such a study is that it could be a form of sexual harassment if a male observer knocked on a ladies rest room to ask if a particular person was therein. A claim might be that the observer "was trying to catch a glimpse of someone undressed." It is not harassment for a designated observer on a designated study to ask someone of the appropriate sex to go into the rest room to inquire. Fortunately, this situation can be avoided nearly altogether if the observer is familiar with the plant and the day's job assignments. With that information, the observer can normally find the person being searched out and leave checking the rest rooms as a last resort.)

19. *On break.* Persons were considered on break from the time they arrived at the break room or at either their desks or the shop area table without work until the time they left. Walking to and from the break room was categorized as *work travel,* as described in Category 2 above. It was also utilized when persons were in the break room or washing up early for lunch or remaining after lunch. At other times during the day, unless due to a delay or interference, such an observation was considered break time, not idle time because some crafts have no set times for breaks.

20. *Other personal allowance.* This category was utilized when persons were conducting what appeared to be personal business not required by their assigned task, such as taking medication (unless in the break room).

21. *Other.* This category was utilized on one study for filling out an accident report.

22. *Other.* This category was never utilized.

Unaccountable

23. *Unaccountable.* This category was used when a person could not be found at the person's assigned work location, or the tool room, storeroom, office, etc., and was only utilized after a 15–20 minute attempt to locate the person was unsuccessful. This category is not intended to indicate that the employee was either working or not working, but simply that the employee could not be located.

(*Note:* This is an area in which it is important for the observer to have a working knowledge of the plant areas and practices, if for no other reason than to be able to find persons. While observations are sometimes not used at the beginning of a study because persons observed have to settle down to their normal behavior, a practical use of beginning observations is to practice finding persons. Unaccountable incidents are useful information if there seems to be certain patterns, such as if the incidents are clustered around the day's end.)

Study Results

The following sections discuss the collection and the subsequent analysis of the observations.

Collection of observation data

During each observation numerous data were collected to allow analysis of the workforce in many ways. This data included employee identification, craft, supervisor, work category, unit status, date, day of week, hour of day, plus comments regarding certain individual observations. Codes used for craft were IT and IA for I&C technicians and apprentices, respectively.

The following list defines the hour periods used in the study. Each period defines the time frame in which the search began.

Period 8 is 7:30 AM to 8:30 AM.

Period 9 is 8:30 am to 9:30 am.

Period 10 is 9:30 am to 10:30 am.

Period 11 is 10:30 am to 11:30 am.

Period 12 is 11:30 am to 12:00 pm and 12:30 pm to 1:00 pm.

Period 13 is 1:00 pm to 2:00 pm.

Period 14 is 2:00 pm to 3:00 pm.

Period 15 is 3:00 pm to 4:00 pm.

Period 16 is 4:00 pm to 5:00 pm.

Period 17 is 5:00 pm to 6:00 pm.

Table G.1, Classification of Observations by Hour, is a checking tool to verify that an equivalent number of observations were made each hour of the day for the study overall to avoid skewing the data. For all persons combined, there is an overall uniformity of about 29 average observations for each period. The fewer than average observations (24) in the 11:00 AM period reflect the weekly crew meeting with the Director of Engineering which was not included nor expected to skew the results. A similar analysis is made throughout the study of examining the observation distribution for specific cases under consideration.

TABLE G.1   Classification of Observations by Hour

Hour period	All	I&C tech	I&C appr
8	30	19	11
9	29	17	12
10	30	14	16
11	24	14	10
12	27	13	14
13	30	13	17
14	29	14	15
15	29	15	14
16	30	13	17
17	30	11	19
Totals	288	143	19

Analysis

The study analyzed the observations in different manners to gather conclusions regarding differences among personnel classifications, differences among different unit status, and differences among different time periods.

Overall.   Table G.2 summarizes the data for all categories during the observation period for this 1993 study. Each category percentage must be considered in light of its margin of error (MOE) shown in Table G.2.

Table G.2 and Fig. G.2, Distribution of Time, show the relative impact of each category's percentage of time. The greatest impact is made by Categories 2, Travel (14.24%), and 4, Work Assignment (11.11%). Table G.3, Minutes per Day for Each Category, shows that these categories take

TABLE G.2   Results for All Personnel

Category	Fri	Mon	Tues	Wed	Thur	Obs.	Percent	MOE,* %
				Work				
1	21	18	30	18	24	111	38.54	6
2	16	10	8	5	2	41	14.24	4
4	6	9	3	8	6	32	11.11	4
5	2	1	1	4	3	11	3.82	2
						Subtotal	67.71	
				Waiting				
6	2	8	3	0	4	17	5.90	3
7	0	1	1	2	1	5	1.74	2
8	2	4	1	5	7	19	6.60	3
9	1	0	0	0	0	1	0.35	1
10	1	0	0	0	0	1	0.35	1
12	0	1	0	0	1	2	0.69	1
						Subtotal	15.63	
				Other				
15	0	1	2	0	3	6	2.08	2
17	0	0	0	0	1	1	0.35	1
19	5	3	7	6	3	24	8.33	3
20	0	0	2	0	1	3	1.04	1
21	0	0	0	1	0	1	0.35	1
						Subtotal	12.15	
				Unaccountable				
23	4	4	2	0	3	13	4.51	2
						Subtotal	4.51	
Totals	60	60	60	49	59	288	100.00	

NOTE: There were no observations in any categories not shown.
*MOE = margin of error.

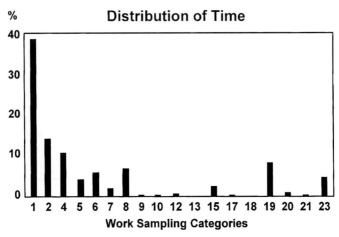

Figure G.2   Time spent in different categories.

TABLE G.3   **Minutes per Day for Each Category**

Category	Name	Percent	Minutes/ day	Comment
1	Working	38.54	231	
2	Work travel	14.24	85	
3	Set-up & take down	0.00	0	Category not used
4	Work assignment	11.11	67	Three daily crew meetings
5	Wrap-up	3.82	23	
6	Waiting for materials	5.90	35	
7	Waiting for tools	1.74	10	Mostly from I&C shop
8	Waiting for instruction	6.60	40	Supvr. or waiting on partner
9	Clearance delay	0.35	2	
10	Interference	0.35	2	Mechanical maintenance
11	Other work waiting	0.00	0	Category not used
12	Waiting for operator	0.69	4	
13	Weather delay	0.00	0	
14	(Category not used)	0.00	0	
15	Meetings	2.08	12	Crew safety
16	Training	0.00	0	Category not used
17	Idle	0.35	2	
18	Rest room	0.00	0	
19	On break	8.33	50	
20	Other personal allowance	1.04	6	
21	Other	0.35	0	Accident paperwork
22	(Category not used)	0.00	0	
23	Unaccountable	4.51	27	
	Total for day		600 minutes	

up 85 and 67 minutes per day, respectively. Categories 6, Materials (5.90%, 35 minutes/day), 8, Instruction (6.60%, 40 minutes/day), and 19, Break (8.33%, 50 minutes/day) also have an appreciable impact.

Reviewing the comments recorded for individual observations in each category allowed accounting for most of the travel by purpose of the travel. Table G.4 and Fig. G.3 illustrate the result of adjusting the percentage of time in major categories by including their associated travel.

TABLE G.4   Comparison of Study with Travel Adjustment

Category	Name	Percent	Travel adjustment	Percent after travel adjustment	Adjusted minutes/day
			Working		
1	Working	38.54	0.00%	38.54%	231
2	Work travel	14.24	−12.12%	2.12%	13
4	Work assignment	11.11	2.78%	13.89%	83
5	Wrap-up	3.82	0.69%	4.51%	27
	Subtotal	67.71	−8.65%	59.06%	
			Waiting		
6	Waiting for materials	5.90	2.78%	8.68%	52
7	Waiting for tools	1.74	1.39%	3.13%	19
8	Waiting for instruction	6.60	0.35%	6.95%	42
9	Clearance delay	0.35	0.14%	0.49%	3
10	Interference	0.35	1.04%	1.39%	8
12	Waiting for operator	0.69	0.00%	0.69%	4
13	Weather delay	0.00	0.00%	0.00%	0
	Subtotal	15.63	5.70%	21.33%	
			Other		
15	Meetings	2.08	0.00%	2.08%	12
17	Idle	0.35	0.00%	0.35%	2
18	Rest room	0.00	0.69%	0.69%	4
19	On break	8.33	2.43%	10.76%	65
20	Other personal allowance	1.04	0.00%	1.04%	6
21	Other	0.35	0.00%	0.35%	2
	Subtotal	12.15	3.12%	15.27%	
			Unaccountable		
23	Unaccountable	4.51	0.00%	4.51%	27
	Subtotal	4.51	0.00%	4.51%	
	Totals	100	0%	100%	601 Min

NOTE: There were no observations or adjustments in any categories not shown.

Figure G.3   Reasons for time spent traveling.

A significant amount of time spent traveling was involved with Work Assignment, Materials, Tools, and Break. After adjusting for travel, the initial categories for greatest impact (Work Assignment, Materials, Instruction and Break) increased from about 3 hours to over 4 hours per day versus less than 4 hours per day in Category 1, Working.

Categories 4, Work Assignment, and 8, Waiting for Instructions, are closely related as they involve sharing information. Together these two categories consume over 2 hours per day. They are the largest opportunity for improvement. The comments made for observations in these areas show the three daily crew meetings consuming the bulk of Category 4 and the time split between getting the supervisor information and watching a partner research files consuming Category 8.

When considering associated travel, Category 19, Break consumes 65 minutes each day. Similarly, Category 6, Materials, accounts for 52 minutes each day. Most job delays waiting for materials involved items kept in the shop rather than the storeroom.

Analyzing the comments for Category 1, Working, gives insight on the type of work done by the I&C craft. Of the 111 Category 1 observations, 39% were of work directly on plant equipment out in the plant. 15% were of work on a computer, 10% were associated with documenting work, and 8% were troubleshooting in files or with other knowledgeable persons. 6% were of work advising others, 6% were on a special assignment to organize the shop, and finally 6% were of work being done in the shop itself on the work benches.

Personnel. The study also analyzed wrench time considering different personnel. The study made classifications by craft skill level and by crew.

Figure G.4, Distribution of Time by Craft Skill, shows the relative percentage of time spent by each level in each category. Tables G.5 and G.6 show the summary data with actual percentages of time, margin of error, and number of observations for each level. Table G.1 allows determination of whether the distribution of observations might have skewed the results for either level.

Figure G.4 shows that technicians have a higher wrench time (Category 1) and Materials delays (Category 6) than apprentices, but these differences are not statistically significant within the margin of error of the study.

On the other hand, at first glance it does appear significant that technicians have a higher amount of time in Category 4, Work Assignment, and a lower time in Category 8, Instructions. The work assignment time is higher only because the observation distributions are skewed, showing markedly more technician observations during the morning meeting time and less technician observations during the last hour of the day. (That skewing is also responsible for showing technicians with a lower wrap-up time, Category 5.) The lower waiting for instruction time is not unexpected as technicians should require less time receiving directions than apprentices.

(Analysis by crew was also made. Initial inspection showed the two crews having different wrench times, one at 44.52% and one at 32.99%. But further analysis showed the low crew's observations were decidedly skewed

Figure G.4   Technicians and apprentices spend time differently.

TABLE G.5   Results for I&C Technicians

Category	Fri	Mon	Tues	Wed	Thur	Obs.	Percent	MOE,[*] %
Work								
1	11	13	12	12	9	57	39.86	8
2	6	6	2	1	1	16	11.19	5
4	4	7	2	2	5	20	13.99	6
5	2	0	1	1	0	4	2.80	3
						Subtotal	67.83	
Waiting								
6	2	4	1	0	3	10	6.99	4
7	0	1	1	1	0	3	2.10	2
8	0	2	0	2	2	6	4.20	3
9	1	0	0	0	0	1	0.70	1
10	1	0	0	0	0	1	0.70	1
12	0	1	0	0	1	2	1.40	2
						Subtotal	16.08	
Other								
15	0	1	0	0	1	2	1.40	2
17	0	0	0	0	1	1	0.70	1
19	2	2	1	2	2	9	6.29	4
20	0	0	2	0	1	3	1.04	1
21	0	0	0	1	0	1	0.70	1
						Subtotal	9.09	
Unaccountable								
23	3	3	1	0	3	10	6.99	4
						Subtotal	6.99	
Totals	32	40	21	22	28	143	100.00	

NOTE: There were no observations in any categories not shown.
[*]MOE = margin of error.

toward more break and meeting time observations, which explained their apparent low Category 1, Working, time as well as high time for travel, work assignment, and break (Categories 2, 4, and 19). Therefore, no graphs or tables were presented. Incidentally, this type of situation was expected where a study of limited observations was conducted. However because the database was constructed to be additive, later studies adding to the observation amount may help analysis of this area.)

Unit status.   The study did not make an analysis by unit event, such as for days when a unit was in outage. There were too few observations for each of the multiple plant conditions observed. Later studies may allow examination of this area.

TABLE G.6   Results for I&C Apprentices

Category	Fri	Mon	Tues	Wed	Thur	Obs.	Percent	MOE,[*] %
				Work				
1	10	5	18	6	15	54	37.24	8
2	10	4	6	4	1	25	17.24	6
4	2	2	1	6	1	12	8.28	5
5	0	1	0	3	3	7	4.83	4
						Subtotal	67.83	
				Waiting				
6	0	4	2	0	1	7	4.83	4
7	0	0	0	1	1	2	1.38	2
8	2	2	1	3	5	13	8.97	5
						Subtotal	15.17	
				Other				
15	0	0	2	0	2	4	2.76	3
19	3	1	6	4	1	15	10.34	5
20	0	0	2	0	1	3	2.07	2
						Subtotal	15.17	
				Unaccountable				
23	1	1	1	0	0	3	2.07	2
						Subtotal	2.07	
Totals	28	20	39	27	31	145	100.00	

NOTE: There were no observations in any categories not shown.
[*]MOE = margin of error

Time.   Finally, the study analyzed work activity over time itself. Figure G.5, Wrench Time Categories by Hour, demonstrates the variance by hour. Category 1, Working, is between 50 and 55% during periods away from day start, day end, lunch, and traditional break or meeting times. However, it drops to between 18 and 30% around breaks, meetings, or lunch time. And while it drops to 20% for the first hour of each day, it is still about 30% for the last hour of each day.

The categories which most vary (indirectly) with wrench time are Category 19, Break, and Category 4, Work Assignment, as expected. Category 2, Travel, is interesting in that most travel appears to be in the afternoon hours.

On the other hand, Category 8, Instructions, seems pretty much constant throughout the day until it drops off for the last 2 hours. (An exception is during the 10:00 AM break time when no instruction observations are made at all.) This constancy indicates a steady work pace.

Figure G.5   Wrench time changes every hour.

Category 6, Waiting on Materials, also seems to be roughly level over the course of the day indicating a steady work rate.

Finally, Category 5, Wrap-up, is only 30% (less than 20 minutes) for the last hour. This time does not include any time spent completing job reports as that was considered Category 1, Working.

In summary of Fig. G.5, it appears that persons work at a steady rate over the entire course of the day, but work is impacted by significant break and meeting times.

Conclusions

The following conclusions are made.

1. A baseline has been established against which to compare future I&C studies.

2. Current wrench time is 38.54%, or almost 4 hours per 10-hour day with all categories summarized in Table G.1. Analysis of the comments and time of day for each observation raises concerns that a significant amount of time is spent sharing information through crew meetings and individual instruction. Together these two areas consume over 2 hours per day.

3. The main difference between technicians and apprentices appeared to be that apprentices experienced more delays in receiving individual instruction.

4. Hour-by-hour wrench time does change throughout the day, but the study suggests that a steady work pace is kept up all day even up through the last hour.

5. Wrench time measured for only 7 individual days (approximately one per week) during the study had an acceptable margin of error, 6%, for overall wrench time. This result suggests that the limited observation effort yielded valid results provided that the days selected are considered typical of the ongoing period.

6. The process of occasional observations over 7 weeks appeared to acclimate the work force to being measured. It should be noted that the I&C personnel had a definite professional nature, and so wrench time might not improve merely as a result of them being observed. But this quality should help the I&C group use measurement results to improve themselves.

Recommendations

The following recommendations are made.

1. Emphasis should be given to addressing communication areas to improve wrench time.

2. A continuing in-house study of observations only one day each week should be conducted to give ongoing feedback on wrench time.

3. Future observation data should be structured to be additive as well as separate from previous observations to allow reduced margins of error to be developed in the analysis of specific groupings such as for supervisors, craft skill level, and special events such as outages and trips.

Attachment A: Procedure for Measuring Work Force Productivity by Work Sampling

This study utilized the following procedure.

1. The Job Planning Coordinator (JPC) and the appropriate managers or supervisors select the workforce personnel and time period to be studied. In addition, work type categories are reviewed.

2. The JPC assigns each person in the study a one- to three-digit, unique number for the purposes of tracking. (The JPC attempts to utilize the same number for persons included in previous studies.) The JPC records each person's name, number, and craft with current skill level in Table G.7. Table G.8 presents codes for craft with current skill level.

TABLE G.7   **Persons in Study**

Name	Number	Craft & level
Davenport	201	IA
Dowenger	202	IT
Fass	203	IA
Gee	204	IA
Henry	205	IT
Hilliard	206	IT
Jorge	207	IT
Lucien	208	IA
Pole	209	IT
Robertson	210	IT
Rou	211	IA
Sandez	212	IA
Sandhill	213	IT
Williams	214	IT

TABLE G.8   **Craft and Skill Designations**

Designation	Craft and skill
MR	Mechanical Trainee
MA	Mechanical Apprentice
M	Mechanic (pre-1994 designation)
W	Welder (pre-1994 designation)
H	Machinist (pre-1994 designation)
MT	Mechanical Technician
P	Painter
CP	Certified Painter
MC	Certified Mechanic
WC	Certified Welder
HC	Certified Machinist
IR	I&C Trainee
IA	I&C Apprentice
IT	I&C Technician
ER	Electrical Trainee
EA	Electrical Apprentice
ET	Electrical Technician
A	Mechanical Apprentice (pre-1994 designation)
T	Mechanical Trainee (pre-1994 designation)

TABLE G.9   **Supervisors of Crews in Study**

Supervisor number	Supervisor
9	Paddington
900	Paddington crew when no supervisor present for the day
901	Anyone working up for Paddington

3. The JPC enters the names of the crew supervisors who have personnel being studied in Table G.9. The JPC assigns each person a one-to three-digit, unique number for the purposes of tracking.

4. For the ministudy, the JPC privately selects observation days from the overall study duration to cover each individual observation time once. The ministudy covers a 40-hour period, but the study is spread over at least 7 weeks. Each afternoon before or morning of a planned observation day, the JPC obtains a schedule or agenda of the day's work with the names of assigned personnel from each supervisor. This helps the JPC to locate persons. It also allows a crew to be on its best performance and less apt to disagree with final study results.

5. Each observation day the JPC utilizes that day's work schedule to select employees to observe and record their work category. The JPC generally selects employees and exact times to observe so as to include as many observations of different employees and times as possible. (The JPC keeps an informal check sheet, such as Fig. G.6, to ensure making an even number of observations each hour of the overall study.) Typically the JPC selects three persons in the same area of the plant to minimize time spent locating the employees. The instant the JPC locates each selected employee, the JPC records that employee's activity in a work category on the Wrench Time Observation Data Sheet (Fig. G.7). (The JPC may occasionally question the employee if it is not certain which category is appropriate for the current activity observed.)

Two sampling methods would ensure representative data, random and sequential. Either method provides nonbiased samples. The observer uses a list of crew employees to create a "sequential" observation strategy. At the beginning of the study, the observer simply starts at the top of the list to select the initial person for the first observation. However, the study also takes advantage of the crew supervisor assigning some employees together as teams and other jobs being in the same area of the plant as the initially chosen person. Because the observer must select three persons each half-hour, the observer selects the initial person and the two supposedly closest persons to that person. Then the observer checks all three names of the crew list. For the next half-hour, the observer goes down the list "sequentially" to select the next unchecked person and then selects the two closest unchecked persons to that person. The observer checks off these three names. When there are not three names left on the list, the observer continues around back to the top of the list. The observer keeps the place on the employee list at the end of a day for the beginning of another day. In this manner, the observer has an equal chance of observing every employee throughout the study and keeps the sample statistically unbiased.

6. After collecting all the study observations, the JPC counts the observations in each category by whichever criteria necessary to make

INFORMAL CHECK SHEET TO BALANCE OBSERVATIONS

Mark the time started to locate persons within beginning half hour (B) or end half hour (E): Hand write a number beside each half hour to designate the week during which the observation occurred.

Weeks:
#1 10/17-10/23   #2 10/24-10/30   #5 11/14-11/19   #7 11/28-12/4

Time Period	Time	Mon	Tues	Wed	Thur	Fri
8	7:30-8:30	B1E1	B5E5	B2E2	B7E7	B1E1
9	8:30-9:30	B1E1	B5E5	B2E2	B7E7	B1E1
10	9:30-10:30	B1E1	B7E5	B2E2	B7E7	B1E1
11	10:30-11:30	B1E1	B5E5	B*E*	B7E7	B1E1
12	11:30-12,12:30 -1	B1E1	B5E5	B*E2	B7E7	B1E1
13	1 - 2	B1E1	B5E5	B2E2	B7E7	B1E1
14	2 - 3	B1E1	B5E5	B2E2	B1E1	B1E1
15	3 - 4	B1E1	B5E5	B2E2	B7E7	B1E1
16	4 - 5	B1E1	B5E5	B2E2	B7E7	B1E1
17	5 - 6	B1E1	B5E5	B2E2	B7E7	B1E1

Special comments: *A special meeting held with the crew and the corporate Director of Engineering this quarter was not included in the study.*

Figure G.6 Informal checks to ensure even observations.

the desired comparisons and contrasts. This study reports this analysis in tables and figures. The JPC calculates the margin of error for each observation category using the formulas in Attachment B.

(*Note:* Manual counting and calculations are adequate for short studies. Appropriately constructed databases or spreadsheets facilitate making the tabulations and calculations.)

WRENCH TIME OBSERVATION DATA SHEET

PAGE ___ OF ___

STUDY _____

PLANT _____

(Cmptr Entry)	Unit Events	Observing Date	Day of Week	Person's Number	Person's Supvr Number	Work Cat #	Time Period	OT?	Any Observation Comments (such as purpose of travel)
		/ /							
		/ /							
		/ /							
		/ /							
		/ /							
		/ /							
		/ /							
		/ /							
		/ /							
		/ /							
		/ /							
		/ /							
		/ /							
		/ /							
		/ /							
		/ /							
		/ /							
		/ /							
		/ /							

Figure G.7   Form to record work sampling observations.

Attachment B: Work Sampling Calculations

The company specifications for the work measurement study state: "The proposed study design will include the methodology for assuring accuracy within ±10% with a 95% confidence level based on the direct work category."

Three study criteria are important to ensuring study accuracy. These criteria are having the observation period span a sufficient portion of the year, having the observations evenly spread out over the course of the shift, and having a sufficient number of observations.

First, the study covered a period of 7 weeks, which should be of sufficient duration to classify as a representative period of typical working conditions and cancel out the effect of most special, limited-duration events that may impact wrench time.

Second, keeping a check sheet ensured making an equal number of observations during each 30-minute block of the entire 10-hour shift by the end of the study. In addition, during the analysis of the results, a check was made of the hourly distribution of observations to determine if any skewing of data might be present. (For example, having a more than average number of observations during the traditional morning break period may cause an artificial decrease in reported wrench time.) Overall there appeared to be an even distribution of observations. In the special cases where there was a slightly uneven distribution, the results are noted as possibly skewed with the possible effect described.

Finally, the margin of error (absolute accuracy) for the direct work category was 6%, well within the 10% required. There were 288 total observations, of which 111 are of direct work, Category 1. The percentage of direct work is 111 divided by 288 which is 38.54%. The margin of error is found by the following equation:

$$a = \{[k^2(1 - p)p]/n\}^{1/2}$$

where a = margin of error
   k = number of standard deviations (k = 2 represents a 95% confidence level)
   p = percent occurrence of the activity or category
   n = sample size (total observations)

Thus,

$$a = \{[2^2(1 - 0.3854)\ 0.3854]/288\}^{1/2} = 5.7\% \text{ (rounded up to 6\%)}$$

This calculation means that 95 percent of the time, any duplicate study done over the same 7-week period and making observations at the same hourly points should have reported a direct work percentage within 6% of the 38.54% reported in this study. Because the study is only repeatable and valid over duplicate conditions, it is clear to see the importance of having a representative study period and an even hourly distribution of observations.

H

Sample Work Sampling (Wrench Time) Study: Full-Blown Study

Two appendices on work sampling, Apps. G and H, contain the reports of actual studies. This type of productivity study is the primary measure of planning and scheduling effectiveness. Consultants typically conduct work sampling studies. Both example studies contain complete procedures for conducting an in-house study. Appendix G shows a streamlined, simple study requiring a minimum of effort. Appendix H contains a more traditional study with many more measurement observations. The traditional study contains a section validating the accuracy of streamlined studies.

This appendix includes the details, results, and analysis of an actual wrench time study, and a procedure for doing an in-house study without the need for a consultant. This appendix is a full-blown study that analyzes work, making an exhaustive number of comparisons. The preceding ministudy, App. G, did not contain enough observations to make such analysis. However, such a ministudy adequately establishes a measure of the so-called wrench or direct work time of a group of craftpersons for improving a planning and scheduling operation.

(*Note:* Although this is a copy of an actual work sampling study, the names have been changed. Minor areas have also been altered to clarify peculiar company jargon, and further notes have been inserted to call attention to particular details.)

**WORK SAMPLING STUDY OF MECHANICAL MAINTENANCE, JANUARY–MARCH 1993.
FINAL REPORT, APRIL 29, 1993**

Table of Contents

Executive Summary

The Maintenance Management Improvement (MMI) program performed a work sampling study for North Station Mechanical Maintenance. The team performed this study to determine the validity of in-house studies and to compare current performance against previous studies. The team also did the study to help accustom station personnel to being measured. A consultant performed the two previous studies of working (wrench) time in 1990 and 1991. In addition to comparisons to the previous studies, analysis of the current study proved the validity of the current study.

The results also suggest future studies can be done with greatly reduced efforts relying on 1-day-per-week observations. These studies should give ongoing feedback of performance and better acclimation of personnel to being measured. It is recommended such future studies be done.

Comparison to previous studies shows wrench time being the same despite changes in other time categories. Current wrench time is statistically unchanged at 35.08% or 3½ hours per 10-hour day with a margin of error of 4%. The previous studies place wrench time at approximately 37% for past years. However, travel has improved from 21 to 15% and work assignment from 5 to 2%. But wrench time apparently does not rise because break times and unaccountable times get worse (as do material and instruction delays to a lesser extent). An analysis of the comments and the time of day for each observation indicate that scheduling and motivational concerns might be associated with these results. (For example, most of the unaccountable observations occur at day start, lunch, and day end.)

There are differences in wrench time among the crafts, with machinists being the highest at almost 51%, or 5 hours per day, and painters being the lowest at about 26%, or 2½ hours per day. There are also distinct differences in the other categories for each craft.

The only event that appeared to affect wrench time significantly was when a unit was tripped or called. Wrench time then fell. Other occurrences such as day of week or loaning of personnel did not affect wrench time significantly.

Future study should help distinguish wrench time per crew, which now appears different among crews, but with overlapping margins of error.

Hour by hour, wrench time changes throughout the day. Overall, there is moderate wrench time in the morning, peak wrench time (over 50%) for 2 consecutive hours after the lunch period, and then less than moderate wrench time thereafter. All of the morning hours have a higher than average delay time waiting for tools and, to a lesser degree, parts and instructions. The initial morning hour has low wrench time associated with high travel and unaccountable personnel. Break and lunch associated periods have only somewhat higher than normal travel. The two periods of peak wrench time also have substantial travel and delays for instructions. The final hour experiences almost no wrench time with high break, wrap-up, and unaccountable percentages.

Introduction

In January 1993, the MMI team initiated a third work measurement study of mechanical maintenance at North Station. The objectives of the study were to (1) analyze results and compare them against two previous studies reported by an earlier paid consultant in September 1990 and July 1991, (2) consider the validity and implications for the future of doing an in-house study, and (3) help station personnel become accustomed to being measured as normal practice for a first-rate maintenance organization.

This study was conducted over 7 weeks during January through March 1993. Observations were taken evenly over 10-hour, first-shift work periods, excluding the lunch period from 12:00 noon to 12:30 PM. Observations were only made of personnel within the plant grounds and did not include personnel going outside the main gate to the fuel oil dock or to other company facilities. Care was taken to duplicate, as much as practical, the practices of the earlier studies. Attachment A, Procedure for Measuring Workforce Productivity by Work Sampling, was developed and utilized for this study.

The measurement focus of this study is on quantifying the amount of time spent by the workforce in various types of activities when the employees are present to work for an entire 10-hour shift. The time spent performing work at a job site is direct work or wrench time. Time spent otherwise, such as in traveling or waiting, is indirect work. From quantifying and analyzing indirect work activities, action can be taken to reduce their amounts and increase the amount of time spent productively performing work at the job site. By conducting followup studies, the effect of such action taken may be determined. The validity of in-house studies may be assessed from comparing the results and methods of this study with previous studies and using the experience of the study analyst and MMI project team to review the results versus expectations.

The Job Planning Coordinator (JPC) performed the duties of observer and study analyst for this study.

It should be mentioned that as indicated by the margin-of-error percentage, results are more statistically significant for larger numbers of observations. This statement means that the accuracy of the results changes from being highly accurate for the overall mechanical maintenance workforce to being less accurate for an individual craft. Likewise, results for an individual employee are not even statistically significant at all due to the low number of observations and are therefore not reported.

Category Definitions

The following are definitions for each of the work sampling study categories. Note that "other" categories are left available to capture special situations that clearly would not fall into set categories.

(*Note:* As studies progressed, occasional activities were observed which did not fall neatly into one of the predefined categories. When necessary, after consulting with either the maintenance supervisor or maintenance manager, a category was selected and an appropriate category clarification was made in these instances. The criteria for altering a classification are usually twofold. First, preference is given to the benefit of the craft technicians, i.e., gives them credit for more wrench time. It is very important to bend over backwards to avoid giving unnecessary criticisms of the study by the crafts themselves. Second, the work sampling categories are mindful not to include in direct work time, wrench time, any activity in which the planner could leverage planning time and avoid a larger delay later in the field.

In addition, as electrical crafts and I&C crafts were included in subsequent studies to the initial mechanical studies, the categories also evolved somewhat. For example, frequently I&C work requires careful study of plant controls schematics; and while a planner finding technical drawing data could probably help a field mechanic save time later, an I&C planner doing too much beyond drawing identification would only be duplicating what the I&C field technician would later have to do. Electrician and I&C research of drawings is considered wrench time, but not for mechanical crafts. To avoid confusion, a consistent set of classifications is used as presented below and represents the most current evolution of the classifications as of 1997.)

Working

1. *Working.* This category was used when persons were performing work (including troubleshooting) physically at the work site or shop. For the I&C and electrical crafts only, this category includes all troubleshooting whether it is at the work site or not. Also considered as this category were completing any job-related paperwork or conferring with a planner to help the planner plan future work. If one or more persons of a work team were working, and other members of the team were idle, then all members of the team were to be judged as working. An example would be the "hole person" who must be outside the boiler when others are working inside. An exception is that an I&C or electrical person away from the work site that is merely waiting for another person to troubleshoot is not in this category. (*Troubleshooting* is defined as determining the cause of a problem rather than finding information such as clearances, etc.)

(*Note:* Category 1 is considered direct work or wrench time. The objective of planning and scheduling is to help keep the craft labor hours in this category. Examining how much time is spent in the other categories facilitates this objective.)

2. *Work travel.* This category includes walking to the work site; walking to the shop, supply room, tool room, operations room, etc.; or traveling by vehicle in connection with work. Traveling within the work task location (e.g., as from one side of the boiler to the other) was not included in this category, but rather was considered as "working," Category 1 above.

(*Note:* It is invaluable to ask and tabulate why someone is traveling because travel is not done unto itself. For example, if getting parts causes most of the travel, then that valuable information guides the improvement of planning or other maintenance processes.)

3. *Set-up and take down.* This category has been infrequently observed and so is not used. This type of work is included in Category 1, Working.

4. *Work assignment.*   Work assignment included any regular crew meeting for sharing information and any occasion involving reassignment to another task, upon completion of a job during the day. This category included being idle between jobs unless in the break room or other break area.

5. *Wrap-up.*   Wrap-up was only used at the end of the day when persons were filling out time sheets, cleaning up after working, and meeting at the end of the shift.

Waiting

6. *Waiting for materials.*   This category was used when a person was waiting at the storeroom for materials, or in the case of a team, when persons were waiting for another person to return with the supplies or materials needed for them to continue their job.

7. *Waiting for tools.*   This category was used when a person was waiting at the tool room (or any other tool location away from the work site), or in the case of a team, when persons were waiting for another person to return with a tool(s) needed by them to continue their job.

8. *Waiting for instruction.*   This category was used when persons performing the job were delayed by the necessity to acquire resolution of questions raised concerning some aspect of the job. It was also utilized as a category for capturing time expended answering telephone calls when a person was paged by another company employee. It was also utilized whenever a person was discussing any work-related question with a co-worker (unless in the case of troubleshooting by I&C or electrical craft) or supervisor.

9. *Clearance delay.*   Clearance delay was employed as a category when a team was delayed from working by the necessity to acquire clearance for a piece of equipment prior to continuing work on their project.

10. *Interference.*   This category was used when a person or team was not able to perform their job until another person or team completed theirs.

11. *Other work waiting.*   This category was never utilized.

12. *Waiting for operator.*   When an operator or engineer was required to inspect or assist in the work and was not available, and therefore created a delay, this category was utilized.

13. *Weather delay.*

14. *Other.*   This category was never utilized.

Other

15. *Meetings.*   Though most meetings are administrative and not included in study time (for company safety meetings or other gatherings), there are meetings such as with operators, supervisors, or vendors

that are brief and usually unscheduled, and therefore are included in the study.

(*Note:* We are only interested in studying persons who are available to work, that is, for persons available to the supervisor to work an entire 10-hour shift, how is their time spent? The planner and supervisor cannot do anything to leverage administrative required time and so that is of no interest here. However, administrative time away (vacations, illness, training) can be a significant management problem worthy of study in itself.)

16. *Training.*   This category was never utilized. (*Note:* See above discussion for Category 15.)

17. *Idle.*   Idle was used usually at the job site when work, tools, equipment, assignment, etc., appeared to be available, the person did not appear to be performing work, and there did not appear to be any obvious interference or delay preventing the employee from performing the job. (*Note:* This is an area where it helps if the study observer has some familiarity with the craft work being done. It is sometimes difficult to tell if a team is idle or being delayed. Questioning without craft familiarity usually does not lead to a confession of idleness.)

18. *Rest room.*   This category was used when persons were in the rest room at times other than traditional break time, lunch, or afternoon meeting. (*Note.* An example of the considerations one has to make in such a study is that it could be a form of sexual harassment if a male observer knocked on a ladies rest room door to ask if a particular person was therein. A claim might be that the observer "was trying to catch a glimpse of someone undressed." It is not harassment for a designated observer on a designated study to ask someone of the appropriate sex to go into the rest room to inquire. Fortunately, this situation can be avoided nearly altogether if the observer is familiar with the plant and the day's job assignments. With that information, the observer can normally find the person being searched out and leave checking the rest rooms as a last resort.)

19. *On break.*   Persons were considered on break from the time they arrived at the break room or at either their desks or the shop area table without work until the time they left. Walking to and from the break room was categorized as *work travel,* as described in Category 2 above. It was also utilized when persons were in the break room or washing up early for lunch or remaining after lunch. At other times during the day, unless due to a delay or interference, such an observation was considered break time, not idle time because some crafts have no set times for breaks.

20. *Other personal allowance.*   This category was utilized when persons were conducting what appeared to be personal business not required by their assigned task, such as taking medication (unless in the break room).

21. *Other.*   This category was utilized on one study for filling out an accident report.

22. *Other.*   This category was never utilized.

Unaccountable

23. *Unaccountable.*   This category was used when a person could not be found at the person's assigned work location, or the tool room, storeroom, office, etc., and was only utilized after a 15–20-minute attempt to locate the person was unsuccessful. This category is not intended to indicate that the employee was either working or not working, but simply that the employee could not be located. (*Note:* This is an area in which it is important for the observer to have a working knowledge of the plant areas and practices, if for no other reason than to be able to find persons. While observations are sometimes not used at the beginning of a study because persons observed have to settle down to their normal behavior, a practical use of beginning observations is to practice finding persons. Unaccountable incidents are useful information if there seems to be certain patterns, such as if the incidents are clustered around day end.

Study Results

The following sections discuss the collection and the subsequent analysis of the observations.

Collection of Observation Data

During each observation numerous data was collected to allow analysis of the workforce in many ways. This data included employee identification, craft, supervisor, work category, unit status, date, day of week, hour of day, plus comments regarding certain individual observations. Codes used for craft were M, P, W, H, A, and T for mechanics, painters, welders, machinists, apprentices, and trainees, respectively.

The two previous studies by the paid consultant measured apprentices and trainees but did not classify them separately. They counted them as part of the craft in which they were working. However, between the time of the last study and this study, several changes occurred in the workforce. First, many of the apprentices were promoted to journeyman status. Second, all the journeymen were promoted provisionally to a newly created multiskill technician status. Finally, the company hired a class of eight new apprentices into the workforce. Yet at the time of this third study, work was still being assigned on the basis of previous craft designation. Because most of the new apprentices were less accustomed to power plant maintenance than were previous apprentices and because work was still assigned by previous designations, a decision was made to use the previous craft designations plus new apprentice and trainee designations to classify the workforce.

Analysis

The study analyzed the observations first on an overall basis with everyone and every circumstance included, then for different subsets of crafts, crews, and other circumstances.

Overall.   Table H.1, Comparison of Studies, compares each work category across the three studies. Each category percentage must be considered in light of its margin of error (MOE) shown in Table H.2, which summarizes the data for all categories during the observation period for this 1993 study.

As can be seen in Table H.1 and Fig. H.1, there is no significant change in Category 1, Working, which was measured at 35.08%.

However, a reduction in time spent is shown in the following categories: 2, Travel; 4, Work Assignment; 10, Interference; 13, Weather; and 15, Meetings. The reduction in travel might be explained by a combination of utilization of the new golf carts and possibly fewer trips in general being made. One new policy in place allows an employee to take a break when it is convenient for the job under way rather than at a specific time each day. In this manner, it is desired that persons take their break in route to a new task rather than take a separate trip. (However, most persons still took their breaks at the traditional times of 10:00 AM and 3:00 PM.) Work Assignment, Interference, and Meetings, presumably are reduced because of the deletion of the morning crew assignment meetings in favor of utilizing posted daily work schedules. The supervisors developed and posted these schedules each day before the work day.

The following categories showed increases: 6, Waiting for Materials; 8, Waiting for Instruction; 19, Break; and 23, Unaccountable. Explanation for Instruction might be in that some discussions regarding work could have been classified as Meetings or Instructions. Break time might have increased because of no longer having set times for break. Unaccountable presumably increased because of methods utilized by the observer. The previous studies made use of specially prepared lists given by the supervisor each morning listing work locations as given at that morning's assignment meeting. But the current study utilized the job schedule completed the day in advance. In addition, where the previous full-time observers allowed a full half hour to locate each set of three persons, the observer for this study made observations in 20-minute blocks to facilitate doing the study on a part-time basis. (*Note:* subsequent in-house studies reverted to the 30-minute allocation as shown in the ministudy, App. G.) Finally, late morning arrivals to work typically would not be measured in the previous studies since the people

Category	Name	1990 study	1991 study	1993 study	Change direction	Comment
			Working			
1	Working	37.45%	37.70%	35.08%	NSC*	
2	Work travel	22.03%	21.46%	15.33%	Down	Carts, no set break time?
3	Set-up & take down	0.00%	0.00%	0.00%		Category not used
4	Work assignment	3.95%	5.14%	1.80%	Down	Three daily crew meetings
5	Wrap-up	6.05%	5.95%	7.87%	NSC	
	Subtotal	69.48%	70.25%	60.08%		
			Waiting			
6	Waiting for materials	2.18%	1.37%	2.76%		?
7	Waiting for tools	2.18%	4.10%	4.83%	NSC*	
8	Waiting for instruction	3.15%	2.81%	3.87%	Up	Some were "meetings"?
9	Clearance delay	0.56%	0.88%	0.69%	NSC*	
10	Interference	2.26%	1.45%	0.28%	Down	Daily schedule meeting helped?
11	Other work waiting	0.00%	0.00%	0.00%		Category not used
12	Waiting for operator	1.86%	0.96%	1.24%	NSC*	
13	Weather delay	0.00%	0.56%	0.00%	Down	No severe weather
	Subtotal	12.19%	12.13%	13.67%		
			Other			
15	Meetings	0.56%	0.40%	0.28%	Down	Some classed as Cat. 8?
16	Training	0.00%	0.00%	0.00%		Category not used
17	Idle	1.86%	2.97%	2.21%	NSC*	
18	Rest room	0.80%	0.80%	1.24%	NSC*	
19	On break	10.90%	10.21%	13.67%	Up	Policy of no set break time?
20	Other personal allowance	0.16%	0.16%	0.97%	NSC*	
	Subtotal	14.28%	14.54%	18.37%		
			Unaccountable			
23	Unaccountable	3.95%	3.05%	7.87%	Up	
	Subtotal	3.95%	3.05%	7.87%		
	Totals	100%	100%	100%		

NOTE: There were no observations or adjustments in any categories not shown.
*NSC = no significant change.

TABLE H.2   **Results for All Personnel**

Category	Fri	Mon	Tues	Wed	Thur	Obs.	Percent	MOE,[*] %
				Work				
1	51	26	61	52	64	254	35.08	4
2	25	14	25	18	29	111	15.33	3
3	0	0	0	0	0	0	0.00	0
4	1	1	5	4	2	13	1.80	1
5	12	4	9	20	12	57	7.87	2
						Subtotal	60.08	
				Waiting				
6	0	1	7	8	4	20	2.76	1
7	6	6	7	7	9	35	4.83	2
8	3	3	8	7	7	28	3.87	1
9	0	0	0	2	3	5	0.69	1
10	0	0	1	0	1	2	0.28	0
11	0	0	0	0	0	0	0.00	0
12	0	0	5	2	2	9	1.24	1
13	0	0	0	0	0	0	0.00	0
14	0	0	0	0	0	0	0.00	0
						Subtotal	13.67	
				Other				
15	1	0	0	1	0	2	0.28	0
16	0	0	0	0	0	0	0.00	0
17	4	2	3	3	4	16	2.21	1
18	2	3	1	1	2	9	1.24	1
19	12	12	19	28	28	99	13.67	3
20	3	1	2	0	1	7	0.97	1
						Subtotal	18.37	
				Unaccountable				
23	9	9	15	15	9	57	7.87	2
						Subtotal	7.87	
Totals	129	82	168	168	177	724	100.00	

NOTE: There were no observations in any categories not shown.
[*]MOE = margin of error.

Figure H.1 Mechanical group with evolving planning department assistance, but no advance scheduling.

arriving late were not yet scheduled to work. But in the current study, these persons were counted as unaccountable if they had been selected to be observed from the advance schedule and did not subsequently get annual leave permission.

The following list defines the hour periods used in the study. Each period defines the time frame in which the search began.

Period 8 is 7:30 AM to 8:30 AM.

Period 9 is 8:30 AM to 9:30 AM.

Period 10 is 9:30 AM to 10:30 AM.

Period 11 is 10:30 AM to 11:30 AM.

Period 12 is 11:30 AM to 12:00 PM and 12:30 PM to 1:00 PM.

Period 13 is 1:00 PM to 2:00 PM.

Period 14 is 2:00 PM to 3:00 PM.

Period 15 is 3:00 PM to 4:00 PM.

Period 16 is 4:00 PM to 5:00 PM.

Period 17 is 5:00 PM to 6:00 PM.

Table H.3, Classification of Observations by Hour, is a checking tool to verify that the observer made an equivalent number of observations each hour of the day for the study overall to avoid skewing the data. For all crafts combined, there is an overall uniformity of about 72 average observations each period. The traditional break times have fewer than average observations and the most observations were in the 2:00 PM

TABLE H.3   Classification of Observations by Hour

Hour period	All	Technicians	Apprentices and trainees	M	P	W	H	A	T
8	69	56	13	26	11	11	8	7	6
9	78	56	22	22	17	9	8	11	11
10	69	56	13	31	9	9	7	7	6
11	72	56	16	27	17	6	6	8	8
12	75	58	17	37	7	9	5	9	8
13	66	55	11	25	15	7	8	9	2
14	82	58	24	35	8	7	8	14	10
15	66	52	14	28	5	8	11	9	5
16	71	56	15	30	11	7	8	13	2
17	76	56	20	34	9	7	6	10	10
Totals	724	559	165	295	109	80	75	97	68

period. These variances may slightly skew the reported results toward showing a higher work time and lower break time than actually exists. A similar analysis is made throughout this study of examining the observation distribution for specific crafts or other cases under consideration.

Figure H.2, Distribution of Time, shows the relative impact of each category's percentage of time. Greatest impact is made by Categories 2, Travel (15.33%), and 19, Break (13.67%). Table H.4, Minutes per Day for Each Category, shows that these categories take up 92 and 82 minutes per day, respectively. Categories 5, Wrap-up (7.87%, 47 minutes/day), and 23, Unaccountable (also 7.87%, 47 minutes/day) also have an appreciable impact.

Figure H.2   Time spent in different categories.

Reviewing the comments recorded for individual observations in each category allowed accounting for most of the travel by purpose of the travel. Table H.5 and Fig. H.3 show the results of adjusting the percentage of time in major categories by including their associated travel. A significant amount of time spent traveling was involved with breaks (and lunch period). Travel dropped to 3.76% (23 minutes/day) and Break increased to 16.70% (100 minutes/day). Wrap-up increased to 8.28% (50 minutes per day). After subtracting the total 30 minutes allowed for breaks and 30 minutes for wrap-up, there was an additional 90 minutes per day spent in these areas (100 + 50 − 30 − 30 = 90). Table H.5 and Fig. H.3 also showed an increase in the relative impact of Categories 6, Materials (4.6%, 28 minutes/day); 7, Tools (7.86%, 47 minutes/day); and 8, Instructions (4.56%, 27 minutes/day) after including their associated travel. Some of the material delays noted in the comments for Categories 2 and 6 were for nuts and bolts that may be difficult to plan or stage. On the other hand, some materials such as drive belts or "all thread" may be easier to anticipate. The most notable of the comments involved tools and appeared to be more schedule than planning related. Many of the observations for Categories 2 and 7 involved tool delays

TABLE H.4   Minutes per Day for Each Category

Category	Name	Percent	Minutes/day	Comment
1	Working	35.08	210	
2	Work travel	15.33	92	
3	Set-up & take down	0.00	0	Category not used
4	Work assignment	1.80	11	
5	Wrap-up	7.87	47	
6	Waiting for materials	2.76	17	
7	Waiting for tools	4.83	29	
8	Waiting for instruction	3.87	23	
9	Clearance delay	0.69	4	
10	Interference	0.28	2	
11	Other work waiting	0.00	0	Category not used
12	Waiting for operator	1.24	7	
13	Weather delay	0.00	0	
14	(Category not used)	0.00	0	
15	Meetings	0.28	2	
16	Training	0.00	0	Category not used
17	Idle	2.21	13	
18	Rest room	1.24	7	
19	On break	13.67	82	
20	Other personal allowance	0.97	6	
21	(Category not used)	0.00	0	
22	(Category not used)	0.00	0	
23	Unaccountable	4.51	27	
	Total for day		600 min	

TABLE H.5   Comparison of Study with Travel Adjustment

Category	Name	Percent	Travel adjustment	Percent after travel adjustment	Adjusted minutes/day
			Working		
1	Working	35.08	0.00%	35.08	210
2	Work travel	15.33	−11.57%	3.76	23
4	Work assignment	1.80	2.48%	4.28	26
5	Wrap-up	7.87	0.41%	8.28	50
	Subtotal	60.08	−8.68%	51.40	
			Waiting		
6	Waiting for materials	2.76	1.93%	4.69	28
7	Waiting for tools	4.83	3.03%	7.86	47
8	Waiting for instruction	3.87	0.69%	4.56	27
9	Clearance delay	0.69	0.14%	0.83	5
10	Interference	0.28	0.00%	0.28	2
12	Waiting for operator	1.24	0.00%	1.24	7
13	Weather delay	0.00	0.00%	0.00	0
	Subtotal	13.67	5.79%	19.46	
			Other		
15	Meetings	0.28	0.00%	0.28	2
17	Idle	2.21	0.00%	2.21	13
18	Rest room	1.24	0.00%	1.24	7
19	On break	13.67	3.03%	16.70	100
20	Other personal allowance	0.97	0.00%	0.97	6
	Subtotal	18.37	3.03%	21.40	
			Unaccountable		
23	Unaccountable	7.87	0.00%	7.87	47
	Subtotal	7.87	0.00%	7.87	
	Totals	100.00	0.00%	100.00	601 min

NOTE: There were no observations or adjustments in any categories not shown.

Figure H.3   Reasons for time spent traveling.

such as the moving of personal tools, the obtaining of small tools that could be anticipated such as shackles or grease guns, or the scheduling of use of equipment such as a crane. Most of the instruction delays were associated with a person-to-person discussion of work-related issues rather than for researching file information. Better clarification of instruction versus meeting categories may be advisable.

Personnel.   The study also analyzes wrench time considering different personnel. The study makes classifications by craft, by days when only one or two crews are present, by supervisor, by when persons are loaned to another plant, and finally by when the apprentices are sent to training. The study compares these classifications to previous study results when the data exists from previous studies.

Figure H.4, Distribution of Time by Craft, shows the relative percentage of time spent by each craft in each category. Tables H.6 through H.11 show the summary data with actual percentages of time, margin of error, and number of observations for each craft. Table H.3 (previously shown) allows determination of whether the distribution of observations might have skewed the results for any one craft.

The first cluster of bars in Fig. H.4 shows Category 1, Working, for mechanics (34.21%), painters (25.6%), welders (33.75%), machinists (50.67%), apprentices (40.21%), and trainees (30.88%). The percentages for the first four of these crafts from the last study (N2) are 34.64%, 26.09%, 37.73%, and 82.17%, respectively.

The only statistically significant change between studies in work time is for the machinist craft which dropped over 30%. This drop is neither

Figure H.4 Separate crafts spend time differently.

within the margin of error for the 1991 study, 9%, nor this study, 12%.
Table H.3 does not indicate a significant skew of the observations.
Between the studies travel rises about 4%, wrap-up about 5%, and
break about 1% for machinists. More remarkable, however, is that in
the last study there is zero percentage (0%) for Categories 6, Materials; 7,
Tools; 8, Instructions; and 17, Idle, as well as for 18, Bathroom; and 23,
Unaccountable. These categories are 1.33, 4.00, 8.00, 4.00, 4.00, 6.67,
and 1.33%, respectively, in the current study. The first five of these categories total 21.33% and it is possible that some observations in these
categories might have been counted as work time in the previous study
for machinists if the previous study observer limited the number of
questions asked of the persons. This difference might explain the results
of machinist work time and give validity to the current study results.

Figure H.4 shows that the highest bars for Category 2, Travel, are
observed in the apprentice (19.59%) and trainee (25.00%) craftpersons.
(Review of the individual observation comments also showed a large proportion of material and tool-related travel by apprentices and trainees
even though many apprentices were not present for most of the study.)
This high travel might be explained by the apparent practice of sending these craftpersons to pick up items when needed by the technicians. Mechanics show the next highest travel at 16.61%. All of the

TABLE H.6   Mechanics

Category	Fri	Mon	Tues	Wed	Thur	Obs.	Percent	MOE,[*] %
				Work				
1	30	6	19	21	25	101	34.24	6
2	11	5	14	9	10	49	16.61	4
3	0	0	0	0	0	0	0.00	0
4	1	1	1	2	1	6	2.03	2
5	4	1	5	10	9	29	9.83	3
						Subtotal	62.71	
				Waiting				
6	0	0	5	2	2	9	3.05	2
7	4	1	3	5	4	17	5.76	3
8	2	0	6	1	4	13	4.41	2
9	0	0	0	2	1	3	1.02	1
10	0	0	0	0	1	1	0.34	1
11	0	0	0	0	0	0	0.00	0
12	0	0	2	1	2	5	1.69	2
13	0	0	0	0	0	0	0.00	0
14	0	0	0	0	0	0	0.00	0
						Subtotal	16.27	
				Other				
15	0	0	0	1	0	1	0.34	1
16	0	0	0	0	0	0	0.00	0
17	2	1	1	1	1	6	2.03	2
18	0	1	0	0	0	1	0.34	1
19	4	5	7	8	12	36	12.20	4
20	1	0	0	0	0	1	0.34	1
						Subtotal	15.25	
				Unaccountable				
23	3	1	2	8	3	17	5.76	3
						Subtotal	5.76	
Totals	62	22	65	71	75	295	100.00	

NOTE: There were no observations in any categories not shown.
[*]MOE = margin of error.

TABLE H.7 **Painters**

Category	Fri	Mon	Tues	Wed	Thur	Obs.	Percent	MOE,[*] %
					Work			
1	2	6	7	4	9	28	25.69	8
2	3	2	2	2	4	13	11.93	6
3	0	0	0	0	0	0	0.00	0
4	0	0	0	0	0	0	0.00	0
5	0	1	0	2	0	3	2.75	3
						Subtotal	40.37	
					Waiting			
6	0	0	0	1	0	1	0.92	2
7	0	0	1	1	0	2	1.83	3
8	0	2	0	2	0	4	3.67	4
9	0	0	0	0	0	0	0.00	0
10	0	0	0	0	0	0	0.00	0
11	0	0	0	0	0	0	0.00	0
12	0	0	0	0	0	0	0.00	0
13	0	0	0	0	0	0	0.00	0
14	0	0	0	0	0	0	0.00	0
						Subtotal	6.42	
					Other			
15	1	0	0	0	0	1	0.92	2
16	0	0	0	0	0	0	0.00	0
17	0	0	0	1	1	2	1.83	3
18	0	1	0	0	1	2	1.83	3
19	1	5	2	13	7	28	25.69	8
20	1	1	2	0	0	4	3.67	4
						Subtotal	33.94	
					Unaccountable			
23	4	4	5	5	3	21	19.27	8
						Subtotal	19.27	
Totals	12	22	19	31	25	109	100.00	

NOTE: There were no observations in any categories not shown.
[*]MOE = margin of error.

TABLE H.8   Welders

Category	Fri	Mon	Tues	Wed	Thur	Obs.	Percent	MOE,* %
				Work				
1	4	3	8	5	7	27	33.75	11
2	2	0	2	0	2	6	7.50	6
3	0	0	0	0	0	0	0.00	0
4	0	0	1	0	0	1	1.25	2
5	2	1	0	2	1	6	7.50	6
						Subtotal	50.00	
				Waiting				
6	0	1	0	2	1	4	5.00	5
7	0	2	1	0	0	3	3.75	4
8	0	0	1	1	1	3	3.75	4
9	0	0	0	0	1	1	1.25	2
10	0	0	0	0	0	0	0.00	0
11	0	0	0	0	0	0	0.00	0
12	0	0	0	1	0	1	1.25	2
13	0	0	0	0	0	0	0.00	0
14	0	0	0	0	0	0	0.00	0
						Subtotal	15.00	
				Other				
15	0	0	0	0	0	0	0.00	0
16	0	0	0	0	0	0	0.00	0
17	0	0	0	0	0	0	0.00	0
18	0	1	0	0	0	1	1.25	2
19	5	1	5	4	1	16	20.00	9
20	1	0	0	0	1	2	2.50	3
						Subtotal	23.75	
				Unaccountable				
23	2	2	2	1	2	9	11.25	7
						Subtotal	11.25	
Totals	12	22	19	31	25	109	100.00	

NOTE: There were no observations in any categories not shown.
*MOE = margin of error.

TABLE H.9   Machinists

Category	Fri	Mon	Tues	Wed	Thur	Obs.	Percent	MOE,[*] %
				Work				
1	6	0	9	11	12	38	50.67	12
2	1	0	2	1	3	7	9.33	7
3	0	0	0	0	0	0	0.00	0
4	0	0	0	1	1	2	2.67	4
5	3	0	1	2	0	6	8.00	6
						Subtotal	70.67	
				Waiting				
6	0	0	0	1	0	1	1.33	3
7	0	0	0	1	2	3	4.00	5
8	0	1	0	3	2	6	8.00	6
9	0	0	0	0	0	0	0.00	0
10	0	0	0	0	0	0	0.00	0
11	0	0	0	0	0	0	0.00	0
12	0	0	0	0	0	0	0.00	0
13	0	0	0	0	0	0	0.00	0
14	0	0	0	0	0	0	0.00	0
						Subtotal	13.33	
				Other				
15	0	0	0	0	0	0	0.00	0
16	0	0	0	0	0	0	0.00	0
17	1	0	1	0	1	3	4.00	5
18	2	0	1	0	0	3	4.00	5
19	0	0	2	1	1	5	6.67	6
20	0	0	0	0	0	0	0.00	0
						Subtotal	14.67	
				Unaccountable				
23	0	0	0	1	0	1	1.33	3
						Subtotal	1.33	
Totals	14	1	16	22	22	75	100.00	

NOTE: There were no observations in any categories not shown.
[*]MOE = margin of error.

TABLE H.10   Apprentices

Category	Fri	Mon	Tues	Wed	Thur	Obs.	Percent	MOE,* %
				Work				
1	9	2	10	8	10	39	40.21	10
2	6	2	1	3	7	19	19.59	8
3	0	0	0	0	0	0	0.00	0
4	0	0	1	1	0	2	2.06	3
5	2	0	3	3	1	9	9.28	6
						Subtotal	71.13	
				Waiting				
6	0	0	1	2	1	4	4.12	4
7	2	0	1	0	1	4	4.12	4
8	1	0	1	0	0	2	2.06	3
9	0	0	0	0	0	0	0.00	0
10	0	0	0	0	0	0	0.00	0
11	0	0	0	0	0	0	0.00	0
12	0	0	2	0	0	2	2.06	3
13	0	0	0	0	0	0	0.00	0
14	0	0	0	0	0	0	0.00	0
						Subtotal	12.37	
				Other				
15	0	0	0	0	0	0	0.00	0
16	0	0	0	0	0	0	0.00	0
17	0	0	1	1	0	2	2.06	3
18	0	0	0	1	1	2	2.06	3
19	1	0	0	2	4	7	7.22	5
20	0	0	0	0	0	0	0.00	0
						Subtotal	11.34	
				Unaccountable				
23	0	1	4	0	0	5	5.15	4
						Subtotal	5.15	
Totals	21	5	25	21	25	97	100.00	

NOTE: There were no observations in any categories not shown.
*MOE = margin of error.

TABLE H.11 Trainees

Category	Fri	Mon	Tues	Wed	Thur	Obs.	Percent	MOE,[*] %
				Work				
1	0	9	8	3	1	21	30.88	11
2	2	5	4	3	3	17	25.00	11
3	0	0	0	0	0	0	0.00	0
4	0	0	2	0	0	2	2.94	4
5	1	1	0	1	1	4	5.88	6
						Subtotal	64.71	
				Waiting				
6	0	0	1	0	0	1	1.47	4
7	0	3	1	0	2	6	8.22	7
8	0	0	0	0	0	0	0.00	0
9	0	0	0	0	1	1	1.47	4
10	0	0	1	0	0	1	1.47	4
11	0	0	0	0	0	0	0.00	0
12	0	0	1	0	0	1	1.47	4
13	0	0	0	0	0	0	0.00	0
14	0	0	0	0	0	0	0.00	0
						Subtotal	14.71	
				Other				
15	0	0	0	0	0	0	0.00	0
16	0	0	0	0	0	0	0.00	0
17	1	1	0	0	1	3	4.41	5
18	0	0	0	0	0	0	0.00	0
19	0	1	3	0	3	7	10.29	7
20	0	0	0	0	0	0	0.00	0
						Subtotal	14.71	
				Unaccountable				
23	0	1	2	0	1	4	5.88	6
						Subtotal	5.88	
Totals	4	21	23	7	13	68	100.00	

NOTE: There were no observations in any categories not shown.
[*]MOE = margin of error.

technician craftpersons show reductions in travel from the previous study which did not separate out the apprentices and trainees.

Category 5, Wrap-up, appears highest for mechanics at 9.83% (margin of error = 3%). This percentage means that on the average, a mechanic spends 59 minutes each day in wrap-up (41 to 76 minutes within the error margin). The apprentices are close behind (9.28%, 56 minutes). In order, machinists are 8.00% (48 minutes), welders are 7.50% (45 minutes), trainees are 5.88% (35 minutes), and painters are only 2.75% (17 minutes). The fact that painters spend less time in wrap-up than allowed might be partially explained by the painters having the largest occurrence in Category 23, Unaccountable, of 19.27%.

Trainees show the highest relative percentage of time in Category 7, Waiting for Tools, at 8.87%.

Painters have a large decrease in Category 8, Waiting for Instructions, from 13.04% (the highest of all craftpersons) in the previous study to 3.67% (among the lowest) in the current study. This decrease might be explained by the work consisting of mainly three large work orders which remained open the entire duration of the current study.

Category 19 shows four crafts over 10% for break time and all crafts over the 5% allowed break time even without including associated travel time. It is possible that the observer did not diligently note if persons in the break room were waiting while being delayed on their assigned jobs or were awaiting job assignment. In order, the break time is painters at 25.69%, welders at 20.00%, mechanics at 12.20%, trainees at 10.29%, apprentices at 7.22%, and machinists at 6.67%. In the previous study these percentages are all lower: 17.39, 11.49, 10.68, and 5.48% for painters, welders, mechanics, and machinists, respectively.

For Category 23, Unaccountable, the painters and welders have the most time with percentages of 19.27 and 11.25%. In the previous study painters had a similar score, but welders had only a 1.25% percentage.

Table H.12 compares the categories for all of the technicians combined and the apprentices and trainees combined. The notable differences are more travel time and less break time for apprentices and trainees. Tables H.13 through H.15 classify work categories by crew.

Each supervisor has a similar percentage of time in Category 1, Working, each within the others' margin of error. Note that Crew 3 has an above average, 20.79%, travel time, but a below average, 6.93%, break time.

The study examined the observation data for possible skewing and crew composition notes. The reported wrench time for Crew 1 might have been a little high due to a skew of having more observations than average in the peak work time periods. There was a fairly even distribution of observations or even trade-offs between opposite effect periods for Crew 2 and 3. It was notable that Crew 1 had all of the painters and Crew 2 had no trainees. Crew 2 also had the majority of the machinists.

TABLE H.12   Comparison of Technicians versus Apprentices and Trainees

Category	Name	Technicians, %	Apprentices, %	Trainees, %
		Working		
1	Working	34.70	36.36	35.08
2	Work travel	13.42	21.82	15.33
3	Set-up & take down	0.00	0.00	0.00
4	Work assignment	1.61	2.42	1.80
5	Wrap-up	7.87	7.88	7.87
	Subtotal	57.60	68.48	60.08
		Waiting		
6	Waiting for materials	2.68	3.03	2.76
7	Waiting for tools	4.47	6.06	4.83
8	Waiting for instruction	4.65	1.21	3.87
9	Clearance delay	0.72	0.61	0.69
10	Interference	0.18	0.61	0.28
11	Other work waiting	0.00	0.00	0.00
12	Waiting for operator	1.07	1.82	1.24
13	Weather delay	0.00	0.00	0.00
	Subtotal	13.77	13.34	13.67
		Other		
15	Meetings	0.36	0.00	0.28
16	Training	0.00	0.00	0.00
17	Idle	1.97	3.03	2.21
18	Rest room	1.25	1.21	1.24
19	On break	15.21	8.48	13.67
20	Other personal allowance	1.25	0.00	0.97
	Subtotal	20.4	12.72	18.37
		Unaccountable		
23	Unaccountable	8.59	5.45	7.87
	Subtotal	8.59	5.45	7.87
	Totals	100	100	100

NOTE: There were no observations or adjustments in any categories not shown.

TABLE H.13   **Crew 1**

Category	Fri	Mon	Tues	Wed	Thur	Obs.	Percent	MOE,[*] %
					Work			
1	21	22	24	17	27	111	31.99	5
2	15	11	7	5	10	48	13.83	4
3	0	0	0	0	0	0	0.00	0
4	0	0	4	0	0	4	1.15	1
5	3	3	4	10	5	25	7.20	3
						Subtotal	54.18	
					Waiting			
6	0	1	1	5	0	7	2.02	2
7	5	6	3	1	6	21	6.05	3
8	0	3	0	4	1	8	2.31	2
9	0	0	0	1	3	4	1.15	1
10	0	0	1	0	0	1	0.29	1
11	0	0	0	0	0	0	0.00	0
12	0	0	1	2	0	3	0.86	1
13	0	0	0	0	0	0	0.00	0
14	0	0	0	0	0	0	0.00	0
						Subtotal	12.68	
					Other			
15	1	0	0	0	0	1	0.29	1
16	0	0	0	0	0	0	0.00	0
17	2	2	0	1	1	6	1.73	1
18	0	3	0	0	1	4	1.15	1
19	2	12	12	18	18	62	17.87	4
20	2	1	2	0	0	5	1.44	1
						Subtotal	22.48	
					Unaccountable			
23	7	8	11	5	6	37	10.66	3
						Subtotal	10.66	
Totals	58	72	70	69	78	347	100.00	

NOTE: There were no observations in any categories not shown.
[*]MOE = margin of error.

TABLE H.14   Crew 2

Category	Fri	Mon	Tues	Wed	Thur	Obs.	Percent	MOE,[*] %
					Work			
1	13	2	16	19	21	71	40.57	7
2	3	1	4	5	8	21	12.00	5
3	0	0	0	0	0	0	0.00	0
4	1	0	1	1	1	4	2.29	2
5	3	1	1	6	3	14	8.00	4
						Subtotal	62.86	
					Waiting			
6	0	0	0	3	3	6	3.43	3
7	1	0	3	3	2	9	5.14	3
8	0	0	0	3	2	5	2.86	3
9	0	0	0	0	0	0	0.00	0
10	0	0	0	0	0	0	0.00	0
11	0	0	0	0	0	0	0.00	0
12	0	0	2	0	1	3	1.71	2
13	0	0	0	0	0	0	0.00	0
14	0	0	0	0	0	0	0.00	0
						Subtotal	13.14	
					Other			
15	0	0	0	0	0	0	0.00	0
16	0	0	0	0	0	0	0.00	0
17	0	0	2	0	1	3	1.71	2
18	1	0	1	1	1	4	2.29	2
19	5	0	6	8	4	23	13.14	5
20	1	0	0	0	1	2	1.14	2
						Subtotal	18.29	
					Unaccountable			
23	0	1	3	6	0	10	5.71	4
						Subtotal	5.71	
Totals	28	5	39	55	48	175	100.00	

NOTE: There were no observations in any categories not shown.
[*]MOE = margin of error.

TABLE H.15   Crew 3

Category	Fri	Mon	Tues	Wed	Thur	Obs.	Percent	MOE,* %
				Work				
1	17	2	21	16	16	72	35.64	7
2	7	2	14	8	11	42	20.79	6
3	0	0	0	0	0	0	0.00	0
4	0	1	0	3	1	5	2.48	2
5	6	0	4	4	4	18	8.91	4
						Subtotal	67.82	
				Waiting				
6	0	0	6	0	1	7	3.47	3
7	0	0	1	3	1	5	2.48	2
8	3	0	8	0	4	15	7.43	4
9	0	0	0	1	0	1	0.50	1
10	0	0	0	0	1	1	0.50	1
11	0	0	0	0	0	0	0.00	0
12	0	0	2	0	1	3	1.49	2
13	0	0	0	0	0	0	0.00	0
14	0	0	0	0	0	0	0.00	0
						Subtotal	15.84	
				Other				
15	0	0	0	1	0	1	0.50	1
16	0	0	0	0	0	0	0.00	0
17	2	0	1	2	2	7	3.47	3
18	1	0	0	0	0	1	0.50	1
19	5	0	1	2	6	14	6.93	4
20	0	0	0	0	0	0	0.00	0
						Subtotal	11.39	
				Unaccountable				
23	2	0	1	4	3	10	4.95	3
						Subtotal	4.95	
Totals	43	5	59	44	51	202	100.00	

NOTE: There were no observations in any categories not shown.
*MOE = margin of error.

There was insufficient unskewed data to allow reporting of wrench time for when there was no supervisor or when someone was working up.

Table H.16 shows the effect of when the work force decreases by loaning personnel to another plant. Category 1, Working, increases by 1 to 36.16% and Category 2, Break, decreases by 1 to 12.52%. All of these changes are within the margin of error of the study. Also, Table H.17 shows there are fewer observations than average in periods 8 and 10 and more observations in period 14. This uneven distribution might artificially raise the reported wrench time and lower the reported break time.

Similarly, Table H.18 shows a decrease of from 1 to 33.89% in Category 1, Working, for the apprentices being sent to training. Category 2, Travel, also decreases by about 1 to 13.93%, but Category 19, Break, increases by over 2 to 16.22%. Again, all of these changes are within the margin of error for each category so they are not statistically significant.

If the change was significant, the decreased travel might be explained by better anticipation of needed parts and tools by technicians who do not have helpers. The lower wrench time and greater break time might be explained by having less peer pressure around to keep on the job.

Table H.19 does not indicate a skewed result due to observation distribution.

The final personnel classification is a repeat of the previous study to examine whether work days with only one or two crews present (usually Monday and Friday) versus the other days with all the crews present have an effect on wrench time. Tables H.20 and H.21 show overall wrench time appears to be higher on "short crew days" (37.82 vs. 33.41%).

Tables H.22 through H.25 show mechanics also had higher wrench time (40.71 vs. 30.22%), but welders had lower wrench time (33.33 vs. 34.00%). These results agreed with the previous study.

Apprentices and trainees (not classified in previous studies) also seem to have higher wrench time on short crew days, but within large margins of error (Tables H.26 through H.29).

Tables H.30 through H.33 show painter wrench time was lower on short crew days than on other days (23.68 vs. 26.76%) as was machinist wrench time (40.91 vs. 54.72%). These results were reversed from the previous study.

With the exception of the mechanics all of the differences between short crew days and other days are within the study margin of error. For the mechanics the data appears to be skewed as shown by Tables H.34 and H.35 by having an unequal number of observations in the 17 (5:00 to 6:00 PM) time period perhaps artificially creating a significant difference. The previous study differences also were within the margins of error for each craft.

It is possible that any actual higher wrench time on short crew days might be explainable by having fewer people to cause disruptions. It might also be explained by the fact that individuals were usually

TABLE H.16   **After Personnel Loaned to Another Plant**

Category	Fri	Mon	Tues	Wed	Thur	Obs.	Percent	MOE,[*] %
				Work				
1	28	26	51	44	56	205	36.16	4
2	19	14	22	11	22	88	15.52	3
3	0	0	0	0	0	0	0.00	0
4	0	1	4	3	0	8	1.41	1
5	7	4	9	13	9	42	7.41	2
						Subtotal	60.49	
				Waiting				
6	0	1	7	7	3	18	3.17	1
7	5	6	6	5	6	28	4.94	2
8	2	3	8	5	7	25	4.41	2
9	0	0	0	2	3	5	0.88	1
10	0	0	1	0	1	2	0.35	0
11	0	0	0	0	0	0	0.00	0
12	0	0	3	0	2	5	0.88	1
13	0	0	0	0	0	0	0.00	0
14	0	0	0	0	0	0	0.00	0
						Subtotal	14.64	
				Other				
15	1	0	0	1	0	2	0.35	0
16	0	0	0	0	0	0	0.00	0
17	2	2	3	3	4	14	2.47	1
18	1	3	1	0	2	7	1.23	1
19	5	12	16	17	21	71	12.52	3
20	2	1	1	0	1	5	0.88	1
						Subtotal	17.46	
				Unaccountable				
23	7	9	13	8	5	42	7.41	2
						Subtotal	7.41	
Totals	79	82	145	119	142	567	100.00	

NOTE: There were no observations in any categories not shown.
[*]MOE = margin of error.

observed more frequently on these days, possibly causing them to be overly sensitive to the observation process. From examining the results with previous studies, it is concluded there appears to be no appreciable difference in actual wrench time overall for short crew days.

Unit Status.   Next the study analyzes unit status.

Table H.36, Days When One or Both Units Are On-Line, shows a Category 1, Working, time of 36.10%. This improvement of 1% versus the overall study average is not significant within the study's margin of error. In addition, Table H.37 indicates a slight skew of observations away from the traditional break times.

On the other hand, for days when either a unit tripped or was called, the results are more interesting. Table H.38 shows a decrease of 3% in Category 1, Working, time to 32.14%. In addition, there is also a decrease in almost all waiting categories and travel. At the same time there are increases in Categories 4, Work Assignment; 5, Wrap-up; 17, Idle; and 19, Break. A possible explanation is that reassignment time to routine start-up and shut-down tasks affects wrench time. The tasks are considered routine because of few delays for parts, tools, or instructions. While there are few observations that cause large margins of error, Table H.39 shows a skew of observations that should cause break time to be underreported. So actual wrench time may be even less. This area may be desirable for collecting additional observations in future studies. Nevertheless, unit trips and calls are not every day occurrences, so lower wrench time might be acceptable.

Time

Finally, the study analyzes work activity and time itself.

Figure H.5, Wrench Time Categories by Hour, demonstrates the variance by hour. Category 1, Working, is between 40 and 53% during

TABLE H.17   Classification of Observations by Hour after Personnel Loaned to Another Plant

Hour period	All	Technicians	Apprentices and trainees	M	P	W	H	A	T
8	49	38	11	15	9	8	6	6	5
9	53	37	16	14	15	5	3	7	9
10	48	36	12	20	6	5	5	6	6
11	60	45	15	21	15	4	5	7	8
12	60	43	17	25	7	7	4	9	8
13	54	46	8	20	15	4	7	6	2
14	72	50	22	28	8	6	8	14	8
15	53	40	13	21	3	6	10	8	5
16	56	43	13	20	11	4	8	11	2
17	62	43	19	23	9	6	5	9	10
Totals	567	421	146	207	98	55	61	83	63

TABLE H.18   **With New Apprentices Sent to Training**

Category	Fri	Mon	Tues	Wed	Thur	Obs.	Percent	MOE,[*] %
				Work				
1	30	22	49	33	29	163	33.89	4
2	10	11	17	14	15	67	13.93	3
3	0	0	0	0	0	0	0.00	0
4	1	0	3	4	2	10	2.08	1
5	9	3	3	13	8	36	7.48	2
						Subtotal	57.38	
				Waiting				
6	0	1	7	3	2	13	2.70	1
7	1	6	6	5	5	23	4.78	2
8	3	3	8	4	6	24	4.99	2
9	0	0	0	1	0	1	0.21	0
10	0	0	1	0	1	2	0.42	1
11	0	0	0	0	0	0	0.00	0
12	0	0	5	2	1	8	1.66	1
13	0	0	0	0	0	0	0.00	0
14	0	0	0	0	0	0	0.00	0
						Subtotal	14.76	
				Other				
15	0	0	0	0	0	0	0.00	0
16	0	0	0	0	0	0	0.00	0
17	2	2	2	2	3	11	2.29	1
18	2	3	0	1	0	6	1.25	1
19	10	12	18	20	18	78	16.22	3
20	1	1	2	0	0	4	0.83	1
						Subtotal	20.58	
				Unaccountable				
23	2	8	8	11	6	35	7.28	2
						Subtotal	7.28	
Totals	71	72	129	113	96	481	100.00	

NOTE: There were no observations in any categories not shown.
[*]MOE = margin of error.

TABLE H.19   Classification of Observations by Hour with New Apprentices Sent to Training

Hour period	All	Technicians	Apprentices and trainees	M	P	W	H	A	T
8	52	46	6	22	11	7	6	2	4
9	58	42	16	17	10	8	7	5	11
10	54	45	9	26	5	8	6	3	6
11	54	44	10	20	16	5	3	3	7
12	51	41	10	27	5	6	3	2	8
13	37	32	5	16	7	6	3	3	2
14	44	33	11	20	5	5	3	2	9
15	34	31	3	19	2	4	6	1	2
16	41	36	5	22	4	5	5	3	2
17	56	44	12	25	9	6	4	3	9
Totals	481	394	87	214	74	60	46	27	60

periods away from day start, day end, lunch, and traditional break times. However, it drops to between 23 and 33% around breaks or lunch. It drops further to 18.84% for the first hour of each day and only 3.95% for the last hour of each day.

The category which most varies (indirectly) with wrench time is Category 19, Break. The morning break consumes 42.03% of the hour between 9:30 and 10:30 (25 minutes) and the afternoon break 33.33% (20 minutes) between 3:00 and 4:00 without even considering associated travel. Similarly, lunch time accounts for 18.87% (11 minutes) of the half hour before noon and the half hour after 12:30 without considering associated travel or the half-hour lunch period itself.

Category 2, Travel, is highest in the first period of the day at 26.09%. So, on the average, 16 minutes is taken by each person just in traveling before job site work begins. Appendix E, Classification of Comments by Hour, shows that this travel is split among getting parts, getting tools, and actually going to the job site. Travel is lowest for the last period of the day (5.26%, 3 minutes). Travel near break time and lunch time appears higher than other times with the exception of the 1:00 to 2:00 PM period. It is noteworthy that the 1:00 to 2:00 period has the highest wrench time, but also the second highest travel.

The next category of note is Category 23, Unaccountable. This category is much higher near day start, lunch, and day end. It is between 11.84 to 17.39% for these periods versus 1.45 to 7.58% for the other periods of the day.

Category 6, Waiting on Materials, seems to be a greater problem, in the morning as does Category 7, Waiting on Tools. Category 8, Waiting on Instructions, appears higher during periods of high wrench time plus during the first hour of the work day.

TABLE H.20   **Monday and Friday (plus Tuesday after Holidays)**

Category	Fri	Mon	Tues	Wed	Thur	Obs.	Percent	MOE,* %
				Work				
1	51	26	27	0	0	104	37.82	6
2	25	14	10	0	0	49	17.82	5
3	0	0	0	0	0	0	0.00	0
4	1	1	0	0	0	2	0.73	1
5	12	4	3	0	0	19	6.91	3
						Subtotal	63.27	
				Waiting				
6	0	1	4	0	0	5	1.82	2
7	6	6	4	0	0	16	5.82	3
8	3	3	4	0	0	10	3.64	2
9	0	0	0	0	0	0	0.00	0
10	0	0	0	0	0	0	0.00	0
11	0	0	0	0	0	0	0.00	0
12	0	0	2	0	0	2	0.73	1
13	0	0	0	0	0	0	0.00	0
14	0	0	0	0	0	0	0.00	0
						Subtotal	12.00	
				Other				
15	1	0	0	0	0	1	0.36	1
16	0	0	0	0	0	0	0.00	0
17	4	2	1	0	0	7	2.55	2
18	2	3	0	0	0	5	1.82	2
19	12	12	7	0	0	31	11.27	4
20	3	1	1	0	0	5	1.82	2
						Subtotal	17.82	
				Unaccountable				
23	9	9	1	0	0	19	6.91	3
						Subtotal	6.91	
Totals	129	82	64	0	0	275	100.00	

NOTE: There were no observations in any categories not shown.
*MOE = margin of error.

TABLE H.21 Tuesday, Wednesday, and Thursday (except Tuesday after Holiday)

Category	Fri	Mon	Tues	Wed	Thur	Obs.	Percent	MOE,[*] %
				Work				
1	0	0	34	52	64	150	33.41	4
2	0	0	15	18	29	111	13.81	3
3	0	0	0	0	0	0	0.00	0
4	0	0	5	4	2	11	2.45	1
5	0	0	6	20	12	38	8.46	3
						Subtotal	58.13	
				Waiting				
6	0	0	3	8	4	15	3.34	2
7	0	0	3	7	9	19	4.23	2
8	0	0	4	7	7	18	4.01	2
9	0	0	0	2	3	5	1.11	1
10	0	0	1	0	1	2	0.45	1
11	0	0	0	0	0	0	0.00	0
12	0	0	3	2	2	7	1.56	1
13	0	0	0	0	0	0	0.00	0
14	0	0	0	0	0	0	0.00	0
						Subtotal	14.70	
				Other				
15	0	0	0	1	0	1	0.22	0
16	0	0	0	0	0	0	0.00	0
17	0	0	2	3	4	9	2.00	1
18	0	0	1	1	2	4	0.89	1
19	0	0	12	28	28	68	15.14	3
20	0	0	1	0	1	2	0.45	1
						Subtotal	18.71	
				Unaccountable				
23	0	0	14	15	9	38	8.46	3
						Subtotal	8.46	
Totals	0	0	104	168	177	449	100.00	

NOTE: There were no observations in any categories not shown.
[*]MOE = margin of error.

TABLE H.22   Mechanics Monday and Friday (plus Tuesday after Holidays)

Category	Fri	Mon	Tues	Wed	Thur	Obs.	Percent	MOE,[*] %
				Work				
1	30	6	10	0	0	46	40.71	9
2	11	5	6	0	0	22	19.47	7
3	0	0	0	0	0	0	0.00	0
4	1	1	0	0	0	2	1.77	2
5	4	1	2	0	0	7	6.19	5
						Subtotal	68.14	
				Waiting				
6	0	0	2	0	0	2	1.77	2
7	4	1	2	0	0	7	6.19	5
8	2	0	3	0	0	5	4.42	4
9	0	0	0	0	0	0	0.00	0
10	0	0	0	0	0	0	0.00	0
11	0	0	0	0	0	0	0.00	0
12	0	0	1	0	0	1	0.88	2
13	0	0	0	0	0	0	0.00	0
14	0	0	0	0	0	0	0.00	0
						Subtotal	13.27	
				Other				
15	0	0	0	0	0	0	0.00	0
16	0	0	0	0	0	0	0.00	0
17	2	1	1	0	0	4	3.54	3
18	0	1	0	0	0	1	0.88	2
19	4	5	2	0	0	11	9.73	6
20	1	0	0	0	0	1	0.88	2
						Subtotal	15.04	
				Unaccountable				
23	3	1	0	0	0	4	3.54	3
		1				Subtotal	3.54	
Totals	62	22	29	0	0	113	100.00	

NOTE: There were no observations in any categories not shown.
[*]MOE = margin of error.

TABLE H.23   Mechanics Tuesday, Wednesday, and Thursday (except Tuesday after Holidays)

Category	Fri	Mon	Tues	Wed	Thur	Obs.	Percent	MOE,[*] %
					Work			
1	0	0	9	21	25	55	30.22	7
2	0	0	8	9	10	27	14.84	5
3	0	0	0	0	0	0	0.00	0
4	0	0	1	2	1	4	2.20	2
5	0	0	3	10	9	22	12.09	5
						Subtotal	59.34	
					Waiting			
6	0	0	3	2	2	7	3.85	3
7	0	0	1	5	4	10	5.49	3
8	0	0	3	1	4	8	4.40	3
9	0	0	0	2	1	3	1.65	2
10	0	0	0	0	1	1	0.55	1
11	0	0	0	0	0	0	0.00	0
12	0	0	1	1	2	4	2.20	2
13	0	0	0	0	0	0	0.00	0
14	0	0	0	0	0	0	0.00	0
						Subtotal	18.13	
					Other			
15	0	0	0	1	0	1	0.55	1
16	0	0	0	0	0	0	0.00	0
17	0	0	0	1	1	2	1.10	2
18	0	0	0	0	0	0	0.00	0
19	0	0	5	8	12	25	13.74	5
20	0	0	0	0	0	0	0.00	0
						Subtotal	15.38	
					Unaccountable			
23	0	0	2	8	3	13	7.14	4
						Subtotal	7.14	
Totals	0	0	36	71	75	182	100.00	

NOTE: There were no observations in any categories not shown.
[*]MOE = margin of error.

TABLE H.24   Welders Monday and Friday (plus Tuesday after Holiday)

Category	Fri	Mon	Tues	Wed	Thur	Obs.	Percent	MOE,[*] %
				Work				
1	4	3	5	0	0	12	33.33	16
2	2	0	0	0	0	2	5.56	8
3	0	0	0	0	0	0	0.00	0
4	0	0	0	0	0	0	0.00	0
5	2	1	0	0	0	3	8.33	9
						Subtotal	47.22	
				Waiting				
6	0	1	0	0	0	1	2.78	5
7	0	2	1	0	0	3	8.33	9
8	0	0	0	0	0	0	0.00	0
9	0	0	0	0	0	0	0.00	0
10	0	0	0	0	0	0	0.00	0
11	0	0	0	0	0	0	0.00	0
12	0	0	0	0	0	0	0.00	0
13	0	0	0	0	0	0	0.00	0
14	0	0	0	0	0	0	0.00	0
						Subtotal	11.11	
				Other				
15	0	0	0	0	0	0	0.00	0
16	0	0	0	0	0	0	0.00	0
17	0	0	0	0	0	0	0.00	0
18	0	1	0	0	0	1	2.78	5
19	5	1	3	0	0	9	25.00	14
20	1	0	0	0	0	1	2.78	5
						Subtotal	30.56	
				Unaccountable				
23	2	2	0	0	0	4	11.11	10
						Subtotal	11.11	
Totals	16	11	9	0	0	36	100.00	

NOTE: There were no observations in any categories not shown.
[*]MOE = margin of error.

TABLE H.25 Welders Tuesday, Wednesday, and Thursday (except Tuesday after Holidays)

Category	Fri	Mon	Tues	Wed	Thur	Obs.	Percent	MOE,* %
				Work				
1	0	0	3	5	7	15	34.09	14
2	0	0	2	0	2	4	9.09	9
3	0	0	0	0	0	0	0.00	0
4	0	0	1	0	0	1	2.27	4
5	0	0	0	2	1	3	6.82	8
						Subtotal	52.27	
				Waiting				
6	0	0	0	2	1	3	6.82	8
7	0	0	0	0	0	0	0.00	0
8	0	0	1	1	1	3	6.82	8
9	0	0	0	0	1	1	2.27	4
10	0	0	0	0	0	0	0.00	0
11	0	0	0	0	0	0	0.00	0
12	0	0	0	1	0	1	2.27	4
13	0	0	0	0	0	0	0.00	0
14	0	0	0	0	0	0	0.00	0
						Subtotal	18.18	
				Other				
15	0	0	0	0	0	0	0.00	0
16	0	0	0	0	0	0	0.00	0
17	0	0	0	0	0	0	0.00	0
18	0	0	0	0	0	0	0.00	0
19	0	0	2	4	1	7	15.91	11
20	0	0	0	0	1	1	2.27	4
						Subtotal	18.18	
				Unaccountable				
23	0	0	2	1	2	5	11.36	10
						Subtotal	11.36	
Totals	0	0	11	16	17	44	100.00	

NOTE: There were no observations in any categories not shown.
*MOE = margin of error.

TABLE H.26   Apprentices Monday and Friday (plus Tuesday after Holidays)

Category	Fri	Mon	Tues	Wed	Thur	Obs.	Percent	MOE,[*] %
				Work				
1	9	2	4	0	0	15	45.45	17
2	6	2	0	0	0	8	24.24	15
3	0	0	0	0	0	0	0.00	0
4	0	0	0	0	0	0	0.00	0
5	2	0	0	0	0	2	6.06	8
						Subtotal	75.76	
				Waiting				
6	0	0	1	0	0	1	3.03	6
7	2	0	0	0	1	2	6.06	8
8	1	0	1	0	0	2	6.06	8
9	0	0	0	0	0	0	0.00	0
10	0	0	0	0	0	0	0.00	0
11	0	0	0	0	0	0	0.00	0
12	0	0	1	0	0	1	3.03	6
13	0	0	0	0	0	0	0.00	0
14	0	0	0	0	0	0	0.00	0
						Subtotal	18.18	
				Other				
15	0	0	0	0	0	0	0.00	0
16	0	0	0	0	0	0	0.00	0
17	0	0	0	0	0	0	0.00	0
18	0	0	0	0	0	0	0.00	0
19	1	0	0	0	0	1	3.03	6
20	0	0	0	0	0	0	0.00	0
						Subtotal	3.03	
				Unaccountable				
23	0	1	0	0	0	1	3.03	6
						Subtotal	3.03	
Totals	21	5	7	0	0	33	100.00	

NOTE: There were no observations in any categories not shown.
[*]MOE = margin of error.

TABLE H.27   Apprentices Tuesday, Wednesday, and Thursday (except Tuesday after Holidays)

Category	Fri	Mon	Tues	Wed	Thur	Obs.	Percent	MOE,[*] %
				Work				
1	0	0	6	8	10	24	37.50	12
2	0	0	1	3	7	11	17.19	9
3	0	0	0	0	0	0	0.00	0
4	0	0	1	1	0	2	3.13	4
5	0	0	3	3	1	7	10.94	8
						Subtotal	68.75	
				Waiting				
6	0	0	0	2	1	3	4.69	5
7	0	0	1	0	1	2	3.13	4
8	0	0	0	0	0	0	0.00	0
9	0	0	0	0	0	0	0.00	0
10	0	0	0	0	0	0	0.00	0
11	0	0	0	0	0	0	0.00	0
12	0	0	1	0	0	1	1.56	3
13	0	0	0	0	0	0	0.00	0
14	0	0	0	0	0	0	0.00	0
						Subtotal	9.38	
				Other				
15	0	0	0	0	0	0	0.00	0
16	0	0	0	0	0	0	0.00	0
17	0	0	1	1	0	2	3.13	4
18	0	0	0	1	1	2	3.13	4
19	0	0	0	2	4	6	9.38	7
20	0	0	0	0	0	0	0.00	0
						Subtotal	15.63	
				Unaccountable				
23	0	0	4	0	0	4	6.25	6
						Subtotal	6.25	
Totals	0	0	8	21	25	64	100.00	

NOTE: There were no observations in any categories not shown.
[*]MOE = margin of error.

TABLE H.28   Trainees Monday and Friday (plus Tuesday after Holidays)

Category	Fri	Mon	Tues	Wed	Thur	Obs.	Percent	MOE,[*] %
					Work			
1	0	9	4	0	0	13	39.39	17
2	2	5	3	0	0	10	30.30	16
3	0	0	0	0	0	0	0.00	0
4	0	0	0	0	0	0	0.00	0
5	1	1	0	0	0	2	6.06	8
						Subtotal	75.76	
					Waiting			
6	0	0	1	0	0	1	3.03	6
7	0	3	0	0	0	3	9.09	10
8	0	0	0	0	0	0	0.00	0
9	0	0	0	0	0	0	0.00	0
10	0	0	0	0	0	0	0.00	0
11	0	0	0	0	0	0	0.00	0
12	0	0	0	0	0	0	0.00	0
13	0	0	0	0	0	0	0.00	0
14	0	0	0	0	0	0	0.00	0
						Subtotal	12.12	
					Other			
15	0	0	0	0	0	0	0.00	0
16	0	0	0	0	0	0	0.00	0
17	1	1	0	0	0	2	6.06	8
18	0	0	0	0	0	0	0.00	0
19	0	1	0	0	0	1	3.03	6
20	0	0	0	0	0	0	0.00	0
						Subtotal	9.09	
					Unaccountable			
23	0	1	0	0	0	1	3.03	6
						Subtotal	3.03	
Totals	4	21	8	0	0	33	100.00	

NOTE: There were no observations in any categories not shown.
[*]MOE = margin of error.

TABLE H.29   Trainees Tuesday, Wednesday, and Thursday (except Tuesday after Holidays)

Category	Fri	Mon	Tues	Wed	Thur	Obs.	Percent	MOE,[*] %
				Work				
1	0	0	4	3	1	8	22.86	14
2	0	0	1	3	3	7	20.00	14
3	0	0	0	0	0	0	0.00	0
4	0	0	2	0	0	2	5.71	8
5	0	0	0	1	1	2	5.71	8
						Subtotal	54.29	
				Waiting				
6	0	0	0	0	0	0	0.00	0
7	0	0	1	0	2	3	8.57	9
8	0	0	0	0	0	0	0.00	0
9	0	0	0	0	1	1	2.86	6
10	0	0	1	0	0	1	2.86	6
11	0	0	0	0	0	0	0.00	0
12	0	0	1	0	0	1	2.86	6
13	0	0	0	0	0	0	0.00	0
14	0	0	0	0	0	0	0.00	0
						Subtotal	17.14	
				Other				
15	0	0	0	0	0	0	0.00	0
16	0	0	0	0	0	0	0.00	0
17	0	0	0	0	1	1	2.86	6
18	0	0	0	0	0	0	0.00	0
19	0	0	3	0	3	6	17.14	13
20	0	0	0	0	0	0	0.00	0
						Subtotal	20.00	
				Unaccountable				
23	0	0	2	0	1	3	8.57	9
						Subtotal	8.57	
Totals	0	0	15	7	13	35	100.00	

NOTE: There were no observations in any categories not shown.
[*]MOE = margin of error.

TABLE H.30   Painters Monday and Friday (plus Tuesday after Holidays)

Category	Fri	Mon	Tues	Wed	Thur	Obs.	Percent	MOE,[*] %
				Work				
1	2	6	1	0	0	9	23.68	14
2	3	2	0	0	0	5	13.16	11
3	0	0	0	0	0	0	0.00	0
4	0	0	0	0	0	0	0.00	0
5	0	1	0	0	0	1	2.63	5
						Subtotal	39.47	
				Waiting				
6	0	0	0	0	0	0	0.00	0
7	0	0	1	0	0	1	2.63	5
8	0	2	0	0	0	2	5.26	7
9	0	0	0	0	0	0	0.00	0
10	0	0	0	0	0	0	0.00	0
11	0	0	0	0	0	0	0.00	0
12	0	0	0	0	0	0	0.00	0
13	0	0	0	0	0	0	0.00	0
14	0	0	0	0	0	0	0.00	0
						Subtotal	7.89	
				Other				
15	1	0	0	0	0	1	2.63	5
16	0	0	0	0	0	0	0.00	0
17	0	0	0	0	0	0	0.00	0
18	0	1	0	0	0	1	2.63	5
19	1	5	0	0	0	6	15.79	12
20	1	1	1	0	0	3	7.89	9
						Subtotal	28.95	
				Unaccountable				
23	4	4	1	0	0	9	23.68	14
						Subtotal	23.68	
Totals	12	22	4	0	0	38	100.00	

NOTE: There were no observations in any categories not shown.
[*]MOE = margin of error.

TABLE H.31   Painters Tuesday, Wednesday, and Thursday (except Tuesday after Holidays)

Category	Fri	Mon	Tues	Wed	Thur	Obs.	Percent	MOE,[*] %
					Work			
1	0	0	6	4	9	19	26.76	11
2	0	0	2	2	4	8	11.27	8
3	0	0	0	0	0	0	0.00	0
4	0	0	0	0	0	0	0.00	0
5	0	0	0	2	0	2	2.82	4
						Subtotal	40.85	
					Waiting			
6	0	0	0	1	0	1	1.41	3
7	0	0	0	1	0	1	1.41	3
8	0	0	0	2	0	2	2.82	4
9	0	0	0	0	0	0	0.00	0
10	0	0	0	0	0	0	0.00	0
11	0	0	0	0	0	0	0.00	0
12	0	0	0	0	0	0	0.00	0
13	0	0	0	0	0	0	0.00	0
14	0	0	0	0	0	0	0.00	0
						Subtotal	5.63	
					Other			
15	0	0	0	0	0	0	0.00	0
16	0	0	0	0	0	0	0.00	0
17	0	0	0	1	1	2	2.82	4
18	0	0	0	0	1	1	1.41	3
19	0	0	2	13	7	22	30.99	11
20	0	0	1	0	0	1	1.41	3
						Subtotal	36.62	
					Unaccountable			
23	0	0	4	5	3	12	16.90	9
						Subtotal	16.90	
Totals	0	0	15	31	25	71	100.00	

NOTE: There were no observations in any categories not shown.
[*]MOE = margin of error.

TABLE H.32   Machinists Monday and Friday (plus Tuesday after Holidays)

Category	Fri	Mon	Tues	Wed	Thur	Obs.	Percent	MOE,[*] %
				Work				
1	16	0	3	0	0	9	40.91	21
2	1	0	1	0	0	2	9.09	12
3	0	0	0	0	0	0	0.00	0
4	0	0	0	0	0	0	0.00	0
5	3	0	1	0	0	4	18.18	16
						Subtotal	68.18	
				Waiting				
6	0	0	0	0	0	0	0.00	0
7	0	0	0	0	0	0	0.00	0
8	0	1	0	0	0	1	4.55	9
9	0	0	0	0	0	0	0.00	0
10	0	0	0	0	0	0	0.00	0
11	0	0	0	0	0	0	0.00	0
12	0	0	0	0	0	0	0.00	0
13	0	0	0	0	0	0	0.00	0
14	0	0	0	0	0	0	0.00	0
						Subtotal	4.55	
				Other				
15	0	0	0	0	0	0	0.00	0
16	0	0	0	0	0	0	0.00	0
17	1	0	0	0	0	1	4.55	9
18	2	0	0	0	0	2	9.09	12
19	1	0	2	0	0	3	13.64	15
20	0	0	0	0	0	0	0.00	0
						Subtotal	27.27	
				Unaccountable				
23	0	0	0	0	0	0	0.00	0
						Subtotal	0.00	
Totals	14	1	7	0	0	22	100.00	

NOTE: There were no observations in any categories not shown.
[*]MOE = margin of error.

TABLE H.33 **Machinists Tuesday, Wednesday, and Thursday (except Tuesday after Holidays)**

Category	Fri	Mon	Tues	Wed	Thur	Obs.	Percent	MOE,[*] %
				Work				
1	0	0	6	11	12	29	54.72	14
2	0	0	1	1	3	5	9.43	8
3	0	0	0	0	0	0	0.00	0
4	0	0	0	1	1	2	3.77	5
5	0	0	0	2	0	2	3.77	5
						Subtotal	71.70	
				Waiting				
6	0	0	0	1	0	1	1.89	4
7	0	0	0	1	2	3	5.66	6
8	0	0	0	3	2	5	9.43	8
9	0	0	0	0	0	0	0.00	0
10	0	0	0	0	0	0	0.00	0
11	0	0	0	0	0	0	0.00	0
12	0	0	0	0	0	0	0.00	0
13	0	0	0	0	0	0	0.00	0
14	0	0	0	0	0	0	0.00	0
						Subtotal	16.98	
				Other				
15	0	0	0	0	0	0	0.00	0
16	0	0	0	0	0	0	0.00	0
17	0	0	1	0	1	2	3.77	5
18	0	0	1	0	0	1	1.89	4
19	0	0	0	1	1	2	3.77	5
20	0	0	0	0	0	0	0.00	0
						Subtotal	9.43	
				Unaccountable				
23	0	0	0	1	0	1	1.89	4
						Subtotal	1.89	
Totals	0	0	9	22	22	53	100.00	

NOTE: There were no observations in any categories not shown.
[*]MOE = margin of error.

TABLE H.34   Classification of Observations by Hour on Monday and Friday (plus Tuesday after Holidays)

Hour period	All	Technicians	Apprentices and trainees	M	P	W	H	A	T
8	27	23	4	12	8	2	1	2	2
9	30	17	13	8	2	5	2	5	8
10	31	27	4	13	4	6	4	1	3
11	32	24	8	12	6	3	3	5	3
12	31	24	7	15	4	5	0	3	4
13	22	19	3	10	3	4	2	2	1
14	30	20	10	14	3	2	1	5	5
15	25	19	6	13	0	3	3	4	2
16	22	18	4	10	0	2	2	3	1
17	25	18	7	6	4	4	4	3	4
Totals	275	209	66	113	38	36	22	33	33

Finally, Category 5, Wrap-up, is not only 61.84% (37 minutes) for the last hour, but 12.68% (8 minutes) of the preceding hour (4:00 to 5:00 PM). This time does not include break time in the last 2 hours, an additional 7.04% (4 minutes) and 9.21% (6 minutes). (Observations made in the beginning 20 minutes of the last hour were normally considered Category 19, Break.) This time also does not include any time spent completing job reports as that was considered Category 1, Working. The result is that on the average, 55 minutes is spent by each person on break or wrap-up in the last 2 hours of the day.

TABLE H.35   Classification of Observations by Hour on Tuesday, Wednesday, and Thursday (except Tuesday after Holidays)

Hour period	All	Technicians	Apprentices and trainees	M	P	W	H	A	T
8	42	33	9	14	3	9	7	5	4
9	48	39	9	14	15	4	6	6	3
10	38	29	9	18	5	3	3	6	3
11	40	32	8	15	11	3	3	3	5
12	44	34	10	22	3	4	5	6	4
13	44	36	8	15	12	3	6	7	1
14	52	38	14	21	5	5	7	9	5
15	41	33	8	15	5	5	8	5	3
16	49	38	11	20	7	5	6	10	1
17	51	38	13	28	5	3	2	7	6
Totals	449	350	99	182	71	44	53	64	33

TABLE H.36   **Days when One or Both Units Are On-Line**

Category	Fri	Mon	Tues	Wed	Thur	Obs.	Percent	MOE,* %
					Work			
1	28	26	33	44	56	187	36.10	4
2	19	14	12	11	22	78	15.06	3
3	0	0	0	0	0	0	0.00	0
4	0	1	4	3	0	8	1.54	1
5	7	4	6	13	9	39	7.53	2
						Subtotal	60.23	
					Waiting			
6	0	1	3	7	3	14	2.70	1
7	5	6	3	5	6	25	4.83	2
8	2	3	4	5	7	21	4.05	2
9	0	0	0	2	3	5	0.97	1
10	0	0	1	0	1	2	0.39	1
11	0	0	0	0	0	0	0.00	0
12	0	0	1	0	2	3	0.58	1
13	0	0	0	0	0	0	0.00	0
14	0	0	0	0	0	0	0.00	0
						Subtotal	13.51	
					Other			
15	1	0	0	1	0	2	0.39	1
16	0	0	0	0	0	0	0.00	0
17	2	2	2	3	4	13	2.51	1
18	1	3	1	0	2	7	1.35	1
19	5	12	12	17	21	67	12.93	3
20	2	1	1	0	1	5	0.97	1
						Subtotal	18.15	
					Unaccountable			
23	7	9	13	8	5	42	8.11	2
						Subtotal	8.11	
Totals	79	82	96	119	142	518	100.00	

NOTE: There were no observations in any categories not shown.
*MOE = margin of error.

TABLE H.37   Classification of Observations by Hour when One or Both Units Are On-Line

Hour period	All	Technicians	Apprentices and trainees	M	P	W	H	A	T
8	47	36	11	13	9	8	6	6	5
9	50	36	14	13	15	5	3	6	8
10	41	31	10	17	6	4	4	5	5
11	53	40	13	17	15	4	4	6	7
12	54	39	15	22	7	6	4	8	7
13	49	41	8	17	15	3	6	6	2
14	66	46	20	25	8	6	7	13	7
15	48	35	13	18	3	5	9	8	5
16	54	41	13	19	11	3	8	11	2
17	56	39	17	20	9	6	4	8	9
Totals	518	384	134	181	98	44	55	77	57

In summary of Fig. H.5, it appears that persons have moderate wrench time in the morning, peak wrench time (over 50%) for 2 consecutive hours after the lunch period, and then a less than moderate wrench time thereafter. All of the morning hours have a higher than average delay time waiting for tools, and to a lesser degree, parts, and instructions (as persons determine what they need?). The initial morning hour has additional low wrench time associated with high travel and unaccountable personnel. Break and lunch associated periods have only somewhat higher than normal travel, but certainly lower wrench time. The two peak wrench time periods also have substantial travel and delays for instructions. The final hour experiences almost no wrench time with high break, wrap-up, and unaccountable observations. Tables H.40 through H.49 contain wrench time observation data for each hour.

Next, the results of the study as the weeks in the study period pass are examined. At this point it must be noted that this study was markedly different from previous studies in the leveling of observations. In the previous studies an equal number of observations were taken each half hour of each date throughout the study (with the exception of skipping one different day each week to allow one full-time observer to observe 10-hour shifts). In the present study, the part-time observer carefully planned for an equal number of observations in each hour and each day of the week for the whole study, but not necessarily for each particular date. In other words, while each hour of the whole current study had a consistent number of observations (about 72), that does not mean that 10 observations were collected in each of the 7 weeks of the study for each hour. Therefore, while the current study was representative of the workforce over the whole period, it was not necessarily so for individual weeks.

TABLE H.38   Days when Unit Tripped or Was Called

Category	Fri	Mon	Tues	Wed	Thur	Obs.	Percent	MOE,[*] %
				Work				
1	0	0	12	16	8	36	32.14	4
2	0	0	8	4	4	16	14.29	7
3	0	0	0	0	0	0	0.00	0
4	0	0	2	1	0	3	2.68	3
5	0	0	6	6	5	17	15.18	7
						Subtotal	64.29	
				Waiting				
6	0	0	0	1	0	1	0.89	2
7	0	0	1	1	0	2	1.79	3
8	0	0	0	1	2	3	2.68	3
9	0	0	0	1	0	1	0.89	2
10	0	0	0	0	0	0	0.00	0
11	0	0	0	0	0	0	0.00	0
12	0	0	0	0	0	0	0.00	0
13	0	0	0	0	0	0	0.00	0
14	0	0	0	0	0	0	0.00	0
						Subtotal	6.25	
				Other				
15	0	0	0	0	0	0	0.00	0
16	0	0	0	0	0	0	0.00	0
17	0	0	1	1	2	4	3.57	4
18	0	0	1	0	0	1	0.89	2
19	0	0	1	6	9	16	14.29	7
20	0	0	0	0	0	0	0.00	0
						Subtotal	18.75	
				Unaccountable				
23	0	0	7	4	1	12	10.71	6
						Subtotal	10.71	
Totals	0	0	39	42	31	112	100.00	

NOTE: There were no observations in any categories not shown.
[*]MOE = margin of error.

TABLE H.39   Classification of Observations by Hour when Unit Was Tripped or Called

Hour period	All	Technicians	Apprentices and trainees	M	P	W	H	A	T
8	16	12	4	4	3	4	1	2	2
9	10	7	3	3	2	0	2	2	1
10	6	5	1	3	2	0	0	0	1
11	12	8	4	4	3	1	0	2	2
12	9	8	1	7	0	1	0	0	1
13	9	7	2	4	0	2	1	1	1
14	11	8	3	3	2	1	2	2	1
15	6	3	3	2	0	1	0	2	1
16	17	14	3	6	5	1	2	3	0
17	6	11	5	7	2	1	1	2	3
Totals	112	83	29	43	19	12	9	16	13

Figure H.5   Wrench time changes every hour.

TABLE H.40 For Period 8 (7:30 to 8:30)

Category	Fri	Mon	Tues	Wed	Thur	Obs.	Percent	MOE,[*] %
				Work				
1	5	2	3	2	1	13	18.84	9
2	5	1	6	3	3	18	26.06	11
3	0	0	0	0	0	0	0.00	0
4	0	0	1	1	1	3	4.35	5
5	0	0	0	0	0	0	0.00	0
						Subtotal	49.28	
				Waiting				
6	0	0	0	0	3	3	4.35	5
7	1	0	1	1	0	3	4.35	5
8	1	0	0	0	2	3	4.35	5
9	0	0	0	1	0	1	1.45	3
10	0	0	0	0	0	1	1.45	3
11	0	0	0	0	0	0	0.00	0
12	0	0	0	2	0	2	2.90	4
13	0	0	0	0	0	0	0.00	0
14	0	0	0	0	0	0	0.00	0
						Subtotal	18.84	
				Other				
15	0	0	0	0	0	0	0.00	0
16	0	0	0	0	0	0	0.00	0
17	1	0	1	0	1	3	4.35	5
18	0	1	0	0	0	0	1.45	3
19	0	1	0	1	4	6	8.70	7
20	0	0	0	0	0	0	0.00	0
						Subtotal	14.49	
				Unaccountable				
23	1	4	6	1	0	12	17.39	9
						Subtotal	17.39	
Totals	14	9	18	12	16	69	100.00	

NOTE: There were no observations in any categories not shown.
[*]MOE = margin of error.

TABLE H.41   For Period 9 (8:30 to 9:30)

Category	Fri	Mon	Tues	Wed	Thur	Obs.	Percent	MOE,[*] %
				Work				
1	8	6	9	7	7	37	47.44	11
2	3	0	4	2	2	11	14.10	8
3	0	0	0	0	0	0	0.00	0
4	0	0	1	0	1	2	2.56	4
5	0	0	0	0	0	0	0.00	0
						Subtotal	64.10	
				Waiting				
6	0	0	2	1	0	3	3.85	4
7	0	4	1	0	2	7	8.97	6
8	0	0	2	2	0	4	5.13	5
9	0	0	0	0	0	0	0.00	0
10	0	0	0	0	0	0	0.00	0
11	0	0	0	0	0	0	0.00	0
12	0	0	2	0	0	2	2.56	4
13	0	0	0	0	0	0	0.00	0
14	0	0	0	0	0	0	0.00	0
						Subtotal	20.51	
				Other				
15	0	0	0	0	0	0	0.00	0
16	0	0	0	0	0	0	0.00	0
17	0	0	1	0	2	3	3.85	4
18	0	0	0	0	0	0	0.00	0
19	0	0	0	2	2	4	5.13	5
20	1	0	0	0	0	1	1.28	3
						Subtotal	10.26	
				Unaccountable				
23	0	0	0	4	0	4	5.13	5
						Subtotal	5.13	
Totals	12	10	22	18	16	78	100.00	

NOTE: There were no observations in any categories not shown.
[*]MOE = margin of error.

TABLE H.42   For Period 10 (9:30 to 10:30)

Category	Fri	Mon	Tues	Wed	Thur	Obs.	Percent	MOE,* %
				Work				
1	5	0	7	1	3	16	23.19	10
2	3	2	2	2	3	12	17.39	9
3	0	0	0	0	0	0	0.00	0
4	1	0	0	2	0	3	4.35	5
5	0	0	0	0	0	0	0.00	0
						Subtotal	44.93	
				Waiting				
6	0	0	1	0	0	1	1.45	3
7	1	0	1	0	2	4	5.80	6
8	0	0	0	0	1	1	1.45	3
9	0	0	0	0	0	0	0.00	0
10	0	0	0	0	0	0	0.00	0
11	0	0	0	0	0	0	0.00	0
12	0	0	0	0	0	0	0.00	0
13	0	0	0	0	0	0	0.00	0
14	0	0	0	0	0	0	0.00	0
						Subtotal	8.70	
				Other				
15	0	0	0	0	0	0	0.00	0
16	0	0	0	0	0	0	0.00	0
17	1	0	0	0	1	2	2.90	4
18	0	0	0	0	0	0	0.00	0
19	2	6	6	10	5	29	42.03	12
20	0	0	0	0	0	0	0.00	0
						Subtotal	44.93	
				Unaccountable				
23	0	0	1	0	0	1	1.45	3
						Subtotal	1.45	
Totals	13	8	18	15	15	69	100.00	

NOTE: There were no observations in any categories not shown.
*MOE = margin of error.

TABLE H.43   For Period 11 (10:30 to 11:30)

Category	Fri	Mon	Tues	Wed	Thur	Obs.	Percent	MOE,[*] %
				Work				
1	9	4	9	5	7	34	47.22	12
2	0	1	2	2	4	9	12.50	8
3	0	0	0	0	0	0	0.00	0
4	0	0	0	0	0	0	0.00	0
5	0	0	0	0	0	0	0.00	0
						Subtotal	59.72	
				Waiting				
6	0	1	2	0	0	3	4.17	5
7	0	1	1	3	1	6	8.33	7
8	2	1	0	1	1	5	6.94	6
9	0	0	0	0	2	2	2.78	4
10	0	0	0	0	0	0	0.00	0
11	0	0	0	0	0	0	0.00	0
12	0	0	2	0	0	2	2.78	4
13	0	0	0	0	0	0	0.00	0
14	0	0	0	0	0	0	0.00	0
						Subtotal	25.00	
				Other				
15	0	0	0	0	0	0	0.00	0
16	0	0	0	0	0	0	0.00	0
17	0	0	0	0	0	0	0.00	0
18	1	0	0	0	0	1	1.39	3
19	1	0	2	2	0	5	6.94	6
20	0	0	1	0	0	1	1.39	3
						Subtotal	9.72	
				Unaccountable				
23	1	0	2	1	0	4	5.56	5
						Subtotal	5.56	
Totals	14	8	21	14	15	72	100.00	

NOTE: There were no observations in any categories not shown.
[*]MOE = margin of error.

TABLE H.44   For Period 12 (11:30 to 12:00 and 12:30 to 1:00)

Category	Fri	Mon	Tues	Wed	Thur	Obs.	Percent	MOE,* %
				Work				
1	3	2	6	5	9	25	33.33	11
2	4	0	0	4	3	11	14.67	8
3	0	0	0	0	0	0	0.00	0
4	0	0	0	0	0	0	0.00	0
5	0	0	0	0	0	0	0.00	0
						Subtotal	48.00	
				Waiting				
6	0	0	0	2	0	2	2.67	4
7	0	1	1	1	1	4	5.33	5
8	0	0	0	0	0	0	0.00	0
9	0	0	0	0	0	0	0.00	0
10	0	0	1	0	0	1	1.33	3
11	0	0	0	0	0	0	0.00	0
12	0	0	1	0	1	2	2.67	4
13	0	0	0	0	0	0	0.00	0
14	0	0	0	0	0	0	0.00	0
						Subtotal	12.00	
				Other				
15	0	0	0	1	0	1	1.33	3
16	0	0	0	0	0	0	0.00	0
17	2	0	1	0	0	3	4.00	5
18	0	2	0	0	0	2	2.67	4
19	3	3	1	1	6	14	18.67	9
20	1	0	0	0	0	1	1.33	3
						Subtotal	28.00	
				Unaccountable				
23	2	0	3	3	1	9	12.00	8
						Subtotal	12.00	
Totals	15	8	14	17	21	75	100.00	

NOTE: There were no observations in any categories not shown.
*MOE = margin of error.

TABLE H.45 For Period 13 (1:00 to 2:00)

Category	Fri	Mon	Tues	Wed	Thur	Obs.	Percent	MOE,[*] %
				Work				
1	9	1	5	9	11	35	53.03	12
2	3	2	3	0	6	14	21.21	10
3	0	0	0	0	0	0	0.00	0
4	0	0	0	0	0	0	0.00	0
5	0	0	0	0	0	0	0.00	0
						Subtotal	74.24	
				Waiting				
6	0	0	0	0	0	0	0.00	0
7	0	0	1	0	0	1	1.52	3
8	0	2	0	2	0	4	6.06	6
9	0	0	0	0	1	1	1.52	3
10	0	0	0	0	0	0	0.00	0
11	0	0	0	0	0	0	0.00	0
12	0	0	0	0	1	1	1.52	3
13	0	0	0	0	0	0	0.00	0
14	0	0	0	0	0	0	0.00	0
						Subtotal	10.61	
				Other				
15	0	0	0	0	0	0	0.00	0
16	0	0	0	0	0	0	0.00	0
17	0	0	0	2	0	2	3.03	4
18	0	0	0	1	1	2	3.03	4
19	0	0	0	0	0	0	0.00	0
20	0	0	1	0	0	1	1.52	3
						Subtotal	7.58	
				Unaccountable				
23	0	0	0	3	2	5	7.58	7
						Subtotal	7.58	
Totals	12	5	10	17	22	66	100.00	

NOTE: There were no observations in any categories not shown.
[*]MOE = margin of error.

TABLE H.46   For Period 14 (2:00 to 3:00)

Category	Fri	Mon	Tues	Wed	Thur	Obs.	Percent	MOE,[*] %
				Work				
1	6	8	8	9	12	43	52.44	11
2	2	3	2	2	1	10	12.20	7
3	0	0	0	0	0	0	0.00	0
4	0	0	1	0	0	1	1.22	2
5	0	0	0	0	0	0	0.00	0
						Subtotal	65.85	
				Waiting				
6	0	0	0	5	0	5	6.10	5
7	2	0	0	1	0	3	3.66	4
8	0	0	6	0	0	6	7.32	6
9	0	0	0	0	0	0	0.00	0
10	0	0	0	0	0	0	0.00	0
11	0	0	0	0	0	0	0.00	0
12	0	0	0	0	0	0	0.00	0
13	0	0	0	0	0	0	0.00	0
14	0	0	0	0	0	0	0.00	0
						Subtotal	17.07	
				Other				
15	0	0	0	0	0	0	0.00	0
16	0	0	0	0	0	0	0.00	0
17	0	0	0	0	0	0	0.00	0
18	0	0	1	0	0	1	1.22	2
19	0	0	0	1	6	7	8.54	6
20	0	0	0	0	0	0	0.00	0
						Subtotal	9.76	
				Unaccountable				
23	3	0	0	1	2	6	7.32	6
						Subtotal	7.32	
Totals	13	11	18	19	21	82	100.00	

NOTE: There were no observations in any categories not shown.
[*]MOE = margin of error.

TABLE H.47   For Period 14 (2:00 to 3:00)

Category	Fri	Mon	Tues	Wed	Thur	Obs.	Percent	MOE,[*] %
				Work				
1	3	2	7	2	5	19	28.79	11
2	3	2	3	3	2	13	19.70	10
3	0	0	0	0	0	0	0.00	0
4	0	1	1	0	0	2	3.03	4
5	1	0	0	0	0	1	1.52	3
						Subtotal	53.03	
				Waiting				
6	0	0	2	0	0	2	3.03	4
7	0	0	0	0	3	3	4.55	5
8	0	0	0	0	0	0	0.00	0
9	0	0	0	1	0	1	1.52	3
10	0	0	0	0	0	0	0.00	0
11	0	0	0	0	0	0	0.00	0
12	0	0	0	0	0	0	0.00	0
13	0	0	0	0	0	0	0.00	0
14	0	0	0	0	0	0	0.00	0
						Subtotal	9.09	
				Other				
15	0	0	0	0	0	0	0.00	0
16	0	0	0	0	0	0	0.00	0
17	0	0	0	0	0	0	0.00	0
18	0	0	0	0	0	0	0.00	0
19	4	2	4	8	4	22	33.33	12
20	0	0	0	0	0	0	0.00	0
						Subtotal	33.33	
				Unaccountable				
23	1	1	0	0	1	3	4.55	5
						Subtotal	4.55	
Totals	12	8	17	14	15	66	100.00	

NOTE: There were no observations in any categories not shown.
[*]MOE = margin of error.

TABLE H.48   For Period 16 (4:00 to 5:00)

Category	Fri	Mon	Tues	Wed	Thur	Obs.	Percent	MOE,[*] %
				Work				
1	3	1	5	11	9	29	40.85	12
2	1	2	2	0	4	9	12.68	8
3	0	0	0	0	0	0	0.00	0
4	0	0	0	0	1	1	1.41	3
5	2	1	2	4	0	9	12.68	8
						Subtotal	67.61	
				Waiting				
6	0	0	0	0	0	0	0.00	0
7	2	0	0	1	0	3	4.23	5
8	0	0	0	2	1	3	4.23	5
9	0	0	0	0	0	0	0.00	0
10	0	0	0	0	0	0	0.00	0
11	0	0	0	0	0	0	0.00	0
12	0	0	0	0	0	0	0.00	0
13	0	0	0	0	0	0	0.00	0
14	0	0	0	0	0	0	0.00	0
						Subtotal	8.45	
				Other				
15	1	0	0	0	0	1	1.41	3
16	0	0	0	0	0	0	0.00	0
17	0	2	0	1	0	3	4.23	5
18	1	0	0	0	1	2	2.82	4
19	1	0	3	0	1	5	7.04	6
20	1	0	0	0	1	2	2.82	4
						Subtotal	18.31	
				Unaccountable				
23	1	1	1	0	1	4	5.63	5
						Subtotal	5.63	
Totals	13	7	13	20	18	71	100.00	

NOTE: There were no observations in any categories not shown.
[*]MOE = margin of error.

TABLE H.49   For Period 17 (5:00 to 6:00)

Category	Fri	Mon	Tues	Wed	Thur	Obs.	Percent	MOE,[*] %
				Work				
1	0	0	2	1	0	3	3.95	4
2	1	1	1	0	1	4	5.26	5
3	0	0	0	0	0	0	0.00	0
4	0	0	1	0	0	1	1.32	3
5	9	3	7	16	12	47	61.84	11
						Subtotal	72.37	
				Waiting				
6	0	0	0	0	1	1	1.32	3
7	0	0	1	0	0	1	1.32	3
8	0	0	0	0	2	2	2.63	4
9	0	0	0	0	0	0	0.00	0
10	0	0	0	0	0	0	0.00	0
11	0	0	0	0	0	0	0.00	0
12	0	0	0	0	0	0	0.00	0
13	0	0	0	0	0	0	0.00	0
14	0	0	0	0	0	0	0.00	0
						Subtotal	5.26	
				Other				
15	0	0	0	0	0	0	0.00	0
16	0	0	0	0	0	0	0.00	0
17	0	0	0	0	0	0	0.00	0
18	0	0	0	0	0	0	0.00	0
19	1	0	3	3	0	7	9.21	7
20	0	1	0	0	0	1	1.32	3
						Subtotal	10.53	
				Unaccountable				
23	0	3	2	2	2	9	11.84	7
						Subtotal	11.84	
Totals	11	8	17	22	18	76	100.00	

NOTE: There were no observations in any categories not shown.
[*]MOE = margin of error.

Figure H.6   Wrench time varies each week.

Figure H.6 shows that as the 7-week study progressed, reported Category 1, Working, time improved from around 32 to 35%. (It must be remembered that unequal hourly observations were made in individual weeks so the different results reflect only those hours as measured in each individual week.)

Figure H.7 again shows wrench time each week, but with events and unit status added.

Figure H.7   Plant events during each week of study.

Figure H.8 compares wrench time with unaccountable percentages. There appears to be no strong correlation.

Finally, seven separate dates were selected from the study period, approximately 1 day per week, where the observations were level with respect to hours of the day. Table H.50, 7 Days, shows every time category being within the margin of error of the complete 7-week study. Table H.51 shows a fairly even distribution of observations.

This study made this analysis to determine if a 1-day-per-week type of observation gathering might be statistically valid. It appears that it is valid.

Validity and implications of in-house studies.   The original company specification for the first wrench time study specifies a 10% accuracy for the overall wrench time. The first two studies achieve margins of error of 3% for Category 1, Working, from over 1200 total observations in each study. This current study achieves 4% with over 700 total observations. Statistically, the in-house study is valid.

However, an important question beyond the study's validity is whether the classification of observations were consistent enough with the previous studies to allow comparison. The fact that many observations had to be carefully examined before classifying would tend to support trying to keep the same observer or group of observers involved in the observation

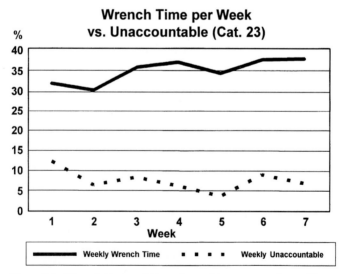

Figure H.8   Wrench time and time charged to persons who could not be found.

TABLE H.50   Seven Days

Category	Fri	Mon	Tues	Wed	Thur	Obs.	Percent	MOE,* %
				Work				
1	13	11	26	18	20	88	31.77	6
2	2	7	13	8	9	39	14.08	4
3	0	0	0	0	0	0	0.00	0
4	1	0	2	1	0	4	1.44	1
5	5	3	3	11	2	24	8.66	3
						Subtotal	155.96	
				Waiting				
6	0	1	7	1	2	11	3.97	2
7	1	4	4	3	3	15	5.42	3
8	1	0	4	3	1	9	3.25	2
9	0	0	0	1	2	3	1.08	1
10	0	0	0	0	0	0	0.00	0
11	0	0	0	0	0	0	0.00	0
12	0	0	2	2	0	4	1.44	1
13	0	0	0	0	0	0	0.00	0
14	0	0	0	0	0	0	0.00	0
						Subtotal	15.16	
				Other				
15	0	0	0	0	0	0	0.00	0
16	0	0	0	0	0	0	0.00	0
17	0	2	2	1	1	6	2.17	2
18	0	3	0	1	2	6	2.17	2
19	7	7	10	12	7	43	15.52	4
20	1	0	0	0	1	2	0.72	1
						Subtotal	20.58	
				Unaccountable				
23	2	7	4	9	1	23	8.30	3
						Subtotal	8.30	
Totals	33	45	77	71	51	277	100.00	

NOTE: There were no observations in any categories not shown. Days included—Jan 20 & 29; Feb 1, 9, 16, 17, & 25.

TABLE H.51   Classification of Observations by Hour for 7 Days

Hour period	All	Technicians	Apprentices and trainees	M	P	W	H	A	T
8	25	21	4	12	3	4	2	2	2
9	28	21	7	5	9	4	3	3	4
10	31	22	9	15	2	3	2	4	5
11	32	25	7	14	4	3	4	3	4
12	30	23	7	12	5	5	1	3	4
13	27	21	6	13	3	3	2	4	2
14	30	23	7	14	2	1	6	4	3
15	22	18	4	11	2	4	1	2	2
16	22	19	3	12	0	5	2	2	1
17	30	21	9	11	4	3	3	3	6
Totals	277	214	63	119	34	35	26	30	33

process. There was a close match of current results to results of previous studies except where changed maintenance practices had explainable impacts. This match appears to suggest that the current study is useful for comparison as well as valid. An example of changed practice is the elimination of the morning check-in meeting and the resulting lower Category 4, Work Assignment, time observed.

Attachment B, Work Sampling Calculations, presents the calculations and considerations involved to ensure that the results are meaningful.

An analysis of only 7 days within the current study achieves 6% accuracy with less than 300 observations. The concern, of course, with a limited observation period is that typical work situations might not be observed. However, since the 7-day result of 31.77% is within the margin of error of the full study, it is felt that a future wrench time study could be conducted on such a limited basis and give meaningful measurement feedback. One different day each week for a period of 1 to 2 months might be used where the observer is careful to make equal observations throughout each hour and each day.

Measurement acclimation.   From Fig. H.9, Wrench Time per Week vs. Number of Observations, it appears there was some improvement in wrench time over the course of the study due to persons being observed. However, as mentioned previously, the study did not attempt to levelize observations to make each week valid for wrench time. The study does not report the initial week of the study (week 0) in the final results to minimize any start-up effects and conscious modifications of persons' efforts. However, although most persons were good-natured about the study, the observer felt that they were becoming a little tired of being observed on a regular basis by the end of the study. Perhaps continual occasional observations in the future might reduce this consciousness.

Conclusions

The study makes the following conclusions.

1. The current in-house wrench time study is valid and is representative of the workforce at this time.

2. Current wrench time is 35.08%, or $3^{1}/_{2}$ hours per day with all categories summarized in Table H.1. Wrench time (Category 1) is statistically unchanged from earlier studies, but travel, work assignment, and interference time has improved. However, break and unaccountable (and to a lesser degree, waiting for materials and instructions) have become worse. Analysis of the comments and time of day for each observation suggests that concerns might be of a work scheduling or motivational nature rather than formal planning of job packages. (However, job planning might help to set job durations and work hour requirements in regard to scheduling.)

3. There are work differences among the mechanical crafts. Table H.52 summarizes the differences with respect to wrench time. In addition, on the whole, break time seems to be greater for technicians than apprentices and trainees combined (15.71 vs. 8.48%) and travel appears to be less (13.42 vs. 21.82%).

4. Crews have somewhat comparable wrench times considering the study accuracy, but it appears that additional observations would make the differences statistically significant.

5. There is little difference in Monday and Friday work with only one or two crews versus other days when all crews are present. There is also little or no difference in the workforce wrench time when personnel are loaned to another plant or the apprentices are sent to training.

Figure H.9 Wrench time and how many observations were made.

TABLE H.52   Conclusion of Wrench Time Differences among Crafts

Craft	Wrench time, %	Hours/10-hour day
Mechanics	34.24	$3^1/_2$
Painters	25.69	$2^1/_2$
Welders	33.75	$3^1/_2$
Machinists	50.67	5
Apprentices	40.21	4
Trainees	30.80	3

6. The only unit status that appears to affect wrench time is when a unit trips or is started. On these days, wrench time becomes worse as persons are reassigned. The data suggests areas where this productivity could be improved, but it may not warrant the effort as these events are not encountered every week.

7. Hour by hour, wrench time changes throughout the day. Overall, there is moderate wrench time in the morning, peak wrench time (over 50%) for 2 consecutive hours after the lunch period, and then a less than moderate wrench time thereafter. All of the morning hours have a higher than average delay time waiting for tools and, to a lesser degree, parts and instructions. The initial morning hour has low wrench time associated with high travel and unaccountable personnel. Break and lunch associated periods have only somewhat higher than normal travel. The two periods of peak wrench time also have substantial travel and delays for instructions. The final hour experiences almost no wrench time with high break, wrap-up, and unaccountable percentages.

8. Wrench time measured for only 7 individual days (approximately one per week) during the study has time percentages for every single category within the margin of error of the whole study. This result suggests that a greatly reduced observation effort yields valid results provided that the days selected are considered typical of the overall period.

9. The process of continual observations nearly every day for 7 weeks did not appear to acclimate the workforce to being measured, but did suggest that a future continual study of occasional observation days may be preferred in this respect.

Recommendations

The study recommends the following.

1. Emphasis should be given to scheduling and motivational areas to improve wrench time.

2. A continuing in-house study of observations only 1 day each week should be conducted to give ongoing feedback on wrench time and to acclimate persons to being measured.

3. Future observation data should be structured to be additive as well as separate from previous observations to allow reduced margins of error to be developed in the analysis of specific groupings such as for supervisors, crafts, and special events.

Attachment A: Procedure for measuring workforce productivity by work sampling

This study utilized the following procedure.

1. The Job Planning Coordinator (JPC) and the appropriate managers or supervisors select the workforce personnel and time period to be studied. In addition, work type categories are reviewed.

2. The JPC assigns each person in the study a one- to three-digit, unique number for the purposes of tracking. (The JPC attempts to utilize the same number for persons included in previous studies.) The JPC provides a number for personnel to affix to their hard-hats. (*Note:* The practice of affixing numbers to hard-hats was discontinued for subsequent studies as some persons considered it demeaning.) The JPC records each person's name, number, and craft with current skill level into Table H.53.

3. The JPC enters the names of the crew supervisors who have personnel being studied into Table H.54. The JPC assigns each person a one- to three-digit, unique number for the purposes of tracking.

4. Each afternoon of the study period, each supervisor (or assistant) provides the JPC a schedule or agenda of the next day's work with the names of assigned personnel.

5. Each day the JPC utilizes that day's work schedule to select employees to observe and record their work category. The JPC generally selects employees and exact times to observe so as to include as many observations of different employees and times as possible. (An informal check sheet such as Fig. H.10 is kept to help make sure an even number of observations is made each hour of the overall study.) Typically three people are selected in the same area of the plant to minimize the JPC's time spent locating the employees. The instant the JPC locates each selected employee, the JPC records that employee's activity in a work category on the Wrench Time Observation Data Sheet (Fig. H.11). (The JPC may occasionally question the employee if it is not certain which category is appropriate for the current activity observed.)

Two sampling methods would ensure representative data, random and sequential. Either method provides nonbiased samples. The observer uses a list of crew employees to create a "sequential" observation strategy. At the beginning of the study, the observer simply starts at the top of the list to select the initial person for the first observation.

TABLE H.53   Persons in Study

Name	Number	Craft and level
Abbington	11	M
Abby	61	H
Andrews	73	W
Brandi	23	M
Brie	15	A
Brown	69	P
Carter, K.	19	A
Carter, S.	67	M
Colter	62	P
Comain	12	T
Cumar	2	W
Dabor	54	M
Douglas	25	M
Eckardt	16	A
Fountain	76	T
Hartness	13	H
Hernandez	4	M
Hobgood	75	A
Hollis	80	A
Jensen	24	T
Jobson, S.	45	P
Jobson, T.	27	T
Johns	63	M
Johnson	17	A
Kent	65	M
Kenny	3	M
Kim	71	A
Lauren	28	M
Marshall	1	H
Morten	10	H
Mott	64	M
Nathaniel	33	P
Noel	68	W
Peek	5	M
Powell	77	M
Roberson	20	A
Rust	70	H
Sanchez	14	A
Spencer	35	M
Strain	74	M
Sunday	7	M
Swanson, R.	9	M
Tien	78	M
Valhalla	79	T
Wall	18	A
William III	72	M
William IV	8	W
Young	46	W

TABLE H.54   Supervisors of Crews in Study

Supervisor number	Supervisor
1	Atwiler
100	Atwiler crew when no supervisor present for the day
101	Anyone working up for Atwiler
2	Hunt
200	Hunt crew when no supervisor present for the day
201	Anyone working up for Hunt
3	Shinsky
300	Shinsky crew when no supervisor present for the day
301	Anyone working up for Shinsky

However, the study also takes advantage of the crew supervisor assigning some employees together as teams and other jobs being in the same area of the plant as the initially chosen person. Because the observer must select three persons each 20 minute period, the observer selects the initial person and the two supposedly closest persons to that person. Then the observer checks all three names off the crew list. For the next 20 minutes, the observer goes down the list "sequentially" to select the next unchecked person and then selects the two closest unchecked persons to that person. The observer checks off these three names. When there are not three names left on the list, the observer continues around back to the top of the list. The observer keeps the place on the employee list at the end of a day for the beginning of another day. In this manner, the observer has an equal chance of observing every employee throughout the study and keeps the sample statistically unbiased.

6. After collecting all the study observations, the JPC counts the observations in each category by whichever criteria necessary to make the desired comparisons and contrasts. This study reports this analysis in tables and figures. The JPC calculates the margin of error for each observation category using the formulas in Attachment B.

(*Note:* Manual counting and calculations are adequate for short studies. As the analysis begins to consider more types of comparisons, appropriately constructed databases or spreadsheets facilitate making the tabulations and calculations. Also, a person performing a study might add the employee craft designation to the observation record to allow the accumulation and analysis of observation data over years of separate studies allowing persons to change crafts and promote up to higher skill levels. One would not have to change the forms, but could add the craft after the observations are complete with a computer "replace" command.)

Attachment B: Work sampling calculations

The company specifications for the work measurement study state: "The proposed study design will include the methodology for assuring

INFORMAL CHECK SHEET TO
BALANCE OBSERVATIONS

Mark the time started to locate persons within beginning twenty minutes (B), middle twenty minutes, or end twenty minutes (E): Hand write a number beside each half hour to designate the week during which the observation occurred.

Time Period	Period Starts	Mon	Tues	Wed	Thur	Fri
8	7:30	B3M3E4 B M E	B4M4E1 B6M6E2	B1M2E* B5M5E6	B2M5E3 B5M6E6	B1M3E2 B6M4E6
9	8:30	B4M3E3 B M E	B2M3E1 B5M6E4	B2M1E5 B6M5E6	B2M3E5 B6M5E6	B2M4E1 B6M6E3
10	9:30	B3M3E4 B M E	B4M1E3 B5M5E6	B2M1E** B6M4E5	B3M3E6 B5M6E6	B1M3E2 B6M6E4
11	10:30	B4M3E3 B M E	B1M5E4 B6M6E5	B**M4E1 B**M5E4	B5M5E3 B6M6E6	B2M1E3 B4M6E6
12	11:30-12 12:30-1	B3M3E7 B M E	B4M1E7 B5M7E5	B4M5E1 B5M6E6	B3M5E1 B5M6E6	B2M3E1 B6M6E4
13	1:00	B4M3E3 B M E	B5M3E3 B7M7E6	B6M1E5 B6M5E6	B1M6E6 B3M6E6	B4M1E3 B6M2E6
14	2:00	B7M3E4 B M E	B3M5E4 B6M6E5	B5M5E1 B6M6E4	B7M7E1 B6M6E6	B3M6E2 B6M7E4
15	3:00	B4M3E7 B M E	B4M6E3 B5M5E6	B1M4E6 B2M6E6	B7M7E3 B1M6E6	B2M6E3 B4M7E7
16	4:00	B3M7E4 B M E	B3M5E4 B6M6E6	B2M5E1 B5M6E6	B1M5E7 B5M6E6	B2M4E3 B6M6E7
17	5:00	B4M3E3 B M E	B4M5E3 B5M6E6	B5M1E6 B6M2E5	B5M5E5 B6M7E1	B4M2E3 B7M7E4

Special comments: *This was a six week study with one make up week. Another initial week taken before the study started was not included.* *Observations not taken during monthly safety meeting. **Several Wednesdays, the VP chatted with the crews. This time was not included for observations.*

Figure H.10   Informal checks to ensure even observations.

WRENCH TIME OBSERVATION DATA SHEET

PAGE ___ OF ___

STUDY _____

PLANT _____

(Comptr Entry)	Unit Events	Observation Date	Circle Day of Week	Person's Number	Person's Supvr Number	Work Cat #	Time Period	OT?	Observation Comments, if any (such as purpose of travel)
		/ /	U M T W H F S						
		/ /	U M T W H F S						
		/ /	U M T W H F S						
		/ /	U M T W H F S						
		/ /	U M T W H F S						
		/ /	U M T W H F S						
		/ /	U M T W H F S						
		/ /	U M T W H F S						
		/ /	U M T W H F S						
		/ /	U M T W H F S						
		/ /	U M T W H F S						
		/ /	U M T W H F S						
		/ /	U M T W H F S						
		/ /	U M T W H F S						
		/ /	U M T W H F S						
		/ /	U M T W H F S						
		/ /	U M T W H F S						
		/ /	U M T W H F S						
		/ /	U M T W H F S						
		/ /	U M T W H F S						
		/ /	U M T W H F S						
		/ /	U M T W H F S						

Figure H.11   Form to record work sampling observations.

accuracy within ±10% with a 95% confidence level based on the direct work category."

Three study criteria are important to ensuring study accuracy. These criteria are having the observation period span a sufficient portion of the year, having the observations evenly spread out over the course of the shift, and having a sufficient number of observations.

First, the study covered a period of 7 weeks which should be of sufficient duration to classify as a representative period of typical working conditions and cancel out the effect of most special, limited-duration events that may impact wrench time. Although personnel were loaned to another plant and personnel were in training at times during the study, these periods were seen as typical circumstances under which the workforce operates.

Second, keeping a check sheet ensured making an equal number of observations during each 20-minute block of the entire 10-hour shift by the end of the study. Then during the analysis of the results, a check was made of the hourly distribution of observations to determine if any skewing of data might be present. (For example, having a more than average number of observations during the traditional morning break period may cause an artificial decrease in reported wrench time.) In most cases there appeared to be an even distribution of observations. In the few instances where there was a slightly uneven distribution, the results are noted as possibly skewed with the possible effect described.

Finally, the margin of error (absolute accuracy) for the direct work category was 4%, well within the 10% required. There were 724 total observations, of which 254 are of direct work, Category 1. The percentage of direct work is 254 divided by 724 which is 35.08%. The margin of error is found by the following equation:

$$a = \{[k^2(1-p)p]/n\}^{1/2}$$

where a = margin of error
k = number of standard deviations (k = 2 represents a 95% confidence level)
p = percent occurrence of the activity or category
n = sample size (total observations)

Thus

$$a = \{[2^2(1-0.3508)\,0.3508]/724\}^{1/2} = 3.5\% \text{ (rounded up to 4\%)}$$

This calculation means that 95% of the time, any duplicate study done over the same 7-week period and making observations at the same hourly points should have reported a direct work percentage within 4% of the 35.08% reported in this study. Because the study is only repeatable and valid over duplicate conditions, it is clear to see the importance of having a representative study period and an even hourly distribution of observations.

Special Factors Affecting Productivity

This appendix discusses additional issues affecting productivity. These issues include crafts already performing at high wrench times without planning and scheduling, utilization of blanket work orders, differentiation between empowerment and scheduling, and underlying causes of low schedule compliance and misused priority systems. The appendix conducts an extensive analysis of low schedule compliance and misused priority systems using "cause maps." The *Maintenance Planning and Scheduling Handbook* addresses these topics in this appendix so as not to distract from the main thrust of the book.

Wrench Time in Exceptional Crafts and Plants

Wrench time may be higher than the industry norm of 25 to 35% in some maintenance organizations without highly developed planning functions. The preceding two appendices showed two such examples. In App. G, the I&C (instrument and controls) group had a 38% wrench time. This I&C group did not have a planning function assisting it. The industry norm more applies to an overall maintenance force than to a specific craft. I&C and electrical crafts without planning typically should be at the top or slightly over industry norm. These groups are typically not at the desired 55% level, but their existing productivity may warrant placing them behind the mechanical group in priority for implementation of planning. In addition, in plants where I&C and electrical groups mostly support the mechanical groups, the improvement of the mechanical group's wrench time through planning improves these other crafts as well. Appendix H shows a mechanical group with a marginal planning effort. With only planning, but no weekly scheduling, the craft is at the top of the industry norm for wrench time. Certain crafts, most

Some crafts might
not need P&S yet.
Plants overwhelmed
with urgent work
need P&S now.

Figure I.1   Exceptional crafts and plants.

notably machinists, achieve 50% wrench time due to the nature of their close-at-hand work assignments.

Another situation, not illustrated by the two included wrench time studies, is that of the plant with extremely high amounts of urgent, reactive work. The craftpersons in these plants have moderate to high wrench times primarily because there is no need to schedule subsequent work assignments. The urgency of the workplace easily directs the resources to the next jobs. There is limited opportunity for idleness or breaks. The plant possesses a productive workforce, but has terrible reliability. Then, as management brings more maintenance personnel to the suffering plant, the workforce is able to catch its breath. The maintenance force grows to where it can keep up with the reactive work and deliver a somewhat reasonable reliability. At this point, wrench time drops as uncertainty sets in as to where to attack the next job. The discussion and all the reasons set forth in Chap. 6, Advance Scheduling, come into play to keep productivity low.

Blanket Work Orders

So-called blanket work orders greatly damage both productivity and record keeping. Rather than accept the expense of writing work orders for every little task, many plants have blankets to which personnel charge time for certain tasks. For example, rather than write a work order, the supervisor may direct a mechanic to hang a bulletin board in the front office and "charge the time to Blanket 103" (miscellaneous mechanical work). The productivity problem created is threefold. One, the supervisor and mechanic have stepped outside of the planning and scheduling process. The supervisor could give the mechanic a time estimate, but probably will not. Two, the extraneous job was not scheduled into the week's allocation of work. As the

Blankets
Smother
Productivity

Figure I.2   Limit blanket work orders.

practice of doing work on blankets expands, the week's allocation becomes meaningless without special attention to schedule compliance. The possibility of doing side jobs on blankets encourages neglect of the weekly schedule. The weekly schedule was based on the work prioritized in the best interest of the plant. These jobs will be delayed by possibly less important blanket work. Finally, the use of blankets invites giving less credibility to the planned time estimates. Instead of assigning a 9-hour job and a 1-hour job from the backlog to a technician on a 10-hour shift, the supervisor may assign only the 9-hour job. The supervisor might direct the technician to do some miscellaneous cleaning up "on the blanket." Likewise, blankets allow the technician to report 5 hours on a planned 5-hour job that took all day. The technician might rather claim the other 5 hours of the day on blanket work rather than having to document all the reasons the planned job took so long. There is a general loss of control, and many plants begin to do an extremely large portion of their work on blankets. One plant could account for less than half of their work hours being spent on specific work orders. The rest was blanket work. Besides productivity, blankets lose vital equipment history. Blanket work leaves no document to store in the paper files or information to record in electronic files. If the plant does some equipment work via work orders and some via blankets, then planners do not have complete history from which to collect delay information or base maintenance decisions going into their plans.

Empowering versus Scheduling

A number of obstacles hinder good scheduling practices in maintenance. One of the most interesting is the notion of empowerment because superior maintenance requires both empowerment and scheduling. Empowering maintenance crews and personnel means allowing them to make decisions within their areas of responsibility. This greatly increases the quality of maintenance work. However, empowering does not mean turning each of the specialized areas of maintenance loose on its own. The maintenance process takes a coordinated team effort to master and an explicit scheduling process is necessary to advance productivity. Superior maintenance requires both empowerment and scheduling.

The following conversation returns to the discussion in Chap. 1 between an owner and technician, but instead this interchange occurs between a maintenance manager and a front line, crew supervisor. Here, the supervisor turns to the notion of empowerment, a common impediment to good scheduling when misunderstood.

MGR: "How did it go this week?"

SUPVR: "We kept the plant running!"

MGR: "Well, how much work did you get done?"

SUPVR: "We did a lot!"

MGR: "Well, how much was that?"

SUPVR: "We did 50 jobs which was more than we ever did before!"

MGR: "But how much was that compared to how much you could have done?"

SUPVR: "You don't understand, we really worked hard! I'm supposed to be empowered!"

MGR: "Yes, empowered to do your job, but maintenance is still a team effort."

This conversation first brings out the issues of effectiveness and efficiency. Note that the crew supervisor correctly understands his job is to keep the plant running and not to exist as a speedy repair service fixing whatever breaks. Note also that there is no argument that keeping the plant running (effectiveness) is the first priority. In addition, whether or not the plant was truly running is self-evident and not even in question. On the other hand, efficiency is a more difficult notion to consider. The supervisor is conscious of how many work orders the crew completed (efficiency). Yet, the conversation implies there exists little or no concrete basis for determining whether the 50 work orders completed was an adequate rate of productivity. Was the week's effort spent productively or not? How would one know? After effectiveness, a plant must consider efficiency.

Secondly, at this point the supervisor raises the issue of empowerment. Management gave the supervisor a crew with which to maintain the plant. Shouldn't he be authorized to do "what it takes" to accomplish this mission? On the surface, this sounds appropriate. If anyone is given a job, surely he should be empowered to carry out those duties.

The empowerment statement arose because the supervisor perceived the manager was interested in some sort of standard or measure of how much work the crew should have done. As it turns out, an advance scheduling process provides such a measure. Crew supervisors generally perceive advance scheduling by an outside group as taking away their control or empowerment.

This apparent conflict or trade-off between empowerment and scheduling vitally concerns maintenance managers because of the impact scheduling makes on productivity.

Probing deeper into the controversy of empowerment and scheduling makes it necessary not only to define the terms empowerment and schedule, but also the precise details of the mission of the maintenance crew and supervisor.

Definitions and details

The dictionary defines EMPOWER as *investing with legal power, authorizing or to enable or permit* and SCHEDULE as *a production plan allotting work to be done and specifying deadlines*. Obviously, these definitions do

Dictionary

◆ Empower

- ▪ *to invest with legal power; authorize to enable or permit*
- ▪ **NOT** *to turn loose on one's own to avoid working with others*

◆ Schedule

- ▪ *a production plan allotting work to be done and specifying deadlines*
- ▪ **NOT** *to specify each and every action of someone else*

Figure I.3 What empower and schedule really mean.

not make the concepts mutually exclusive, that you cannot do one if doing the other. A maintenance crew clearly does not exist as an island isolated from the rest of the organization. In addition, just because a crew receives an advance schedule does not mean that it is hopelessly constricted and hindered from doing its job. The problem is that maintenance crew supervisors have *traditionally* exercised an extraordinary amount of freedom in selecting work activities and dictating productivity. Furthermore, while current management circles have rightly favored the concept of empowerment, they have sometimes allowed the term to carry a life of its own even to an unjustified extreme. Finally, advance scheduling requires more coordination and accountability with crew supervisors. Consequently, when management implements an advance schedule program in a modern environment promoting empowerment, crew supervisors usually perceive a loss of "control."

Empowered to do what?

Listing a number of necessary components of the maintenance process helps one define the proper area of responsibility of a maintenance crew and supervisor and one deals with the concern of empowerment. First, the maintenance process makes considerable use of the work order system. The work order itself is the vehicle by which requesters of work identify needs; maintenance planners predict parts, tools, skills, and hours; and the crew executes work. Clearly, the crew supervisor does not "lose empowerment" by various groups processing the work order. Next, the established plant priority system aids communication on the importance of individual jobs. In fact, this system really should drive to a large degree which job the crew tackles next. Again, the crew supervisor sees this system helping, not

Empowered to Do What?

- ♦ Work order system?
- ♦ Plant priority system?
- ♦ Hiring, training?
- ♦ Equipment information, CMMS?
- ♦ Tools, tool room?
- ♦ Spare parts, Storeroom?
- ♦ Payroll?

Figure I.4 How much empowerment do crew supervisors need?

hindering. Similarly, the crew supervisor does not see it necessary to have complete control over hiring, training, tools, spare parts, and payroll (Fig. I.4).

Beginning with choosing exactly what job to execute, the crew supervisor begins to feel a little more ownership in what lies more completely within his control. Yet, did not the plant priority system help decide this? In addition, the culture itself, if not management, should desire adequate preventive maintenance (PM) to prevent failures. The plant itself should also promote predictive maintenance (PdM) to head off problems and project work to improve equipment. These are not just the concern of the crew supervisor. Plus, coordination with production crews, even if just to clear equipment for maintenance, naturally would take a crew supervisor outside of any isolation. Yet, most crew supervisors would agree that even these areas do not infringe on their "empowerment (Fig. I.5)."

Now consider actually assigning work to individuals. Certainly, the crew supervisor is most knowledgeable about which individual persons

Empowered to Do What? (cont.)

- ♦ Choosing work?
- ♦ PM, PdM, projects?
- ♦ Coordination with production?
- ♦ Assigning work?
- ♦ Reacting to emergencies?
- ♦ Outages?
- ♦ Nonoutage times?

Figure I.5 How much empowerment do crew supervisors need?

work best together and are best suited for specific assignments at specific times. In addition, he is in the best position to direct the sudden reaction of even an entire crew to handle emergencies. These are areas most within the area of responsibility of a crew supervisor.

Next, consider two times to apply the preceding concepts. First, think about an outage period requiring a company to take an entire unit or plant out of production for maintenance. Different groups come together to ensure there are sufficient spare parts and labor for the anticipated work. A schedule in the form of a work scope lists all this work. Companies execute outages with great efficiency in part due to this advance schedule. Few persons would insist on lessening the coordination involved for the sake of "empowerment." Second, consider a maintenance period where no outages take place, just routine maintenance. This is the time that suffers from low productivity. Why should a company have different applications of empowered responsibility between these two times?

Proper empowered responsibility between planning and crew supervision

Job planning provides work order time estimates needed for scheduling. The crew supervisors could provide these individual job estimates, but a separate planning group truly frees and helps empower crew supervisors to function "in the present." Without having to prepare for future work, supervisors can give more attention to today's work quality and crew performance. The planning group functions "in the future" giving attention to preparing job plans, not only with time estimates, but also with details to avoid anticipated delays. See Fig. I.6.

The planning group works in the future. Planners develop job plans with time estimates. They develop the weekly schedule looking at the entire plant work backlog. The planning group uses crew forecasts of overall labor without regard to individual names. The planning group executes its portion of the maintenance process for the overall benefit of the maintenance team and maintenance manager. This is the proper area of responsibility for the planning group.

The crew and its supervisor work in the present. The crewmembers execute assigned work, empowered to concentrate on today's work without regard to organizing details for future work. The supervisor monitors today's work and assigns tomorrow's work. The supervisor develops the daily schedule using the weekly schedule allocation, but also considering any reactive work that cannot wait until next week. It is okay to break the schedule for urgent reactive work. The supervisor works with individual technicians. The crew and its supervisor execute their portion of the maintenance process for the overall benefit of the maintenance team and maintenance manager. This is the proper area of responsibility for the crew and its supervisor.

The Maintenance System (Team)

Planning Department	Crew Supervisors
Work in the future	Work in the present
Develop job plans with estimates	Assign jobs & monitor jobs-in-progress
Develop weekly schedules	Develop daily schedules
Use whole backlog	Use weekly + urgent
Look at overall labor	Look at individuals
For manager & team	For manager & team

Figure I.6   Maintenance is NOT separate teams, each doing everything.

The supervisor must be empowered to execute his part of the process. The supervisor is in the best position to handle the current day's work by assigning names to tasks, coordinating resources and clearances, and handling emergencies or other urgent work that cannot wait. See Fig. I.7.

Together with an advance schedule from the planning group, the supervisor directs the daily execution of work toward an allocated goal of work tailored to his crew's abilities for the week.

The result of proper empowerment

The plant with advance scheduling and empowered crews results in discussions between management and supervisors that lead to information that can improve plant maintenance. Such information becomes

Crew supervisor must be empowered to:

- Handle the current day's work
- Assign names to tasks
- Coordinate resources and clearances
- Handle emergencies

Figure I.7   Proper areas to empower supervisors.

There is no trade-off between empowering crews and working toward a week of scheduled work. Do both.

Figure I.8   Schedule and empower together.

available because advance scheduling provides a standard that leads to specific questions. See Fig. I.8.

MGR: "How did it go this week?"

SUPVR: "We did a lot!"

MGR: "Well, how much of the work scheduled for this week is done?"

SUPVR: "Most of it."

MGR: "Let me see the jobs not yet started . . . I see we did not start about 100 hours of work. What happened?"

SUPVR: "Let's see. On three jobs, we didn't have parts in stock so we had to place an order for next week. On one job we did complete, the estimate was too short by half even though the job went smoothly. Another job was delayed when we ran out of solvent and I authorized a field order to buy some so we could finish."

MGR: "That's okay. On some weeks, we have things we just have to deal with. I know you were working hard because I was on the shop floor several times. I am concerned a bit about three planned jobs not having parts available as well as running out of solvent. I'll talk to the planners about the parts and the tool room about the solvent."

Accomplishing productive maintenance takes a systematic approach, not isolated groups of "empowered" crews working as best they can. Many different elements make up the effective maintenance process. An important element in the effective maintenance process is empowerment at the crew level to promote effective maintenance. There is no trade-off between empowering crew supervisors and working toward a week of scheduled work. The maintenance force should do both.

Schedule Compliance

Figure I.9 shows a cause map for low schedule compliance following the Mark Galley format of ThinkReliability.com. The cause map reads from left to right with causes and effects in boxes. Each box states an *effect* with *causes* of that effect stated in boxes connected by arrows from the right.

Figure I.9   Overall cause map for low schedule compliance.

Each arrow leads to an effect box from a cause box. Because each cause can also be an effect, each cause can have proceeding cause box(es) leading to itself. One of the chief values of a cause map is that it clearly shows the interconnecting relationships of all the issues leading to an ultimate result or effect. The Fig. I.9 cause map identifies and discusses most of the likely root causes of low schedule compliance and the following discussion reviews portions of the map and suggests likely solutions to the lower level underlying or root causes. The readers can compare the map with their own plants and experience to help resolve their own low schedule compliance situations.

To further explain how a cause map works, example Fig. I.10 shows the *effect* of a man tripping was *caused* both by a rake on the ground and by a man walking. Furthermore, the rake on the ground is both a cause and an effect. The rake on the ground is an effect caused by someone not putting the rake away earlier. Note also that the final effect had two causes, both of which were mandatory to cause the effect. One significance of knowing that an effect had multiple mandatory causes is that if *any one* of these multiple causes had been eliminated, the final effect would have been avoided.

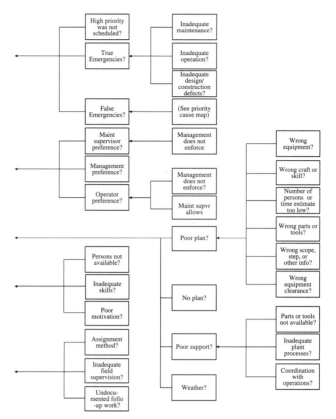

Figure I.9 *(Continued)*

This logic aids the investigator in finding economical solutions to plant problems mapped out in cause maps. The investigator seeks to identify all the mandatory causes including their preceding causes. With the map showing how all the causes fit together, the investigator can generate solutions to the most easily avoided causes that would eliminate entire branches leading to the final effect. Then the investigator can compare the different possible solutions and implement the more economical or practical solutions to avoid the final effect. In the example cause map, a solution to have persons put away rakes after use might be more practical than attempting to keep persons from ever walking in the general area.

A cause map usually deals with a specific occurrence such as a single week having low schedule compliance. In such maps, the investigator maps all the contributing causes in their mandatory relationships. In contrast, Fig. I.9 shows an overall perspective of different reasons a week *might* have low schedule compliance. Therefore, those possible causes that may or may not exist for every instance have question marks after cause statements. To find the actual causes for a specific plant or specific week, the plant should gather data to support which questionable causes are significant.

Figure I.10 Simple cause map showing a middle box to be both an effect and a cause.

The following discussion examines the cause map for low schedule compliance in Fig. I.9 by reviewing each portion of the map one at a time. The review begins with the final effect (low schedule compliance) caused by three mandatory causes, then follows each cause back through its preceding mandatory and possible causes. Solutions are often obvious to the lowest level causes as discussed below. What the investigators need to do is determine which possible causes apply to their particular plants.

Major causes

The first portion of the map shows the three major mandatory causes of scheduled work not being done that lead finally to low schedule compliance. The top box shows that for scheduled work not done, obviously there must have been work scheduled. Management's directive with productivity in mind caused this effect. The other two causes leading to scheduled work not done are more interesting. First, the crew did not do all

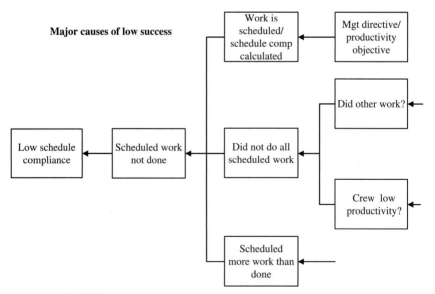

Figure I.11 Cause map portion for major causes.

the scheduled work and second, there was more work scheduled than was done. These two causes might seem a duplication of each other, but they are not. See how each leads from a different rationale of causes. The effect of a crew not doing all the scheduled work must be the result of either a cause of doing other work instead of the scheduled work or a cause of having low productivity (or both). On the other hand, considering the effect of having more work scheduled than was done leads from another set of possible causes altogether as pointed out by Fig. I.12.

Overloaded schedule

Figure 1.12 shows that scheduling more work than is being done could result from a number of possible causes. The top box suggests that perhaps time estimates on planned work have been too low in general. This would cause a scheduler to schedule too much work because the scheduler matches planned hour estimates against the forecasted craft hours available for the next week. The scheduler would put too many jobs in the weekly schedule. This is a very real possibility because companies should train planners to plan jobs for "good" technicians with "no unanticipated" delays. Many, many jobs are not assigned to "good" technicians or do experience unanticipated problems. Nonetheless, this is not a "problem." The purpose of measuring schedule compliance is to help management find opportunities to improve productivity. Management wants to know how many jobs did not run smoothly and why. The low schedule compliance score should lead to management questions. The questions in this case might lead to identifying that too many technicians do not have "good" skills and need training or that too many jobs run into

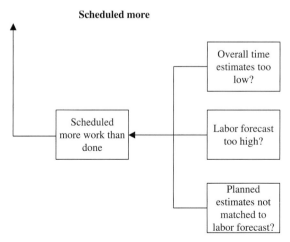

Figure I.12 Cause map portion for scheduling more work than is done.

unanticipated problems. These problems might involve a lack of parts, tools, or other necessities to complete jobs in a timely manner. A low schedule compliance score helps point managers in the proper direction for resolving them. Continuing to measure schedule compliance gives management an indicator of success in finding and dealing with such problems. Please note that in such cases, a plant does not want planners to add extra time for lesser-trained craftpersons or unanticipated delays as these practices would defeat the purpose for having standards, scheduling work, and measuring compliance against a schedule.

Moving on in the cause map, the middle box suggests that the schedule contained too much work because the labor forecast was too high to begin with. If the crew supervisor did not account for vacations, training, meetings, or other absences in the forecast, then the resulting schedule would include too much work. The crew supervisor should carefully consider all known instances when technicians would not be available for work the next week on an hour-by-hour basis. If the crew normally has an hour-long crew meeting each Wednesday after lunch, the forecast should reflect an hour not available for each crewmember. Otherwise, the scheduler would try to schedule an hour too much work for each crewmember resulting in a lower schedule compliance score. The purpose of the schedule compliance score is to lead management to find problem areas for productivity. Crews having weekly meetings for valid reasons should not cause schedule compliance to suffer. The forecast should have included the set apart time to begin with. The forecast should account for all known unavailable time. On the other hand, unexpected absences, short-notice training, or short-notice meetings are not known in advance and thereby "appropriately" lower schedule compliance. Management needs to know if there is a significant problem with absenteeism, training sessions without notice, or excessive last minute meetings. The schedule compliance score does its job in directing management to knowing what is going on with its workforce. The bottom box in the cause map leading to more work scheduled than done is that of the scheduler not matching planned estimates to labor forecasts at all. Many plants use a standard or average forecast. Using such forecasts avoids having to work with the supervisors and speeds the scheduling effort. However, the actual craft-hours available for individual crews vary greatly from week to week as training and vacations arise. These events must not exert undue influence on the schedule compliance score and obstruct management's view of actual maintenance problems. Management should ensure that the crew supervisors always provide valid forecasts to the schedulers. Indeed, a low schedule compliance score might lead management to consider an abnormally high absenteeism rate.

Crew not making it

Figure 1.13 shows that a crew not doing all of its scheduled work could be caused by either the crew doing other work instead of the scheduled work

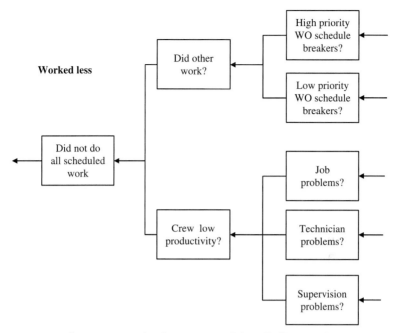

Figure I.13 Cause map portion for a crew not doing all of its scheduled work.

or by the crew not working as productively as expected (or both). In turn, the top box of doing other work is obviously caused by schedule breakers. These schedule breakers could be high priority work or low priority work. Figures I.14 and I.15 examine each of these possibilities. Figure I.13 also shows that the secondary bottom box for crew low productivity could be caused by problems with the job itself, or with the technicians or supervisors. Figures I.16, I.17, and I.18 examine each of these possibilities.

Schedule breakers

Beginning with Fig. I.14, this figure evaluates the effect of a crew doing other work than scheduled work due to high priority schedule breakers. Considering the top box, while it is obvious these schedule breakers were by definition not scheduled, management should see if the work was already known at the time of schedule creation. Schedulers must include all known higher priority work when they create schedules. Not including known urgent work makes the schedule irrelevant. Schedulers must include the highest priority work still in the backlog when creating schedules. The other two causes of high priority schedule breakers are each only possibilities. The schedule breakers might be true emergencies or false emergencies.

The case of true emergencies or urgent work is the ultimate purpose of tracking schedule compliance. Crews cannot comply with the schedule

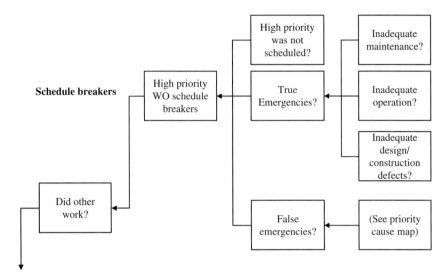

Figure I.14   Cause map portion for high priority schedule breakers.

when urgent work arises. As expressed in Chap. 4, the crew must break the schedule to react to urgent work. The ultimate reason for *having* the measure of schedule compliance is to gauge the degree of the reactiveness of the plant. The lower the schedule compliance, the more reactive is the plant in the nature of its workload. This is management's success indicator for implementing proactive maintenance strategies. Is the maintenance force controlling the plant, or is the reactive nature of the plant controlling the maintenance force? A high schedule compliance score should indicate the former and a low score the latter. In the case of these true emergencies, several possibilities exist. Obviously, true emergencies are caused by inadequate maintenance, inadequate operation, or inadequate equipment (design or construction).

First, inadequate maintenance results in emergencies when PM does not properly prevent equipment problems, PM inspections or PdM detection does not catch work in time, or previous maintenance work was poorly executed. PM's should be in place to lubricate and otherwise service equipment to reduce failures. Some PM's should also be in place to catch problems before they cause high priority work orders. PdM should also catch problems before they affect plant reliability. In addition, just because a plant performs preventive and predictive maintenance, does not mean that they are the right PM and PdM tasks. Having the proper PM's and PdM tasks in place is the purpose of reliability centered maintenance (RCM). Poor execution of maintenance work leading to rework and other premature failures also creates high priority schedule breakers.

Second, inadequate operation could also cause excessive amounts of reactive maintenance. Does management adequately train or staff operators to operate the plant correctly? Do the operators catch and report problems as soon as they notice them? Do the operators have the mind-set to wait on reporting minor problems until they become emergencies?

Third, inadequate equipment in place could ultimately cause the low schedule compliance through either design or construction. Investments in better plant design frequently improve reliability and schedule compliance by reducing emergency work.

Thus, the schedule compliance score gives management its ultimate indicator of the health of the plant. The manager interprets the score to gauge the reactiveness of the plant. Following up the precise causes of the emergencies causing low schedule compliance allows the manager to make changes in the plant proactive maintenance strategy through better maintenance, operation, or equipment. The plant manager uses the weekly schedule compliance scores to measure and track subsequent improvement.

The lowest box in Fig. I.14 considers the possible cause of false emergencies causing high priority schedule breakers. In other words, crews break the planned weekly schedule to attack new highly prioritized work, when in fact the work could have waited and not caused the crew to break its schedule. The work was not that important, but was given a higher priority than it deserved. This is such a pervasive problem that this appendix dedicates an entire separate cause map to investigate abuse of the plant priority system (beginning with Fig. I.19). This appendix presents and discusses the priority system cause map following the completion of the discussion of this map for low schedule compliance.

Figure I.15 covers another circumstance altogether for breaking the schedule with other work, that of low priority work orders breaking the

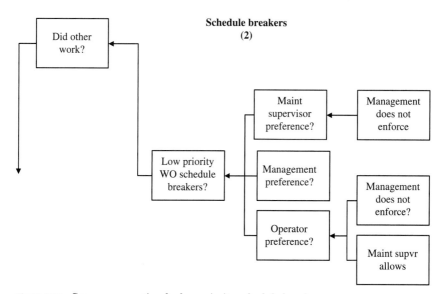

Figure I.15   Cause map portion for low priority schedule breakers.

schedule. In other words, the crew abandoned its planned weekly schedule for work that was not urgent. John E. Day, Jr. (1993) points out that individuals naturally prefer to work on things they enjoy, things they are good at, or things they think are important before they would work on things that the plant considers important. This is one of the purposes of having a published schedule of work, to align everyone to focus on agreed work orders. This could happen on a technician basis, but the top box of this portion of the cause map considers that the supervisor controls the crew's choice of work. In general, the underlying cause of supervisors breaking the schedule for low priority work is inattention on the maintenance manager's part. Management must make the schedule important to the supervisor. This is easier said than done, because plants should not tie schedule compliance directly to supervisor pay. Supervisors must feel free to break the schedule for urgent work. Nevertheless, managers should expect the supervisor to account for schedules not completed and explain other work that broke the schedule. A manager should recognize low priority schedule breakers and coach the errant supervisor. Some plants use a complementary report to the schedule compliance indicator. This report lists all the scheduled jobs done, all the scheduled work not done, and all the nonscheduled work done. The report lists all of the work priorities with the work orders and managers look for low priority work orders showing up in the last portion of the report when the scheduled work was not all completed.

In addition, management allowing the maintenance supervisors to break the schedule for low priority work, management itself sometimes orders that low priority work break the schedule. Management sets a bad example in such cases and encourages others such as engineers not to think ahead and request that special work be included in the next week's schedule. Hopefully, management would be leading the reliability and productivity efforts in the plant. But if not, some of the best counters to such a problem is including management in the weekly scheduling process and publishing both the weekly schedule and the schedule compliance scores. Publishing the schedule reminds management that work wanted should be on the schedule. Publishing the schedule compliance score reminds management that the success of the schedule is important. Chapter 3 Schedule Principle 3 also gives a case for the supervisor protecting the schedule by requiring the interrupting party to prove its case for why the work cannot wait. This cause of management interrupting schedules is very dangerous because it cannot be easily avoided and signals that management may not be firmly behind true schedule success.

The bottom box in this portion, operator preference, is also a potential problem. Many plant operators feel that it is their job to tell the maintenance force what to do each day. Fortunately, this can only be a problem if the supervisor allows it. Nevertheless, many plants have operators empowered by plant management to do exactly that. Management orchestrates the operations led interruption of scheduled work, usually through morning meetings. Management must first buy in to the concept of higher productivity from working to a weekly schedule and then management must not allow operations personnel to dictate low priority jobs for immediate work. Of course, management and maintenance must assure operations that high priority work gets quick attention while insisting that low priority work must wait. Morning meeting can be a real problem if not properly managed. This meeting should be to assess proper priorities and assure recognition of urgent work, not facilitate a routine time for operations to dictate which jobs maintenance should do today. Operators might also dictate routine schedule interruptions for lower priority work because they distrust the work order system. Are jobs done "now, tomorrow, or never" in the minds of the operators? Look for ways to allow operators to see the status of their jobs in the computerized maintenance management system (CMMS). Involving the operations group in the weekly schedule process also allows them a proper time to push lower priority work in the context of all the plant work. Publishing the weekly schedule batch of work also reminds the operators of the process and success of maintenance completing lower priority work. Above all, keep in mind that pushing a weekly schedule increases maintenance productivity and ultimately, the completion of more low priority jobs. Again, Chap. 3 Schedule Principle 3 presents a case for supervisors protecting the schedule.

Low producing crews

Figure I.16 addresses a familiar problem leading to low producing crews, namely unanticipated job problems. They are unanticipated by definition because planners should include time in their estimates to deal with expected delays such as getting known parts out of the storeroom. Jobs run into problems when the plans are wrong, the jobs do not have a plan, the plant has poor support processes, or even when the crews experience weather problems.

First, jobs can run into problems when the job plan is wrong. The job might have had the wrong equipment indicated and so technicians took longer than expected to straighten out the mistake. (They may have even worked on the wrong equipment, but that would not have necessarily delayed their productivity in working on scheduled work.) The planner might also have specified the wrong craft. Sending out a mechanic for a certified welder job naturally delays jobs and interrupts productivity. The mechanic soon finds out that the job requires a welder and returns to the supervisor. By then, the welder is already elsewhere and the day's schedule of work is disrupted. These matters of reassigning work create even more severe problems when they involve extra coordination with operations. Planners specifying too few persons also worsen productivity. The supervisor assigning two persons to a job planned for

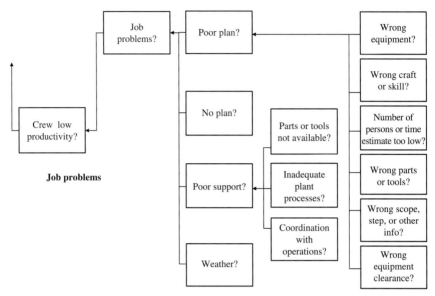

Figure I.16   Cause map portion for job problems leading to low crew productivity.

one, cuts productivity by half. Similarly, a job planned for too few hours impacts productivity by definition: the actual hours take longer than the estimate. Planners might also specify incorrect parts or tools requiring the technician to spend extra time finding the correct items. The overall thrust of the planned job might also be incorrect. The technician after opening a pump casing might find the planned work of replacing an impeller to be woefully inadequate. The pump might need new bearings and a refurbishment of the casing itself. Even if the planner made the right call on the scope of the work, individual steps of the procedure might lead a technician astray and require extra time to correct. Finally, a planned job might incorrectly assess the clearance requirements of a job. Clearing a piece of equipment that did not need clearance for repair, or vice-versa, might cause confusion and take extra time to remedy. All of these problems with job plans are not catastrophic or even a problem necessarily. Proper planning expects technicians to find and report problems with job plans. Planning Principle No. 2 explains that maintenance forces work on the same equipment repeatedly over the course of several years. Instead of expecting to provide perfect plans, proper planning expects to receive technician feedback and be a faithful library service to avoid repeating the same mistakes on the same jobs. Management should ensure the feedback loop is working and not get overly concerned with mistakes on single jobs. On the other hand, repeated poor plans might indicate a more serious problem with planner qualification. Chapter 10 explains the seriousness of having qualified planners. Thus, a number of wrong elements in job plans might lower productivity, but management should primarily concern itself with insuring that planners are improving plans as problems occur.

Second, plants experience lower productivity on unplanned work than planned work. Because many schedule breakers are unplanned, they worsen schedule compliance in two ways. As discussed above, they take the place of scheduled work. Nevertheless, they also carry additional productivity destroying activities of forcing the assigned technicians to solve nearly all the problems that might have been headed off by prior planning. The operator might have identified the wrong equipment. Now instead of a planner having the chance to waste time straightening out the discrepancy, an entire crew might sit idle (or even worse, a plant could be idle) while assigned technicians scramble. Equally damaging to productivity might be wrong guesses about the number of persons and skills to assign or if the equipment requires clearance. The technicians also have no planner head start from looking at history for parts, tools, scope, or procedures. The worse the schedule compliance because of schedule breakers, the more impact unplanned work might have on productivity and schedule compliance in turn.

Third, planners plan for "smooth" jobs to set a standard for good performance and so that technicians are more likely to recognize and report delays in feedback for jobs that take too long. If the maintenance force has genuinely poor support, then productivity suffers. The planners might plan for such problems as anticipated delays and they may increase many job time estimates, but poor support situations are so serious they significantly jeopardize overall productivity. The top box showing a possible underlying cause for poor support is with obtaining parts and tools, whether they are obtained from storerooms and tool rooms or they are specially procured for tasks. Management inquiry into low schedule compliance may quantify the effect of having an inadequate storeroom or tools. Planning might help bypass some of these problems by anticipating them before they are scheduled and assigned. However, if planning must spend an extraordinary amount of time on too many jobs to provide the most basic parts and tools, the planners either cannot plan all the work or must severely restrict their efforts planning many jobs. Some of the unplanned jobs end up leading to becoming unplanned schedule breakers. Some of the jobs planned with less-than-desired attention lead to problems when put into progress. Consequently, poor plant support for parts and tools leads to low productivity and ultimately low schedule compliance. The next box showing a possible underlying cause for poor support is that of other inadequate processes other than for parts and tools. Perhaps tools and parts are available, but processes to check tools in or out or processes to obtain the parts or return unused parts are cumbersome. Perhaps the CMMS is unwieldy and not everybody on each crew can find or use information in it. These inadequate plant processes are candidates for continual improvement

efforts. The last box considered a possible underlying cause of poor support, that is coordination with operations. This is a significant concern
in many plants. Operations must act on advance notice to have equipment ready for maintenance service. Needlessly having a group of technicians wait for operations to clear equipment wrecks productivity.
Management can often see problems in this area from wrench time
studies where operator delays form a category. On the other hand,
repeated technician and supervisor complaints should reach management well ahead of any sophisticated studies.

Fourth and finally, in the underlying causes for job problems is the
environment, usually weather. A planner would not typically anticipate
an entire week of inclement weather that drastically slows work. Yet
weather does influence maintenance and sometimes it affects entire
weeks. This is a good opportunity to present the case that management
should not overly concern itself with a single week of low schedule compliance. Management should principally concern itself with trends and
the identification and correction of system problems. This is true of any
data set and statistical analysis. If there is an outlier, see if a special
cause deserves special attention, otherwise go on.

After unexpected job problems slowing productivity, Fig. I.17 examines problems with technicians themselves. Persons might not be available for a number of reasons. The weekly forecast expected their
presence, but excessive absenteeism could cause continual problems.
Perhaps management subjects the crews to many unexpected meetings. Does the plant suddenly assign technicians to unexpected training? Supervision and management should become aware if these issues
are influencing schedule compliance. Next, consider that even if persons
are available, they may not have the correct skills. The supervisor must
carefully fill out the weekly schedule forecast form so that the scheduler
is aware of what skills are available. Management should investigate
a frequent mismatching of skills with an eye toward the proper use of

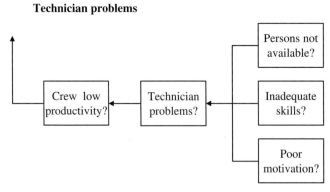

Technician problems

Figure I.17   Cause map portion for personnel problems leading to low
crew productivity.

the forecast and job matching routines within the weekly scheduling process. Finally, poor motivation is a very real underlying cause for technicians causing low crew productivity. Poor motivation could stem from any number of reasons that management must uncover and address. The scheduling process at least sets forth the expectation of how much work is required allowing management to quantify productivity problems to some degree. Without a set goal of work through a schedule, management do not even know that productivity had suffered.

Figure I.18 examines the last portion of possible underlying causes of low crew productivity, namely those due to supervisor problems. Supervisors might cause low productivity through their assignment methods, their field supervision, or their allowing undocumented follow-up work. The top box in Fig. I.18 suggests that supervisors might cause low productivity through their assignment methods. The entirety of the scheduling process endeavors to help supervisors assign the proper amount of work. Chapter 6 and 7 address assignment methods that may cause low productivity. The middle box points at field supervision as a possible cause for low productivity. Many plants tie up their supervisors in endless administrative or clerical tasks. The primary tasks of supervisors should revolve around being in the field with their technicians they supervise. As obvious as this sounds, many supervisors cannot find the time to be in the field with their crews helping them overcome obstacles. Managers must facilitate crew supervisors being with their crews to facilitate them in being productive. Finally, undocumented follow-up work upsets productivity. This is similar to low priority work breaking schedules. Many times technicians find additional work during their assignments. If they can easily execute the additional work in the course of their current assignments without greatly altering their schedules, all is well. However, if the technicians proceed with follow-up work to the

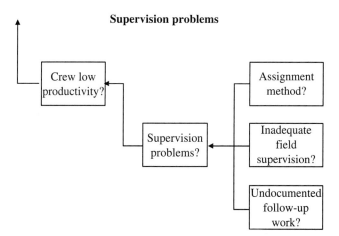

Figure I.18 Cause map portion for supervision problems leading to low crew productivity.

neglect of the overall schedule, schedule compliance suffers. This problem occurs most noticeably on inspection oriented PM's. As technicians inspect equipment and uncover problems (as intended), technicians should make minor adjustments and equipment corrections as they go along. Nevertheless, if the problems uncovered would take more than a specified moment to correct, the technicians should write work orders for their correction later. The chief concern is that a technician becoming overly concerned with correcting a problem here and there might not complete the entire PM inspection and leave other problems unfound altogether. This problem extends to routine maintenance beyond PM's when technicians find problems. There is a point where future work orders should address the problems instead of corrections on the spot.

Priority Systems

Figure I.19 shows a cause map for setting false work order priorities. This map examines why persons might abuse the plant priority system. Similar to the cause map for low schedule compliance, readers can compare this map

Figure I.19 Overall cause map for incorrect priorities.

with their own plant experiences for a head start to resolve priority system problems. Many different possibilities exist for valid priority systems. Earlier, Fig. I.10 explained how a cause map works. The priority system cause map in Fig. I.19 identifies the issues that any adopted system faces.

The following discussion reviews and discusses the priority system cause map in portions offering solutions to specific causes as appropriate. The basic causes are evident, but the in-depth review of the underlying causes beyond these basics is necessary to find specific areas for management to address to correct priority system abuse. Solutions are often obvious to the lowest level causes as discussed below. What the investigators need to do is determine which possible causes apply to their particular plants.

Major causes

The first portion of the map (Fig. 1.20) shows the basic causes of how plants assign incorrect priorities to work. Cause maps run from left to

Figure I.19   (*Continued*)

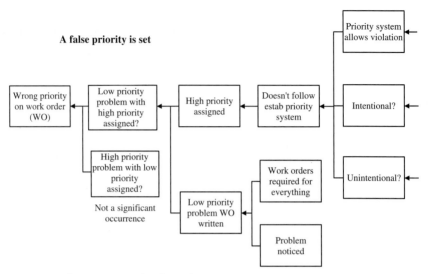

Figure I.20   Cause map portion for major causes.

right and this analysis starts by looking at the first box on the left. The first box states the ultimate effect, namely a work order with the wrong priority assigned. Two immediate causes lead to this ultimate effect. Both high priorities assigned to low priority problems or low priorities assigned to high priority problems can cause incorrect priorities on work orders. The cause map immediately dismisses plants giving low priorities to important problems. The format of a cause map places data supporting the cause underneath the box. The statement underneath the box of high priority problems being assigned low priorities simply states that this is not a significant occurrence. The other box, low priority problems with high priorities assigned, has two supporting causes. The top cause box shows the originator assigned a high priority and the bottom cause box shows the originator wrote a work order for a low priority work order. The bottom effect box resulted from two mandatory causes. First, management requires work orders for maintenance work and second, the originator has (notices) a problem needing maintenance. The cause map provides this line of logic for completeness. Certainly, a plant does not solve the whole priority assignment problem by stopping the writing of work orders or stopping the noticing of problems.

The majority of the cause map follows the logic of the causes leading to low priority problems being assigned high priorities. Obviously, the incorrect priority did not follow the established priority system of the plant. But three preceding causes lead to this nonadherence to the plant priority system. First is the mandatory cause of the plant priority system allowing violation. That is, it is *possible* to assign an incorrect priority. Second is the possible cause that the violation may have been intentional and third is the possibility that it was unintentional. The following sections discuss each of these portions of the cause map.

Choice

Figure I.21 shows the top box and the first major division of the basic underlying causes of incorrect priorities. The system under evaluation allows choice in priority. Some systems do not. Those systems might tie priority solely to a predetermined criticality of equipment, a measure of equipment that might affect plant or unit reliability. Those systems might well place higher priorities on reliability issues, but many times an operator or other originator or work sees instances of true urgency regardless of so-called equipment "criticality." This is especially true of safety issues. A walkway on the thirteenth floor of the boiler might not contribute directly to plant availability, but it might sure contribute to death if allowed to rust away. The plant for which Fig. I.20 is typical allows choice by the first writer of the work order. The system also allows choice by reviewers of the work order who may change the priority based on their superior knowledge of the overall backlog or relative plant system requirements. In other words, other work orders in the backlog truly are more important or the true functional requirements of the plant make the work order more or less important than originally thought. Nevertheless, the priority system allows the originator of the work, the person on the scene who first identified the problem to choose a priority and the work process system allows reviewers to modify the priority later. The CMMS allows this choice making by persons rather than a predetermined equipment criticality formula. The plant considers that the originator can make the best initial assessment. If the plant considers this area to be the most significant problem in setting correct priorities, management could authorize changing the CMMS to set priorities automatically, but this is not recommended. Some CMMS systems actually programmatically override the originator and review committee assignment of priority when the planner attaches a job plan. It is considered unwise for a priority in saved job plan from a databank to erase a priority set by an originator that was

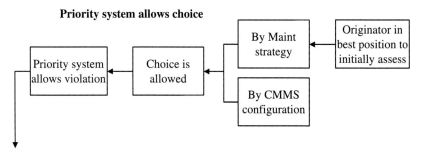

Figure I.21   Cause map portion for priority system allowing violation.

aware of unique plant conditions at the time of writing the work order. It is true that the CMMS may do a better job of remembering relative importance of different equipment and systems set in advance by plant engineering. Nonetheless, the CMMS should suggest, rather than dictate, criticality into a priority selection. A better solution might be for the plant to review its priority system choices rather than not allow choices. Several of the following branches of the cause map discuss complexity and relevancy of the priority system choices.

No priority system in reality

Skipping to the bottom box of Fig. I.20, Fig. I.22 shows another major division of the basic underlying causes of incorrect priorities. This box's statement has "Unintentional?" with a question mark to mark it as a possible cause of not following the priority system. An originator might unintentionally not follow the established priority system. Looking at the uppermost boxes in Fig. I.22, the originator might not be aware a priority system exists. Perhaps the plant has a paper work order system with a simple blank for the work order priority. The paper form might not list the alternate choices and there may be limited or training or advertisements around the plant that priorities exist. Presumed here

Figure I.22 Cause map portion for setting false priorities by accident.

is the originator placing a high priority on the work order. The CMMS might not require a priority and somehow a high priority is assigned or the priority field might default to a high priority. Some systems default to a high priority depending on the system configuration. CMMS systems should not default to a high priority.

The lower boxes of Fig. I.22 show the causes associated with an originator being aware of, but not understanding the priority system. The priority system might be too vague because the criteria are not well defined. Solutions to this situation might be giving examples or rewording individual priorities for better definition. Perhaps the plant has never trained operators and maintenance personnel who write work orders on the priority system. They may not need formal training but discussion on the system. The plant might provide notices, reminders, or even feedback on selections already made. On the other hand, originators might not understand the system because it simply is not relevant. The following section for Fig. I.24 discusses relevancy further.

In all of these situations where the originator might set a false priority unintentionally too high, the supervisor must protect the schedule against false emergencies.

Gaming the priority system

Figure I.23 is the beginning of the cause map section for "gaming" the priority system, *the heart of why persons put higher than appropriate priorities on work orders.* The priority system dictates a low priority, yet they put a high priority on the work order. They have a desire to get the work done faster than the priority system demands. Moreover, the plant management does not enforce the priority system allowing the originator to place a higher priority than justified. Without proceeding further, management could resolve this inappropriate prioritizing by enforcing the priority system. One method could be through the much maligned

Gaming the priority system

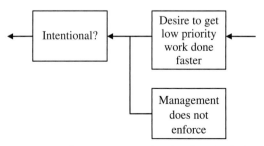

Figure I.23 Cause map portion for "gaming" the system, setting false priorities on purpose.

"morning meeting." Many companies discontinue the regular morning meeting because it simply degenerates over time into an opportunity for operations to direct the current day's maintenance work. Instead, this meeting could be an opportunity to ensure that work orders have the proper priorities and that maintenance has already started work on true emergencies. Nevertheless, rather than dictate a solution by management decree, the overall cause map further develops the underlying causes of intentionally setting higher priorities, specifically the effect of someone desiring to get low priority work done faster.

Figure I.24 examines the case of the work order originator not agreeing with the priority system before Fig. I.25 addresses the case of the originator agreeing, but not following the priority system anyway. A person might disagree with the priority system and set a falsely high priority system for two reasons. One, the person thinks the system is simply not relevant. The system might be too complex or too simplistic. If the system is too complex, the originator simply errs on the side of getting the work done. Writers of work orders might not want to wade through endless cases and explanations for setting different priorities. The complexity of the priority system is not helpful for the operator simply saying how much of an emergency exists. Endless cases presented as examples in a priority system are sometimes hard to match with the specifics of a current situation. On the other hand, some systems are too simplistic. These systems might have no allowance for human comfort: the priority system only addresses machine reliability. However, humans do have needs. Priority systems should allow originators to write work orders for some creature comfort tasks. The originator might also perceive the system as "urgent or maybe never." The system might not expressly say this, but in practice, many plants address work orders either as work needed or work not needed. The maintenance

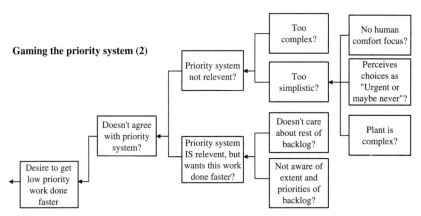

Figure I.24 Cause map portion for gaming the system "because" (of disagreement).

forces sit back on their heels waiting for those urgent work orders and allow the other work to sit in the backlog. The obvious solution to such situations is to establish scheduling so that the plant selects all the work that it can do in order of priority, but not stopping after the urgent work. The plant might be too complex for the plant's simple priority system. Does the system allow the originator to distinguish relative priorities among work for the utility's most important unit versus other units? If the plant has several product lines, does the originator know which is most important? Therefore, the plant's priority system might be too complex or too simple. The plant might have tried to prepare an adequate system for a complex plant, but ended up making it too complicated to be practical. Alternatively, the plant might not have made the system sophisticated enough. With such trade-offs, the perception of "now or never" becomes evident in practice. Management must provide a priority system that addresses concerns other than strict consideration of plant reliability. Plants must operate safely with humans within environmental and legal constraints. Priority systems must allow for all of these aspects.

Make the priority system as simple as possible, but not too simple. Use at least five choices, each containing a few key words to describe its relative importance. Examples of work for each choice should be readily available, but not necessarily built into the choice description. Having as many as ten choices could be helpful in encouraging originators to make choices away from the most urgent priorities; the extra choices allow a wider spread of priorities. Suppose that in a choice of ten priorities, no one ever chooses less than a seven (with ten being the least important). This might be a very successful system because in a system with only seven choices, no one might ever choose less than a five. The extra three choices in the system of ten allow a greater spread. The spread is what schedulers value because they can determine the relative value of work.

Time based systems might also be more relevant. These systems might assign maximum times for maintenance to respond or complete work. Consider the following:

Priority 0   Emergency. Respond within 1 hour.

Priority 1   Urgent. Complete within 48 hours.

Priority 2   Serious. Complete within 14 days.

Priority 3   Routine. Complete within 28 days.

Priority 5   Routine. Requires special arrangements. Complete within 6 months.

The above system dictates time requirements. The danger in such a system is when maintenance suddenly has more work than it can complete in 2 weeks. Does that automatically authorize overtime or the use of contactors or does management allow the maintenance force to complete them as best as it can? Does management merely track that the average completion time of priority-2 jobs is 20 days? If management routinely allows taking longer than the priority-dictated target, does that damage the credibility of the priority system? Does it encourage originators to choose priority 1 even if they could wait 14 days (but not 20 days)? Furthermore, how can an originator claim with certainty that maintenance *must* complete the work within a certain timeframe? In truth, an engineered economic analysis might establish a timeframe. Such an analysis might say that the lost revenue of the equipment is worth hiring outside labor for the repair in "10 days as a breakeven point." This is not very practical for day-to-day maintenance even though in theory it could be set up in advance on a CMMS (back to the criticality approach). Wouldn't it be more prudent simply to allow the maintenance group to work on the highest priorities first? This would let the amount of work in each priority dictate how long it will take to complete any work. In actual practice, operator knowledge of the relative importance of problem is probably practical. Management could supplement this approach by allowing operations or maintenance to intervene specially and identify certain work that maintenance *must* complete by certain deadlines.

The most relevant ("best") priority systems probably take advantage of multiple facets of the maintenance process rather than count on a single listing of priority selections. Originators make a best guess at the overall priority of the work among 5 to 10 choices. A brief morning meeting including key operations persons reviews priorities. Planners may question priorities as they plan. Supervisors protect their schedules daily against false emergencies. Management supports maintenance in not letting supervisors abuse the priority system. Supervisors also review the proposed weekly schedules for appropriate work orders. The operations group reviews the proposed weekly schedules for appropriate work orders. Just as maintenance planning is not the "silver bullet" for maintenance, a list of priority choices on its own cannot fully make a system "relevant."

Moving on, the originator might agree with the priority system, but wants maintenance to do this work faster than the priority system allows. The originator might not care about the rest of the backlog. This selfish approach says, "Do my work now." Then again, the originator might not be aware of other work in the backlog and simply want the work done. Making the backlog work more visible might help originators lower their priority selection.

Finally, Figure I.25 probably addresses more significant reasons why originators favor "gaming" the priority system. The originator agrees with the priority system, but thinks that maintenance simply might not respond in time to low priority work. The originator might think there is a very unpredictable time for maintenance completing low priority

Gaming the priority system (3)

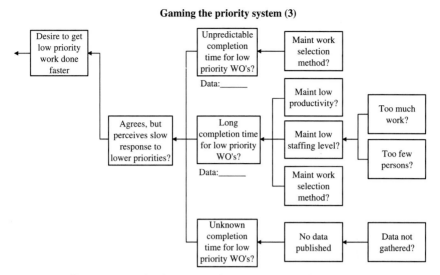

Figure I.25   Cause map portion for gaming the system "anyway" (in spite of agreement).

work. Maintenance might complete it tomorrow, next week, next month, or next year without any prediction of the actual time. Maintenance might have a work selection method that is very unpredictable. It might pick a low priority job or it might not. Under such a scenario, the originator would pick a higher priority than necessary for work that is not urgent but does need attention before it is too late. Then again, the originator might not perceive an unpredictable completion time; the originator might know it will take a long time to complete any low priority work. Maintenance might have a hard time getting to lower priority work because it has a low productivity, it has too few persons, or it simply never selects low priority work. In the case of too few persons, the mandatory underlying causes are too much work for too few persons. In these cases, management must recognize when sufficient productivity is present, but maintenance resources are exhausted. Management might not have replaced persons who retired and it has reached the point where it does not have enough persons. Management cannot adjust a priority system to compensate for not having enough persons. Alternatively, instead of a "known unpredictability" or a known long time for completion, the originator simply might not have any idea how long it takes to complete low priority work. Originators never know if maintenance ever does nonemergency work because maintenance never publishes (or gathers any data to publish) any information about routine repairs. Perhaps maintenance does not return a copy of completed work orders to originators or there is no CMMS to query. Originators might feel that maintenance might not ever complete any work not prioritized.

The modern CMMS allows more access to work completion data, but management might go further and publish expected completion times for each priority level of work. It might also report to originators notice of completion of individual work orders to overcome the perceptions of unpredictable, long, or simply unknown completion times for routine work.

Figures I.19 through I.25 cover the most probable reasons for setting false priorities on work orders. From allowing choice on priority selection to understanding unintentional and intentional violations, management has a number of options to manage the priority system. Perhaps the most pervasive problem is operators "gaming" the priority system. The *Maintenance Planning and Scheduling Handbook* contends that improving maintenance productivity to complete all work orders, including those of low priority, helps make the issue of setting priorities irrelevant. With a highly productive maintenance workforce, operators need not game the work order system to have maintenance address low priority work.

Summary

One sees that there are additional issues and situations affecting how planning and scheduling affect productivity. One of these is higher wrench time experienced under certain conditions without significant planning and scheduling. Certain crafts have somewhat higher wrench times than industry averages for overall maintenance. However, even these craft are helped by maintenance planning. Plants experiencing extreme trouble also often do not have poor wrench time. Yet as these plants try to increase their reliability, they must employ planning and scheduling to maintain a high productivity. They must maintain higher productivity to allow completion of more proactive work. Another issue is the interference that blanket work orders cause the planning and scheduling effort. Management should eliminate or greatly restrict the use of blankets. Finally, management should understand and address, if necessary, empowerment and common failings of schedule success and priority systems.

Work Order System and Codes

Starting a work order system is the most important improvement one can make to a maintenance program. The work order system is the process by which the maintenance manages all plant maintenance work. The system assists the plant in keeping track of, prioritizing, planning, scheduling, analyzing, and controlling maintenance work. The plant must have a viable work order system as a foundation to planning.

This appendix illustrates a typical manual a company would use to document its work order process. Notes have been inserted to call attention to particular details. Example work orders throughout this book have used the codes from this appendix. The manual should primarily illustrate the work order form, the work order process, and identify the codes used in the system. As a computerized maintenance management system (CMMS) becomes developed, it is not unreasonable to expect that the CMMS could contain the essentials of the work order manual and replace it.

This example company uses a paper work order system for persons that originate work orders. A clerk enters the work orders into the CMMS system. The plant is currently considering creating a CMMS screen to resemble its work order form and allowing persons to enter their own work orders directly into the system.

It is a good idea to have a document that sets forth the rules of using work orders. It is not necessary that the document be exceedingly thick.

COMPANY WORK ORDER SYSTEM MANUAL

Table of Contents

Introduction

Work Flow

Introduction

The work order system is the process which the Maintenance Department uses to manage all plant maintenance work. The work order system assists the plant in keeping track of, prioritizing, planning, scheduling, analyzing, and controlling maintenance work. (The terms *WO* and *work order* have the same meaning.) A major purpose of using work orders for plant equipment is to be able to track its history. Using blanket work orders or having several pieces of equipment on the same work order destroys the process of keeping history.

The *work flow diagram* shows the steps of the cycle that occur from initiating a work order to work completion.

The plant uses the work order form as the document to record information associated with executing the work request.

After work completion, WO forms which have historical value are filed to assist future work. Codes are used with each work order to allow various sorts and analysis of maintenance work. For example, the outage code allows sorting out all work orders that must be done during a unit outage.

The CMMS (computerized maintenance management system) allows computer tracking and analysis of work orders as well as plant equipment data. The system is on the personal computer (PC) network and may be accessed from any plant.

The WO numbering system provides for assigning each separate work order a unique number to allow keeping the work done under that number separate from other maintenance work.

A *WO System Manual* is distributed to each person on the manual distribution list who is responsible for keeping it up to date with all issued revisions. The Job Planning Coordinator coordinates and distributes all revisions to the *WO System Manual.*

Work Flow

Work flow diagrams show the steps of the cycle that occur from initiating a work order to work completion. Figure J.1 shows the normal steps of the work process. Figure J.2 shows the steps taken during emergencies. In emergencies, action begins with verbal instructions and the paperwork follows later.

Work Order Form and Required Fields

The WO form is the document used to record information associated with:

1. an identified problem, need, or work request
2. the work authorized to be done
3. the plans, details, and schedules necessary to perform the work
4. the results of that work

After work completion, the planning department files work order forms which have historical value to assist future work.

Figure J.3 illustrates the work order form this company uses. Figure J.4 identifies who is responsible for completing each section of the form.

The following information is used by the Job Planning Coordinator when ordering new work order forms from the printer. The JPC gives this information to the company graphics department, printer, or forms order company along with a reproducible original.

General information

All parts are printed in black.

Carbonless paper is used.

Note that the original is reduced in size!

Copy all four parts; enlarge them to have 1/4-inch left and right margin.

Numbering should be in red ink at the top of the form.

No additional numbers should appear anywhere on the form or tab.

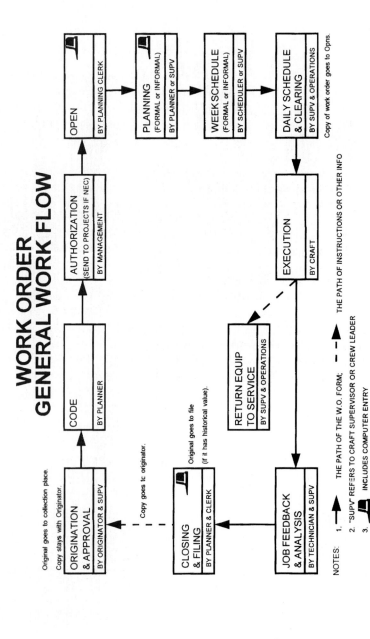

WORK ORDER
GENERAL WORK FLOW

Original goes to collection place.

Copy stays with Originator.

ORIGINATION & APPROVAL
BY ORIGINATOR & SUPV

CODE
BY PLANNER

AUTHORIZATION
(SEND TO PROJECTS IF NEC)
BY MANAGEMENT

OPEN
BY PLANNING CLERK

PLANNING
(FORMAL or INFORMAL)
BY PLANNER or SUPV

WEEK SCHEDULE
(FORMAL or INFORMAL)
BY SCHEDULER or SUPV

DAILY SCHEDULE & CLEARING
BY SUPV & OPERATIONS

Copy of work order goes to Opns.

Copy goes to originator.

RETURN EQUIP TO SERVICE
BY SUPV & OPERATIONS

EXECUTION
BY CRAFT

Original goes to file
(if it has historical value).

CLOSING & FILING
BY PLANNER & CLERK

JOB FEEDBACK & ANALYSIS
BY TECHNICIAN & SUPV

NOTES:
1.   ⬆ THE PATH OF THE W.O. FORM;   - - ▶   THE PATH OF INSTRUCTIONS OR OTHER INFO

2. "SUPV" REFERS TO CRAFT SUPERVISOR OR CREW LEADER

3.   ▟   INCLUDES COMPUTER ENTRY

4. THE PROJECTS GROUP RECEIVES REFERRED WORK AS A CRAFT

5. "INFORMAL" REFERS TO CRAFTS WHERE THE SUPV DOES THE PLANNING OR SCHEDULING

Figure J.1   Company work flow diagram.

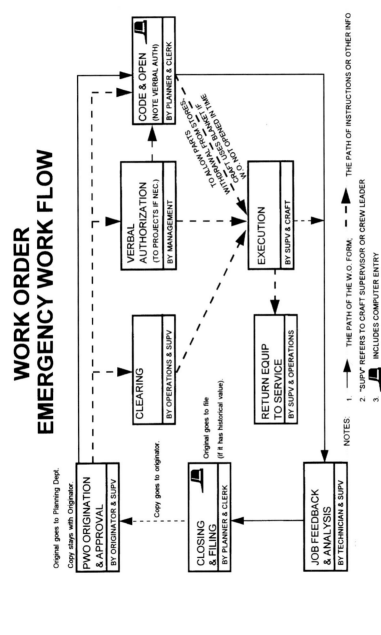

WORK ORDER
EMERGENCY WORK FLOW

Original goes to Planning Dept.

Copy stays with Originator.

PWO ORIGINATION & APPROVAL

BY ORIGINATOR & SUPV

Copy goes to originator.

Original goes to file
(if it has historical value).

CLOSING & FILING

BY PLANNER & CLERK

JOB FEEDBACK & ANALYSIS

BY TECHNICIAN & SUPV

CLEARING

BY OPERATIONS & SUPV

VERBAL AUTHORIZATION
(TO PROJECTS IF NEC.)

BY MANAGEMENT

CODE & OPEN
(NOTE VERBAL AUTH)

BY PLANNER & CLERK

TO ALLOW PARTS
WITHDRAWAL FROM STORES;
BLANKET USES BLANKET IF
CRAFT OPENED IN TIME
W.O. NOT

EXECUTION

BY SUPV & CRAFT

RETURN EQUIP TO SERVICE

BY SUPV & OPERATIONS

NOTES:

1. ——▶ THE PATH OF THE W.O. FORM; ─ ─▶ THE PATH OF INSTRUCTIONS OR OTHER INFO

2. "SUPV" REFERS TO CRAFT SUPERVISOR OR CREW LEADER

3. ⬛ INCLUDES COMPUTER ENTRY

4. THE PROJECTS GROUP RECEIVES REFERRED WORK AS A CRAFT

5. "INFORMAL" REFERS TO CRAFTS WHERE THE SUPV DOES THE PLANNING OR SCHEDULING

Figure J.2 Company emergency work flow diagram.

WORK ORDER

REQUESTER SECTION   Priority __ __

Equipment _____ Tag # _____

Problem or Work Requested: Def Tag #_____

Outage Req? Y/N   Clearance Req? Y/N   Confined Space? Y/N

By: Date & Time: **APPROVAL:**

PLANNING SECTION   Assigned Crew:_____   Attachment? Y/N

Description of work to be performed:

Labor requirements:

Parts requirements:

Special tools requirements:

By: Date & Time: **Job Estimate:**   **Actual:**

CRAFT FEEDBACK (Modify plan sections above: actual labor, parts, & tools)

Work performed including equipment changes & any problems or delays:

Date & Time Started:_____   Date & Time Completed:_____

By: Date :   **APPROVAL:**

CODING

Figure J.3   Company work order form.

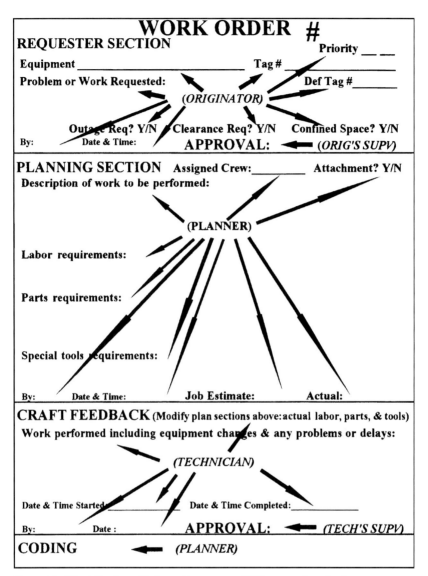

Figure J.4   Identification of information responsibility on work order form.

Numbering sequence should be as follows:

Lot 1: A000 through A999	Lot 2: B000 through B999
Lot 1: C000 through C999	Lot 2: D000 through D999
Lot 1: E000 through E999	Lot 2: F000 through F999
Lot 1: G000 through G999	Lot 2: H000 through H999
Lot 1: J000 through J999	Lot 2: K000 through K999
Lot 1: L000 through L999	Lot 2: M000 through M999
Lot 1: N000 through N999	Lot 2: P000 through P999
Lot 1: Q000 through Q999	Lot 2: R000 through R999
Lot 1: T000 through T999	Lot 2: W000 through W999
Lot 1: X000 through X999	Lot 2: Y000 through Y999

Note: Do not use letters i, o, s, u, v, or z. Package in nearly equal amounts (between 50 and 200); no specific amount is required.

Part 1 information: White stock, $8^{1}/_{2} \times 11$ inch, print the front and back (see sample).

Part 2 information: Pink stock, $8^{1}/_{2} \times 11$ inch, print the front only (see sample).

Part 3 information: Blue stock, $8^{1}/_{2}$ inch $\times$ cut at bottom line, print the front only (see sample).

Part 4 information: Goldenrod stock, $8^{1}/_{2}$ inch $\times$ cut at bottom line, printing front only (see sample).

Contacts: (Job Planning Coordinator's name and phone number.)

CMMS Instructions for Plant-Wide Use

(*Note:* This section should cut down the 1-inch-thick computer manual that came with the CMMS to a few pages of key instructions for non-planning group persons such as most plant supervisors, technicians, operators, and managers.

Codes

Codes are used with each work order to allow various sorts and analysis of maintenance work. For example, the outage code allows gathering all work orders that must be done during a unit outage. Here is an example set of codes for an oil and gas fired electric power station that has both steam and gas turbine generating units. These codes illustrate what codes are used for and look like.

Priority

(*Note:* Priority codes allow ranking work orders in order of importance to know which to handle first. See Scheduling Principle 2 for more discussion of their importance and how incorrect usage can hinder productivity. Organizational discipline that comes through education, communication, and management commitment helps ensure correct usage.)

Work order priority codes.   First digit—priority type. (*Note:* The first digit does not indicate priority preference; each type has equal weight.)

S   Safety

H   Heat rate

E   Environmental or regulatory

R   Reliability or availability

G   General

Second digit—priority order. (*Note:* The second digit indicates priority preference.)

0   Emergency Conditions
    Loss of unit
    Immediate or imminent loss of unit capacity
    In violation of environmental regulations
    Loss of unit load control
    Emergency safety hazard

1   Urgent Conditions
    Significant potential for loss of unit or unit capacity
    Major loss of heat rate
    Significant potential for violation of environmental regulations
    Urgent safety hazard

2   Serious Conditions
    Condition that could cause serious damage to critical equipment
    Serious loss of heat rate
    Condition that could cause violation of environmental regulations
    Serious safety hazard
    PM's

3   Noncritical Maintenance on Production Equipment

4   Noncritical Maintenance on Nonproduction Equipment

(*Note:* Therefore, work orders with "1" priorities should be handled ahead of work orders with "2" priorities and so on regardless of the letter priority type. Having the letters associated with the priorities helps persons better understand the nature and reasons behind the work orders and their numerical priorities. If one carefully examines the point criteria examples, safety is indeed given more weight in that it is easier for a safety (S) work order to receive a 1 score than it is for heat rate, a unit efficiency measure. In a well functioning plant, 0 and 1 work orders should be infrequently received. It is maintenance's objective to have mostly 2 level preventive maintenanace (PM) work orders and 3 and 4 level work orders as the maintenance organization goes about its task of maintaining, not recovering. There is a valid argument that such a short range between 0 and 4 tends to encourage persons to classify work orders as 1 (urgent) and 2 (serious) unnecessarily. For that reason many plants have an extended range, perhaps 0 to 10. They do not get many 10 work orders, but they do get more 3's and 4's and less false 1's and 2's.)

Status

(*Note:* These codes are solely for the CMMS, the simple reason being that the placement of paper work orders determines their status. For example, paper work orders waiting for parts are kept by the purchaser, work orders in progress are kept by the technicians, etc. The CMMS allows the rest of the plant to ascertain where the work orders are in the work order process.)

Work order status codes

WAPPR—Waiting for approval. This status is the initial status of a WO before plant management has authorized the work request.

APPR—Approved. Management has authorized the work request and it is ready for planning.

WSCH—Waiting to be scheduled. The work order has been planned (if planning was necessary) and it is ready to be scheduled for work.

HOLD-MATL—Waiting for material or tools. Materials or tools are unavailable to either start or continue work.

HOLD-OTHER—Waiting for other reason than materials or tools. Work is waiting on engineering or management decision to either start or continue work.

PROJ—Project. This work is being undertaken by the corporate project group. Plant maintenance forces are not involved at this time.

SCHED—Scheduled. Work has been included on the weekly schedule.

ASSGN—Assigned. Work has been assigned or included on the next day's schedule.

INPRG—In progress. Work has already begun.

COMP—Completed. Work has been completed, but the finished work and documentation have not yet been reviewed or analyzed. WO has not yet been closed

COMP-DWGS—Completed, waiting on drawings. Work has been completed, but required drawings have not yet been revised. WO has not yet been closed.

COMP-OTHER—Completed, waiting on other. Work has been completed, but some specific requirement has not yet been submitted. WO has not yet been closed.

CLOSE—Closed. All work and documentation have been completed satisfactorily. WO is closed.

CAN—Canceled. This WO is not considered necessary. WO's may be canceled for a variety of reasons such as the WO may be a duplicate of another WO, the need for this work no longer exists, or plant management has decided not to approve the work order for economic, budget, or other reasons.

Department and crew

(*Note:* These codes when assigned to work orders make it possible to determine and track who originated the work and who is responsible for executing the work. Backlogs and work already performed are easily sorted into crafts such as mechanical and electrical as well as into crews for responsibility.)

Department and crew codes

First digit—Department. Second digit—Crew.

DEPARTMENT 1. Mechanical (Cost Center 1004)
  CREW 0: Overall craft superintendent
  CREW 1
  CREW 2
  CREW 3
  CREW 4
  CREW 5

DEPARTMENT 2. Instrumentation & Controls (CC 1006)
  CREW 0: Overall craft superintendent
  CREW 1
  CREW 2

DEPARTMENT 3. Electrical (CC 1005)
  CREW 0: Overall craft superintendent
  CREW 1
  CREW 2

DEPARTMENT 4. Operations (CC 1002)
  CREW 1: A shift
  CREW 2: B shift
  CREW 3: C shift
  CREW 4: D shift
  CREW 5: Relief crew
  CREW 6: X crew

DEPARTMENT 5. Plant Administration (CC 1001)

CREW 1: Plant Manager
CREW 2: Plant Assistant Manager
CREW 3: Plant Engineers
CREW 4: Administrative Support
CREW 5: Training Department
CREW 6: Planning Department

DEPARTMENT 6. Chemistry and Water Support Division (CC 1012)

CREW 1: Administration
CREW 2: Results Laboratory
CREW 3: Central Laboratory

DEPARTMENT 7. Performance Engineering Division (CC 1032)

CREW 1: Administration
CREW 2: Predictive Maintenance
CREW 3: Diagnostic Engineering
CREW 4: Production Analysis
CREW 5: Equip Database, Tagging, and Schematics

DEPARTMENT 8. Support Outside

CREW 1: Outage Management & Contractors (CC 1021)
CREW 2: Project Group & Contractors (use craft CC)

(*Note:* Thus, a work order coded as written by Crew 4-1 and assigned to Crew 1-3 would be a work order originated by the A shift crew in operations and it was for mechanical type work falling into the Crew 3 area of responsibility in the mechanical maintenance department.)

Work type

(*Note:* The work type coding allows backlogs and work already performed to be sorted to determine how the plant is performing. Is most of the work trouble and breakdown type work or is most of it preventive maintenance? Is there any corrective maintenance to correct situations from turning into trouble and breakdown type problems? The company trying to have a maintenance program and not just a repair service considers this type of information.)

Work type codes

1. *Spare equipment (noninstalled).*   Maintenance activity performed on noninstalled (spare) plant equipment. Rebuilding rotating spares is included.
2. *Structural.*   Maintenance activity performed to maintain general plant structural integrity. This includes associated support work such as surface preparation, painting, structural weld repair, and insulation renewal. Insulation and lagging removal and replacement necessary to

perform an inspection or repair of a covered or protected component would not be considered as structural.

3. *Project.* Maintenance activity performed on equipment or plant involving a modification, upgrading, or improvement and performed on a one-time basis, whether or not a unit outage is involved.

4. *Buildings and grounds.* Maintenance activity to maintain general plant appearance such as trash hauling, janitorial work, gardening, material handling, safety inspections, and general housekeeping efforts. *Exceptions:* Additional effort required due to extensive plant structural repairs is structural work.

5. *Trouble and breakdown.* Maintenance activity required to return a unit or plant equipment to normal operating conditions due to an equipment breakdown or failure to operate properly (well enough for its intended function). Failed or improperly operating equipment discovered or worked during a planned outage is this work type only if the particular item is preventing the return to service of the unit at the time of discovery.

6. *Overhaul.* Maintenance activity performed for general overhaul inspection or rebuild of one or more major equipment groups during a scheduled outage shown on the 1-Year Outage Plan. An overhaul is intended to keep equipment in proper operating condition or restore capability that has gradually diminished over time. All work performed in preparation of, during, in cleanup from, or otherwise due to the overhaul is considered to be this work type. The work is not project work because an improvement over the original design capability is not intended. An overhaul is beyond the scope of normal preventive maintenance. The overhaul occurs before specific trouble or breakdown has happened and before predictive maintenance is recommended.

7. *Preventive maintenance.* Maintenance activity performed which is repeated at a predetermined frequency and is scheduled from a preventive maintenance listing or program. The frequency established need not be related to calendar time but can be required based on predetermined service time or events such as the number of service hours or starts. Work which is performed during an overhaul outage, but would be performed anyway if the outage was postponed, is included in this work type.

8. *Predictive maintenance.* Maintenance activity performed based on a prediction that future equipment breakdown or failure to operate properly will occur. The prediction is based on trend analysis of diagnostic data collected by techniques such as vibration or lube oil analysis. The need for maintenance would not normally have been apparent without analysis of the diagnostic data.

9. *Corrective maintenance.* Proactive maintenance activity performed based on a prediction that future equipment breakdown or failure to operate properly will occur. The intent of corrective maintenance is to maintain equipment optimum performance. The need for maintenance is normally established by inspection. The prediction is *not* based on a

predictive maintenance type technology or trend analysis, nor is it based on a preventive maintenance type predetermined frequency.

How found

(*Note:* These codes are very valuable for evaluating the ability of plant programs for finding problems in time. The PM and predictive maintenance (PdM) programs should be finding problems before equipment breaks. Although valuable, this code is considered too much for a paper form. As the plant adds more and more codes, the ability of a paper work order becomes cumbersome. This code can best be selected by the originator of the work. A paper work order would have to list the choices for the originator to review. Thus, the examples in this handbook do not use this code because the handbook illustrates a paper system for the origination of work. The handbook purposely illustrates a paper system to establish the fundamentals of planning before a company automates or facilitates planning with any computer. Nevertheless, the system in this book is only a step away from having its originators directly enter their requests on a CMMS screen. One of the first additions a company should make with direct screen work requests is the addition of a field with "drop-down" choices for how the originator discovered the work.)

How found codes

OPN-OP—Opns found while operating. Operations personnel encountered need for work while operating equipment.

OPN-RND—Opns found while making rounds. Operations personnel discovered need for work while making scheduled rounds.

OPN-PM—Opns found while performing PM. Operations personnel discovered need for work while performing PM work.

OPN-OTH—Opns found otherwise. Operations personnel discovered need for work, but not while operating, making rounds, or performing PM.

MAINT-PM—Maint found while performing PM. Maintenance personnel discovered need for work while performing PM work.

MAINT-WK—Maint found while performing non-PM work. Maintenance personnel discovered need for work while performing non-PM maintenance work.

MAINT-OTH—Maint found otherwise. Maintenance personnel discovered need for work, but not while performing work.

PDM-RT—PdM found while performing route. PdM personnel discovered need for work while performing a PdM route.

PDM-OTH—PdM found otherwise. PdM personnel discovered need for work, but not while performing a PdM route.

ENGRG—Engrg found work. Engineering personnel discovered or decided need for work.

OTH-FND—Other found work. Other personnel than listed above found or requested the work.

OTH-SYST—Other system generated work. The work was not found or requested. The system generated the work order work established by plant programs such as PM (or operator rounds if performed by work order).

Plan type

(*Note:* The proactive and reactive part of the plan type code is similar to work type but is a broader category. In general, all the work types except number 5, trouble and breakdown, are proactive. The maintenance department wants to be working on proactive work to avoid later trouble that would inconvenience the plant operation and require excessive maintenance effort. By definition, proactive work is not urgent, so the planning department wants to spend more time planning these jobs to be most effective and efficient when executed. The planning department also wants to recognize the reactive work orders so they can be quickly passed on to the maintenance department for timely resolution. Minimum versus extensive maintenance coding also helps the planning department recognize how much effort to expend on planning the work. Small jobs are frequently not worth much time spent on planning, though some time is still necessary.)

Plan type codes.   First digit—Reactive or Proactive

R   Reactive

Equipment has actually broken down or failed to operate properly.

Priority-1 jobs are defined as urgent and so they are reactive.

P   Proactive. Proactive work heads off more serious work later. If the damage is already done, the work is reactive.

Work done to prevent equipment from failing.

Any PM job.

Generally Predictive Maintenance initiated work that if done in time will prevent equipment problems.

Project work.

Second Digit—Minimum Maintenance or Extensive Maintenance

M   Minimum maintenance. Minimum maintenance work must meet *all* of the following conditions:

Work has no historical value.

Work estimate is not more than 4 total work hours (e.g., two persons for 2 hours each or one person for 4 hours).

While parts may be required, no ordering or even reserving of parts is necessary.

E   Extensive maintenance. All other work is considered "extensive." (*Note:* "RE" would be an example plan type code for a reactive job that is also extensive.)

Outage

(*Note:* These codes are not the same as work order status codes. If the maintenance group can only perform a certain work order during a unit outage, it is critical that this limitation be recognized as soon as possible. This is regardless of whether the work order has yet been approved or planned. The planning group must be able to recognize these work orders in order to plan them more expeditiously than other work. There may or may not be a sudden window of opportunity to execute the work. If there is an unexpected outage for some other reason, it is wise to have planned these work orders in time. *Outage* refers to the condition of the entire plant unit, not the individual piece of equipment.)

Outage codes

1. *Forced short outage.*   A component failure or other condition which requires the unit to be removed from service immediately or at any time prior to the end of the upcoming weekend.

2. *Scheduled short outage.*   A component failure or other condition which requires the unit to be removed from service, but not necessarily before the end of the upcoming weekend. The work does not have to be done during a major outage.

3. *Major outage.*   Work which can only be done during a major outage such as a major rebuild or overhaul inspection or work on one or more major equipment groups. The outage is normally shown as a planned outage on the 1-Year Outage Plan. Testing after a major outage at limited loads is included.

4. *Not Used.*

5. *Forced derating.*   A component failure or other condition which requires the maximum capability of the unit to be reduced immediately or at any time prior to the end of the upcoming weekend.

6. *Scheduled derating.*   A component failure or other condition which requires the maximum capability of the unit to be reduced, but not necessarily before the end of the upcoming weekend. The work does not have to be done during a major outage.

7. *Potential outage, redundant installed equipment.*   A trouble condition which requires a piece of redundant equipment to be taken out of service without affecting the unit capability; but if the backup equipment failed an outage or derating would result.

8. Not used.

9. *Potential outage, noninstalled spare equipment.* Work performed on noninstalled spare equipment (such as rebuilding rotating spares) where if the unit suddenly needed the spare, the spare would not be available and a more serious than normal unit outage or derating would occur.

U. *Starting up.* Work which requires the unit to be in a start-up mode where the unit is being brought on-line. Equipment tests or adjustments that can only be done during a start-up condition are included.

D. *Shutting down.* Work that can or should only be done when the unit is on line, but Dispatch no longer needs the unit and the unit is about to be shut down. Typically, equipment tests or adjustments that may trip the unit are included.

(*Note:* Thus, a work order coded as a 2 could only be done when the unit is down on outage for some reason. If a short notice outage work (SNOW) outage occurs and the outage will take 48 hours, this SNOW work order might be a candidate for including in the outage work scope. This work order could only be included if it is already planned and ready for scheduling. One might not know if the work order could be done if the job was not furnished with an expected duration via planning, say 24 or 78 hours of duration.)

Plant and unit

(*Note:* The plant and unit coding begins the equipment coding system. The company has an "intelligent" numbering system. The equipment number identifies the plant, unit, and system of the equipment being discussed.)

Plant and unit codes. First digit—Plant. Second and third digits—Unit.

PLANT N—North Generating Station
    UNIT 00—Steam Plant Common
    UNIT 01—Steam Unit 1
    UNIT 02—Steam Unit 2
    UNIT 03—Steam Unit 3
    UNIT 04—Steam Unit 4
    UNIT 30—Gas Turbine Plant Common
    UNIT 31—GT Unit 1
    UNIT 32—GT Unit 2
    UNIT 33—GT Unit 3
    UNIT 34—GT Unit 4
PLANT S—South Generating Station
    UNIT 00—Steam Plant Common
    UNIT 01—Steam Unit 1
    UNIT 02—Steam Unit 2
    UNIT 03—Steam Unit 3
    UNIT 04—Steam Unit 5

(*Note:* Thus, a code of N02 would indicate North Station Unit 2.)

Equipment group and system

(*Note:* Each equipment group is shown with its composing systems. For example, Group A, Air, is followed by the systems within the Air Group, System I, Instrument Air, and System S, Service Air. Following this listing is a comprehensive definition of each system.)

Group and system codes.   First digit—Group. Second digit—System.

A   Air
   I   Instrument
   S   Service

B   Boiler
   A   Air Flow for Combustion
   B   Boiler Tubes and Steam Generating Section
   C   Controls
   E   Air Preheater Heat Exchanger
   F   Burner Front
   G   Gas Flow from Combustion
   I   Aspirating Air
   J   Casing and Structure
   P   Convection Pass
   S   Sootblowers
   T   Seal Air
   V   Vents and Drains
   W   Wash Drains

C   Condensate
   D   Feedwater Heater Drains
   F   Flow
   P   Polishers
   R   Recovery
   S   Supply
   V   Vacuum Supply

D   Feedwater
   A   Auxiliary Feedwater
   D   BFP Fluid Drive
   F   Flow
   O   BFP Lube Oil
   P   Boiler Feedwater Pump
   T   BFP Steam Turbine

E   Electrical
   C   Communication
   D   120-250 Vdc
   I   Control Room Instrumentation
   L   Plant Lighting and Distribution

    M   Miscellaneous
    S   Switchgear, Load Center, and MCC
    Y   Switchyard

F Fuel
    B   Fuel Oil Burner Supply and Return
    C   Fuel Oil Service Pump
    D   Diesel Storage and Transfer
    H   Fuel Oil Heaters
    I   Ignitor Fuel Supply
    N   Natural Gas
    O   Fuel Oil Storage and Transfer
    P   Propane Gas
    V   Vehicle Fuel Storage
    W   Wharf Facility for Ship Unloading

G Generator
    D   Diesel Generator
    E   Exciter
    H   Hydrogen
    I   Isolated Phase
    O   Seal Oil
    P   Protection Circuit
    R   Rotating Field
    S   Stator
    V   Voltage Regulator
    W   Stator Oil and Cooling Water

H Water
    A   Treatment Plant Acid
    C   Treatment Plant Caustic
    D   Demineralizer
    P   Potable Water Supply
    R   Raw Water Supply

I Intake
    C   Canal
    F   Fish Protection Traveling Screens
    R   Intake Chemical Treatment

J Cooling Water
    C   Condenser Cleaning
    I   Circulating Water
    L   Closed Cooling
    R   River Water
    T   Cooling Tower
    X   Condenser

K   Fire Protection

    D   Dry Chemical
    P   Portable
    U   Gas Turbine
    W   Water and Foam

N   Environmental

    A   Air Quality
    W   Water Quality

R   Reagent and Chemical

    A   Acid Cleaning for Boiler
    C CO_2 and H_2 Supply
    D   Condenser Discharge Chemical Injection
    M   Miscellaneous
    N   Nitrogen Supply
    O   Fuel Oil Treatment
    P   Condensate pH Chemical Injection
    W   Boiler Wash

S   Steam

    A   Auxiliary Piping
    E   Extraction Piping
    M   Main Steam Piping
    P   Primary Superheating Section
    Q   Secondary Superheating Section
    R   Reheat Piping
    S   Reheat Superheating Section
    X   Auxiliary Boiler

T   Steam Turbine

    A   HP and IP Section
    B   IP Section
    C   LP Section
    F   Front Standard
    H   HP and IP Turbine Control
    I   Monitor Supervisory Instrumentation
    J   Turbine Controls
    L   Steam Seals
    O   Lubricating Oil
    P   Pedestal
    S   Structure
    V   Vents and Drains
    X   Crossover Pipe

U   Gas Turbine

    A   Accessory Station
    B   Bearing and Coupling

C   Compressor Section
D   DEH Control
E   Exhaust
G   Generator
H   Housing
I   Combustor Section
J   Cooling Water
O   Lubricating Oil
P   Protection and Control
R   Starting
S   Supports
T   Turbine Section
U   Turning Gear
W   Water Wash
Z   Atomizing Air

V   Ventilation

A   Air Conditioning
D   Equipment Dry Layup
M   Miscellaneous Vents and Exhausts
S   Steam Heating

W   Waste

F   Waste Fuel Collection
H   Water and Boiler Wash Collection
I   Water Instrumentation
L   Liquid Waste
S   Solid Waste Disposal
W   Sewage Treatment

Y   Blanket Accounts

A   Operations
B   Maintenance
C   Administration
D   Engineering
E   Structural
F   Production Equipment
G   Facilities
H   Computer

Z   Miscellaneous

B   Buildings and Grounds
C   Turbine Deck Bridge Crane
E   Electric Tools and Equipment
F   Freeze Protection
H   Hydraulic Tools and Equipment
L   Laboratory

M   Machine Shop Equipment
O   Other Tools and Equipment
P   Pneumatic Tools and Equipment
S   Security
V   Plant Vehicles

Group and system definitions.   Except as noted in the individual system definitions, several general definitions apply. Each system includes its related gauge work, instrumentation and controls, power and control wiring, and breakers. *Piping* means the complete piping system, including the pipe itself, insulation, lagging, hangers and supports, steam tracing lines, valves, safety relief valves, orifices, strainers, drains, and traps associated with that pipe. All electrical, instrumentation, and mechanical devices which exclusively serve a major component are accounted for in the same system as that component.

A   AIR

AI   Air Instrument

Provides compressed air to pneumatic controls and instrumentation. Includes compressors, dryers, coolers, separators, air tanks, and piping. Instrument Air can also be described as piping coming from the service air compressors when the compressors serve both systems.

AS   Air Service

Provides compressed air for general plant service (house air) such as air motors and air hose connections. Includes compressors, dryers, coolers, separators, air tanks, and piping.

B   BOILER

BA   Boiler–Air Flow for Combustion

All combustion air ducting from the air intake ports to the windbox. Includes all related hangers and supports, dampers, damper controls, expansion joints, orifices, and the windbox. Includes FD fans, fan casings, inlet vanes, motors, and foundation. Includes the overfire air system. Does not include preheaters (BE) or seal air system (BT).

BB   Boiler–Boiler Tubes and Steam Generating Section

Steam drum and its internals, mud drum, furnace tubes (all passes), downcomers, risers, tube headers, boiler circulation pump, and drum safety valves. The blowdown lines, drain lines, and flash tank are part of system BV. The convective pass tubes are found in system BP. The economizer is part of the feedwater flow in system DF.

BC   Boiler–Controls

Controls for the fuel and air supply to the burners. Includes boiler master controls and associated signal transmitters and analyzers. Includes electronic control systems, transmitters, and terminals.

BE   Boiler–Air Preheater Heat Exchanger

Regenerative or tube type air preheater. Bearings, baskets, casing, seals, lube oil system, drive motors (air and electric), and gears. Includes wash lances. Also includes the steam coil air preheater. Includes coils, inlet and outlet valves, piping from the steam source, condensate hotwell, and any condensate drain piping to condensate recovery. Does not include sootblowers which are in BS.

BF   Boiler–Burner Front

Oil and gas burners, ignitors, air register, and flame controls. Steam used to purge and clean the burners and fuel oil headers at the burner decks. Includes piping downstream of the auxiliary steam header.

BG   Boiler–Gas Flow for Combustion

All combustion gas ducting from the economizer plenum to and including the exhaust stack and the gas recirculation ducting. Related hangers and supports, dampers, damper controls, expansion joints, and the exhaust stack. Includes ID and GR fans, fan casings, ID fluid drives, motors, fluid drive oil coolers, the lube oil piping system, and foundation. Includes any turning gear drive systems associated with the fans. Stack gas monitoring equipment are in NA. Does not include the air preheaters (BE) or the seal air system (BT).

BI   Boiler–Aspirating Air

Blowers and piping that provide aspirating air to the burner front flame scanners. The same system may also provide combustion air to the propane ignitors. Does not include seal air (BT).

BJ   Boiler–Casing and Structure

Outer casing, insulation, inner casing, structural steel, access doors, gas seals, expansion joints, deck plates and grating, hand rails, stairs and ladders, access doors, and furnace buckstays. Includes any penthouse and header enclosure structures. Includes ash hoppers. Includes any penthouse blowers. Includes any boiler seal skirt which wraps around the bottom hopper of the boiler furnace. Does not include steam piping or headers (Group S).

BP   Boiler–Convection Pass.   Includes convective pass wall tubes and inlet and outlet headers.

BS   Boiler–Sootblowers

Sootblowers, drive motors, steam supply, and air supply piping. Includes control panel interlocks and wiring. Does not include the boiler caustic wash piping (BW).

BT   Boiler–Seal Air

Piping that provides seal air from the combustion air system to the induced draft and gas recirculation fan seals. Piping that provides cooling air to the gas recirculation fan. For positive pressure boilers such as South Unit 3, this system also provides seal air to sootblower and view port openings.

BV   Boiler–Vents and Drains

Waterside drain lines and all vent lines necessary for boiler filling and draining. All boiler blowdown and drain piping and valving. Blowdown (flash) tank, blowdown tank drain pump. Does not include the headers being serviced by the vents and drains. Does not include the nitrogen supply which is in system RN.

BW   Boiler–Wash Drains

Wash drain piping and collection troughs which serve the furnace, air preheater, and economizer sections of the boiler and includes piping prior to the sumps and trenches in system WH.

C   CONDENSATE

CD   Condensate–Feedwater Heater Drains

Drain piping from all feedwater heaters. Includes heater drain pumps, high- and low-pressure flash tanks, and piping up to the condensate flow system (CF). Includes turbine water induction protection system. Also includes manual vent and drain piping.

CF   Condensate–Flow

Condensate piping beginning at the condensate pump outlet through to the deaerator, condensate booster pumps, low- and intermediate-pressure feedwater heaters (high-pressure feedwater heaters are in system DF), the deaerator, and the deaerator storage tank. Does not include the condensate (hotwell) pump which is part of the condenser system (JX).

CP   Condensate–Polishers

Serving vessels, regenerator tanks, resin, backwash pump, sluice pump, and piping.

CR   Condensate–Recovery

Collects high temperature condensate from various parts of the system. May collect boiler water for transfer to storage tank. Includes condensate recovery flash tank, cooling condenser, condensate recovery transfer pump, hotwell drain pump, and piping.

CS   Condensate–Supply

Includes condensate storage tanks whether the water is polished and treated water tanks or only demineralized. Includes pumps and piping to and from the demineralizer system. Also includes the boiler fill pump.

CV   Condensate–Vacuum Supply

Vacuum pumps, exhaust silencers, hogging jet, primary and secondary steam air ejectors, moisture collection tanks, oil drain tanks, vacuum breaker, and piping.

D FEEDWATER

DA   Feedwater–Auxiliary Feedwater

Auxiliary feedwater pump, feedwater heaters, and piping which supply the reboilers (SA). Does not include the auxiliary boiler feedwater system which is included with the auxiliary boiler (SX).

DD   Feedwater–BFP Fluid Drive

Fluid drive unit to the boiler feedwater pump and fluid cooling system. Includes oil supply and return piping to BFP bearings.

DF   Feedwater–Flow

Feedwater booster pumps, high-pressure feedwater heaters, economizer. Piping, including piping after the economizer up to the steam drum and piping after the deaerator storage tank. Does not include the feedwater pumps which are in system DP. For North Unit 1, this includes the piping from the economizer outlet header up to but not including the furnace first pass inlet headers, which is in system BB.

DO   Feedwater–Lube Oil for BFP and BFP Steam Turbine

Lube oil storage tank, lube oil pumps, oil coolers, and piping.

DP   Feedwater–Boiler Feedwater Pump

Boiler feedwater pump, motor, feedwater piping up to and including suction and discharge valves. Includes seal water piping. Does not include the fluid drive (DD) and lube oil piping (DO). Does not include a drive turbine (DT), if present.

DT   Feedwater–BFP Steam Turbine

Turbine, inlet and outlet steam piping, including the steam governing valves, condensate drains. Lube oil is covered under system DO. Steam seals are included in system TL.

E ELECTRICAL

EC   Electrical–Communication

All in-house and outside telephone systems. Public address system.

ED   Electrical–120–250 Vdc

DC inverters and motor-generator sets used for charging the relay batteries and any other dc applications. Includes the batteries.

EI   Electrical–Control Room Instrumentation

Controls and instrumentation for specific devices are part of the system in which the device resides. For example, the boiler feed pump controls belong in the boiler feed pump system. Otherwise, this system contains all other panel controls, recorders, gauges, displays, etc., located in the boiler or turbine-generator control rooms. Nonelectric controls are usually specific enough to place them in their respective systems.

EL   Electrical–Plant Lighting and Distribution

All general illumination fixtures, including their wiring and switches. Includes aviation lighting and portable lighting.

EM   Electrical–Miscellaneous

Reserved for any equipment that would not be in any of the other designated systems. Examples are cable trays, portable power centers, and grounding grid. Also includes electrical freeze protection circuitry if not specific to a system.

ES   Electrical–Switchgear, Load Center, and Motor Control Center

Includes main buses and housing. Tie breakers, feeder breakers, and related transformers. Does not include individual breakers which belong to the motors of another system.

EY   Electrical–Switchyard

Switchyard breakers, power lines, and gantry work. Main transformer, startup transformer, auxiliary transformer, etc.

F   FUEL

FB   Fuel–Fuel Oil Burner Supply and Return

Oil supply piping downstream of fuel oil heaters, high-pressure strainers, constant differential pumps, piping serving individual burners, burners in and out controls, and oil piping upstream of the return line to the valve farm. Includes oil purge system. Does not include oil burners which are in system BF.

FC   Fuel–Fuel Oil Service Pump

Fuel oil service pumps, strainers, piping up to but not including the fuel oil heaters.

FD   Fuel–Diesel Storage and Transfer

No. 2 oil storage tanks, retaining dikes, transfer pumps, and piping to and from the gas turbines. Does not include the unloading dock (FW).

FH   Fuel–Fuel Oil Heaters

Fuel oil heaters and related piping. Steam supply piping from auxiliary steam (SA). Not included are the condensate drains, which are part of the heater drain system (HK).

FI   Fuel–Ignitor Supply

Equipment and piping leading from the natural gas (FN) and propane (FP) systems which uniquely supply the ignitors of the main burners of the boiler. Includes the natural gas trip valves, burner ignition valves, and vent piping.

FN   Fuel–Natural Gas

Metering station, scrubbers, vaporizers, and piping (including vents and drains) to the burners. Does not include gas burners which are in system BF. Does not include ignitor supply as defined in system FI.

FO   Fuel–Fuel Oil Storage and Transfer

Fuel oil tanks, steam coil heaters and transfer pumps including their reduction gearing and piping, including the valve farm, retaining dikes, and oil sumps. This would also include any fixed oil and water separating equipment. Also includes the supply piping between the valve farm and unit fuel oil strainers (FC) and the return piping (FB). Includes steam supply and condensate return for steam tracing.

FP   Fuel–Propane Gas

Propane storage tank, vaporizer, common piping up to but not including the gas ignitor fuel supply FI. Does not include propane to the auxiliary boiler, which is part of system SX.

FV   Vehicle Fuel Storage

Separate diesel or gasoline storage tank for vehicle supply. Propane for the forklifts is tapped off the main storage tank in system FP.

FW   Fuel–Wharf Facility for Ship Unloading

Marine arms, unloading pumps, stripping pumps, piping, and dock structure, including the tanker-man house. The oil piping is included up to the fuel storage tanks.

G   GENERATOR

GD   Diesel–Engine and Generator

Includes any reserve generator driven by a piston engine (diesel or gasoline). The rest of the systems in G group are for turbine-driven generators.

GE   Generator–Exciter

Generator exciter. Includes the casing, rotating and static elements, and field breaker. Includes any exciter cooler and closed cooling piping exclusive to the exciter.

GH   Generator–Hydrogen

Hydrogen coolers, hydrogen control panel, hydrogen and carbon dioxide manifolds, liquid detector, purity meter, gas dryer, and hydrogen and carbon dioxide piping. Does not include the hydrogen or carbon dioxide bulk storage which are in system RC.

GI   Generator–Isolated Phase

Bus assembly from the generator lead box to the main transformer, ducting, heater and cooler, and generator disconnect switch. Main generator breaker in the switchyard is in system EY.

GO   Generator–Seal Oil

Main seal oil pump, emergency seal oil pump, recirculation pump, vacuum pump, vacuum tank, detraining tank, shaft seals, seal drain header, and piping. For a unit that shares a common system for generator seal oil and turbine lube oil, use system TO.

GP   Generator–Protection Circuit

Various relays, transmitters, recorders, panel gauges, switches, and interlocks used to protect the generator from damage.

GR   Generator–Rotating Field

Generator rotor and coils, collector ring and brushes, rotor couplings, any rotating blades, and field rheostat.

GS   Generator–Stator

Main stator assembly, generator casing, support assemblies, end bells, generator bearing assemblies, fan shrouds, hydrogen diffuser baffle, bus bar enclosure, and main lead connection box and bushings.

GV   Generator–Voltage Regulator

Voltage regulator assembly, excitation transformer, control panel.

GW   Generator–Stator Oil and Cooling Water

Stator cooling oil tank, stator cooling pumps, coolers, vacuum pumps, and piping. Includes any stator cooling water deionizer, heat exchanger, pumps, and piping.

H   Water

HA   General Water–Treatment Plant Acid

Acid tank, feed pump, and related piping.

HC   General Water–Treatment Plant Caustic

Caustic transfer pump, caustic mixing tank, and related piping.

HD   General Water–Demineralizer

Cation, anion, and mixed bed tanks, resins, carbon purifier, degasifier or aerator, silica analyzer, transfer pump, and related piping.

HP   General Water–Potable Water Supply

Plant water piping downstream of the chlorination source. This could be a city water tie-in or an on-site chlorinator. Includes the chlorinator. Includes general use fixtures such as domestic water heaters, sinks, commodes, etc. Includes any storage tanks and service pumps. Does not include piping exclusively dedicated to fire protection or fire hydrants (system KW).

HR   General Water–Raw Water Supply

Well pumps, raw water booster pumps, aerator tank, and related piping which is used strictly for raw water. Does not include plant chlorination equipment, which is the beginning of system HP.

I   INTAKE

IC   Intake–Canal

Intake canal, log screens, bubble buster, trash rakes, cathodic protection, tide gate, and intake well. Includes stationary cranes.

Mobile cranes would be in ZV. Does not include the circulating pumps, which are in system JI.

IF   Intake–Fish Protection Traveling Screens

Used for filtering of debris, fish, etc., from the intake canal water before it enters the circulating pumps. Includes the traveling screen assemblies, fish return troughs, and screen wash pumps. South Stations has a screen wash pump with each screen assembly. North has a screen wash pump that supplies all screen assemblies.

IR   Intake–Intake Chemical Treatment

Sodium hypochlorite or chlorine tank and piping, including the injector pump and piping to each intake well.

J   COOLING WATER

JC   Cooling Water–Condenser Cleaning

Condenser cleaning system.

JI   Cooling Water–Circulating Water

Circulating water pumps and intake piping leading to the condenser water boxes and condenser discharge piping.

JL   Cooling Water–Closed Cooling

Closed cooling provides a heat sink for various bearings, lube oil systems, and other processes throughout the plant. The system is isolated from other water systems in the plant, hence its name. The closed cooling heat exchanger discharges this heat into the river water system. Includes closed cooling heat exchangers (river water heat exchangers), closed cooling pumps and booster pumps. Includes supply and return piping. Does not include the supplied heat exchangers, each of which belongs to its respective system.

JR   Cooling Water–River Water

River water is taken from the condenser inlet headers and sent to the closed cooling heat exchanger to provide a cool source for the accumulated heat of the closed cooling water system. The warmed river water is then sent to the condenser outlet header to be sent to the discharge flume or cooling towers via the circulating water piping. River water booster pumps, strainers, and piping used for providing river water between the condenser (JX) and closed cooling heat exchangers (JL).

JT   Cooling Water–Cooling Tower

Cooling tower structures and components receive water from and return water to the circulating system (JI). Does not include air conditioning cooling towers.

JX   Cooling Water–Condenser

Condenser shell, tubing, water boxes, condensate drain nozzles into condenser, and condensate pumps. Includes exhaust shroud between LP turbine and condenser. Includes the water box vacuum pumps used to assist the filling of the circulating water system. Does not include the vacuum pumps in system CV.

K FIRE PROTECTION

KD   Fire Protection–Dry Chemical

Fixed dry chemical systems exclusively dedicated to fire protection. Includes associated alarms and controls.

KP   Fire Protection–Portable

All types of portable extinguishers. Water, CO_2, Halon, potassium nitrate (Purple K), etc.

KU   Fire Protection–Gas Turbine

Fixed systems self-contained in the gas turbine unit. Includes carbon dioxide and Halon systems.

KW   Fire Protection–Water and Foam

Fixed dry and wet water piping systems exclusively dedicated to fire protection. Includes fire hydrants feeding from potable water supply. Includes protected areas such as burner fronts, transformers, lube oil reservoirs, and offices. Includes associated alarms and controls. Includes any exclusively dedicated fire pumps. Potable water itself is listed as system HP.

N ENVIRONMENTAL

NA   Environmental–Air Quality

Atmospheric discharge monitoring equipment. Weather monitoring equipment. Opacity monitor, NO_x monitor, SO_2 monitor, CO monitor, ambient temperature and humidity probe, transmitters, analyzers, and other computer equipment.

NW   Environmental–Water Quality

Groundwater monitoring wells and instrumentation. Cooling water and liquid waste discharge monitoring are included in JI and WI, respectively.

R REAGENT AND CHEMICAL

RA   Chemical–Acid Cleaning for Boiler

Piping (temporary or permanent) and piping connections for acid cleaning of the boiler waterside surfaces. See system RW for caustic washing of the fireside surfaces. For the boiler wash drains, see system BW.

RC   Chemical–CO_2 and H_2 Supply

This system provides bulk gas storage for the generator. Hydrogen is used in the generator as a heat removal medium. Carbon dioxide is used to purge the generator of hydrogen for fire prevention purposes. This is not considered a fixed system to be included in

KD. Includes tanks, manifolds, regulators, and piping. Does not include the hydrogen panel in system GH. Does not include CO_2 supply for gas turbines or chemical waste treatment systems.

RD   Condenser Discharge Chemical Injection

Chemicals such as sodium bisulfate which are used to dechlorinate the condenser discharge water.

RM   Chemical–Miscellaneous

Chemical storage not specified in other systems.

RN   Chemical–Nitrogen Supply

Nitrogen is used primarily for blanketing the waterside of the boiler and sometimes feedwater heaters during reserve shutdown. It is usually tied into the equipment vents. Includes tanks, manifolds, regulators, and piping.

RO   Chemical–Fuel Oil Treatment

Chemicals which are added to fuel oil in order to induce an effect. Includes magnesium oxide day tanks, feed pumps, and piping into fuel oil supply.

RP   Chemical–Condensate pH Chemical Injection

Chemicals which are used to neutralize the pH condition of condensate and feedwater. Includes tanks, pumps, and piping for ammonia and hydrazine injection into the condensate flow system. Also includes phosphate injection system into the steam drum. Does not include sodium hypochlorite or chlorine which are in system IR.

RW   Chemical–Boiler Wash

Caustic solution used to clean the firesides of the boiler and other equipment subjected to flue gases. Includes soda ash storage tank, mixing and service tanks, related pumps, piping up to the sootblower piping (BS), and boiler wash hose connections. Includes any piping that is subjected to the boiler wash caustic solution, even if it can also be used to supply raw water. See system RA for waterside acid cleaning. The air preheater wash lances are included in system BE.

S   STEAM

SA   Steam–Auxiliary Piping

All auxiliary steam piping not specified in other systems. Includes desuperheater supply.

SE   Steam–Extraction Piping

Steam extracted from various stages of the superheated steam flow used primarily as a heat source for feedwater heating. Extraction steam piping begins at the first field weld at the extraction nozzle (either in the turbine or steam piping) up to the final field weld at the feedwater heaters. Includes extraction drain lines up to the condenser. Does not include feedwater heaters (CF and DF) or condenser (JE).

SM   Steam–Main Steam Piping

Main steam piping from the first field weld of the secondary superheater outlet header to the final field weld before the main steam stop valve at the high pressure turbine. Includes drain piping.

SP   Steam–Primary Superheating Section

Primary superheater, including inlet and outlet headers, and piping to the secondary superheater section (SQ). Includes any flow control valves, hangers, and tube supports. Includes attemperator and attemperator supply piping.

SQ   Steam–Secondary Superheating Section

Secondary superheater, including inlet and outlet headers. Includes flow control valves, hangers, and tube supports.

SR   Steam–Reheat Piping

Hot and cold reheat piping between turbine and reheat superheater. Includes attemperator and attemperator supply piping.

SH   Steam–Reheat Superheating Section

Reheat superheater, including inlet and outlet headers, hangers, and tube supports.

SX   Steam–Auxiliary Boiler

Auxiliary boiler(s), fuel supply, water supply, air supply, exhaust ducting, and common steam headers.

T   STEAM TURBINE

TA   Steam Turbine–HP and IP Section

Outer and inner casings, all internal components including HP and IP rotor, blade rings (diaphragms), stationary impulse diaphragm, balance piston, reheat seal, and nozzle block and chest. Does not include steam seals (TL). For units with separate HP and IP sections, use system TA for HP section and system TB for IP section.

TB   Steam Turbine–Intermediate Pressure (IP) Section

Outer and inner casings, all internal components including IP rotor, blade rings (diaphragms), stationary impulse diaphragm, balance piston, reheat seal, and nozzle block and chest. Does not include steam seals (TL). For units with combined HP and IP sections, use system TA.

TC   Steam Turbine–LP Section

Inner and outer casings and all internal components including LP rotor and blade rings (diaphragms). Includes multiple LP sections on same unit. Includes any low-pressure exhaust hood sprays and supply piping. Does not include steam seals (TL). Does not include the crossover pipe(s) which is listed in system TX.

TF   Steam Turbine–Front Standard

Casing, pedestal, thrust bearing and adjuster, main speed governor and overspeed governor, and instrumentation board assembly. The shaft driven lube oil pump is part of system TO.

TH   Steam Turbine–HP and IP Turbine Control

The steam side of the turbine controls. Main steam stop valves and actuators, governor valves, main steam bypass valves, warm-up valves, steam chest, piping to nozzle block, and intercept and reheat valves. Includes any drain lines from these devices.

TI   Steam Turbine–Supervisory Instrumentation

Vibration monitors and controls, bearing temperature monitors and controls, cylinder expansion monitors, rotor position monitor, eccentricity monitor, speed monitor, various control valve position indicators, steam pressure and temperature indicators and transmitters, and no-load trip devices.

TJ   Steam Turbine–Turbine Controls

Includes electric-hydraulic, digital-electric-hydraulic, and mechanical-hydraulic controls. Any related hydraulic pumps, piping, coolers, etc.

TL   Steam Turbine–Steam Seals

Steam packing exhauster, shaft seals and casings, supply and drain piping up to the turbine casing, associated desuperheaters, regulator, and controls. Includes steam seals to any BFP turbines.

TO   Steam Turbine–Lubricating Oil

Main and auxiliary pumps, lube oil reservoir, filter tank (bowser), storage tanks, coolers, and piping. For a unit that shares a common system for generator seal oil and turbine lube oil, the system is in TO.

TP   Steam Turbine–Pedestal

Concrete pedestal, sole plates or pedestal cover, bearings, oil wipers and baffles, and the couplings. For all HP, IP, and LP sections. Includes the turning gear assembly which includes the bull, pinion, and reduction gears, turning gear motor, clutch, lube oil supply, and return piping.

TS   Steam Turbine–Structure

Decking, stairs, handrails, enclosures, foundation, and pedestals not otherwise listed. The turbine building itself would be accounted for in buildings and grounds (ZB).

TV   Steam Turbine–Vents and Drains

Drain piping and valves from the various stages of primary steam turbines to their drain destinations.

TX   Steam Turbine–Crossover Pipe

Includes crossover piping leading into the LP turbine. Includes bolting and thrust protection.

U GAS TURBINE

UA   Gas Turbine–Accessory Station

Motor control center, fuel oil pump. Does not include starting motor, which is in system UR.

UB   Gas Turbine–Bearing and Coupling

Thrust bearing loader and unloader, labyrinth seals, bearings and casings, oil wiper seal ring, and coupling.

UC   Gas Turbine–Compressor Section

Provides compressed air for fuel atomization and combustion as well as combustion gas cooling. Includes compressor rotor, blade rings, and casings.

UD   Gas Turbine–Digital-Electric-Hydraulic Control

Hydraulic pumps, tubing, and DEH controls.

UE   Gas Turbine–Exhaust

Exhaust hood, air cone, bearing heat shield, and expansion joint.

UG   Gas Turbine–Generator

Generator section of gas turbine.

UH   Gas Turbine–Housing

Air inlet and exhaust housing, control, accessory, excitation, and gas turbine compartment housings.

UI   Gas Turbine–Combustor Section

Combines no. 2 oil with combustion air and ignites the mixture. Includes combustion chamber assemblies, transition ducting, cross-fire tube, fuel manifold and fuel nozzles, chamber casing, and diffuser case.

UO   Gas Turbine–Lubricating Oil

Provides lubricating oil to all bearings, gears, and control system. Includes oil reservoir, lube oil pumps, pressure regulators, oil coolers, oil filters, and piping.

UP   Gas Turbine–Protection and Control

Temperature control circuit, fuel control and monitoring, governor and speed control, sequencer, vibration monitors and trips, overspeed trips, combustible gas detector, flame detector, relays, relay batteries, etc. Includes "false start" drain sumps, used for protection to allow fuel drainage in case of an aborted start. Does not include the DEH controls which are in system JD.

UR   Gas Turbine–Starting

Starting motor, accessory gearbox, and torque converter.

US   Gas Turbine–Supports

Alignment devices and pedestal foundation.

UT   Gas Turbine–Turbine Section

Combustion gases expand through the turbine to provide shaft power to turn the generator and compressor sections. Includes the turbine rotor, nozzle guide vanes (blade rings), and casings.

UU   Gas Turbine–Turning Gear

Prevents warping of the turbine rotor by turning it at slow speed while the unit is cooling down from operation. Includes drive assembly, jaw-type clutch, and turning gear motor.

UZ   Gas Turbine–Atomizing

Air Starting air compressor, cooler, separator, tank, and piping required to provide atomizing air to the gas turbine fuel nozzles.

V VENTILATION

VA   Ventilation–Air Conditioning

Includes central systems and window units.

VD   Ventilation–Equipment Dry Layup

Includes dehumidification equipment and piping for short- and long-term "cold" storage.

VM   Ventilation–Miscellaneous Vents and Exhausts

Includes any exhaust fans, hoods, etc., not associated with other systems.

VS   Ventilation–Steam Heating

Steam or hot water heating systems used for personnel space heating needs.

W WASTE

WF   Waste–Waste Fuel Collection

Waste oil tank and piping to the fuel oil service pump strainers. Piping from the various burner front collection funnels.

WH   Waste–Water and Boiler Wash Collection

Sumps, sump pumps, drain trenches, including storm drains, and drain piping to chemical waste treatment system or percolation ponds (WL).

WI   Waste–Water Instrumentation

Water quality monitoring equipment at the chemical waste treatment system (WL).

WL   Waste–Liquid Waste

Chemical waste treatment system or percolation ponds. Includes the grit bed pumping station as it receives influent from system WH, surge tanks, sludge settling ponds, and percolation ponds. Includes any associated clarifiers. Includes the lime storage bin, rotary lime feeder and slaker, work tank, slurry pump, rapid and slow mix tanks, piping, and controls. Includes the CO_2 storage tank, refrigeration and vaporization equipment, piping, and controls. The monitoring wells surrounding a basin are included in system NW.

WS   Waste–Solid Waste Disposal

Fixed systems such as a landfill.

WW   Waste–Sewage Treatment

For on-site treatment, includes the sewage collection piping, lift stations, discharge piping, barscreen, aeration tank, blower, settling tank, digester, clarifier tank, and chlorine and aluminate feed systems. For discharge to the city sewage system, includes the sewage collection piping and any pumping stations.

Y   BLANKET ACCOUNTS

Used for plant expense accounting for work that is not equipment specific or otherwise accounted for in another of the equipment systems. These systems are to be used sparingly and are not to be used as an easy substitute for more detailed work order coding.

YA   Blanket Accounts–Operations

Operations-related expenses.

YB   Blanket Accounts–Maintenance

Maintenance-related expenses.

YC   Blanket Accounts–Administration

Miscellaneous administrative expenses as clerical, guarding, and grounds keeping.

YD   Blanket Accounts–Engineering

Engineering expenses.

YE   Blanket Accounts–Structural

Maintenance of steam structures expenses.

YF   Blanket Accounts–Production Equipment

Maintenance of power production equipment.

YG   Blanket Accounts–Facilities

Miscellaneous buildings expenses.

YH   Blanket Accounts–Computer

This would include maintenance of mainframe and PC computer networks. Use this code for time and monies expended on a CMMS.

Z MISCELLANEOUS

ZB Miscellaneous–Buildings and Grounds

Building structures not otherwise listed in other systems. Also includes items as roads, parking lots, elevators, lawns, and picnic areas.

ZC Miscellaneous–Turbine Deck Bridge Crane

Entire crane assembly, including the brakes and gears, power wires, track, controls, and supports. Does not include portable hoists (ME), shop hoists (ZE), or mobile cranes (MV).

ZE Miscellaneous–Electric Tools and Equipment

Various fixed and portable electric driven tools and equipment. Arc welders, metallizing machines, bolt heaters, portable fans and space heaters, hand drills, electric chainfall hoists, etc.

ZF Miscellaneous–Freeze Protection

Various systems, not otherwise specified, used to protect plant equipment from freeze damage.

ZH Miscellaneous–Hydraulic Tools and Equipment

Various fixed and portable hydraulic tools. Includes presses, bending brakes, lifts, porta-power devices, and wrenches.

ZL Miscellaneous–Laboratory

Laboratory analysis equipment for fuel oil, water, and steam sampling analysis. Sample points would be listed in the involved system.

ZM Miscellaneous–Machine Shop Equipment

Drill presses, lathes, milling machines, grinders, etc.

ZO Miscellaneous–Other Tools and Equipment

Other tools, usually unpowered, not covered by the other tool systems. Includes specialized tools for turbines, oil spillage control, and safety. Includes nonproduction equipment such as ice makers. Measurement tools such as micrometers are also included.

ZP Miscellaneous–Pneumatic Tools and Equipment

Various fixed and portable pneumatic driven tools. Includes sandblasters, spray paint equipment, impact wrenches, and portable air compressors.

ZS Miscellaneous–Security

Perimeter fences and gates, guard houses, surveillance cameras, intruder alarm systems, etc.

ZV Miscellaneous–Plant Vehicles

Forklifts, mobile cranes, bulldozers and other heavy equipment, plant assigned cars, trucks and boats, etc.

(*Note:* An example of using the plant, unit, group, and system codes would be N02-CP to indicate North Station Unit 2's condensate polishing system.)

Equipment type

(*Note:* These codes are useful for analyzing problem areas in the plant. For example, the plant could segregate work orders and determine how much plant work was being expended on pumps, equipment type code 01. Planners place these codes on the work orders during the coding process, but the codes are not part of the equipment component tag number. Equipment type codes could be included in the equipment number itself, but the numbering system would be stretched to the limit. Adding the equipment type codes to the equipment number itself may be beyond the point of diminishing returns. It can become confusing and frustrating to require a tremendous string of numbers to be manipulated. Certainly, additional intelligent equipment coding beyond these numbers would be difficult.)

These codes are to be used consistently throughout the various equipment systems. For any system described, these code numbers will be unique to the equipment described. For example, 01 means pump regardless of in what system the equipment lies. A pump in the condensate polishing system would be coded as 01 and a pump in the intake basin would be coded as 01 for the equipment type code. Certain major pieces of equipment such as the boiler feed pumps and the steam turbine generators are complex enough to merit their own Group and System designations. In these cases, major pieces may be given their own major code numbers, e.g., high-pressure steam turbine inner casing.

(A later section discusses using these codes as "problem classes.")

Equipment type codes

00. *General:*   Intended to represent the system in general. When a specific piece of equipment is involved, this code is not to be used.

01. *Pump:*   Device intended for the movement of liquids. Does not include vacuum pumps.

02. *Compressor and Vacuum Pump:*   Device intended to change the pressure of the gas involved. Device should provide a pressure differential greater than 35 psi.

03. *Fan and Blower:*   Device intended to increase the flow volume of the gas involved. Device should provide a pressure differential less than or equal to 35 psi.

04. *Hydraulic Coupling:*   Fluid drive unit located between a driven object and its driver. (Simple couplings are considered a part of the driven object.)

05. *Gear Set:*   Separately cased gear set, reduction gear, or turning gear located between a driven object and its driver. (Gear sets integral to an object are considered a part of that object. Simple couplings are considered a part of the driven object.)

06. *Pressure Vessel:*   A container designed to continuously hold a pressurized fluid.

07. *Vented Vessel:*   A container which is vented so that it does not hold more than static pressure head of the substance contained. The vent need not necessarily go to the atmosphere.

08. *Heat Exchanger:*   Includes open and closed types such as tube-and-shell and plate-type heat exchangers, condensers, and deaerators.

09. *Piping or Ducts:*   Piping which is used for transferring fluid from one point to another. Includes flexible piping and hose. Insulation, simple hangers, supports, flange work, expansion joints, etc., are considered part of the piping or duct to which it is attached. No distinction is made on operating pressure. Does not include open troughs (see item 50).

10. *Hanger Assembly:*   Includes large hanger mechanisms as found on the major steam piping. (Simple hangers are considered a part of the piping under item 09—Piping.)

12. *Control Valve and Actuator:*   Includes all valves with remote actuation (except solenoid valves). The valve and its actuator are considered as one item, regardless of how attached they are to each other.

13. *Solenoid Valve:*   Any valve which is solenoid actuated. Includes the solenoid.

14. *Manual Valve:*   A valve which must be manually operated. Includes check valves. Includes backflow preventer assemblies. Does not include safety valves.

15. *Safety Valve and Others:*   Includes safety valves, relief valves, rupture disks, vacuum breakers. Can also include trip valves.

16. *Flow and Restriction Orifices:*   Includes the orifice plate (Venturi or other), and upstream and downstream pressure taps.

17. *Steam Trap Assembly, Moisture Trap Assembly, Float Trap Assembly:*   Includes trap, strainer, isolation valves, bypass valve, drain valve, and any test valve. Continuous drains are listed here.

18. *Filter and Strainer.*

19. *Regulator.*

21. *Switchgear:*   Circuit breakers, including the primary disconnects and operator with auxiliary equipment such as the switch, trip device, solenoid and arc chute assembly. For voltages of 460 and higher. Smaller voltages are typically considered a part of the equipment being served, or are listed as part of a distribution panel in systems such as EB. Breakers such as metal-clad, oil-filled, air blast and air magnetic types are included.

22. *Electric Motor, Single-Phase.*

23. *Electric Motor, Three-Phase, under 500 V.*

24. *Electric Motor, Three-Phase, 500 V+.*

25. *Electric Motor, dc.*

26. *Power and Control:*   Cables, control devices, etc., between the switchgear and electrically driven device, typically a motor.

28. *Motor Starter and Contactor:*   Includes the toggle switch, solenoids, relays, and timers.

29. *Transformer.*

30. *Structure:*   Girderwork, stairs, floors, walls, ceilings, containment walls, etc.

31. *Air Dryer:*   Includes both refrigerant and desiccant type dryers. Includes the dehumidifiers.

33. *Motor Control Center:*   A grouping or combination of motor starters in a single cabinet or enclosure. Consists of lead-in cables, molded case circuit breaker, starter contactor, control transformer, and fuses.

34. *Instrumentation:*   All types of instruments including devices used to record information automatically on paper and flow meters. Does not include flow orifices, which are listed as major code 16. Includes rotometers and level indicator vessels.

35. *Auxiliary Driver:*   Diesel engine and its accessories such as air starter, fuel pump, heat exchanger, etc., that drives the emergency fire pump. Also included is the steam engine that drives the fuel oil transfer pump.

41. *Sootblower Assembly:*   Device used to remove accumulated soot and debris from a heat transfer surface. Typically found in boilers and air preheaters. Can include permanent wash lances.

43. *Attemperator:*   Includes the spray head, thermal liner, inlet nozzle, and casing. Does not include the control valve, which would be coded as 12.

44. *Lubricating Assembly.*

45. *Burner Assembly:*   Includes oil and natural gas burners—the nozzle, gun, diffuser, canes, etc. Does not include the air register, which would be coded as 47.

46. *Generator:*   Small generators, ac or dc, such as motor-generator sets. Does not include prime mover generators, which are covered in Group G.

47. *Damper or Register Assembly:*   Device for regulating gas flow, including its driver and actuator (if applicable).

48. *Exhaust Stack, Exhaust Silencers.*

49. *Traveling Screen Assembly:*   Does not include the screen wash pump or troughs.

50. *Trough, Trench, Ditch, Sump Pit.*

51. *Air Motor:* A rotating pneumatic device used to provide power to a driven device such as a pump or generator.

52. *Gas Ignitor:* Natural gas and propane ignitors.

99. *Other:* Equipment not listed above.

(*Note:* Notice that code 12 includes both actuators and their valves and that there is no code 11. It became impractical to classify valves as 11 separate from actuators. It was frequently difficult to tell which was the problem area, the valve or its actuator. Early planning strategy at this plant was to have the planner be alert to change the code after the job was completed to have the exact equipment identified. In actual practice, however, it became apparent that the first equipment number placed on the work order would stick and not be changed. It made life much easier to combine these devices. Later as the planning system matured and computer "help" was brought in, the computer required that everything be separate to list the manufacturer with the equipment on a one-to-one basis. Now it was easy with a paper file to list two manufacturers and scores of suppliers for any particular valve and actuator combination. At this plant, the computer was modified to allow continuation of the existing practice.)

Problem class, problem mode, problem cause, action taken

Using problem class, problem mode, problem cause, and action taken codes helps future reliability analysis by reliability groups. By analyzing work orders with these codes, reliability groups determine the most common types of equipment problems with their most common modes and causes. With this knowledge, reliability groups can dictate changes to PM or PdM programs to prevent or to provide early detection for common plant problems.

The purpose of the codes is *not* to reduce written feedback received from technicians. The purpose of the codes is *not* to specify ahead of time the actions for the technician to take. Operators and planners make their best estimates of the proper class and mode and perhaps the cause code, rarely the action code. Technicians have the final word on what actually happened to what, why it happened, and what the technicians did about it. Planners verify that technicians selected the proper codes for what the technicians finally decided and did.

Problem class codes should be the same as equipment type previously shown. Each different type of equipment class has only a limited number of principal problem modes. The mode is how (not why) the problem occurred. For example, a problem class of VALVE might have ten principal ways a valve could have a problem. These might include WILL NOT

OPERATE OR SLOW MOVEMENT, LEAKING EXTERNALLY, and LEAKING THROUGH INTERNALLY. It is important to always list PM and OTHER as choices for problem modes. The choice of PM is necessary because the work might not be a "problem," but scheduled PM work. Listing OTHER allows the list to be limited to 10 or fewer choices for easy use without damaging the credibility of the choices. Persons complain if management insists they must select a code and no appropriate code exists. OTHER allows selection of an appropriate code. Persons occasionally find other modes than the principal ones, but the list becomes too cumbersome if listing every conceivable possibility. Engineering might occasionally include more modes in the list if persons frequently select OTHER for certain problem classes. Listing PM and OTHER for every problem mode allows management to require using the codes.

Similarly, each problem mode for a specific equipment class (problem class) has only a limited number of principal causes. The cause is the why of the mode. Why did the valve (problem class) leak externally (problem mode)? The cause might be looseness of fasteners or corrosion and the listing of principal causes would list LOOSENESS and CORROSION among principal causes for the mode of LEAKING EXTERNALLY. Again, OTHER should always be a choice. SCHEDULED is also listed to allow selecting a cause for a mode of PM.

Finally, each cause has limited number of actions a technician could take to remedy the specific cause. For a cause of LOOSENESS, the technician might select TIGHTENED or REPLACED. OTHER is always listed as an option. PERFORMED is listed to allow selecting codes for preventive maintenance.

Planning manages the collection of these codes. In a paper system, the planner enters the correct problem class and failure mode based on the originator's description. With CMMS, the computer would automatically populate the problem class field upon the originator's selection of equipment. The originator could change the class from a drop-down list if needed. The originator could then select from a drop-down list that only shows the short list of problem mode choices specific to that problem class. In a paper system, the planning or engineering department would issue small booklets listing all the plant's problem classes to technicians they could keep in their toolboxes. Then under each separate class, the booklet lists principal problem modes. Each problem mode, in turn, lists principal problem causes. Finally, each problem cause lists principal choices for action taken. The technicians look in their booklets for the problem class and mode already listed on the work order. Under the mode area, the technicians select appropriate codes for problem cause and then action taken. In a CMMS where technicians enter feedback on the computer, they could have specific drop-down lists for each choice. However, even with a CMMS, the *Maintenance Planning and Scheduling Handbook* prefers technicians use paper work orders in the field. It is still preferred that they use their booklets to select codes and the planners or planning clerk updates these on the CMMS during job closing.

(*Note:* Plants using these codes experience several significant problems. The principal problem is the choices frequently overwhelm persons. Some plants abandon the use of the codes because they have straight lists of all possible modes, all possible causes, and all possible actions. With each list numbering well over 100 items, persons typically favor using about 10 to 20 codes for any job. In addition, each person usually favors a different set of 10 to 20 codes. Management can alleviate some of this problem by providing a very structured breakdown list or tree structure and limiting the choices to principal ones using "other" and "PM." Yet, even with a tree structure, if an equipment class has subcomponents such as a valve may have an actuator, distinguishing between modes and causes becomes difficult. Another significant problem with these codes is that as management refines the tree structure, especially with a CMMS, usually only one choice is allowed at each step. In other words, a valve might have a problem with looseness and corrosion. The resulting work order might clean the corrosion and tighten the fasteners. Another example might be technicians executing a PM route as scheduled and they tighten a number of loose fasteners. Coding system usually address this issue by requiring that persons select the "most correct" code or the code that "pertains to the majority of the work." CMMS systems that allow multiple selections are preferred, but uncommon. A paper system might be the best collection method for multiple selections. On the other hand, reliability groups typically need this information entered into some sort of database or spreadsheet to allow analysis. Even if a paper system allowed multiple choices, the plant would need to exercise some foresight to accommodate data entry for analysis. Another significant problem is that plants are rarely advanced enough to have a reliability group that looks at the result of such codes. There then arises the problem of the "chicken and the egg." Should the plant go through the trouble of collecting the data in the expectation of possible future use or should the plant wait to begin collecting the data until a reliability group wants it? Many plants abandon collecting the data after determining that "No one wants it, anyway." This handbook advocates collecting the data for future use, but would prefer an engineering group establish the codes in a tree structure and booklets. The planning department could establish the codes in the tree structure and make the booklets, but this would be a lower priority for the planning department.)

The following example illustrates a tree structure for a single problem class. Imagine the immenseness of this structure if completed for all equipment classes at a plant. A booklet can become daunting and a computer becomes preferred to limit the irrelevant codes for each class.

Problem class, problem mode, problem cause, and action taken codes example

Problem class: VALVE

Problem mode	Cause	Action taken
Will not operate or slow movement	Actuator problem	Cleaned
		Adjusted
		Lubricated
		Repaired or replaced broken part
		Replaced actuator
		Other (See remarks)
	Power loss	Replaced fuse
		Reset or replaced breaker
		Repaired ground
		Repaired relay
		Repair wiring
		Repaired transformer
		Other (See remarks)
	Controls	Adjusted
		Replaced card
		Other (See remarks)
	Valve body problem	Cleared debris
		Cleaned corrosion
		Repaired parts
		Replaced parts
		Replaced valve
		Other (See remarks)
	Other (See remarks)	Other (See remarks)
Leaking externally	Looseness	Tightened fasteners
		Replaced valve
		Other (See remarks)
	Packing, o-ring, gasket	Tightened
		Replaced packing, o-ring, gasket
		Other (See remarks)
	Worn	Repaired part
		Replaced part
		Replaced valve
		Other (See remarks)
	Corrosion	Cleaned
		Replaced valve
		Other (See remarks)
	Other (See remarks)	Other (See remarks)
Leaking through internally	Debris	Cleaned
		Other (See remarks)
	Looseness	Tightened fasteners
		Replaced valve
		Other (See remarks)
	Corrosion	Cleaned corrosion
		Replaced parts
		Replaced valve
		Other (See remarks)

(Continued)

Problem mode	Cause	Action taken
	Worn	Repaired parts
		Replaced parts
		Replaced valve
		Other (See remarks)
	Other (See remarks)	Replaced parts
		Replaced valve
		Other (See remarks)
Noisy	Looseness	Tightened fasteners
		Replaced valve
		Other (See remarks)
	Corrosion	Cleaned corrosion
		Repaired parts
		Replaced parts
		Replaced valve
		Other (See remarks)
	Worn	Repaired parts
		Replaced parts
		Replaced valve
		Other (See remarks)
	Other (See remarks)	Repaired parts
		Replaced parts
		Other (See remarks)
Appearance	Corrosion	Cleaned corrosion
		Cleaned and painted
		Replaced
		Other (See remarks)
	Other (See remarks)	Other (See remarks)
PM	Scheduled	Performed
		Other (See remarks)
Other (See remarks)	Other (See remarks)	Other (See remarks)

Work Order Numbering System

(*Note:* The WO numbering system provides for assigning each separate work order a unique number to allow keeping the work done under that number separate from other maintenance work. The work order number arrangement at many plants evolves in form such as its length and inclusion of alphabet characters. Having a record of how the numbering system works and has worked is very useful. Keeping this record pays dividends in the future if there is commitment to keeping it current.)

Current numbering system

Each WO is numbered with a separate, unique seven-digit code.

The first digit is the plant code (as defined in the Plant and Unit Codes). The second and third digits are the last two digits in the calendar year when the WO is originated (e.g., "93"). The originator writes in the first three digits (plant and year) when initiating a WO. The last four digits are made up of one letter followed by three numbers. Each separate WO is given a unique combination of last four digits, which may be from A001 to Z999. The last four digits (unique) of the WO are already preprinted on the blank WO forms.

For example, WO number N93G457 is a unique WO number for a work order at North Station written in 1993.

The WO number is also used for the CMMS system as well as the mainframe system.

Blanket work orders are kept in the mainframe system exclusively. A WO form is not required for a blanket work order. The Maintenance Department Mainframe System Administrator assigns blanket work order numbers using the same seven-digit numbering sequence as for regular WOs except that the second and third digits are the unit codes (as defined in the Plant and Unit Codes) and the last four digits are all numbers from 8000 to 9999. The Mainframe System Administrator keeps a log to maintain consistent blanket numbers among plants and to avoid assigning duplicate blanket numbers for different activities.

For example, blanket WO numbers N009132, S009132, and K009132 are for maintenance planning activities for each plant.

The use of blanket work orders is not encouraged for equipment maintenance. The creation of a new blanket WO must be authorized by the Assistant Plant Manager.

Previous numbering systems

Prior to mid-December, 1993, the first digit (plant) was preprinted on the form and the second and third digits were the unit codes (as defined in the Plant and Unit Codes). The originator wrote in the Unit Code when initiating a WO. (This system was changed to the current system to allow using the same form at all plants and because there were still too few available WO numbers.)

Example: WO number N01G257 would be a unique WO number for a work order on North Unit 1.

Prior to 1991, the last four digits of regular WOs were 0001 to 7999 and blanket WOs were 8000 to 9999. (This system was eventually changed because there were too few available WO numbers.)

Prior to about 1990, each plant had separate WO numbering systems and conventions such as using NS at the beginning to designate North Station.

Notes

These are general notes for numbering work orders. Using years instead of unit for the second two digits will cause a potential conflict with old

WOs and blankets in the year 2000 that will need to be addressed eventually. The conflict arises because the year 2000 and unit common both would use 00.

Manual Distribution

(*Note:* The plant should keep a list of who has the manuals so that it can update their documents. It is also valuable to list actual names along with titles so that all crew supervisors, managers, engineers, planners, and even computer support persons can be included in having the work order system documentation. Normally this listing is published along with the manual in this back section. A plant with general access to an internal intranet Web site should consider discontinuing the maintenance of a hard copy work order system manual in favor of the more plant-wide access to a Web document. Alternately, a plant with a CMMS with plant-wide access may have the computer system contain the bulk of the documentation of the work order system and codes. In either case, this paper document would be discontinued because of its more limited access and the trouble of keeping it current.)

K

Equipment Schematics and Tagging

This appendix sets forth guidelines for equipment identification and tagging which are essential for setting up the filing system used by planning. Planning uses the equipment tag numbers in its filing system. Nearly every piece of equipment should be tagged with a number if maintenance work is ever performed on it. The repetitious nature of maintenance work demands that records be kept identifying the circumstances of past work. The past work cannot be improved upon if records are not kept. Records cannot be kept if there is no numbering system to allow arrangement of the files or computer records. It is also not enough to have an equipment number for each item. That number must be clearly visible on the piece of equipment. With this arrangement, it is possible to utilize equipment numbers practically to assist finding equipment and keeping records. A plant might set guidelines if not all equipment is to have an equipment number or tag. All lines and devices critical to processes should have numbers and tags. It may be permissible to exclude certain drain lines smaller than 2 inches. Appendix J equipment type codes recommended giving control valves and their actuators a single number. Similarly, a plant should seriously consider giving a single number to an entire drive-train such as a pump, its fluid drive, and its motor. Components internal to other equipment such as boiler tube section inside a furnace may be impractical to tag, but numbers can be established for files to track different sections of tubes. Valves contained within the protective structure of a steam turbine might be numbered, but not tagged directly. A placard may be set near the turbine identifying such concealed valves and their designated numbers. The numbers might be placed on labels in the control room near appropriate controls or switches.

The following sections establish practical guidelines that may be used to establish equipment numbers and place appropriate tags on the equipment.

Equipment Tag Numbers

There are several issues necessary to mention when discussing equipment numbers. Included among these issues are the uniqueness of the numbers, the use of intelligent numbers, and the use of equipment schematics.

First, equipment numbers must be unique. A number used by one piece of equipment cannot be used for any other equipment. Because the planners will keep individual records for each piece of equipment, each number must indicate a single repository for information. If the plant utilizes existing numbering systems supplied by vendors, it must avoid accidental duplication.

Second, intelligent numbers aid maintenance work because they help identify involved equipment and facilitate filing or computer classification. For example, an equipment number that begins with B for all boiler equipment would allow all boiler files to be kept together. Figure K.1 shows an example of an intelligent equipment number for the valve used in the work order example given in Chap. 5. Although intelligence in equipment numbers is helpful, one must take care not to put so many digits into the equipment number so that the number becomes awkward. Excessively long strings of digits make it more difficult to record the correct number when writing work orders or otherwise using the number. The length and arrangement of the example shown in Fig. K.1 seem to present a sufficient amount of information without being too lengthy. It also helps if the characters have some recognizable aspect in themselves.

Figure K.1   Intelligent equipment number for North Station Unit no. 2 Fuel Oil Service Pump Control Valve Strainer B.

For example, the plant character of the example is N for North, the unit characters are 02 for Unit 2, and the group number is F for Fuel. All of these characters make sense and avoid the problem of having a long string of characters making little obvious sense to most plant personnel. Another consideration with an intelligent number is whether to use process location or physical location (or both) within the equipment number. For example, should a number for a process line valve designate that it is in the XYZ process or that it is in the ABC building on the first floor? This issue is whether one should use function or location within the intelligent coding. Either arrangement might be satisfactory. Additional codes can always be placed elsewhere in the work orders or computer record to retain the information not given by the equipment number. The coding structure in App. J and Fig. K.1 uses a process function intelligence for the equipment numbers. A third choice is using the specific equipment without regard to either its function or location. For example, a motor might be numbered as MOT-53. Regardless of where the plant utilizes or moves this motor, it will always retain its same number. Moving a motor from one process or building in the previous arrangements would cause problems trying to keep perfect equipment records. The problems would occur because the motor would change equipment numbers to fit in a new process or location. On the other hand, both maintenance work and plant operation are usually focused around plant processes or systems. Therefore the most helpful equipment numbering arrangements appear to be those utilizing process, not location or type of equipment. This touches on the last issue dealing with intelligent numbers. What then should be done to track rotating spares or other equipment that moves from place to place? *In actual practice, many plants have few pieces of equipment that move from place to place.* Maintenance supervisors or planners can track the location of rotating spares or particular motors by serial numbers or other easily developed spreadsheets. *The intelligence of the equipment number is probably not the place to track migrating equipment.*

Third, if schematics exist, they can greatly assist the plant develop an equipment numbering system. If the schematics exist with a previously developed numbering system on some systems, but not a plant-wide numbering system, the previous numbers can be incorporated into the new numbering system. For example, if there had been existing schematics with a previous numbering system with numbers such as CVS1, CVS2, etc., then Fig. K.1 might have been given the number N02-FC-CVS2. The plant uses the previous vendor number for the unique part of the equipment number. If the plant is creating the schematics and the equipment numbers at the same time, it is a good opportunity to utilize both efforts together to develop a logical numbering arrangement.

Figures K.2 and K.3 show how a plant can use an existing schematic with no existing number system to develop equipment numbers. The planner or

Figure K.2 Example of an equipment schematic for North Station Unit no. 2 Fuel Oil Service Pump System.

Figure K.3 Example of use of a schematic to give unique equipment designations for the last digits of equipment tag numbers.

647

other person responsible for choosing the numbers simply goes around the process flow diagram adding sequential unique numbers for the end of the equipment code. Only the unique last numbers might be shown for each equipment to avoid cluttering the drawing. The plant would use the schematic with the balloon numbers to create a list of complete equipment numbers with equipment names for the FC system. The plant could send this list to someone who makes the tags. The tags could be made all at once, then hung as maintenance is performed, or a more extensive initial effort could be made to hang all the tags. The problem with this latter approach is having to locate the equipment when it is not already identified in the field for maintenance work with a deficiency tag. The schematics with the balloons help finding the equipment as well.

If the plant uses schematics, there should be a designated responsibility with resources to create and maintain them. This effort is normally beyond the scope of the planners and possibly beyond the scope of the planning department itself. The planning department could acquire an engineer, designer, or drafter to lead the effort. This effort is most appropriately placed in either the plant or corporate engineering support groups.

Without any schematics, the planners might easily mount an effort to hang tags that do not contain equipment names. The planners would have perhaps 100 sequential tags made for each system, such as N02-FC-001 through N02-FC-100. Then, as the planners scope work orders for each system, they would take the next available tag for each system being scoped and hang it on the equipment. For example, a planner is about to scope a job within the Unit 2 fuel oil service system. The planner takes a tag off the top of the N02-FC stack and hangs it on the equipment during the initial visit to the equipment. An appropriate job for a college, technical major intern or summer student might be the hanging of all tags within systems. Preferable, however, is the hanging of tags with numbers and names by a first rate effort of the plant engineering staff.

Equipment Tag Creation and Placement

This section addresses two issues with the tags themselves. What should the materials and other physical arrangements of the tags be and how should they be placed on the equipment?

First of all, there are an extraordinary diversity of tag materials and companies specializing in equipment tags. Each plant should evaluate its own circumstances to select the tags that make sense. Many companies purchase thousands of small metallic tags with sequential numbers for asset accounting by the financial department. These numbers may be satisfactory for maintenance use as well but they lack the advantages of an intelligent numbering system. Many plants have found engraved plastic laminate tags to be satisfactory and this section gives guidelines with their usage in mind. Many of these guidelines are general in nature, applicable to other types of tags.

Typically, plastic tags use two-color, laminate material. The first color is the top or overlay material and the second color is the underlying color that will show through as the letters after the engraving. "Black and white" designates a black tag that will have white letters. "White and black" designates a white tag that will have black letters after the engraving. $^1/_6$-inch thickness is adequate for most applications. One should normally specify standard stock materials. Special colors may dictate special orders for office areas. Normal material is usually adequate, but there is material that is ultraviolet safe and continual sunlight in the outdoors might require this material. There is also sturdier material for higher ambient temperature, but usually hanging tags by wire will keep a normal tag safe enough.

A tag that is easy to see is desired. A plant person looks for the tag rather than the letters. In a low light situation one might want to look for a white tag that has black letters than try to find a black tag.

2×3-inch makes a good size tag for most applications. Most equipment should have a tag that is readable at arm's length. Letters should be proportionally in height and width to the tag. This means that after the equipment number, there might be only two lines for equipment name on a 2×3-inch tag. The engraving machine might make only a single pass with the engraver point. Letters cannot be made too bold or they will run together and become illegible.

Certain tags such as for large tanks may have to be read far away. Tags can be made in any size. For a 6×8-inch tag, the engraver point would make several passes to create wide individual letters. In any case, one wants the letters visible, but not running together.

Rounded corners are possible. These may be practical in a high traffic area where a person's clothing may catch on the tags. The engraver company would take the extra step and individually stamp each corner of each tag to provide rounded corners.

A local trophy shop should be able to make most plastic tags. Appendix C gives suggested specifications and names one possible supplier. The creation of the tags requires someone competent to place complete thoughts on each line. For example, below the tag number one would want this:

CONDENSER PRE-BOILER
CLEAN UP RUPTURE DISC

and not this:

CONDENSER PRE-BOILER CLEAN
UP RUPTURE DISC

What does "*UP RUPTURE DISC*" mean? In addition, the engraver may have to make abbreviations such as shown in the sample tag for Fig. K.4 to avoid having letters too small.

Figure K.4 Example of 2 3 3-inch equipment tag for North Station Unit no. 2 Fuel Oil Service Pump Control Valve Strainer B.

Second, several comments regarding the placement of the tags are appropriate and, while obvious, should not go without mention. Tags should be hung vertically, not face up. Even if the equipment is lower than eye level, hanging the tag tilting upward only invites dust accumulation and illegibility. For neatness, tags should be of a uniform size and color. It also looks neat if they are facing in the same direction. On the other hand, using wire to hang the tags allows reattachment by the technicians after maintenance. This is better than the neatness of precise tags all glued and facing in the same direction. Tags should also be consistently placed if possible. For example, tags might be attached with silicone caulk on tanks under nameplates or on the same sides. Obviously, tags should be hung where visible, not obstructed by pipes or conduit. Tags might be normally wired to valve handles. Wiring tags might use two tag holes to allow facing the tag toward an aisle to make it readable without handling. Appendix C identifies wire that has great ability to be twisted repeatedly without becoming brittle and breaking.

Summary

In summary, the identification of equipment with unique numbers is of vital importance to the planning operation. It is to be hoped that the planners can make use of existing numbers to create unique equipment files. If equipment numbers must be created, intelligent numbers should be employed since they help facilitate the filing and other maintenance operations. Several considerations must be made when establishing intelligent numbers. The plant must also go beyond assigning equipment numbers and actually attach tags with assigned numbers to the equipment in the field. The tags themselves offer a variety of choices. Simple plastic tags may be successfully used and customized. Tags allowing easy reattachment by technicians are preferred.

Computerized Maintenance Management Systems

This appendix gives additional information on implementing and using a computerized maintenance management system (CMMS) to assist planning, as referenced in Chap. 9. The main chapters did not include this information to avoid detracting from the book's main purpose of explaining how to do planning. This appendix mentions some common features and aspects a planner would like to have in a CMMS and then discusses different types of CMMS software projects such installing a new system or simply upgrading an existing system. While not providing a project management guidebook, the appendix also covers major problems of software projects and how to try to avoid them. Furthermore, take any discussion points below that seem derogatory toward the Information Technology (IT) group primarily as a caution.

Planning Principles versus Using a CMMS

It is appropriate to begin this appendix by distinguishing plainly between planning and using a CMMS. They are not the same. Planning is a tool to increase productivity. A CMMS is a tool to apply information. They are overlapping tools.

Examine the six planning principles of Chap. 2 against this information tool. The CMMS does not dictate that planners should be *separate* from crews. It does not dictate that planners should concentrate on *future* work. The CMMS does require *component-level* equipment designations for recording information. The CMMS does not emphasize the technical *skill of the planner* in setting scopes and estimates. The CMMS also may not count on *craft skills*. The CMMS certainly cannot measure *wrench time* as only statistical work sampling is adequate.

On the other hand, one intent of the CMMS is the same as the planning system, to help reduce delays.

Examine the six scheduling principles of Chap. 3 against the CMMS. The CMMS does not dictate that plans should be made for the *lowest possible skill* level. The CMMS may presume that *priorities and schedules* are important, but may not dictate that schedules should not be interrupted nor enforce organizational discipline against setting false priorities. The CMMS does become involved in *forecasting* availability of personnel. The CMMS may allow sensitivity to examine scheduling work hours for 80%, 100%, or 120% of a crew's available hours. On the other hand, the CMMS would not dictate the principle of *100% as a weekly goal*. Neither would it normally dictate the principle of working persons below their skill levels to match job hours to crafts available. Instead, the CMMS may blend the weekly and *daily scheduling* routines into setting weekly schedules on a daily basis by the computer operator who may be a planner, scheduler, or crew supervisor. The CMMS may help measure *schedule compliance*, but it could not on its own define what activity would constitute compliance.

In addition, regarding reactive work explained in Chap. 4, the CMMS may not differentiate planning's different responses to different degrees of *job reactiveness*. One sees the principles that embody a successful planning program do not come from the CMMS. Largely, a CMMS is a database manager where planning may seek information. The CMMS contains information; it should not dictate planning strategy.

With this distinction made, the following section lists helpful computer characteristics for maintenance, in particular, the planning group.

Helpful Features for Planning and Scheduling

The following features are more common for a CMMS to include (whereas, Chap. 9 mentioned other more advanced features that a CMMS might not commonly support). The following may be common-sense, but having a list and discussing them to some degree is worthwhile. The list includes aspects such as system performance that readers should never take for granted. The list also includes simple fields easy to add to make the system more useful.

User friendly

If the CMMS is not helpful, it will not help maintenance. If it is not useful, users will not use it. One would think that all CMMS software is useful and helpful, but many systems are not. Many systems are painful to use and cause resentment in maintenance toward management that inflicted the system upon them to collect information. Many times management is not even aware of the pain because the IT group managed the project. Rick Baldridge (2005) of Cargill advocates that a CMMS "should be fast and painless."

What are some aspects of so-called "user-friendliness?" Perhaps most important to the maintenance person is using single screens mimicking the printed work order form. Users should be able to fill in different aspects of the work order form on the computer. However, the computer can provide help such as drop down choices and tab orders that take users to the next field they would normally use. If a single screen would be too cluttered, the main screen might have limited tabs to bring up new subscreens with additional information. Above all, the maintenance persons should know where they are in the overall CMMS without becoming confused as to which screen to choose to go where they want. David Berger and Bruce Tognazzini have also each described excellent points about user-friendliness.

David Berger (2003) who writes well-respected articles on CMMS usage for *Plant Services* magazine pins down several points on software being simple and easy to learn or use. As mentioned above, the software should follow how the user would fill out a work order rather than simply provide data fields. Bubble help would provide inexperienced users information on fields. The software should group menus logically, consistently, and with a minimum of nesting. Clear, action words help (such as Chap. 5 recommends for job plans). The CMMS should allow the company to configure the screen for how the company wants to use it, not just how the CMMS vendor thinks it ought to be. Error messages should be clear and suggest a solution. Finally, dangerous actions should contain warnings along the lines of "Are you sure you want to do this? All changes will be lost if you continue!"

In addition, Bruce "Tog" Tognazzini (2005) of the Nielsen Norman Group (www.nngroup.com) writes web software articles including the referenced article on *First Principles of Design*. This article wonderfully specifies terms of user-friendliness in IT language. He confirms many of David Berger's points. Among additional specifics he proposes are the following. The system continuously saves the users work with the "option for the user to undo any activity at any time." Status mechanisms should keep the user informed (such as tiny schedule icons filling up with planned jobs). Use colors carefully and with other visual clues as many persons have some color deficiency. Defaults should help users not have to fill in every field, but the user should be able to use something other than the default. The user should have well-marked roads to follow with landmarks to do their job, but they ought to be able to go off-road if they know what they are doing. "Use larger objects for more important functions." Make objects look like their function. For example, many CMMS packages use an oilcan icon to direct users to preventive maintenance (PM). Users should never lose their work because of something they did by accident. Use large-enough font sizes. Finally, users should be able to take up where they stopped. This is similar to adding a bookmark to a book. It should not be a chore to take up where one leaves off. Tog also says that the only way to know what the users expect is to do user testing. There is no substitute.

Be sensitive that some CMMS packages are not user friendly. Do not accept programs that cause a hardship. Do not accept that IT understands the maintenance need for ease of use.

Speed is everything

Along the lines of user-friendliness is speed. Speed is everything. Consider the following:

> You arrive home, flip the light switch and wait 30 seconds for the light to come on (log onto the CMMS). You go to the living room, flip the light switch, and wait 20 seconds for the light to come on (open work order module). You flip the TV switch and wait 20 seconds for the TV to start (submit query to find certain work orders). You push the channel button and wait 5 seconds for each channel to change (view each work order). After 45 seconds, you have been able to page through nine channels (work orders). You ask the electric and cable TV companies to speed their service (ask them to speed up the CMMS), but both ask you to present a "business case" why you cannot wait a measly 20 seconds for any one task. You change electric and cable TV providers (change CMMS vendors and outsource IT to someone who cares).

Paper work orders *never* slow down the user. As soon as the user writes down certain information, the user can proceed to the next blank or give the completed work order to whoever needs it. A computer can exasperate the user with endless processing delays. Many computers are slow and frustrate users past the point of reasonableness. Maintenance must demand a speed standard if the IT group will not recognize what is reasonable. David Berger specifies that any function should have a response time of 2 seconds or less. Tog Tognazzini further specifies that the computer should acknowledge all commands within 50 milliseconds. An hourglass should appear for anything taking longer than $1/2$ seconds. A message should estimate the time remaining for anything that takes longer than 2 seconds. An animated progress indicator (such as a scroll bar) should communicate how long the task might take. If anything takes longer than 10 seconds, a beep or something should indicate the finish. As a minimum, the computer should allow the user to do something else while it finishes a long task. *Tog says, "Make it faster...Be ruthless."*

Anything in seconds sounds fast, but for a user making repeated key strokes and operations on each work order, waiting for even a few seconds every time is a productivity killer.

Reliability is second

Next is reliability. If a company commits itself to a CMMS, will the CMMS crash unexpectedly, continually? Some vendors have offered near-beta versions of their products in an attempt to stay ahead of the market with "bleeding edge" technology. Because they have not thoroughly tested what real users might do, common mistakes of users lead to server crashes and system freezes. Do not be quick to use a new CMMS or CMMS major new release.

In addition, as mentioned in Chap. 9, if the system does crash, will the crash corrupt data? Will the system lose collected information? Pay attention to system parameters such as continual backup or duplication of information.

Inventory help

Moving from overall characteristics of a good system to more specific ideas, consider the inventory module. Many storerooms are not under the control of maintenance and their portion of the CMMS usually offers limited rights to maintenance personnel. Nevertheless, some CMMS packages allow adding a single field in the inventory module to which any planner or technician can add free text comments. For example, if the standard inventory name of something were not very helpful, technicians could add a comment that the item is "used on unit 3 boiler feed pump."

More obvious and already discussed in Chap. 9 is automatically collecting a record of everything ever used on a piece of equipment. The CMMS might also print this list at the end of any printed work order as a helpful reference for the technician. This is not a waste of paper.

Is this a modification? Rework? Call out?

Three helpful fields help assessment of the maintenance program. These can each be a simple check off or yes/no field.

A field asking, "Is this a modification?" can tip off engineering to look for design changes. It might also alter the planning process to seek engineering approval on certain tasks. A modification is not replacing a defective valve, repairing a leaking flange, or overhauling a pump. A modification is changing the design basis of the plant such as adding a new strainer in a piping system, changing a motor that increases the horsepower, or changing a piping system from cast iron to stainless steel. The originator can check the box, but planning can later verify the choice. Technicians working on a job can also check the box and the planner can finalize the choice during closeout of the work order.

A similar field asking, "Rework?" allows operators to voice their assessment of a particular task. Perhaps the operators feel that the equipment repaired only a week ago should not have a new problem so soon. Perhaps the work completed by maintenance last time did not fix the problem. The operators write a new work order and designate it as rework. In addition, a single work order where maintenance incorrectly repairs something might stay open until the problem is resolved. Although rework is involved, there is never a subsequent work order. In this case, the technicians, supervisor, or planner can check the rework box. Anyone that feels a task involved any rework can check the box. An issue arises that perhaps someone might "uncheck" the box for one of two reasons. First, a person might disagree with the original assessment. In this case, the rework choice generates healthy communication. The plant might have a policy that once the box is checked, no one can uncheck it. Alternately, the plant might have an audit field capability that allows following changes to the field. It might be the plant's choice to let the planning department have rights to uncheck the box. Many plants also favor allowing the operators to claim rework that stays with the record even if the work later turns out not to be so. The plant wishes to use the rework field to gauge the opinions of the operators. Allowing the operators the final say on rework gives this customer of maintenance the right to voices its opinion for the record. Second, the supervisor or technician might fear penalties for having rework. In this case, Chap. 9 insists the plant should never use the results of the CMMS for discipline or else data will never be entered accurately.

Finally, a field asking, "Call out?" allows the operator or technician to record whether the work involved calling anyone out to the plant outside of normal work hours. This is different from simply recording overtime. Later reliability analysis might use the rate of call outs. If neither the operator nor technician checked this box when appropriate, the planner might check it during project close out.

Deficiency tag

Chapter 8, Resources, and App. J, Work Order System, discuss the usefulness of a deficiency tag. The CMMS should allow the originator to enter the deficiency tag number for cross-referencing.

Outage and clearance versus status

An *outage* is taking an entire unit such as a steam plant or entire assembly line out of production for maintenance work. A *clearance* is making a single piece of equipment available for maintenance that may not prevent product production (due to redundancies or excess capacity built into overall units). Work order *status* refers to whether the work is approved, planned, or completed; not whether it takes an outage of a unit or clearance of equipment to execute the work.

Work orders must identify work that maintenance can complete only during outages as soon as possible and this identification must remain visible on the computer regardless of the work order status. Identifying outage requirement separately from status allows planners to see how many work orders requiring an outage have not yet been planned or schedulers to see how many work orders requiring an outage are ready to work for the upcoming weekend outage. Some CMMS packages only designate outage as a status such as "Waiting for Outage." This keeps management from reviewing outage work before the work is planned and ready to work. Keep a separate field in the CMMS to denote whether work is outage work. Similarly, use an extra field to denote whether work needs equipment clearance from operations whenever it is worked. Both outage and clearance need to be independent of work order status.

Priority

Appendix I on scheduling dynamics sets forth the extensive case that originators set valid priorities that equipment criticality should not easily over-write. Ensure originators have a field to select priority. Do not let the CMMS dictate priorities solely based on equipment or existing job plans. A CMMS suggestion might be acceptable.

How found

Appendix J on work orders and codes suggests some helpful codes on "how found." Knowing whether an operator found the problem on a round or a maintenance technician found the problem helps engineers evaluate the reliability program of the plant. How are most problems discovered? Is predictive maintenance (PdM) doing its job? Are maintenance PMs effective? Add a field with a drop-down list to capture this origination information.

Attachment or link

If a CMMS allows attaching or linking electronic documents, their existence should be obvious to the person accessing the work order. A person with a paper work order sees that a staple attaches an alignment checklist. The person on the CMMS should also notice any attached forms. This is a common failing of CMMS packages where the user must hunt for suspected attachments.

Equipment module

For the equipment module, imagine a paper file and being able to see all the drawings and original purchase information including warrantees. Planners and technicians should be easily able to access drawings and schematics from the CMMS or at least see their numbers so they can go fetch them from paper drawing systems. The CMMS should display model numbers, serial numbers, and other specification information. CMMSs should have most of this information automatically print at the end of printed work orders for reference.

Types of Projects

Software projects take several forms. CMMS vendors often release patches to existing systems to help companies correct "bugs" found by the user community. Less frequently, vendors release new versions of the software requiring users to upgrade. These new versions often take advantage of new technology and processes. Sometimes, plants change their existing CMMS vendor altogether. Commonly, this is due to a large company trying to standardize the different CMMS systems used across various plants or units. Finally, there is the company without a formal CMMS that underlies the supposed basis for this handbook, doing planning and scheduling without a CMMS. This company seeks its first system.

Patches

Vendors try to minimize the release of patches and honestly try to test systems before release. Yet there is no substitute for actual user testing. CMMS vendors test alpha versions of new software in-house and usually partner with several companies to test beta versions. Nevertheless, hundreds of final users across different platforms and final configuration discover new problems. Some problems affect only individual companies while others are obvious defects for every user. The vendor attempts to help individual companies find easy work-a-rounds, but sometimes must issue patches to the user community.

The receipt of a patch usually allows a company to evaluate whether it wants to go ahead and install the patch. If the patch does not appear to include any problems pertinent to the specific company or if the company IT group is too busy, the company might wait to install any individual patch. However, at some point, a company usually installs several patches through the IT group (presuming a normally networked CMMS requiring IT involvement). The preferred method of patch installment involves the IT group implementing the patch on a QA or test server and allows the system administrators in the maintenance group to test the overall functioning of the system before going into production.

Upgrades

Upgrades to new versions of the CMMS software are much more serious and risky. Standard industry wisdom is to wait for others to upgrade first and obtain experience. Maintenance is about reliability, not about being the first. One incentive to move to a new version is that the CMMS vendor cannot "support" the old version forever. This means that while the vendor will answer the help line and explain how the system works; the vendor will soon stop developing patches to fix common bugs discovered in the old software. Nevertheless, this is not enough incentive to make a company move too rapidly to an unproven new technology with more bugs.

Companies should approach major upgrades as formal projects heeding the cautions identified in the remainder of this appendix.

Changing systems

Changing systems because of company mandates or simple exasperation with an old system is not uncommon. However, beware that "the grass is always greener on the other side of the fence." You are aware of your software problems and only see the salesperson promotion of the new software, if you are not careful. Again, treat these as formal projects.

New systems

New systems often cause the least frustration for CMMS users. The CMMS user does not have a baseline for judging what is worse about the new system. The users often have a better idea of what they want out of the system in terms of automating the paper process. Users converting from other systems often lose this sense of automating the maintenance process and focus on the processing problems of the new system versus the old system.

Nevertheless, these companies are most at risk for underestimating the risk of software problems interrupting their processes. Any substantially-sized company not installing a simple patch should approach a CMMS as a formal project. Formal projects should implement the best CMMS programs where they leverage the proper balance of maintenance and IT expertise.

Big Glitches in Real Systems

The following listed examples supplement the few presented in Chap. 9 to illustrate the seriousness of IT problems. All of these "glitches" are true problems in actually released software in wide usage by major CMMS vendors. While it might be mentioned that none of these problems would occur in a paper system, there is another more serious point to make. Even when found, these problems were not easily resolved. These problems do not go away soon after they are encountered and reported. In addition, frequently fixing one problem introduces another problem. Many major upgrades planned for 12 months continue past the

24-month mark still testing and fixing without going live. If projects go live anyway with significant problems, they damage the CMMS and IT's credibility even they do not cripple planning.

- You cannot order more of an item from the storeroom unless ordering more than ordered the first time. In other words, if you ordered eight widgets the first time, you must at least order nine this time.

- The entire server crashes when selecting the "move equipment" command accessed through the work order module.

- The entire server crashes if you try to create a new work order after changing the status of several other work orders.

- The server crashes at least once per day for no apparent reason.

- Although technicians can enter their hours on timesheets, these hours do not come back to the referenced work orders correctly.

- You cannot print more than one work order at a time if any of them has an attachment.

- The PM "next due date" never updates in certain instances, which effectively cancels the PM forever.

- The new upgrade version is slower than the old version.

- You can only see six steps of a 100-step job plan at a time and you have no indication that any of the step fields have long descriptions. Some might, some might not. You would not know unless you click on each individually.

- You can put as much information as you want into a long description field without warning, but the final work order report will only print out a limited amount.

- Spell check locks up the computer.

- You cannot restrict inventory searches to items where the balance is greater than zero.

- If an equipment number has more than 16 characters, you cannot drill down to see or select any of its children.

- If an equipment has more than 1000 children or grandchildren, you cannot assign it to a new parent itself.

- It takes 4 minutes to save a work order with any changes.

- Error message simply says, "Boolean field is blank."

- Selecting the "cancel" button from a drop-down list locks up the computer.

- After selecting a piece of equipment from a hyperlink, it takes 20 seconds to return it to the work order.

- The upgrade conversion mapped all the jobs in the old CMMS with a clearance requirement of "Yes" to the new CMMS with a clearance requirement of "No."
- Occasionally, the order process will print out 10 pick tickets in the storeroom for an item when only one was requested.

Death March Projects

Edward Yourdon (2004) wrote a wonderful book about large software projects, *Death March, Second Edition*. This sounds extreme, discussing a "death march" when a company is simply trying to use a computer to facilitate maintenance. Is maintenance that complicated? Are computers that complicated? For whatever reason, this discussion is not extreme because many software projects do go awry and recovery is not so simple. As the *Maintenance Planning and Scheduling Handbook* presents planning in Chaps. 2, 3, and 4, it is much easier to start out with proper planning than to recover from a poor planning program. Maintenance planning seems easy in theory, yet most planning organizations are frustrated with planning. Using a CMMS presents the same scenario. Seemingly simple computer software causes a lot of frustration in industry. A word to the wise beforehand is appropriate. Knowing the extreme of what can go wrong is invaluable for anticipating possible problems. Being familiar with warning signs might allow problem avoidance or timely recovery action. Therefore, it is entirely appropriate to consider extreme "death march" software projects, what they are, why they happen, and key points to survival. The following points all come from Edward Yourdon (with page references). The selected points are only the tip of the iceberg of primary interest to maintenance persons. Yourdon's book offers a much more detailed analysis of these issues, especially from an IT standpoint.

What they are

Death march projects are defined by likelihood of failure in greater or equal to 50% or by having one or more of the following (pp1-3):

- $1/2$ schedule needed
- $1/2$ staff needed
- $1/2$ budget needed
- 2 times features as normally required

Death march projects are the norm, not the exception (pxvi).

Why they happen

The following points help explain why death march projects occur.

- Naïve promises made by marketing, executives, or others (p6).
- Claims such as "Unless it is available in the next six months. The competition will grab the entire market" (pxvii).
- Naive optimism of youth: "We can do it over the weekend" (p6).
- Attempt to use "bleeding edge" technology (p6). Thought processes such as "If the new technology does work, it will be worth it" (pxvii).
- Management ignoring new situations until it is almost too late (p6).
- "Methodology gives those with no ideas something to do." Mason Cooley, *City Aphorisms*, Eleventh Selection (p115). (This suggests a company might implement a CMMS simply because it is supposed to be a good idea.)
- Death march projects magnify politics (p46).
- One of the problems with a death march project is the owner is higher in the organization than for a normal IT project and the project manager does not have much contact (p45).
- Upper management dictates tight deadlines and budgets on large software projects because it is tired of numerous past software projects where IT made its own schedule and budget, then delivered nothing, ever (p68).
- Software project managers tend to underestimate the importance of "soft" factors in a project such as morale and perceived quality (p152).
- Persons joining and leaving the project team during the course of its work have a huge impact on the project (p160).
- Brooks' Law: Adding more persons to a late software project just makes it later. The addition of new persons requires extra time for existing persons to train them (p159).
- The persons who quit due to frustration in the middle of a software project are frequently the most productive. The average productivity of the group drops and frustration increases (p154).
- Project dysfunctionality. Grading individual project members on individual tasks during the project harms the overall project. Most companies reward team members for individual accomplishments because of a sense of unfairness to tie their rewards significantly to the project success. While the schedule go-live date is easily measured, the success in terms of quality such as user usefulness is not. In addition, both overall schedule and quality are interdependent among project team members. Management may feel that not rewarding individuals for

independent performance is unfair. Nonetheless, human nature makes rewarding small tasks unwise. Punishing a person who is late on an individual task even though the overall project succeeds, may seem like a method to ensure team members finish individual tasks for the good of the overall project. This is unsound theory. In reality, teams members rewarded thusly make every effort to complete individual tasks regardless of the overall project. In addition, sometimes the identification of proper individual tasks is subject to uncertainty. If the organization rewards a team member for the completion of Task X and the team member finds that Task X is unnecessary, the culture might cause the team member to complete the task anyway. In addition, the success of individual tasks does not ordain overall success because the interfaces among the multitude of individual tasks are often as important as the tasks themselves. Team members are smart enough to choose their success above the project success. Projects fail if team members know that project success does not reward them (p167).

- Time slippage from key stakeholders not making timely decisions (p183).
- No one will say the emperor is wearing no clothes (p178).

Key points to survival

Yourdon includes the following recommendations for surviving a death march project.

- Death march projects *must* have a sponsor that can stop the nonsense (p51).
- The difference between success and failure on a death march project often lies in the project team's ability to identify the critical features of the system (pxviii).
- Quality is also a project outcome and a team may have to decide what "good-enough" software is (p76).
- Do not design for anticipated future needs (p136).
- Keep the owner happy (p45).
- Realize that the IT managers are the problem, not the developers (p33).
- Do not put in a safety factor. Putting a safety factor in a project simply delays the project as work expands to fill the available time, "Parkinson's Law" (p154).
- The project manager must document and publish key decisions being delayed and the slippage of the project schedule. As soon as the project manager is ordered to stop publishing such information, the project manager must at least identify each delay as a project risk (p183).

- Be aware of the 90% done illusion. (It is not done until it is done) (p189).

- The project manager must insist on an absolute right to veto anyone adding an unacceptable person to the project team (p91).

- Do not allow transfers off the project until it is complete. End-time problems need the experts on the team for proper resolution. End-time problems may delay the project and make its end not so near (p159).

- Realize the term "man-month" is a myth. Adding twice as many persons to a project will not cut the time in half. Slight exchanges may be linear; management insisting on a 10% decrease in schedule might compensate with a 10% increase in staff. If more than 10%, square the tradeoff: if management wants to cut the schedule in half, instead quadruple the staff or quadruple the budget (p71-76).

- Provide reasonable office conditions. Persons that are able to close doors and avoid interruptions are over twice as productive (p107).

- Avoid project dysfunctionality. Make the overall success of the project the basis of individual evaluations as much as possible. Team members are smart enough to choose their success above the project success. Projects fail if team members know that project success does not reward them (p167).

Planning a CMMS Project

This section creates a general "plan" for a CMMS project that could be applicable to implementing a major upgrade for an existing CMMS or changing over from another existing CMMS. The plan could also be applicable installing a new CMMS from scratch, but with special modification to add a system selection phase. Chapter 9 addresses a typical strategy for selecting a system. The plan below presumes the company knows which CMMS it is going to use. This presumption addresses the common plight of many companies already having CMMS systems but facing major upgrades or being directed by a parent company to change to another system for company-wide standardization. In the case of multiplant efforts, the company would have a larger team involved with the project, but apply the same concepts. In addition, anytime a CMMS faces a major upgrade could be a time to consider other systems. Therefore, just as the maintenance planner plans a work order for maintenance, this handbook plans a work order for a CMMS. This plan uses the feedback loop concept to identify special areas of concern based on previous industry experience with similar industry software projects. This information should help a team on its current project. (The plan does not consider minor patches. The discussion earlier in this appendix primarily recommends testing for functionality.) Just as in a Chap. 5

job plan, this is not a complete project management guide and presumes that the project team has some skill in managing projects.

Work Request for a CMMS

Senior management has agreed with maintenance to authorize a new CMMS. Management expects a project team to identify the proposed benefits and cost with a project plan.

Planning for a CMMS

The CMMS plan will cover the project team, a scope, a project plan, hardware besides the software needed, a general procedure, and the estimated cost.

Requirements must be established for the project. (This presumes a need for the entire project can be established.) However, because maintenance is only now evolving as a science, the maintenance group and higher management often find it difficult to express requirements in terms useful to the IT group. Even worse is that IT persons have less of an understanding of what maintenance is, but they *think* they do. They say, "Maintenance means fixing things when they break." If the IT group does happen to realize it does not know the proper requirements and has trouble understanding the maintenance group's interpretation of what it needs, it often turns to consultants. While they *might* understand modern maintenance, many consultants may not understand the valid practices of the current maintenance group. Instead, they might introduce too many alternate practices ("to bring the maintenance group into the 21st Century"). Alternately, the consultants or IT itself may simply deem that the proposed CMMS has all the "right" practices. In either case, they are *not* automating the current maintenance program.

This matter of the difficulty of translating maintenance requirements leads to two remarkable project needs. These requirements involve the selection of the project manager and the finalization of the project scope.

First, the manager of the project *must* come from the maintenance side of the house. Maintenance professionals commonly acknowledge that most IT-led CMMS projects are grossly inadequate. Day-in and day-out, a maintenance professional as project manager can properly resolve the seemingly minute details that are critical to software usefulness. The IT group becomes overly concerned with schedule and budget at the end (and frequently throughout) the project. The IT group does not understand that if the CMMS is not useful to the users, it is useless. Only a maintenance professional can make proper judgment calls along the way to make the project successful.

Second, while most projects seek to eliminate scope creep and lock project requirements, the difficulty of expressing maintenance requirements together with the uncertainty of the capabilities of the new CMMS makes it prudent to allow the project to clarify its requirements throughout the project. This is a truly unique, but understandable condition. Scope creep destroys many projects and companies frequently mandate "design freezes" after the initial project charter or as soon as the project runs into any kind of technical, schedule, or budget difficulty. "No changes" IT project management declares. Yet, this undermines the very reason for undertaking the project. The purpose of the project according to IT is "to finish it." Yet, the true purpose of the project is to improve plant reliability through better maintenance. Only a maintenance professional can make proper judgments to make this happen. The project can freeze requirements at the functional level, but not at the lower level that IT desires. For example, consider a project requirement that maintenance be able to plan jobs and keep master plans for continual improvement. During development and testing, the maintenance user leads find that a planner can only see six steps of a job plan at a time and cannot easily scroll down the screen to see the rest of a job plan. The IT group might claim that there was no requirement to see a certain amount of the job plan at a single time and resist any change. Yet, only when planners review the trial job plans on the CMMS can they even tell if the CMMS plan capability is practical. Certainly, IT persons cannot tell what is practical from a planner viewpoint.

Some CMMS programs do not adequately meet many common maintenance expectations and needs because the vendors are primarily software companies at heart. They hire expert programmers to integrate many fast changing technologies of computer hardware and software. On the other hand, these companies see maintenance as much less sophisticated. They may not even have a true maintenance professional on staff, feeling they can interview potential clients for maintenance needs. Even if they do have a maintenance professional on staff, the company may not let this person make the critical decisions that would make the product most useful. When they do let the maintenance professional participate, they may bring various department heads together and vote. In a vote among five computer group leaders and a single maintenance person, decisions swing toward computer solutions over maintenance preferences.

General disrespect in the IT group for maintenance knowing what it wants and the expectation that a large CMMS provider or consultant does know leads toward the maintenance department not receiving a good product.

Keep project requirements at the "performance specification" level and insist on professional determination of what meets proper performance. Support from the project sponsor may be necessary to stop nonsense about specific requirements.

Staffing

Typically, a work order plan develops a scope before establishing the labor requirements, but in this case, a project team will help develop the scope and the rest of the plan.

Member: Project manager assigned from the maintenance workforce, typically an engineer or other professional with project management experience and familiarity with the planning group.

Member: Planning supervisor.

Member: Plant system administrator.

Member: IT programmers and support persons involved with the CMMS.

Resource: IT programmers and support persons involved with other company software.

Resource: Maintenance manager.

Resource: Maintenance planners.

Resource: Consultants typically from the vendor CMMS or ones certified by the CMMS vendor. These persons will do the more intricate configuration that requires more familiarity with the system.

The time commitment for the project manager is near half time for most of a year. The other members also will have a heavy time commitment that will affect their other responsibilities.

Scope

The scope of the project identifies specific overall areas of benefit for the CMMS and identifies the expected value. The project team should also capture some of this information in a formal project charter recommended in the special tools section of this project plan.

Areas of benefit.   The scope of the project should identify the major benefits expected from the project. Chapter 9 discusses the key benefits that should help maintenance planning including standardizing work processes, inventory control, information for metrics and reports, finding work orders, linking information to equipment, having a common database, helping with scheduling, and helping with PM generation. (The reader should not include any of these areas that are not significant drivers for the reader's situation.) A project charter should capture the scope as described below in the special tools section.

It would be helpful to include the word "additional" when discussing features of the CMMS project. The existing maintenance process and existing CMMS have certain features that genuinely help maintenance. Users often take many of these features for granted. For example, the overall speed of the existing CMMS may be adequate, but the project team might not realize that the proposed newer system might be configured to slow down some processes to an unsatisfactory speed. The old system allowed scrolling through work order screens each in a fraction of a second. The new system might take over 4 seconds for each record, making it almost impractical to scan through a large amount of work. The maintenance group never insisted on a "speed specification" because it did not imagine that a new system would be worse in this critical regard. Specifying that the new CMMS would provide the "additional" benefits over the old system helps the maintenance group discuss details with the IT group. It is not acceptable for the new CMMS to cripple existing satisfactory capabilities. On the other hand, new systems might well compromise certain old features in some respects, but maintenance would be overall better off going forward to the new system. (Careful CMMS selection up front can help reduce the likelihood of selecting altogether unsatisfactory systems.) A final note on specific features such as speed, system stability, and user-friendliness: If any desired features are specific reasons a plant is upgrading or changing a CMMS, the project should explicitly list them among the overall benefits expected. The existing system might simply be too slow. It might lose data consistently or be unavailable too many times. It might simply require too much computer expertise to use. The specified benefits of the project should certainly include these areas.

Value of benefits. COPQ stands for Cost of Poor Quality. Management wants the CMMS project to deliver an overall benefit. This means there exists a present opportunity to improve. The gap between the most improvement possible and the current situation represents waste and inefficiency known as "poor quality." COPQ represents this poor quality, usually in monetary terms. COPQ does not represent the benefit value of the CMMS project, but the total wasted dollars in pertinent areas that are not perfect.

The project team should determine COPQ for each area and the potential value for the CMMS to help. Benefits include effectiveness such as plant availability and efficiency such as workforce productivity. Plant availability usually carries far more value than productivity, though each is significant and interdependent.

For example, a power station might have an 85% equivalent availability factor. The 15% wasted availability might be worth $4.5 million/year to a 1000 MW power station in power transaction capability and $270 million in capital construction cost for the 150 MW "lost" by the low availability. The CMMS will not save *all* of this amount for the utility, but it does establish a basis for how much waste is available for improvement. The project team must decide how much of this COPQ the CMMS might help. It might be reasonable to expect that a new CMMS that improves plant areas with evident problems (listed as benefit areas above) could expect to take credit for a certain percentage increase in availability with a specific value.

The project team might decide how much process standardization and better planning information would contribute to maintenance productivity. For a 90-person workforce with average wages of $25/hour and 35% wrench time, the COPQ is $90 \times 25 \times 2080$ (hr/yr) $\times$ (100% − 35%)/35% = $8.7 million. Nothing will improve a workforce wrench time from 35 to 100%, but it would be reasonable that having a good CMMS over a poor CMMS or no CMMS might be responsible for 5% of an improved wrench time for a company with a good planning program. The planning program can be expected to help improve wrench time from 35 to 55%, but 5% of this could be due to other system factors including a good CMMS. It would be reasonable to attribute 5% to the CMMS as $90 \times 25 \times 2080$ (hr/yr) $\times$ 5%/35% = $668,000 per year.

If better inventory control is an expected benefit, the project team might enlist the maintenance supervisors, purchasing agents, and storeroom personnel to consider the COPQ of the current process and how much improvement the CMMS might yield.

Similarly, the project team should consider each area of expected benefit to develop a COPQ and estimate how much the proposed CMMS might help reduce COPQ. Reports, metrics, PM generation, and similar benefits might claim a portion of the overall availability COPQ. Finding work orders, scheduling, and similar benefits might claim a portion of the productivity COPQ.

Obviously, the potential value of a CMMS depends on the current COPQ of a particular plant and how a CMMS might help specific areas. Knowledge is power and the better a project team knows its individual situation, the better it can estimate the impact of the CMMS. Following a structured approach that breaks down each potential benefit into COPQ's can help sort through information and build a more credible business case for the value of the CMMS.

Project plan

The *project plan* includes the schedule and communication elements of the overall plan.

Schedule.   Develop a high-level schedule for major parts of the project such as develop scope, design, construct, test, train, go live, and follow-up. Major software projects such as upgrades easily can be expected to take 12 months if a team thoroughly participates in reviewing the new

CMMS and creating a helpful design document to guide configuration. Testing and practice implementations on quality assuranace and testing (QA) servers take time. Scheduling training before going live can also take weeks or more.

Do not expect to consume testing or training time as contingency if the project runs into trouble. Problems might have to *extend* testing time to ensure product usefulness.

Communication.   Plan a communication strategy for all stakeholders including approval authorities, team members, resources, and interested parties including:

- Have weekly meetings with team members.
- Email reports to sponsor.
- Plan updates for approval authorities and resources.
- Plan information updates other stakeholders especially in their areas of interest.

Parts

In a normal job plan, this section would list parts from the storeroom or specially purchased from vendors. For the CMMS project, this section mainly includes the software itself and computer hardware.

The software itself is only a portion of the CMMS cost. Typically, CMMS vendors have standard prices for the system itself, but these prices are also negotiable depending on the size of the company. Especially negotiable are the licensing fees per user. Cost for vendor help to install and configure the system is included in the labor or special tools section. Plan for the longer-term vendor licenses and support of the system where appropriate.

All computers in the work locations must be capable of handling the new software. Depending on the maintenance plan, consider whether all maintenance persons need computers. The planners need the fastest computers possible.

Printers should also be considered. Each work area should have printers fast enough to print numerous work orders and reports. In addition, servers to handle modern CMMS software can be very expensive. Not only does the company need a "production" server to handle the application, the company should also consider having separate servers for "development, "QA", and "training." Some companies use the QA server for training to save expense, but this takes a little more coordination when soon going live with a system. The testing environment needs to become stable with limited changes so trainers can predictably set up test work orders and other scenarios for students. During this time, the system testers can use the development server, if needed, to check out final adjustments by the programmers.

Special tools

In a normal job plan, this section would list special tools not normally found in a technician toolbox. For the CMMS project, this section includes consultants as a cost, special documents that will help the project team manage certain items, test labs, and programmer facilities.

Consultants.   Consultants help implement the software, especially configuration where the in-house IT persons need special programming assistance. On the other hand, it might be entirely practical for the in-house IT persons to handle all of this for less complex commercial systems.

The CMMS vendors or other consultants certified by the CMMS vendor can give price quotes for different arrangements of providing assistance. Be aware that as in the case of a major pump overhaul, jobs sometimes do not run smooth and additional work is required to maintain the equipment properly. Do not expect that the CMMS project will run completely smooth and the CMMS consultant would never need additional funding. The team should plan for additional help for problems. Also, ensure any arrangements include post-going-live support to resolve startup issues. Although the testing group thoroughly reviews the system ahead of time, there is no substitute for actual usage in finding problems and issues.

Documents.   Several special documents will be special tools to help the CMMS implementation team manage the project. These special tools include a project charter, a design document, an issues log, a test script, and a conversion script.

The *project charter* briefly describes the project as developed in the scope of this plan. It should list primary reasons management authorized the project including the overall benefits expected with some specific measures of improvement expected. The charter should also specifically mention what is included and what is excluded as brief lists or statements. For example, "the project will include the software, hardware, initial training, inputting equipment nameplate information," etc. "It will not include purchasing hand tools, evaluating the inventory current stocking levels," etc. The idea of *included/excluded* lists is to help everyone focus and not get distracted. List obvious items the project includes. List items that might not be obvious that the project includes. List items that might not be obvious that the project does not include. The charter should also list principal team members, sponsor, approval authorities, and resources by name with their roles. The charter should also show a general schedule for major parts of the project such as develop scope, design, construct, test, train, go live, and follow-up.

This charter will provide an overall focus for the project team. Having the overall benefits and schedule helps remind the team not only does it have a schedule, but why. The schedule is not to implement just anything; the schedule is to implement a desired product. The charter constitutes an agreement among the team members and sponsor. The charter might be two or three pages of text.

The *design document* is more of a living document to keep track of important detail decisions. The document should cover each of the CMMS modules or major areas listing particular important features in each. It should list fields expected in each area. Developing the design document requires the team have an in-depth knowledge of the new CMMS capabilities, the existing CMMS (if any) capabilities, and the existing maintenance process. The team develops the design document through a series of meetings. The meetings should compare how the new CMMS expects the maintenance process to work versus its existing CMMS and maintenance process. Hopefully, the CMMS can accommodate the existing process wherever possible. If not, the team has to make and document how the design will work.

The design document must include a section entitled "Operating Definitions." Operating definitions explicitly capture agreements where terminology is critical, customization versus configuration for example. Companies should resist customizing systems that affects the source code and hinders later upgrades or service. On the other hand, virtually no CMMS can be used "out of the box." (Think of eating pancake mix "out of the box.") The company must make some tailoring or configuring to add fields or change screen layouts where necessary and allowed by the software. Capture operating definition agreements in this section as they come up during the process. Later, when the schedule is tighter, the IT group has a tendency to declare *any* change is a customization and not allowed.

The project team also employs an *issues log*. This is an action items list for unresolved issues. During the design phase, the team might not be able to come to an agreement about the use of specific job statuses. There may be more fundamental issues such as using the inventory module of the CMMS versus interfacing with other company software. Not every unresolved issue goes on the issues log. Early during the design phase, the team might simply leave blanks for space holders in the design document for areas needing information. However, later as problems arise and potential disagreements need resolution, the project must record these issues to keep track of them. The primary use of the issues log is usually in conjunction with the testing.

The *test script* allows a testing group to proceed through the CMMS in a logical manner to check functionality and performance. The project team needs to develop a list for each module of the CMMS. The test script would be a list of items to check in each module. The following list illustrates a good generic starting point for a work order module. The checklist looks at the overall module and screens and then attempts routine operations. Then the checklist attempt unusual operations. Finally, the checklist notes overall speed and reliability.

Work order module test script

- Check the overall screen for adequate layout and inclusion of all the fields. (Specifically list certain fields if they are new fields or might easily be overlooked.)
- Check that all fields are sized correctly to display enough data without scrolling.
- Check that all fields can hold the proper amount of database characters even with scrolling.
- Check that all fields have been mapped properly from the existing CMMS such as physical location in the old CMMS populating the physical location in the new CMMS.
- Check that all fields have correct dropdown lists.
- Check all field drill-downs work correctly.
- Check all field hyperlinks work correctly.
- Create a work order.
- Check that proper field auto-populate.
- Do required fields work?
- Change status on the work order to allow planning.
- Attach a job plan.
- Attach documents.
- Link documents.
- Print the work order with and without attachments.
- Put the work order in progress.
- Charge time to the work order.
- Charge parts to the work order.
- Charge tools to the work order.
- Add feedback comments to the work order.
- Complete the work order.
- Close the work order.

- For any operation that involves interfaces to other company software, check those functions in the other software as well.
- Run and check reports for single and multiple work orders.
- Approve multiple work orders.
- Run queries to find work orders.
- Create another work order using invalid accounts.
- Attempt to add invalid data to fields.
- Try to approve a work order with insufficient computer rights.
- Try to remove a job plan.
- Judge overall speed and performance
- Judge overall system reliability.

The test script would have a separate column to enter whether each item passed, partially passed, or failed. The script would also have a column for comments. The script might also have another comments column to include expectations. The team uses test scripts to allow the programmers to correct problems. The issues log seems a bit redundant to the test script, but using both can keep minor test script comments off the issues log. In addition, some issues are too broad to fit on the test script document. It is okay if an item is on both documents.

Note that just because something is not on the test script does not mean the CMMS does not have to accommodate it. The test script is a tool, not a specification. Even if the specification does not explicitly exist, that does not mean the CMMS is adequate. In the world of maintenance, a job plan does not have to specify "good workmanship" for a supervisor to reject a repair job because of "poor workmanship." This is why the project team should be as much as possible under the direction of maintenance to ensure a satisfactory product.

On the other hand, the test script should be as helpful as possible. The intent is to help find problems. Finally, for completeness, *conversion scripts* should be mentioned. These scripts are simply the detailed process steps that the programmers follow when making a cutover (practice or final). Of interest to maintenance is that the scripts contain agreed data mapping and configurations. For example, the old CMMS had an "extra" field used for crew number. The new CMMS has a specific field for crew number and the conversion should map all the data in the old extra field over to the new specific field. The programming members of the team maintain these scripts rather than the project manager.

Test labs.   The project team should come together to test the new software together in a common lab, if possible. In this manner, the testers can consult each other and leverage their collective experience to make testing more thorough and faster. The programmers and consultants can also be on hand to assist the learning curve. The project might arrange to have a number of computers available in a common room for this purpose for an extended period. If the CMMS involves network issues, the testing will eventually have to test out access and performance at remote sites.

Programmer facilities.   As recommended by Yourdon above, having special facilities for programmers where they can avoid needless interruptions helps them become more productive.

Procedure

This section identifies and reviews critical project steps including system selection, design, construction, testing, signoff, going live, and training.

Select.   Chapter 9 generally covers *system selection*. If the plant has to include this task, the project must include this in the project schedule and charter. The team should still assemble and complete the job plan to this point assessing benefits and value before selecting the system. In addition, the following section on design should be started (especially tasks involved with selecting critical desired features) before starting to select a system. Add at least 6 months to the project schedule for selecting a potential system.

Seriously consider using the maintenance portion of an existing enterprise major software already used in the company for financials. Although the system might not specialize in maintenance, the risk in interfacing with systems is great enough to consider a decrease in supposed CMMS expertise for an easier interface.

Design.   The *design phase* of the project consists of defining the current process, determining critical factors, defining exact solutions, and documenting agreements and clarifications.

The project team should first define the overall flow of the maintenance process if there is not already an adequate flowchart. The process map should be at an overall high level, such as in Chap. 5 Fig 5.1, and can be refined for more detail if necessary. This map should encompass all of the areas affected by the expected benefits identified in the project scope. In other words, if inventory was an expected benefit, the map should include the storeroom. Using more than a single map is acceptable.

Next, using this map, examine each step and list all the factors required for that step. For example, for a process step for a technician to receive a storeroom item, factors might include:

- Job needs a part
- Planning knows what part

- Technician knows what part
- Part is in stock
- Storeroom is open
- Storeroom person available
- Storeroom person can find part
- Storeroom safely stored part
- Technician has authority to request part
- Maintenance funds available to buy part
- Work order is approved

The project team repeats this for every step to obtain a large list of factors that affect maintenance around the areas where benefits are expected. The team should also conduct focus groups among key stakeholders including planners, accountants, technicians, and managers to identify potential factors and their relative importance on the ultimate expected benefits. The team then must review each of these factors to determine which have a critical affect where the CMMS is involved. In the above example, the team might decide that "planning knowing what part" and "part is in stock" are *critical* to better performance for inventory where the CMMS is involved.

The list of these critical items gathered from all the process steps are known as "critical factors." *Critical factors* determine the success of projects. A project team *cannot* by decree "make inventory more effective." However, a project team *can decree solutions to underlying causes of specific critical factors.*

The team next takes this list of *critical factors* and analyzes them to determine probable problems and probable underlying causes of each. For example, the team through investigation and interviews determines that planners can generally find parts in the current CMMS, but they are limited to searching by keywords and then can only attach a single item to a work order at a time. The team specifies that the planners should be able to must be able to query inventories by key words, manufacturer part numbers, and past equipment usage. In addition, planners should be able to select multiple items in the inventory screens and return them easily to a job plan. The team also investigates and determines that purchasing is not very quick at replenishing exhausted stores. The team decides that the CMMS should provide reports on how many items are below reorder points and how many of these items do not have purchase orders to replenish them.

In this manner, the team identifies critical factors, probable underlying causes of probable problems, and incorporates appropriate solutions into the design document.

In addition, the project team should take a step back and ensure the critical factors include items such as hardware and interfaces with other company software such as timesheets, financials, and inventory. Potential solutions to interfaces should use the vendor standard interface packages whenever possible.

This is the preferred method for the project team: the team should identify the critical 20% of the CMMS project that will lead to ultimate project success for a CMMS to provide a benefit. The design document should be a compilation of critical factors and selected design solutions. This general list should be the starting point of preparing the detailed design document for the proposed CMMS. This list should also be starting point for selecting a new CMMS.

The project team should also change the design document with specific clarifications as the project progresses. The team must record items in the design document as it encounters them and make them visible so that everyone associated with the project can see the current best description of project requirements. Items that cannot be easily resolved should be listed in the issues log. The project manager needs to review items on this list to ensure they are necessary clarifications and not superfluous distractions. The project manager from the maintenance side of the house must continually decide what constitutes "good enough" software. The project team can never deliver "perfect software" by any deadline.

Construct.   The *construction phase* of the project consists of the project team and consultants installing and configuring the agreed design in a development environment. There is limited maintenance involvement except to resolve issues that arise where the programmers feel that the design might not be adequate for user satisfaction. At some point, the programmers will want the users to test the product as installed and configured and will move the development product into the QA environment. Here the project team will run through the test script.

Test.   In the *test phase*, the project team begins using the test script to evaluate the CMMS installation. The team runs through the test script noting problems and may enter issues in the issues log. After going through the QA system as much as possible, the programmers react to issues and problems to correct the development environment and update conversion scripts.

After the programmers feel they have successfully addressed all the concerns and the testers have evaluated the changes, the programmers perform a practice cutover (also known as a mock cutover) converting the existing CMMS into the development environment. This conversion must also address "in-flight" transactions. For example, if the new CMMS uses new statuses, work orders coming over from the existing CMMS must be converted. If the new CMMS uses a different process for purchase order approval, the conversion must deal with purchase requisitions awaiting approval in the existing system. If successful, they convert the development environment into the QA environment for testing. Again, the testers run through the scripts noting concerns and issues. This iterations of testing, correcting, and testing may repeat until the programmers can deliver a practiced cutover without significant problems or concerns. It is not enough for programmers to declare "the last cutover had problems, but at least we know what to do about it." There must be a near-perfect practice mock cutover.

As a special note, the team must test the function of the data backup or security system that would safeguard data in the event of system problems.

Finally, in the case of a new CMMS, the team has an entirely different approach available. It does not have to count on a conversion or practice cutover from an existing CMMS. The program has the option of going live for the exclusive use of data input persons for equipment data or PM's. It can later allow planners to start making job plans. It could even wait longer for originators to enter work requests directly.

Signoff.   The *signoff* moment occurs when maintenance formally accepts the product from the IT group. However, even after a successful practice cutover, the project team should be allowed to experiment and use the QA system for a brief period, perhaps a week or so. In this period, the team practices to ensure usefulness and begins to demonstrate the CMMS to users such as technicians and managers. These persons should have had some earlier involvement to validate requirements, but chiefly the project team handled most issues up to this time. At this time, these users may uncover fundamental problems that might cause another iteration of programming and testing. The project team may call for another practice cutover if the team suspects the changes are severe enough to affect system performance. The team might sign its acceptance of the project anytime during this period. Nevertheless, this signoff does not authorize the IT group to abandon the project. The signoff essentially indicates an agreement to go live.

Train.   *Training* takes place before going live, but also continues into the future. Training is a method of control for ensuring the users can use the software well enough for the company to benefit. This idea of control permeates the entire project, not just this stage at the end. The project team should think of training in this context. Plants train users for a reason, to ensure correct results. There are other methods to ensure correct results. Control means comparing and keeping an activity on track to a desired standard. Control measures range from awareness to full automation.

Awareness is the least effective and full automation is the most. For example, consider charging work orders to a correct general ledger account. As a minimum, the plant could send out emails to tell originators how the account system works and distribute charts of correct accounts for different crews and activities. The plant could also have the CMMS provide pop-up menus with helpful information.

More effective than awareness in terms of reducing errors for incorrect accounts would be *training* where persons receive specific instruction and practice selecting correct account codes. Beyond training would be *statistical process control* (SPC) where some analyst reviews the selection of proper accounts and reports problems to appropriate managers. The manager would use this information to correct problems with individuals or groups that do not select proper codes. The plant could employ awareness, training, and SPC methods together to help ensure proper account selection.

Nevertheless, there are more effective methods for ensuring proper code selection. The first is *simplification* where the CMMS simply removes the decision from work order originators and allows a small group of specifically trained planners to make proper account selections as part of their coding duties. This simplifies having to make hundreds of users aware, trained, or controlled to select correct codes.

Error proofing is the next most effective of the control strategies. In the case of the general account codes, a company might decide to use a single general ledger account for all maintenance at a single plant so that the CMMS does not even allow selection.

At the top of the control hierarchy is *full automation* where the account code fills in automatically and no choice is necessary. Many plants use a strategy for account codes where selection of the equipment brings over a predefined account code making the account code specific and useful, but also reducing errors from persons. In actual practice, the planners can also change these codes as necessary.

Thus, training is merely one aspect of control. Throughout the project, the project team should be implementing ways to control the final use of the CMMS. The team might publish process maps for awareness. They might hold classes for crews for training. The project team should develop include fields and develop reports to allow SPC. Some CMMS packages automatically generate graphs or other KPI's. The project team should develop appropriate indicators. From the first, when developing process maps, the project team should simplify processes to avoid problem areas altogether. The team should also put hard barriers in place to prevent errors. Finally, the CMMS is a perfect tool to automate choices. Instead of allowing infinite paper entries, the CMMS can fill in fields such as physical location and general ledger account simply upon the selection of a piece of equipment.

The following is a sample of a control device. Making learning aids such as laminated wallet cards can greatly increase awareness and help users with a CMMS. Conduct training classes and hand out "How to Cards." The following is a sample of a "How to Card."

How to order a storeroom part

1. After accessing the work order in the work order module, click on the material order tab.
2. Enter the part # you need in the item field.
3. To find a part #, right click in the next empty item box and hyperlink to Inventory. Search by key words and return to work order with your selection.
4. Enter the number of items you need in the quantity requested field.
5. For each part, tab over to the allocate/pick column, click the dropdown menu, and select PICK.
6. Repeat steps 2 to 5 for each different type of part needed.
7. Save the work order to activate the picks.
8. The pick ticket will print in the storeroom for you to pick up your part.

Cutover and going live.   After initial training of the users, the project team converts the old CMMS into the new CMMS. The project team should all be available as much as possible to resolve sudden urgent problems at this point.

For the next week or so, the project team must be available to address issues and concerns as the maintenance workforce uses the new CMMS. It cannot be emphasized enough that the project team including any consultants or other key programmers should be available for near-instant support.

Estimated job cost

A plan always ends with the estimated cost. For a normal work order, this allows engineering to conduct reliability studies and replace equipment that is too expensive to maintain. In the case of this CMMS plan, management needs a realistic idea of what the project will cost. CMMS projects can become very expensive. Management must balance this cost against the calculated project value developed in the project scope. Maintenance is expensive. Nevertheless, poor maintenance is more expensive. An enlightened maintenance management can employ a CMMS to great advantage. Nevertheless, all project team members, maintenance managers, IT managers, and above all, the project sponsor should understand the concept of a sunk cost. It does not matter how much a company has already spent on a project. If the project looks like it will not turn out a helpful CMMS, it should abandon the project. If the software ends up upon closer inspection looking like it will not deliver the benefits expected, management should evaluate the revised expected benefits against the revised remaining costs. This allows the project team or management to make a simple investment decision whether to continue the project or not. A team should not complete a project simply because it started it. The project effort to date might have become a learning experience and not completing the project makes the best financial sense.

Ongoing Support

The implementation of a CMMS does not end after a successful startup. Ongoing license fees and patches involve the CMMS vendor throughout the life of the software. In addition, user leads (usually the system administrators) at each plant should continually coordinate with each other to share best practices and police consistency of coding practices. Many user leads at different plants within a company meet regularly once each month. These user leads also attend user group meeting conducted by the vendor each year. Management should encourage user leads and IT programmers to stay up-to-date with the software in terms of conferences and training just as any industry professionals should stay up-to-date with their crafts.

Perspective

Finally, above all, keep your sense of perspective when working with the IT group. Do not lose your sanity. The following contribution from Bob Anderson, outage manager extraordinaire, kept our team from taking it personal in our last CMMS upgrade.

Meeting to Review Screen Design

Attendees: Doc, Bob, (and Fred: He showed up very late and did not contribute any to the meeting, although he did complain a lot about something.)

The meeting attendees agreed that the following information and only this information will be included on the work order request:

- Description (limited to 12 characters): This will avoid specifics and allow the person who does this job to do anything they want.

- Work Type: We added a work type to our current list with the designation of UW—Unnecessary Work.

- Crew: This will automatically fill in with "NONE" and cannot be changed. Nobody wants to be responsible anyways.

- Deficiency Tag: Will have a drop down list with three choices: There was a tag but I removed it. I wrote a tag, but don't know where I hung it. I pulled a tag from another location and used it for this job.

- Equipment description: This way you might get some idea what might need to be fixed. We don't know how you will fill this in since it is a "read only" field.

- Target start: the computer will automatically fill this field with a randomly generated date. The only condition we stipulated was that the date could not be earlier than 3 months from the report date.

- Reported by: We have to blame somebody so it might as well be you.

- Action taken: We always know how to fix something even though we don't know what the problem is.

- General ledger account: You cannot choose your own; otherwise, your managers will have a heart attack because they are running out of money.

- Permit #: We added a drop down list: Hunting. Fishing. Burning.

- Area and functional system have been combined into one field called dysfunctional area. This will automatically be filled in with "everything."

- Fields not included on the new screen that were on the old screen: Equipment number was not included so there is no possible way to identify the specific piece of equipment. The description was removed because as I mentioned before, we don't really need to know what should be fixed. We just fix whatever we feel like. Status was not included because nobody agrees with them. Problem was not included because it is usually just a figment of someone's imagination.

- We added additional functionality to the screen: Upon saving the new work order, a little trash can will display on the screen, the lid will open, and your work request will be tossed into the can. No matter how many mistakes you make, you will not get any error messages.

If you wish to add or remove something from these requirements, leave us a message or send us an e-mail. Both of us have had extensive training on how to use the delete key.

M

Setting Up and Supporting a Planning Group

This appendix covers how to go about establishing a planning group in actual practice. It follows the mechanics of establishing the organization with a minimum amount of philosophical discussion. The appendix begins by making a straight run through starting planning in an existing industrial plant that previously has not had planning. Following sections explore more precisely the location of planning within the overall organization, the selection of persons to be planners, the physical workspace layout for planners, and the management of the initial and ongoing process. The appendix then adapts this basic approach for plants that need to modify an existing planning operation and for plants that are under construction or have nontraditional maintenance structures. The appendix concludes by presenting an aids and barriers analysis for each of the major areas of planning and key issues involving planning.

Setting Up a Planning Group in a Traditional Maintenance Organization for the First Time

Consider a traditional maintenance organization in an industrial wastewater plant. Figure M.1 shows the existing organization with separate crews for each main craft. Under the maintenance superintendent, there are 23 maintenance technicians under the control of crew supervisors. The maintenance manager believes planning is essential to having high maintenance productivity on a routine basis. The maintenance manager decides to implement maintenance planning.

The manager first discusses the concept with the maintenance superintendent and reaches an understanding that planning is desired. Together, they learn about the concepts of planning and gain the support

Figure M.1   A typical industrial maintenance organization contemplating planning.

of the plant manager. They decide to select a single mechanical planner because the bulk of the maintenance work is mechanical.

The manager and superintendent have two or three meetings with the crew supervisors to explain the vision of planning. They invite the operations manager and superintendents as well as the superintendent of the pump stations to attend these meetings. They express the ability of planning to use file information to help technicians move up the learning curve from past jobs. They express that planning gives supervisors job time estimates and skill requirements to help them assign work more intelligently. They express the management desire to measure the success of the maintenance crews being able to schedule their time on equipment rather than equipment failures directing the efforts of the crews. They explain the high cost of delays in plants without planning, and how planning reduces delays and improves time spent on the equipment. They assure the supervisors that planning will not delay the start of urgent, reactive work. They lead the discussion of starting planning for the mechanical crew, the machine shop crew, and the painter and helpers of the building and grounds crew. The electrical and instrument supervisors also desire to have the planner process their work orders to gain the benefit of file information. These two supervisors want to try the overall scheduling concept and agree to help the planner set time estimates on the plans. The superintendent of the remote pump stations plans to discuss including that area and whether an additional planner would be needed at a later date.

The maintenance manager and superintendent then help each supervisor explain the changes to their crews along with the importance of giving adequate feedback on jobs to help future work. They explain that the planner is committed to being their personal file clerk as well as helping their supervisor establish estimated times and skill levels. They insist that quality remains more important than meeting any time estimate the planner may assign. The planner is merely giving the crew a head start, and the recognition of the skill of the crew is vital to the success of the planning operation. How could the crews or the supervisors complain about management wanting a system that should help reduce delays so that technicians can spend 55% of their day on their jobs?

The maintenance manager and superintendent then advertise among the existing technicians for the new planner. They had been able to arrange with Human Resources to place the position at the supervisor pay level and they had only to interview for the position. They had several candidates who were good technicians in their own right and had good communication skills. Some were surprised to hear about the amount of filing that a planner would do. Two candidates not only understood the filing requirements, but endorsed the idea for the right reasons. Both had the respect of their peer craftpersons and would have made good planners. The manager and superintendent selected one of these persons to become the next planner based largely on their perception that this person was the better "self-starter" of the two.

The maintenance manager and maintenance superintendent personally spend time with the new planner planning a number of work orders in the backlog. These initial work orders are not difficult to plan because there is no file history to research. The technicians will be doing most of the materials information hunting from scratch. On the other hand, the planner must create a new component level file for each job. One of the superintendent's primary concerns is that the planner plans the preventive maintenance (PM) work orders that the clerk issues. The maintenance clerk will continue to enter work orders into the computerized maintenance management system (CMMS). One of the plant engineers will continue to direct the implementation of the CMMS under the guidance of the maintenance manager and maintenance superintendent. The superintendent helps the planner set up a new office or cubicle and obtain files and other office supplies. The superintendent's foresight has already established a suitable, convenient filing area.

Fortunately, an equipment numbering system already exists at the plant with every piece of equipment bearing a six-digit, sequentially numbered tag. Otherwise, a major task of the maintenance manager and maintenance superintendent would have been to work with the plant engineering group to develop a numbering system and place tags on all the equipment.

The maintenance manager and maintenance superintendent also agree on indicators of the program's success. The maintenance manager wants to see an indication of the percentage of crew work spent on planned jobs. The plant manager and maintenance manager have been tracking work orders completed per month and expect this number to rise. They do not necessarily expect the backlog to drop because they are emphasizing the writing of more proactive work orders from the PM program which keys on inspections. The managers expect the extra work orders completed to consist of more proactive work and activities to head off failures. The manager also wants to see a schedule success score.

One month after selecting the planner, the manager and superintendent agree to start the scheduling effort on a specific Thursday 2 weeks from the present date by which the planner should have the entire backlog planned. The group reviews the scheduling process. The planner will obtain a weekly schedule forecast of work hours from each crew supervisor on Thursday morning. The maintenance superintendent working with the operations superintendent will give the planner an overall idea of any urgent plant needs that may not be addressed by the work order codes. Then the planner will allocate work for each crew from the entire plant backlog. The crews will use this work as their backlog from which to select work during the week unless other urgent work arises. As always, the crew supervisor assigns and directs all daily work.

Before the first schedule Thursday arrives, the maintenance manager checks with the planner regarding planning the entire backlog and reminds the supervisors that the planner will be requesting their forecasts soon. On the morning of the first schedule Thursday, the manager keeps tabs on the forecasts and schedule process. The manager leaves the success of the planning operation more and more up to the maintenance superintendent as time passes.

Organization and interfaces

Without formal ties, planning interfaces with nearly every other plant department. The planner resolves work scope questions with operations, engineering, technicians, supervisors, superintendents, and managers. The plant may also have the planners work with a corporate project group.

Figure M.2 shows that the example organization placed the planner parallel to the first line supervisors. This plant kept the planner in the union, but made the pay grade and reporting line equal to the first line supervisors. This is the *minimum* acceptable organization position. The goal should be a choice of promoting a technician to a first line comparable position or promoting a supervisor to a level equal to or slightly less than the superintendent. The planner must deal with supervisors on a peer level or from a higher planner position. Planning brings a structure to the maintenance department that active management support must

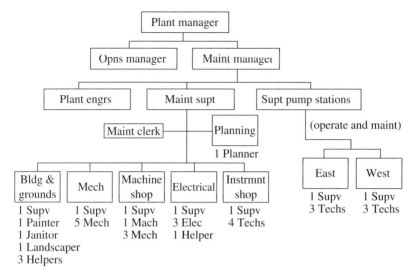

Figure M.2   The minimum placement of the planning group within the maintenance hierarchy.

maintain. *The first step in holding this structure together is to make
planning a highly sought after position.* Planning does not receive the
"buy in" and loses value when planners are lower than crew supervisors.
Management cannot straddle the fence and plan to promote technicians
serving as planners after planning proves itself. Without the support,
planning will not succeed.

Depending on the size of the maintenance staff, the planning group may
have an organization of its own. Two or three planners may work with a
clerk for the superintendent. Beginning with the third planner, management may designate one of the planners to be a lead planner. More
planners may encourage management to have a planning supervisor. If
there is a planning supervisor, the plant may have a problem making the
planners equal to the crew supervisors. The planning supervisor must be
high enough in the organization to avoid this problem. The plant may
make the planning supervisor equal to the superintendent for this large
a maintenance organization. The planners must be kept together in a
group and not separated to report to each crew supervisor. Besides the
abandonment of planning duties this encourages, separating the planners
leads to inconsistent planning. It is acceptable and even desirable to
assign planners specific crews to plan for, but they must report to the planning organization. It is acceptable and even desirable that planners attend
crew meetings, but they must have cubicles or offices so as to work out of
the planning office. Other assignments make sense when having multiple planners. A specific planner may work on long-term outage plans. A
specific planner may interface with the corporate project group.

Maintenance must not overextend the responsibilities of the planners. A well-defined experienced planner may be able to plan for 30 technicians, but management is wise to limit a new or struggling planner group to have closer to 20 technicians for each planner. In many cases, 15 technicians is the desirable number of technicians per planner depending on the amount of plant bureaucracy and purchasing the planner encounters. For example, does the planner have to fight the bureaucracy to buy parts? Will the planners have to attend technician training classes? The ability of a maintenance clerk to assist the planner dealing with the logistics of handling work orders also makes a difference. If there is a considerable amount of equipment tagging or CMMS development requiring planner input, management should keep the ratio of technicians to planners low. Having at least two planners in the above example organization would greatly help the plant during vacation periods and with sick days. Because the potential number of technicians to benefit from planning is almost 30 with the pump stations included, the ratio would be close to 15 technicians to each planner. Because management understands the magnitude of the savings from a single planner, a second planner may be a possibility.

Finally, what type of hours should the planners work? The answer to this question implies an adequate understanding of the planners' mission to focus on future work. The planners do not have to work the same hours as the crews for which they plan. As long as some overlap exists in the work hours among crews and planners, planners may work hours convenient to themselves. The most preferred shifts would have the planners present for some time each weekday to handle new work orders, but the specific hours are not that important.

It may sound quite easy to establish the planning department. However, be aware of the tendency for too many plants to give it much less attention than it deserves. An engineer responsible for leading the planning effort may feel that the program is not technical enough to give it the development it deserves. The plant manager may feel the planning effort is too low in the organization to give it personal attention. The subtleties of the planning system require resolution of all the issues identified by the planning and scheduling principles to implement an effective program. The maintenance manager may lack the knowledge, skills, or willingness to introduce such a program to alter the current modes of responsibility and control. These reasons have left planning as an undeveloped opportunity of tremendous importance.

Planners

Having the right person as planner is the single most critical factor governing the success of the planning program. Almost the entirety of Chap. 11, Control, points to the selection and development of the planner as the primary control of the planning program. Management must pay careful attention to all of the issues surrounding the planner position including selection, wages, training, and evaluation of the person's performance. The planner is not the technician who did not work out and so is shifted to a less important job. The planner is the person having a direct bearing on nearly every job that maintenance executes. Each experienced planner should increase the effectiveness of 30 persons to the level of 47. The planner must possess people skills as well as technical craft knowledge and skill in handling files and data. At the same time, the planner must continually deal with a myriad of different equipment situations in rapid succession. The planner is a major component of the organization.

Selecting.   Planners must be skilled in dealing with people, machines, and data. They must be highly motivated and self-starters. They must have a solid understanding and belief in the purpose of planning and why and how planning helps.

Planners should have good machine skills. Planners should have a good understanding of the plant systems and equipment. Planners must have a background of technician level, practical experience. They should have a thorough knowledge and appreciation of safety rules. Planners must be able to assess equipment problems correctly and develop appropriate solutions quickly. Planners must be able to anticipate job problems and delays and plan appropriate measures to avoid them. They must be able to determine appropriate skill levels of crafts to assign. Planners must be able to estimate reasonable time requirements for the execution of job plans. Planners must be top craftpersons to command the respect of technicians who receive their plans.

Planners should also have good analytical skills in handling data and information. They must be able to read schematics, drawings, and manuals. They should be able to operate computers. Planners must be able to file information and retrieve information from files to improve job plans. Planners must be able to put together the elements of planned work including work scopes, materials, tools, time and craft estimates, and other requirements together into a coherent job plan. Planners must be able to put together work orders into proper schedule allocations. Planners should easily be able to handle the math involved in adding estimates and ordering materials.

Planners should also have good communication and people skills. Planners help coordinate work involving more than one craft. They should be able to order materials and supplies. Planners should be able to communicate and maintain extremely effective working relationships with all levels of plant personnel. Planners should be humble while maintaining a strong self-esteem. Planners need to accept being in a file clerk role if they are being paid enough.

Planners should be able to lead and follow at the same time. Planners should be self-motivated. They should have a low tolerance for the status quo and bureaucracy. The planners must be able to question supervisor authority. The planners should not be afraid of authority, and not be prone to shirking responsibility. The planners must be able to follow their supervisor's directions, but not let other supervisors frustrate their efforts. The planners must follow the planning mission and be able to lead the other personnel out of the traditions of the past. Planners should be diplomatic, hard workers, who do not care about what other persons think about them departing from tradition.

After hearing this personality requirement, one superintendent remarked that he had an idea who might make a great planner. It was the person who had tested the superintendent's new shoes with a hammer to determine if they had steel toes!

Appendix N gives an example of a formal job description. However, *the industry wisdom is to handpick the proper person from the workforce who would make a good planner and then arrange the job to make that person want it.* Many times the existing persons who meet these criteria are in line for promotion to supervisor or are already supervisors. Frequently, a company's existing supervisor promotion criteria help determine the proper person. The company could add "planning and scheduling duties as assigned" to the existing supervisor job specification and promote the right person to be a planner. The company could also reassign a good supervisor to become the planner. Using a supervisor also allows rotation if needed to obtain the best planner. One cannot rotate technicians in and out the planner position because the position must be above the technician level. One should also be wary of using a supervisor that has "retired" into the supervisor position. One should also avoid allowing the planner selection to be made solely by seniority among technicians desiring the position. More than a single company has selected planners from among technicians by seniority and then later abandoned planning. Do pick someone that wants the job, but do not pick someone just because that someone wants the job.

Managers should take their time and select the right persons as planners. They should not quickly select a planner and then expect to manage the wrong person into the right planning behavior. The majority of the managing with respect to planning is to select the right person.

When interviewing persons that have been planners before in other companies, ask if they feel technicians could somehow get more work done. Determine if they know the purpose of planning is to improve productivity. If they do know that, it would be an advantage. If they do not, that is okay because they will soon.

Wages.   Making the pay grade if not the position equivalent to the first line supervisor is only part of the equation. What if different first line supervisors make different wages? If so, pay the planner at least at the level of the lowest first line supervisor, but definitely ahead of the highest field technician. A mechanical planner should make above the highest paid instrument technician, for example. If the planner position remains in the union, it should also definitely be a stepping stone to a supervisor's position.

The market issue also comes into play. The planner is valuable to other companies as well. If one had 20 electricians, one could risk losing one or two before realizing that the market commands a higher pay rate for the most capable technicians. Reacting in time to prevent the loss of the only electrical planner one has would not have the benefit of such advance warning.

Consider also that if the plant moves technicians straight to planner positions, those persons will start drawing *less* wages. This is because in many plants, technicians do make a significant amount of overtime money each year. The planner position makes it difficult to allow the planner to work most overtime jobs. Much overtime comes from working jobs that run over and cannot be stopped simply because the shift ends. Keeping the same technicians on the job for overtime allows the job to continue without interruption. The planners do not work jobs and so are not considered for this work. Overnight emergencies are also not practical for planners to work as technicians or supervisors because the plant cannot afford planners to miss the following day of planning future work for entire crews. Making a technician into a planner without any wage adjustment would punish the one person management felt was qualified enough for the job and who wanted to do it. In actual experience, this situation does not attract the top technicians who do not mind getting dirty in the field. The situation practically shuns the very persons desired. The top technicians who volunteer for overtime and do not mind getting dirty and working long hours to make additional money for their families will not apply. The persons who have a healthy self-esteem and do not care too much for authority do not need an office job that pays less. These are the persons who frequently work through break to finish a job. Conversely, the situation attracts the less competent technicians who want a clean job at a desk. The situation attracts persons who are willing to make less money to do less.

The plant must make the planner wages sufficient enough to attract and retain the right personnel who can properly execute the planner's duties. Because a planner is worth 17 persons, the company has the funds to make the planner pay adequate. Management also wants to send the signal that it values the importance of planning. This gives the planners the respect and support to make things happen.

Training. Two issues arise with respect to the training of maintenance planners; namely, learning planning techniques and retaining craft skills.

Retaining craft skills is usually not a problem. The planner's close involvement with the jobs keeps the knowledge and usually the actual skills fresh enough. This is the similar case for the crew supervisor. On the other hand, the planning techniques are all new and must be acquired.

The selection of the right persons as planners makes these persons capable of acquiring the planning skills. The first task at hand is to share the vision of planning with these persons and then help them pick up the necessary techniques of planning. Several companies offer excellent classes to acquaint new planners with the purpose and techniques of planning that can be directly applied at home. Caution must be exercised because many training companies also have differing visions and related techniques about what planning should be. Management should determine if the course objectives match close enough with the vision adopted by management. The management or other specific person driving the adoption of planning may be the best source to instill the vision of planning to the planners. This person may be the best person to attend different outside training classes and bring home the necessary "profound" knowledge. The best training on techniques comes from OJT or on the job training. An existing planning operation can easily train a new planner, but a new planning organization does not have the luxury of letting a new planner tag along with an experienced planner. Again, the person driving planning must spend time coaching and advising a new planner doing actual planning and scheduling work. For example, the coach may help the planner gather the appropriate unplanned jobs to begin a field inspection tour. The coach may advise that the planner is taking too much time to add unnecessary job details to a particular plan. The coach may help the planner gain confidence in being able to assign a time estimate to a job. The coach may help the planner realize that a short 1- or 2-hour job need not take into account time for breaks, but a longer job should. The coach may remind the planner to make new files or remember to check existing file information. The coach can later look in on or check back with the planner to see how it is going. This one-on-one interaction in actual planning situations has been proven to be among the most effective for training. This is an appropriate investment of time for a manager, superintendent, or other person. Proper control of planning comes from selection and training of planners, not continual meetings, not direct supervision, not rules, and not indicators.

Evaluating. On an overall basis, management evaluates the performance of the planning and scheduling operation with the amount of increased productivity it perceives, possibly with a wrench time measure. On an individual basis, management may wish to evaluate individual planners.

A person must show evidence of certain traits to qualify as a living human being. Among other things, a person must breathe and eat at various times. Similarly, a planner simply must perform certain actions to qualify as a planner. This is not an attempt to second guess the extreme care with which management selected the persons with the right potential. These traits are not the subjective criteria that went into selecting the planner. These traits are the simple, objective actions of planning. Principal among these are the following. A planner must be able to code new work orders accurately. A planner must create a component level file for nearly every new job where a file does not yet exist. A planner must use the equipment number to create the file. A planner must place planned jobs in the correct waiting-to-be-scheduled files or properly change the CMMS status depending on the degree of plant computerization. A planner involved in purchasing must place the associated work orders and documents in the correct files and change any computer as appropriate. A planner must understand to plan jobs for the lowest craftperson classification that could execute the work without regard for who the planner thinks the supervisor probably will assign. A planner must understand to plan for a good technician without any unanticipated delays. A planner must make a copy when necessary and not send the only copy of an equipment manual out into the field. A planner must file job feedback. A planner must check file history for all but the most obvious reactive, minimum maintenance tasks. A planner involved in scheduling must gather work hour forecasts from each crew supervisor. A planner involved in scheduling must be able to create the weekly allocation. A planner involved in scheduling must be able to calculate the schedule compliance measure.

These evaluation criteria demonstrate that after the planners were selected, the situation became "are they planning or not" rather than one of "how well are they planning." Management should avoid the problem of the "crack battalion" when evaluating individual planners. The problem of the crack battalion is this. On regular maintenance crews, there are always one or two vastly superior maintenance technicians. These technicians normally always receive the higher evaluations and raises. In military terms, these persons might be the best shooters or "crack shots." If management decided to create a crack battalion, it might take all of the top technicians to make a super crew for extremely difficult work. The military might take all the best shooters and make a sharp shooter division. The evaluation problem occurs when suddenly the technicians who always received above-satisfactory marks are now entirely among their peers of ability and skill. Should management rate them relatively as average or from below to above satisfactory? Management should be willing to rate everyone in the crack battalion above satisfactory that does not evince an absolute dereliction of duty. That is the purpose of the above planner actions being offered for evaluation.

Management should ensure that the right persons are in planning and continue to reward them for being this type of person. The management does better by evaluating planners on a supervisor basis than a technician basis if any subjectivity is to be included. A person is a planner or is not a planner. If a person is not a capable planner, management must see if another position is more suitable for this person and bring in another planner.

Workspace layout

The maintenance manager or superintendent should personally see to helping the new planner establish an office or cubicle. The issues involved with this topic include the size and location of the planner office, the arrangement of the files, and miscellaneous office equipment and supplies.

The issue of office size is one of the few areas where the planners might not rank as high in privilege as the crew supervisors. Crew supervisors frequently need private offices to hear private personnel issues with members of their crews. There are no hard and fast rules on specific size or dimensions, but the planners should not have large offices. Small offices or cubicles encourage the planners not to keep files within their offices. A cubicle arrangement frequently gives the planners sufficient privacy, but brings them close to shared files. The primary advantage of the cubicles is that they may be kept small enough to discourage keeping equipment files. On the other hand, management may decide that in order to demonstrate its support of planning, the planner should have an actual office.

Files must be kept out and accessible to technicians working jobs in progress or to other planners who would otherwise interrupt the planner continuously. Chapter 8 discusses the various types of files and resources the planners utilize. Open files whose folders can be seen at a glance are preferred to encourage persons to use file history information. This also helps planners and clerks file information. Appendix C identifies the style of file folders preferred to use with open files. Shelves set up along walls near the planners' cubicles are satisfactory. One of the best arrangements the author has observed has been the use of rolling files where several planner cubicles occupied one side of a large room and the rolling files occupied the other side. File security was a concern addressed earlier in this book and might be best handled by having the files where persons would at least have to walk past the planner, maintenance clerk, or maintenance superintendent. A sign mandating that "files are not to be taken from the room, but left on the table for the planners to return to the shelves" is helpful. A table near the files helps encourage persons to use materials there rather than take original materials from the area. A copier should be close at hand for similar reasons.

There should be a small desk or table near the files set up as a File Creation Station with all the files necessary to create component level files.

The location of the planning operation including the equipment files should be close to the work. It is to be hoped that this is also close to the other crew supervisor offices. The intent is to have the files accessible to technicians so they can research jobs in progress as needed. File security may dictate a specific office location, perhaps in a main maintenance building with the superintendent.

The planners require standard office equipment such as a copier, fax machine, and personal computer with email and Internet access. The copier should preferably have two-sided copying and 11 × 17-inch capabilities. Its speed should be at least 20 copies per minute. Within each planner's cubicle besides a desk, the planner should have a file cabinet and book shelf. A telephone with two extensions and voice mail on the primary line is helpful. Beyond identifying the preferred type of file folders and tag material, App. C contains suggestions for miscellaneous office supplies such as a desktop calculator. The maintenance superintendent or manager should take care to outfit the planning office as a first class operation with a comfortable chair, desk, and other furniture and supplies. The planner is an important member to welcome aboard the team!

Management and control

After the selection of planners and the assurance that everyone understands the vision and basic process, management's role reverts to more of an insistence on overall performance from maintenance. Specifically, management should expect that the maintenance crews spend an increasing amount of work hours on planned work. There should also be an increase in the number of work orders completed as well as the amount of hours spent on proactive work. Management should expect the reporting of schedule compliance. Management might want to measure wrench time if it does not have a feel that there is less standing around or other delay time. Management has a right to expect improvement in plant availability or reliability as the maintenance force completes more work.

On a longer-term basis, as with most programs after successful implementation, someone must sponsor planning and remember why it is successful. This continued attention ensures the creation of a culture that will reinforce the program. It also hinders gradual changes that unwittingly undermine the success of maintenance planning. For example, will someone remember why the planners are organized apart from the crews in 5 years and not try to place a planner on each crew? Will someone remember why planners do not place too much detail into job plans and not task planners with a new requirement to add more detail? Management has five basic jobs: to plan, organize, staff, direct, and control. Management planned to improve productivity and organized a planning department. Management staffed it well. Management directed the initial implementation, and not only productivity improved, but reliability improved as well. Management must remember why planning works to control the future success of the maintenance planning program. Appendix P offers helpful probing questions for management.

Redirecting or Fine-Tuning an Existing Planning Group

Think of an existing plant with a maintenance planning operation that is not contributing to the effectiveness of maintenance. From the start, technicians are promised that they should never have to look for parts. The planners have never been able to fulfill this promise, and this is part of the reason for an atmosphere of distrust. Not helping this situation is the presence of at least one planner that no one ever respected as a competent craftperson. None of the planners consult past jobs when planning new ones. The old jobs are copied onto microfilm and conveniently stored away by work order number by the document control group. The planners could access the work orders, but see no need even if they have the time to request the records. There is not much time anyway because of the time it takes to thoroughly write out a decent job plan. The planners profess to research history if they know maintenance has performed a lot of work on the equipment. Each planner plans differently. One planner says that there is no need to write even known inventory item numbers on the work order because "any technicians worth their salt can find the numbers themselves." There is a lack of clarity in the planner purpose. No one understands the concept of improving the amount of time technicians spend on the equipment. There is a mass of guidelines and rules written over the past years to cover isolated incidents and situations. Equipment tagging is a concern. Tagging should have proceeded on higher maintenance items first. There is no scheduling other than a daily schedule prepared by each supervisor. Management has taken a wait-and-see attitude to decide if planning is really necessary before supporting it. None of the managers wants to risk supporting a losing cause.

Finally, one manager directly supervises a difficult pump refurbishment and insists that the technicians record the information developed on the job, such as the exact clearances and steps involved in the rebuild. Some months later the manager encounters another crew in the middle of an almost identical job. It comes as a surprise that the crew has none of the information from the previous job. After reviewing the situation, the manager discovers that the old information had never even made it back to the files where the planner could have found it. In fact, this kind of thing appeared to be pretty much of a normal mode of operation. The manager undertakes to learn what planning should be about and then wishes they could start over again without preconceived ideas. The manager resolves that planning must improve, but discovers a widespread emotional attachment to the existing system. The persons involved appear to be capable of executing the planning system needed, but actively resist change. The planners have long fought the maintenance crews to defend their concept of planning. Asking them to change now involves losing face by everyone involved.

As one can see, it is very difficult to back up and start anew with maintenance planning. It is much easier to implement a new planning operation than to revise an existing one in many cases. Incorrect concepts about planning can greatly hinder the acceptance of the correct concepts. Unlearning improper planning is difficult.

To redirect a planning operation with significant problems requires that managers work closely with supervisors to ensure adherence to a planning mission statement and set of principles. Managers themselves must follow up specific problems in areas of concern. For example, listing parts on a planned package is a carrot to make persons want planned jobs. However, the initial implementation of planning may have involved telling the troops that they would never have to hunt for parts again. Management must now explain to everyone that this is not a valid concept, but rather that planners should promise to be faithful file clerks to help retrieve any parts that are discovered on past jobs. Management must obtain commitments that the plant wants 55% wrench time and that the planning and scheduling tool is essential to take the plant there. Management must obtain commitments that the plant wants weekly scheduling and that it wants certain indicators to drive it, such as schedule compliance and planned coverage. Management must ensure there are qualified personnel in planning and give them the necessary support. Management must insist on creating files and utilizing them for every job. *The plant culture largely derives from the first line supervisors. Management must take extraordinary measures to enlist their commitment to make the changes to make planning work.*

On the other hand, many maintenance planning functions are not achieving success due to only a few misunderstandings of critical points. One or more of the planning and scheduling principles may address these areas entirely. Perhaps the organization is planning without scheduling. Perhaps planners place too much detail into the plans and cannot plan all the work. Perhaps the filing system is not component specific. Perhaps the planners are not kept separate and are taken for extra help on crews. It is hoped that the principles expressed in this book have focused on specific issues that the maintenance practitioner can grasp and apply to fine-tune a maintenance planning operation.

Considerations

There are other considerations with regard to establishing a planning operation. The thrust of this book so far has been implementing a planning function in an existing plant with a traditional maintenance group organized along craft lines. Mention should be made of newer plants, even plants under construction. Mention should also be made of less traditional organizations such as those with area responsibilities or self-directed work teams.

Older facilities versus newer facilities

Newer facilities have several advantages favoring the implementation of maintenance planning. For one thing, there is less momentum in the current culture to oppose major changes to the work process. For another thing, maintenance manuals from the construction of the plants are more available for collection into a proper file system. In addition, the plant may not yet have begun to experience the occurrence of many reactive events. A maintenance group has now a better opportunity to establish a first class PM program to keep the plant from such deterioration. The plant can establish a culture of proactive maintenance without having to turn around a reactive maintenance mentality. All of these factors work toward the success of a maintenance planning group. On the other hand, one of the few disadvantages of a newer plant is that an unseasoned management staff may make unwise decisions regarding maintenance staffing. It may not see the need to maintain an adequate maintenance staff or to implement a process as planning to maintain a high productivity.

In a newer plant, the planning staff should be quick to establish a proper work process including planning and scheduling. It should include a comprehensive PM program. The staff should be alert to the need to collect manuals. In addition, because the plant is new, there should be more opportunities for the planners to propose projects both large and small as design problems become evident. In an older plant, projects come about more from replacing worn out equipment with equipment based on newer technology.

Facilities under construction

Plants so new as to be still under construction should establish a planning organization without waiting for first operation. Planners should be involved in establishing equipment numbers, tagging equipment, and collecting manuals and design documents for files.

Centralized versus area maintenance considerations

In large, spread out facilities, companies might have centralized or area maintenance functions. Centralized groups would dispatch technicians from a central location. Perhaps a school board would have a central yard and dispatch maintenance personnel to various schools. An area arrangement might have each school with a small maintenance staff of its own. The advantage of the centralized staff is the ease of implementing a planning system and controlling productivity. The advantage of the area arrangement is that it fosters ownership and quicker reaction time to urgencies. Some companies have a combined system of a small resident staff at various locations even if not outlying and a larger centralized maintenance staff.

The planning operation would certainly function for the centralized staff. It is usually not a problem for planners to be wide ranging in their travel to scope and plan jobs. Depending on the size of the area groups, planning may or may not be appropriate, especially if the areas have only a single person or two. Many more persons allow the planners to organize covering several close areas with a single planner even if the technician-to-planner ratio is low. Remember that a theoretical break-even point exists so that out of every three technicians, one could be made a planner. Ratios of six technicians to one planner might be appropriate to justify adding a planning and scheduling boost to productivity. Without much question, large area groups would merit planning. The planners may be geographically spread out, but would still maintain a separate function accountable only to the overall superintendent or manager.

Another variation of area teams would be the plant that has divided its craft among the various crews. There might be a single crew complete with mechanics, electricians, and instrument personnel to handle all boiler work. There might be another crew complete with all three crafts to handle all river water work. There would be other crews for various process areas of the plant. This type of organization does not much affect the planning organization. An electrical planner would probably still plan all the electrical work for all the crews. The mechanical planners would each have various crews for which to plan the mechanical work.

Traditional versus self-directed work teams

Self-directed work teams are a special case. The organization of these teams and their responsibilities vary widely from company to company. The individual plant must decide whether there is an opportunity for planning assistance. In their truest sense, some of these groups both operate and maintain their equipment. It would be difficult to schedule work for any crew that both operates and maintains its equipment. For one thing, it is difficult to establish a meaningful forecast for work hours available to maintain apart from operate. The self-directed team not only operates and maintains, but by definition does not utilize much outside support. The sense of ownership and pride drives such a team to keep its equipment available and efficient. In such organizations management may wish to offer the services of a planner to help the team keep useful records for future reference on tasks. The diplomatic planner could explain the repetitious nature of equipment maintenance and the advantages of using work orders. The team may develop a member with an interest in filing. The planner can also suggest work order selection for repair periods.

On the other hand, many organizations with such teams do have a separate maintenance organization that does much of the maintenance work after all. The teams actually function more as operators with a highly developed sense of preventive maintenance. The teams perform many of the minor maintenance operations and adjustments to keep the machines operating well. The teams write work orders to the main maintenance group to handle repairs or other procedures. In these cases, planning can proceed as traditionally envisioned. The planners develop work scopes, estimates, and other job requirements to allow improvement on past jobs and scheduling of current work by the main maintenance group.

Aids and Barriers Overview

This section of this appendix, Setting up a Planning Group, identifies and analyzes aids and barriers for the planning and scheduling program. As shown in Fig. M.3, the overall planning and scheduling program can be divided into five broad areas with major objectives. These areas are *organizing* a planning group or system in the first place, then *planning* enough work for a crew for a week, then *scheduling* the right amount of work for a crew for a week, then *executing* that scheduled work and giving feedback, and finally keeping the program *ongoing* and successful.

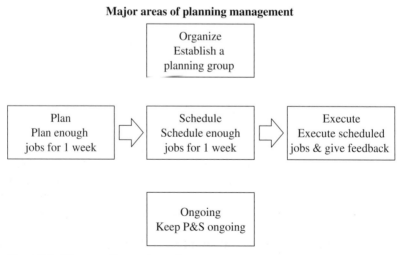

Figure M.3 The overall areas to analyze.

The successful management of the planning and scheduling program meets each of these objectives in turn.

This appendix section first considers the major objectives of the overall program and develops an aids and barriers analysis for each. Then the section selects key aids and barriers from among these analyses and defines them as separate objectives for which to analyze, deriving both aids and barriers for achieving those objectives. Finally, the section considers and analyzes aids and barriers for several special circumstances for planning and scheduling.

Aids are resources and include what is needed or what might help the particular objective being considered. Many of the aids should exist as mandatory prerequisites before this objective can be met. An example is the requirement that planners should have an adequate filing or records system before they begin planning jobs. Another example is that planners need a backlog of planned jobs before they can work toward the objective of establishing a weekly schedule. On the other hand, many aids are not necessarily mandatory, but simply helpful. This includes the aid of having helpful forms to assist the scheduling process. While not essential, this item is listed as an aid because it might help.

Barriers are constraints or impediments; things that might hinder the objective being considered. A barrier critical to the objective to keep planning and scheduling ongoing is not scheduling. A barrier that would certainly hinder, but not destroy, the ongoing success of planning and scheduling would be getting inadequate feedback on executed work. An overwhelmingly large portion of the success of planning and scheduling comes from the scheduling process and its effect on maintenance productivity. On the other hand, the scheduling process can take place without the individual improvements feedback can make to individual jobs.

Why should a program consider aids and barriers? Recognition of these areas can lead the supporters of the program to take action. Such action can ensure the success of planning and scheduling. This action may be to establish or reinforce aids or to head off or overcome barriers. Recognition of potential aids and barriers are crucial to taking action steps to deal with them. Planning and scheduling problems may be caused by aids that do not exist or are inadequately in place. Other problems may be caused by barriers that do exist or may come into existence. In each analysis, certain aids and barriers may not be a problem in the reader's particular plant. For example, supervisor support, critical to several objectives, may already be in place. However, early recognition of the problems that do exist can lead to the establishment of plans and action items to overcome them. The readers should consider which of these may be issues in their facility and establish action items and plans for them where possible.

The appendix uses the format illustrated by Fig. M.4 to perform the aids and barriers analyses. The rectangle in the middle identifies the

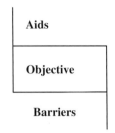

Figure M.4 The basic aids and barriers analysis map.

specific *objective* to be met in the specific analysis. The analysis attaches *aids* to either side of vertical line ascending from the top of the rectangle. The analysis attaches *barriers* to either side of vertical line descending from the bottom of the rectangle. The analysis is essentially a discussion session to identify likely aids and barriers. For each analysis the figure contains only key words or phrases for each potential aid or barrier while the text following the figure not only expounds on certain aids and barriers, but also identifies where the issues are discussed throughout the handbook.

A number of observations should be mentioned about the identification of aids and barriers in these analyses. These include using this analysis method versus root cause mapping, duplicating some aids and barriers for different objectives, stating the negative of an aid as a barrier, listing the same item as an aid as well as a barrier, listing questionable aids and barriers, and finally, the brainstorming nature of the analyses.

Ultimately, every aid could be traced back and laid at the feet of management understanding and support and every barrier could be blamed on the lack of the same. However, this appendix conducts an aids and barriers analysis and not a cause map or similar technique asking "why?" five times for each issue. This appendix does not attempt to trace each aid and barrier back to its systemic root, but simply lists and briefly explains (where warranted) significant areas for the key driver of the planning program to be aware.

Considering duplication, sometimes, a particular aid or barrier might be important to consider for different objectives. The person trying to assist the planning and scheduling program must address each objective within the overall program separately. Therefore, for the most part, the analysis for each objective lists the aids and barriers for that objective regardless of other objectives.

In most cases, an area of concern could be stated positively for the aid (or barrier) and negatively for the barrier (or aid). For instance, management support might be listed as an aid while the lack of management support might be listed as a barrier. Rather than do this for most items, the aids and barriers analysis for a single objective usually lists the concern only in the area it is most likely to be addressed. The most critical items are more often listed twice. Sometimes the same item might be listed as both and aid and a barrier to the same objective such as having a CMMS to the objective of planning jobs. In many of these cases where different aspects of the same item could be considered as a potential aid or barrier, a question mark (?) follows the item in both sections. The text section addresses the finer details and distinctions of meaning. Other items might be followed with a question mark (?) such as *labor agreements* listed as a possible hindrance to selecting a planner. This does not mean that it is always a hindrance, but could be a hindrance for some plants in some instances. The idea of listing the item is not to identify it as a hindrance, but to identify it as a potential hindrance. Indeed, a labor agreement might be an aid to the effective communication necessary for a management to show its support of planning. The thought behind listing *labor agreements* as a barrier is the typical potential to slow the process of creating a planning position. By anticipating this barrier, the supporter of planning can address this concern in a timely manner to the satisfaction of all parties.

Some items such as the aid *hide like an elephant* and the barrier *lack of support up and down* for the objective *selecting the right planner* reflect the real world nature of the development of most of these analyses. They were derived in brainstorming and discussion workshops conducted by the author around the world and are many times presented as they were received, especially when they reflect the sentiment intended.

Major Areas of Planning Management

The five aids and barriers analyses in this section address the management of the planning and scheduling program. Management is trying to establish and maintain a successful ongoing program.

Organize—establish a planning group

The principal objective of this area is beginning a new planning program. Aids or barriers meriting their own analysis later as issues with subordinate aids and barriers include management support, having the right planner, and planner training as shown in Fig. M.5.

Aids

Which plant. Company-wide efforts must decide where to start. Planning requires a dedicated person at the plant site (App. A) to make it happen as well as management commitment at the top. Which plants have a person that is "bought in" to planning? Also, companies may consider a limited area to prove

Figure M.5 Aids and barriers for organizing the whole program.

the concept. On the other hand, a greater number of plants could help each other as they learn and improve. In any case, this is a decision that must be made.

Management appreciation of WIIFM and support of planning (Chap 1, Chap 4, Chap 12, App. A, App. P).   Planning cannot exist without the active support of management. This handbook exists for the management as much as for the planning department. What should management expect from planning if they support it? Chapter 12 covers WIIFM, What's in it for me.

Driver or key supporters (App. A, App. F).   Should a project manager be responsible for implementing a planning organization? The person responsible must believe in the concepts of planning, be intimately involved with maintenance, and have the authority from management to complement the responsibility.

Knowledge of planning and scheduling (Handbook).   Training

Knowledge of own plant and preferences. The planning program may approach resolving the issues presented by the principles differently based on personalities or particular characteristics of the plant. For instance, some supervisors may be more adaptable to change and should be the focus of early efforts to gain support. Or budgeting may be more critical in some plants than others.

Communication (Chap. 1, App. A). With and among management, supervisors, all personnel.

Supervisor support (Chap. 1, Chap. 3, Chap. 4, Chap. 12, App. A). Other than budgeting and position allocation, management support is primarily needed in order to garner supervisor support and willingness to follow a planning and scheduling system.

Which crafts. Mechanical crafts lend themselves most to maintenance planning, but planning can boost other crafts as well. The plant may wish to proceed with mechanical crafts first.

Budget. Not only personnel (technician promotion to planner), but also facilities such as file system and office equipment. A CMMS, if employed, might be overwhelming. The IT support personnel aspect of a CMMS may require future budgeting.

Placement of planning within the organization. (Chap. 1-Principle 1, App. M).

Creation of planner position(s) and planner selection (Chap. 11, App. M). The "wrong" planners simply cannot be left alone, "worked with", governed by rules, or measured with indicators enough to "make them make planning work." The art of planning demands a qualified planner. The supporter of planning must have the right planners in place to achieve successful planning and scheduling.

Creation of other positions? (Chap. 5, App. F, App. M). Clerk? Purchaser? Parts expediter? Designated operations coordinator or contacts?

Performance indicators. A "before" snapshot –. Wrench time study? (Chap. 2-Principle 6, Chap. 11, App. A, App. G, App. H). The plant needs indicators to manage change. A wrench time study is very intrusive, but may illustrate the low productivity of the current workforce.

CMMS? (Chap. 9, App. L)

Barriers

WIIFM. What in it for me? Unclear (Chap. 12).

Lack of management support (Chap. 1).

Lack of supervisor support. The supervisors have a lot of concerns and not much readily apparent benefit to embrace planning.

Lack of others' support (Chap. 12). If the supervisors do not support planning, the technicians probably will not either.

Poor culture – not ready for this (App. A, App. M). The supervisors are the keepers of the culture of the plant and their attitude is driven by their relationship with plant management

Change, Inertia. (App. M).

Budget. (Chap. 1).

Unclear vision. (Chap. 1).

Wrench time study? (Chap. 2-Principle 6). Wrench time will improve after the plant begins weekly scheduling. The productivity will be evident by an increased number of work orders competed each month. The plant may not want to embark on a formal wrench time study which some technicians may resent.

CMMS lack of IT support (Chap. 9, App. L). Once planning has committed to a CMMS, IT support is critical. Reductions in IT maintaining the CMMS itself will greatly hamper planning.

CMMS – limitations (Chap. 9, App. L). Many CMMS systems simply cannot reproduce the easy steps of manually planning jobs. For instance, whereas before the planner attached documents with a stapler, the new system may have a less than clear method of linking files and only for electronic documents.

Plan—plan enough jobs for one week

The principal objective of this area is to plan enough work for crews to be able to work planned jobs and to enable the schedule program in the next objective. Aids or barriers meriting their own analysis later as issues with subordinate aids and barriers include management support, equipment tags, files, purchasing, work order system, CMMS, urgent breakdowns, and technician interruptions as shown in Fig. M.6.

Aids

Management support of planning and appreciation of WIIFM (Chap. 1, Chap. 4, Chap. 12, App. A, App. P).

Work order system (Chap. 5, App. J). Obviously a prerequisite for having a planning system.

Backlog of prioritized work orders (Chap. 5). The planners do not generally write work orders, but take previous written work orders to plan in order of plant priority (but also considering reactiveness).

Priority system (Chap. 5, App. I, App. J). Appendix I considers "gaming the priority system."

Enough planners (Chap. 2-Planning Principle 1, Chap. 5). The typical planner to technician ratio is 1:20-30.

Equipment tags hung on equipment (Chap. 2-Planning Principle 3). An almost mandatory prerequisite.

File system. Records, past work orders, files with equipment history (Chap. 2-Planning Principle 3, Chap. 5, Chap. 8). A mandatory prerequisite.

CMMS (Chap. 9, App. L). A CMMS might greatly assist planners store and retrieve information for job plans.

Planner with craft experience (Chap. 2-Planning Principle 4, App. M). Plant knowledge, craft skills, communication skills, data skills. Knowledge of supervisor preferences.

Knowledge of skills of craft-persons (Chap. 2-Planning Principle 5). Planners plan for skilled craft-persons, but they should be cognizant of what the crafts are capable.

Storeroom
Parts catalog
Processes for stock.
Processes for nonstock.
Vendor information for nonstock.
Processes for special tools
Planner facilities
Already created standard jobs
or booklets of key information

Resources: What is needed or might help.

Management support
Work order system
Backlog of prioritized work orders
Priority system
Enough planners
Equipment tags
File system
CMMS?
Planner with craft experience
Knowledge of skills of craft-persons
Knowledge of crew available hours per week
Field inspections of jobs
Knowledge of when operations can typically make equipment available

Plan
Plan enough jobs
for one week

Planners taken away or given other work
Dumping of odd jobs on planners
Poor or no work order system
Lack of information on new work orders
Lack of information and feedback on history work orders
Inadequate authority or processes to get parts or make decisions
Lack of equipment information, O&M manuals
Plant reliance on blanket work orders
Lack of processes for stores, purchasing, tools
Lack of training on processes
Lack of planner training and understanding of planning fundamentals
Highly reactive plant maintenance conditions
"Back door" jobs being done instead of routing through planning
Extensive duplication of work orders

Constraints and impediments: What might hinder.

Reluctance to write work orders
Lack of planning facilities and tools
Technician interruptions
Operations unavailability
Inadequate storeroom
Overwhelming backlog
Wrong or untrained planners
No driver of system
File security
Library hassle
CMMS?

Figure M.6   Aids and barriers for planning enough work.

Knowledge of crew available hours per week.   This would simply give the planner an idea of how much to plan to stay ahead.

Field inspections of jobs (Chap. 5).   Planners should almost always go inspect jobs if practical before planning them. This may not be possibly geographically for spread out operations.

Knowledge of when operations can typically make equipment available.

Organized and stocked storeroom (App. A).   The more parts available, the better from a planning standpoint. However, planning may be able to optimize use of plant funds by being able to order nonstock parts more efficiently and ahead of time than a plant with no planning function. This could lead to more intelligent reductions in what to carry for stock.

Accurate parts catalog or system (Chap. 5).   The planner would use this information on job plans.

Adequate processes for identifying and reserving storeroom stock (Chap. 5).   The plant needs a defined process for how to identify parts and plans and how to reserve them.

Adequate processes for purchasing nonstock (Chap. 5). The plant needs a defined process for the planner to obtain parts the storeroom doesn't carry.

Vendor information for nonstock (Chap. 5, Chap. 8). – who, where, when, how.

Adequate processes for identifying and reserving special tools (Chap. 5, App. A). This might consist of a tool room or area where tools not normally carried in tool boxes can be acquired by technicians. The plant needs a process for how to identify tools on plans and possibly how to reserve them. These would be tools not normally carried in technician tool boxes.

Planner facilities (Chap. 8, App. M, App. C). This would include all the necessary conveniences planners need in an office environment including a filing system, copiers, printers, and computers.

Already created standard jobs or booklets of key information (Chap. 2-Principle 5, Chap. 5, Chap. 7). Although planning strives to create a cycle where it can accumulate and develop better plans and information over time, many plants have recent or soon-to-be retirees (or others) establish initial plans with quite a bit of detail for critical equipment or commonly performed maintenance operations. This gives the planning department more of a head start.

Barriers

Planners taken away or given other work (Chap. 2-Planning Principle 1). There must be planners to plan.

Dumping of odd jobs on planners (Chap. 2-Planning Principle 1). Planners must be allowed to plan. The planning department becomes an easy area to assign odd jobs such as taking surveys, responding to engineering requests, or other tasks that management does not have the will to abandon or hire other staff to handle.

Poor or no work order system (App. J). Not having work orders to attach plans to would really cripple planning.

Lack of information on new work orders (Chap. 5). Planning slows when planners have to collect basic information. Work order forms with sections, information fields, and check boxes can help encourage originators to provide information. Be careful with having too many "required" fields in a CMMS or a process for "rejecting" incomplete work orders that may discourage persons from writing work orders.

Lack of information and feedback on history work orders (Chap. 2-Planning Principle 3, Chap. 5). The more information coming back from the field on what the technicians actually did, the better. Chapter 5 covers a number of useful ways to increase getting good feedback.

Technician interruptions (Chap. 2-Planning Principle 2). Not all jobs will go as smooth as planned, but technicians should not ask the planner to resolve problems for jobs already in-progress.

File security (Chap. 8). The files should be accessible to persons other than planners so that planners do not have the only means of solving problems for jobs-in-progress. However, more accessibility means more likelihood of files disappearing.

Library hassle (Chap. 8). If the planners do not control the plant files, there is likely to be hassle in quickly accessing and storing information. Sometimes the engineering department has control of all the O&M manuals as well as the plant drawings and schematics.

Operations unavailability (App. F). The planners must have quick access to ask questions when necessary of the operations or production group. Many times the operator who wrote the work order is not on shift when the planner is planning it.

Inadequate storeroom (App. A).

Inadequate authority or processes to get parts or make decisions. Planners may need authority or the ability to get approvals when planning jobs.

Lack of equipment information, O&M manuals (Chap. 8). This will slow planning, but the planning department will accumulate file information through the cycle of planning maintenance actions for the same equipment over time.

Plant reliance on blanket work orders (App. I). This is as bad as not having a work order system. Planners base their plans on discrete work orders.

No driver of system (App. A). Planning needs someone in authority and someone intimately involved in maintenance both keenly interested in making planning work.

Lack of processes for stores, purchasing, tools (Chap. 5). How do planners designate and reserve items for plans?

Lack of training on processes. Many times these processes are not clearly understood. Planners need to understand.

Lack of planner training and understanding of planning fundamentals (Chap. 2, Chap. 3, Chap. 4, Chap. 5). Planning has a lot of room for serious error in overall direction.

Highly reactive plant maintenance conditions (Chap. 4). The plant is in a poor condition with a high amount of urgent or emergency work.

"Back door" jobs being done instead of routing through planning (Chap. 3, Chap. 4, Chap. 6). Many plants plan a lot of work, but it's not the work being done. The planned work sits on the shelf.

Extensive duplication of work orders (Chap. 5). A potential source of confusion. Deficiency tags generally head off duplicate work orders.

Reluctance to write work orders. This leads ultimately to only emergency work being verbally reported directly to crew supervisors and takes planning out of the loop.

Lack of planning facilities and tools (Chap. 8). This might cause planning to slow. Planning benefits from a somewhat modern office environment.

Overwhelmingly large backlog. This may make the planning challenge seem impossible. However, only enough work needs to be planned for crew to complete in 1 week. If the backlog is creditably prioritized, the planners can start with the highest priority jobs not yet worked and plan them. They could also plan the entire backlog in a fairly short time if they give identification of craft and hours on each job on a first pass and make a second pass to add details and better scope definition. The first pass alone can be used when it is time to make the weekly schedule.

Wrong or untrained planners (Chap. 11, App. M). Planners who are not self-starters and cannot quickly plan in accordance with the general principles of planning. Planners with inadequate craft skills, inadequate communication skills, or inadequate ability to work with data, files, and computers.

CMMS? (Chap. 9, App. L). A CMMS might overly restrict and frustrate the task of planning jobs. An example of this frustration is if planners come to rely on the system and then the IT department allows the network to become overloaded and the CMMS speed dramatically slows to a crawl. Another example of a CMMS hindrance might be a problem with linking documents or printing attachments to work orders if planners have a number of standard forms and reference documents that were formerly easily attached to work orders in a pure paper system.

Schedule—schedule enough jobs for one week

The principal objective of this area is to set the goal of work by creating a weekly schedule. This schedule is a simple list of work orders that matches the crew's labor hours and skills available for the next week. The list takes relative work order priorities into consideration. Figure M.7 shows aids or barriers meriting their own analysis later as issues with subordinate aids and barriers include management support, supervisor support, and urgent breakdowns.

Aids

Management support of planning and appreciation of WIIFM (Chap. 1, Chap. 4, Chap. 12, App. A, App. P).

Supervisor support of planning and appreciation of WIIFM (Chap. 12). Supervisor going along with the scheduling is mandatory even if it must be insisted upon by plant management. However, as keepers of the plant culture, supervisors open acceptance goes a long way toward helping planning.

Planned work orders identifying labor hours and skills needed (Chap. 3-Scheduling Principle 1, Chap. 6). This is a mandatory prerequisite because the process of scheduling requires knowing how many hours and for what craft skill for each work order.

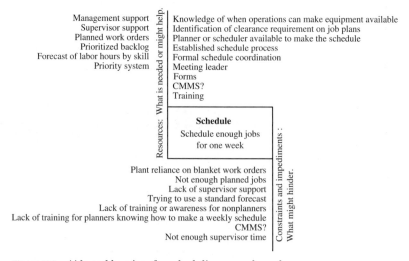

Figure M.7   Aids and barriers for scheduling enough work.

Prioritized backlog of work orders (Chap. 3-Scheduling Principle 2, Chap. 6).   This is a mandatory prerequisite because the scheduling process considers priorities.

Priority system (Chap. 5, App. I, App. J).   Appendix I considers "gaming the priority system."

Forecast of labor hours that need work for the next week identifying skill levels available (Chap. 3-Scheduling Principle 3, Chap. 6).   This is a prerequisite because the scheduling process allocates backlog hours according to the forecast. Scheduling without the forecast is possible, but not recommended.

Knowledge of when operations can make equipment available.   Not necessarily required, but it helps if planners have a general feel for what equipment can be cleared for work during a week even though the originator of the work should note if an entire system outage or merely equipment clearance is required.

Identification of clearance requirement on job plans.   This is a help. This is not necessarily the details of how to clear equipment, but a yes or no identification of if the equipment will need clearance.

Planner or scheduler available to make the schedule (Chap. 3, Chap. 6).   Someone has to make the schedule. The actual process of making an advance schedule (a simple list of jobs to do) can be made needlessly complex.

Established process or procedure on how to develop the weekly schedule (Chap. 6). It is required that planners create the schedule according to a set process, one that most everyone understands.

Formal schedule coordination or meetings to give input or accept schedule (Chap. 6). It is helpful to incorporate specific opportunities for others to give schedule input.

A meeting leader.   Charging a specific individual with the responsibility to lead any of the scheduling meetings helps to ensure that they do take place and are more effective.

Forms to help the forecast and scheduling processes (Chap. 6).   This helps standardize the process.

CMMS? (Chap. 9).   It would seem that a CMMS could easily create the required weekly allocation of work, but in practice many CMMS systems over-complicate the process.

Training for the planner, supervisors, and key operators on the scheduling process. The driver of the system needs to spend time walking each of the participants of scheduling through the process and then monitoring its usage because scheduling is so important to productivity.

Barriers

Plant reliance on blanket work orders (App. I).   Blankets rob the plant of an ability to schedule sufficient work.

Not enough planned jobs (Chap. 4).   If planners cannot plan enough work, this will seriously hamper the scheduling process.

Lack of supervisor support and supervisors won't give adequate crew forecast (Chap. 12, Chap. 6).   A communication effort from plant management must convince supervisors why the plant and the supervisors need planning. The planner may have to go to the supervisor to help collect the forecast information.

Trying to use a standard forecast (Chap. 6).   Crew labor hours vary tremendously from week to week and the schedule must reflect expected labor to have enough credibility. Reasons for not meeting the schedule would otherwise concentrate too heavily on why the standard forecast was inadequate for that week.

Lack of training or awareness for nonplanners.   Supervisors mostly need to understand the concepts of the scheduling process and approach.

Lack of training for planners knowing how to make a weekly schedule (Chap. 6).   The planners making nonsense schedules would hinder the productivity gain possible and damage planning's credibility.

CMMS? (Chap. 9, App. L).   A CMMS might needlessly over-complicate the scheduling process.

Not enough supervisor time to help develop labor forecast.   This is symptomatic of the plant supervisors and possibly plant management not being serious about planning. The planners should track down the supervisors and help them work out the forecast.

Execute—execute scheduled jobs and give feedback

The principal objective of this area is to work on the goal list of jobs established by the weekly schedule. This schedule guides the crews toward the appropriate list of jobs out of the backlog considering the relative job priorities and the available craft skill hours. The crew does not have to choose jobs from the entire backlog and should only deviate from the culled backlog list for new urgent work that cannot wait until the following week. Figure M.8 shows the aids or barriers meriting their own analysis later as issues with subordinate aids and barriers include supervisor support, technician support, and urgent breakdowns.

Aids

Management support of planning and appreciation of WIIFM (Chap. 1, Chap. 4, Chap. 12, App. A, App. P).   Supervisor support of planning and appreciation of WIIFM (Chap. 12). Necessary because the supervisors must assign from the allocated schedule goal.

Technician support of planning and appreciation of WIIFM (Chap. 12).   Preferred to get good feedback on job plans.

Manpower available as forecasted.   During the week, the supervisor uses the labor resources forecasted.

Effective daily schedule (Chap. 7).   This is outside planning and weekly scheduling, but supervisors greatly benefit from creating a daily schedule. Planners can help support this daily effort by supplying forms and training.

Daily schedule form (Chap. 7).   Chapter 8 shows this helpful form.

Coordination with operations for equipment availability to service (Chap. 7).   This coordination occurs daily as the supervisors line up work for the next day.

Staging? (Chap. 6).   Staging or kitting is not required for effective planning or productivity increase by maintenance. However, staging could help productivity even more for certain items and jobs.

Training on the planning and scheduling process.

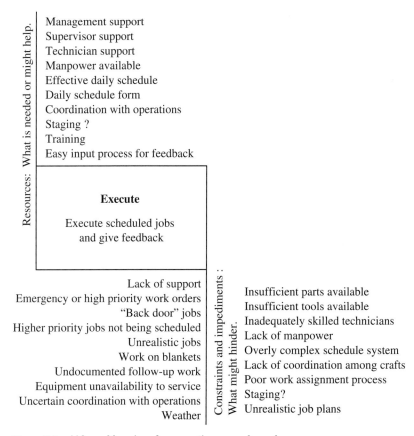

Figure M.8   Aids and barriers for executing enough work.

Easy input process for feedback (Chap. 5).   The easier it is for technicians to provide feedback, the better. Easiest methods involve simply marking and handing in a paper work order.

Barriers

Lack of support from management, supervisors, or technicians. Planning will not survive as a grassroots effort.

Emergency or high priority work orders (Chap. 4).   This could be used as an excuse to derail planning, but is probably a reason the plant should utilize planning.

"Back door" jobs being done instead of scheduled work (Chap. 4).   Supervisors ignore the planning and scheduling process and accept work requested without work orders.

Higher priority jobs not being scheduled.   The schedule did not include the important jobs.

Unrealistic jobs being scheduled.   The schedule included jobs that could not be cleared, had dramatically wrong estimates or anticipated parts, or were all lower priority jobs.

Work on blankets (App. I).   There is so much work on blankets that the schedule becomes meaningless.

Undocumented follow-up work.   The technicians do more work than required and ignore the rest of the schedule. For example, the technicians find extensive damage during the course of a PM inspection and instead of writing work orders for the work, they abandon doing the rest of the scheduled work.

Equipment unavailability to service.   This could be due to operators not clearing and tagging out equipment when needed for two reasons. One, the equipment cannot be taken out of service and two, there is not an adequate process for obtaining timely clearance and tagging from operations (or it is not being followed). The planners must obtain information with new work orders or clarify with operations whether equipment can be cleared with adequate notice or cannot be cleared without a plant outage. This information must be noted clearly on the work orders.

Uncertain coordination with operations.   It is difficult to coordinate when operations will clear equipment for technicians to service.

Weather.   Weather conditions prevent planned work.

Insufficient parts available (App. A).   The parts planned on jobs were not available after all or the storeroom is inadequate for any parts not planned ahead of time.

Insufficient tools available (App. A).   Special tools planned on jobs were not available after all or the tool room is inadequate for any tools not planned ahead of time.

Inadequately skilled technicians (Chap. 2-Principle 5, App. A).   Planning should not be expected to take the place of technician skill. Planning should plan for skilled technicians and the lack of technician skills would lower the workforce's ability to completely execute the work on the schedule.

Lack of manpower.   If enough persons call in ill or were not correctly forecasted to be on vacation or in training, the schedule would contain more work than could be executed.

Overly complex schedule system (Chap. 3, Chap. 6).   Some schedule systems are too complicated to assist the supervisor in efficiently assigning work.

Lack of coordination among crafts (Chap. 3).   Some jobs require inter-craft assistance and not using the advance weekly schedule to communicate might delay jobs.

Poor work assignment process.   Work not given to technicians each day. Technicians expected to find own work.

Staging? (Chap. 6).   Staging might be a hindrance if it requires more work overall to stage the parts in advance than is saved obtaining the parts only when they are needed.

Unrealistic job plans.   Job plans might not be possible if the planner incorrectly scopes the work.

Ongoing—keep planning and scheduling ongoing

The principal objective of this area is to maintain the successful planning and scheduling program itself. This management effort controls planning by keeping the aids and barriers necessary for long term success under control. Figure M.9 shows the aids or barriers meriting their own analysis later as issues with subordinate aids and barriers include

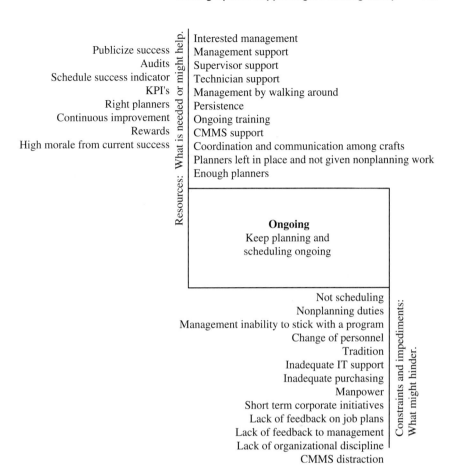

Figure M.9 Aids and barriers for keeping the whole program working.

management support, supervisor support, technician support, having the right planners, and having trained planners.

Aids

Interested management.   Management support of planning and appreciation of WIIFM (Chap. 12, Chap. 4, Chap. 1, App. M, App. P).

Supervisor support of planning and appreciation of WIIFM (Chap. 12).

Technician support of planning and appreciation of WIIFM (Chap. 12).

Management by walking around.   Management needs feedback from the field.

Persistence (App. M).   Planning "goes against the grain" and would soon drift into disarray without active control.

Ongoing training for planning and scheduling including CMMS if used. Refresher training is a good aid.

CMMS support. If the company uses a CMMS, active support from the IT department is mandatory.

Coordination and communication among crafts. The coordination available from knowing each other's weekly schedule can help crafts work together and aid the viability of planning.

Publicize success. Prove and keep the benefit of planning in view.

Audits by management (App. P). Management can help its control of planning by establishing regular audits of practicing the principles.

Schedule success indicator (Chap. 3, Chap. 11). This indicator drives a lot of good behavior. Schedule success cannot be measured if the crews are not using a schedule. A schedule cannot be created if enough jobs are not being planned. Management's insistence of seeing Schedule Success scores help keep the other practices active.

Key performance indicators (KPI's). Measurement systems (Chap. 11). Ongoing, before and after snapshots. relevant and visible. Schedule success, planned coverage, backlog, work orders completed, wrench time?

Right planners (Chap. 11). Keeping the right planners is a mandatory aid to keep planning working.

Planners left in place and not given more and more nonplanning work. Management must guard against planning gradually taking on so many nonplanning duties that reduces actual planning.

Continuous improvement in P&S performance. Active management working with planning to improve would reduce falling into bad habits.

Rewards for achievement. Pay structures, wages, incentive awards should all continue past the initial implementation of the planning program.

High morale from current success.

Barriers

Not scheduling (Chap. 3). Drifting into not scheduling drops productivity gains.

Nonplanning duties given to planners. These gradually add up.

Management inability to stick with a program, short term memory. Forgetting vision and WIIFM.

Change of personnel. Change of management. Loss of key drivers and supporters. Lack of succession planning. Loss of key planners.

Tradition. Tendency to revert to the good old days. A very powerful barrier.

Inadequate IT support (Chap. 8, App L). The IT department considers the CMMS as a program to implement and not have to maintain. IT runs out of budget to support CMMS. IT will not make continuous improvements or any changes maintenance wants.

Inadequate purchasing processes or support. The processes do not exist or are too difficult to use by planners or technicians.

Manpower availability. Planners are not replaced or the key supporters are no longer in place.

Short term corporate initiatives. Cost cutting. Personnel cutbacks.

Lack of feedback on job plans (Chap. 5). This is a significant hindrance, but not vital.

Lack of feedback to management on P&S success.   Management loses focus and does not allocate resources to ongoing planning.

Lack of organizational discipline. Noncompliance allowed.

CMMS distraction (Chap. 9).   The ability of a CMMS project to distract the real focus of planning is significant.

Key Aids and Barriers

Some of the aids are so important for meeting the analyzed objective and some of the barriers are such a problem that a separate barriers and aids analysis is shown for each of those aids and barriers. For example, the obvious aid of supervisor support is such a critical aid (and lack thereof, a critical barrier) to several objectives that supervisor support as an objective in itself is analyzed. Figure M.10 shows the aids and barriers that for which the following section analyzes subordinate aids and barriers.

Management support—sponsor a P&S system

This aid appears in most of the major areas of planning management. The handbook addresses management support and communication in a number of areas including Chap. 1, Chap. 4, Chap. 12, App. A, and App. P, as shown in Fig. M.11. Chapter 1 gives management the overall vision. Chapter 4 provides guidance through the controversies of reactive maintenance. Chapter 12 clearly defines WIIFM, What's in it for me, for management. Appendix A discusses the leadership and communication aspects of management support as a necessary prerequisite of planning. Finally, App. P sets forth honest questions for management to pursue to gauge the plant's interest and effectiveness in pursuing effective planning.

Aids

Current poor maintenance performance.   Not effective or not efficient. Plant desperation. Willingness to try something new.

Cost benefit analysis (Chap. 1).   Management needs to see the bottom line.

Key Aids and Barriers

- ◆ **Management Support**
- ◆ **Supervisor Support**
- ◆ **Technician Support**
- ◆ **Right Planner**
- ◆ **Planner Training**
- ◆ **Urgent Breakdowns**
- ◆ **Technician Interruptions**

- ◆ **Equipment Tags**
- ◆ **Files**
- ◆ **Purchasing**
- ◆ **Work Order System**
- ◆ **CMMS**

Figure M.10 Supporting areas for the overall program objectives.

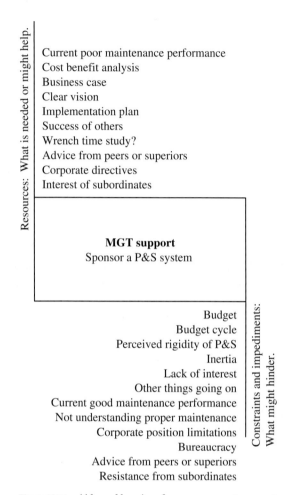

Figure M.11   Aids and barriers for management support.

Business case (Chap. 1).   Management needs to see the bottom line.

Clear vision of planning.   Close to mandatory is management really understanding the purpose of planning.

Implementation plan.   This would be helpful for management to see the steps involved in implementing planning. Preferably one with least disruption,

Success of others.   Other plants with planning success stories would be helpful.

Wrench time study? (Chap. 2-Principle 6) Actually seeing the existing wrench time as a measure usually convinces management to do something about it.

Advice from peers or superiors.

Corporate directives.

Interest of subordinates.   Planning is such a change of maintenance practice that interest by the persons affected may soften management reluctance to implement the changes.

Barriers

Budget. Planning facilities are biggest expense unless a CMMS is involved. The cost of the planner salary should not be a large item because existing positions can be converted.

Budget cycle. Implementation may have to wait.

Perceived rigidity of P&S not being able to react to plant needs. Management may wrongly perceive emergencies as having to wait on planning.

Inertia. Changing means having to do something.

Lack of interest. Management may not be convinced.

Other things going on. Management must allocate scarce resources among competing programs.

Current good maintenance performance. Plants are often "victims of their own success."

Not understanding proper maintenance. Maintenance should be preventing plant disruptions, not merely reacting to them quickly. Management may be satisfied with the fast reaction to emergencies approach of maintenance and not want planning.

Corporate position limitations. Management might have restrictions on creating new positions.

Bureaucracy. It takes a monumental effort to change anything.

Advice from peers or superiors. These may be against the concept of planning because of industry-wide frustration with planning.

Resistance from subordinates. Management may give in to subordinate objections.

Supervisor support—follow a P&S system

Supervisor support critically affects the success of planning as it enters the scheduling and work execution phases. The handbook addresses supervisor support in a number of areas including Chap. 3, Chap. 4, Chap. 12, and App. A, as shown in Fig. M.12. Chapter 1 gives management the overall vision. Chapter 3 covers culling the backlog for the supervisors and not holding schedule compliance as a weapon against them. Chapter 4 provides assurance that planning will not hold up crews from attacking urgent work. Chapter 12 clearly defines WIIFM, What's in it for me, for supervisors. Appendix A discusses the supervisors as keepers of the plant culture.

Aids

Management support. Usually required.

Clear vision (Chap. 1). Supervisors need to understand why.

Better control of individual jobs (Chap. 5, Chap. 12). Crew supervisor now has hour and craft skill estimates before assigning work.

Service of getting a culled backlog (Chap. 6). Scheduler goes through entire backlog for the supervisor selecting jobs that make sense.

Peer relationship between supervisors and planners (App. M). Supervisors are more likely to support persons "on their level."

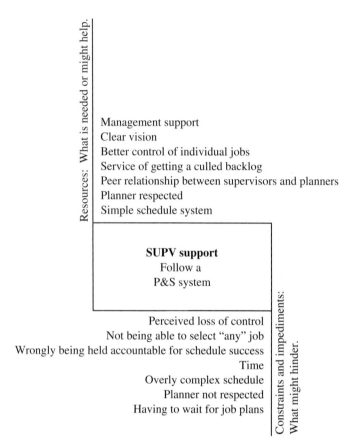

Resources: What is needed or might help.

Management support
Clear vision
Better control of individual jobs
Service of getting a culled backlog
Peer relationship between supervisors and planners
Planner respected
Simple schedule system

SUPV support
Follow a
P&S system

Perceived loss of control
Not being able to select "any" job
Wrongly being held accountable for schedule success
Time
Overly complex schedule
Planner not respected
Having to wait for job plans

Constraints and impediments:
What might hinder.

Figure M.12   Aids and barriers for supervisor support.

Planner respected among the crafts (Chap. 11, App. M).   Supervisors are more likely to support persons that their crews respect.

Simple schedule system (Chap. 6).   The schedule should be a simple list or package of work to start the week with. Forms can help. Not having a schedule that must be adjusted each day helps. Don't correct the weekly schedule once issued. Simply use it as a starting point and work in urgent jobs that can't wait until the next week.

Barriers

Perceived loss of control.   The supervisors are accountable to a schedule.

Not being able to select "any" job.   The supervisors are restricted in selecting work to execute.

Wrongly being held accountable for schedule success.   Schedule success tied to pay. Supervisors do not affect schedule success as much as does the overall condition of the plant. Reactive work will interrupt the schedule and supervisors must react to that work instead of meeting a set schedule. In fact, schedule success is more of an indicator of the overall plant condition than it is the supervisor's willingness to work towards a set schedule. In addition, if plant management allows operations to declare false emergencies, schedule success suffers. Managers should hold supervisors accountable for starting each day with a daily schedule and supporting the planning department in developing the weekly schedule.

Time to develop labor forecasts.   The supervisor already has a lot to do and the system is adding another task.

Overly complex schedule (Chap. 6).   Having to spend too much time working with the schedule.

Planner not respected among the crafts (App. M).   Supervisors are reluctant to support such an intrusive program if they do not respect the planners chosen.

Having to wait for job plans (Chap. 4).   The proper planning system should not make supervisors ever wait for planned work. The supervisor should be free to start an urgent job anytime. However, the planners can quickly plan reactive work under Chapter 4 guidelines.

Technician support—follow a P&S system

Technician support affects the success of planning mostly as it enters the work execution phase. The handbook addresses technician support mostly in Chap. 2 which sets forth planning as a library service and which holds that technicians are not wholly accountable for wrench time apart from the maintenance system, as shown in Fig. M.13. Chapter 12 also defines WIIFM, What's in it for me, for technicians. This appendix, App. M, discussed the needing respect of planners among the technicians in an earlier section for selecting the right planners.

Aids

Supervisor support of planning.   The technician usually support planners heartily endorsed by the supervisors.

Planners at peer level with supervisors (App. M).

Planner respected among the crafts (App. M).

Planners attending crew meetings (Chap. 5).   Planners can reinforce support of planning with interaction.

Not having to select the next job.   Technicians should not have to select their own work. This leads to peer pressure to slow down productivity.

Not having to determine how many jobs.   Technicians appreciate being assigned specific quantities of work.

Having a measure of daily success in terms of jobs finished versus jobs assigned. Everyone should go home at the end of the day knowing they've won or lost.

Getting a head start on jobs from a respected planner.   Technicians don't like reinventing the wheel and get a heads up from an experienced planner.

Library service of past job knowledge (Chap. 5).   File information history, parts, tools, special problems.

Job time estimate (Chap. 5).   Technicians get a standard against which to work.

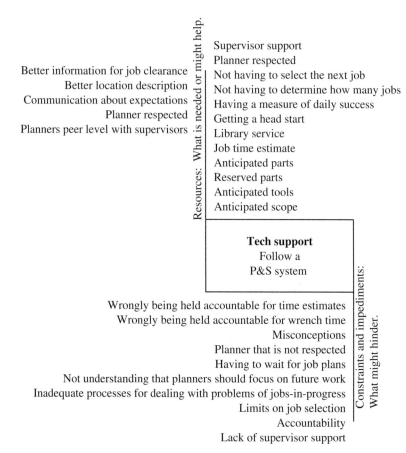

Figure M.13   Aids and barriers for technician support.

Anticipated parts (Chap. 5).   Technicians get an advance estimate of parts needed.

Reserved parts (Chap. 5).   Technicians get more parts help.

Anticipated tools (Chap. 5).   Technicians get an advance estimate of tools needed.

Anticipated scope (Chap. 5).   Technicians get an idea of the job from an experienced craft-person.

Better information for job clearance (Chap. 5).   The planner may provide a heads up to avoid time wasted in the past.

Better location description.   The planner finds the job and can reduce hunting by the technician.

Communication about expectations.   Craft persons should be expecting a head start based on a respected planner looking over the job and guessing what is needed based on his experience and file information. The craft-persons must expect to add to this file information over time. The planner cannot do perfect research and perfectly plan any job. The planner and craft-person are part of a system for perfecting a job plan for that equipment over time and repeated maintenance cycles.

Barriers

Wrongly being held accountable for time estimates (Chap. 5).   The uncertain nature of any maintenance makes time estimates on individual jobs too uncertain against which to hold technicians accountable. Holding them accountable introduces fear into maintenance and ill will toward planners.

Wrongly being held accountable for wrench time (Chap. 2).   Craft-persons do not affect wrench time as much as the overall system does. If the system does not assign enough work, wrench time suffers. If false emergencies interrupt jobs-in-progress, wrench time suffers. If craft-persons have inadequate information, parts, and tools, wrench time suffers. If craft-persons have not been adequately trained, they have to delay work seeking information, and wrench time suffers.

Misconceptions.   Expecting perfect plans (scope, parts, tools, times) instead of expecting a head start.

Planner who is not respected among the crafts.

Having to wait for job plans.   Crews should not have to wait on planning to begin jobs.

Not understanding that planners should focus on future work.   Planners should not be in the business of helping jobs already in-progress or they cannot plan future work. Then a vicious cycle starts where all jobs-in-progress need extensive help because planners never had time to plan them.

Inadequate processes for dealing with problems of jobs-in-progress.   Craft-persons must have ready access to peers, O&M manuals, supervisors, purchasing assistance, and other resources for resolving problems without having to interrupt planners focusing on future work.

Limits on job selection.   Not being able to select any job. A planning and scheduling system makes the job selection process more structured and limits the ability of technicians and supervisors somewhat from selecting pet projects or other inappropriate jobs.

A sense of accountability over the long run for jobs finished versus assigned.

Lack of supervisor support.

Right planner—create positions and select the right planners

This aid so critically affects planning success that Chap. 11 and App. M (in an earlier section) cover selecting the right planner extensively, as shown in Fig. M.14.

Aids

Respect from craft-persons.   Mandatory.

Planners at peer level with supervisors (App. M).   Very helpful. This makes the planner position much more sought after by technicians.

Good craft experience and skills.   Mandatory.

Good communication skills.   Mandatory.

Good data skills.   Mandatory.

Self starter.   Mandatory.

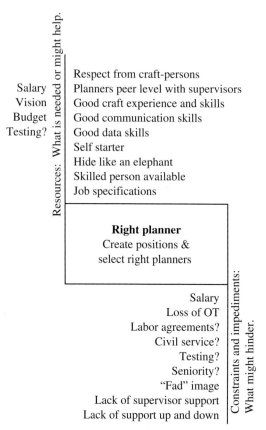

Figure M.14   Aids and barriers for selecting the right planner.

Hide like an elephant.   Thick skinned, but diplomatic. The successful planners need enough self-esteem where they feel good about what they are doing in spite of resistance from some supervisors, some craft-persons, and even some managers. On the other hand, they may need to "lay low" occasionally and exert their obvious presence diplomatically.

Skilled person available.   It helps if the supervisory ranks or craft ranks already contain persons that meet the mandatory requirements.

Job specifications (App. N).

Salary (App. M).   Must be competitive to make the job desirable for qualified persons.

Vision (Chap. 1).   Planners need to be instructed in the purpose of planning.

Budget.   Budget changes may be necessary to establish the planner position.

Testing?   Companies such as civil service may have to arrange testing for new positions. Tests should be structured toward measuring communication and data skills as well as machine skills. Interviews may be necessary to assess communication skills and self initiative.

Barriers

Salary (App. M).   If planner take home pay is too low, the skilled technicians in the crafts will not be interested.

Loss of OT (App. M).   Making technicians into planners for a modest salary increase and then not allowing them overtime may give them a net pay reduction and cut interest among top technicians. The plant should make the salary rate more than compensate for lost overtime.

Labor agreements?   This could be a potential to slow the creation of a planner position and should be addressed in a timely manner.

Civil service? (App. M).   There might be a large bureaucracy in place that needs timely action to address.

Testing? (App. M).   Testing might not be accurate for professional positions such as planning.

Seniority? (App. M).   Not all mechanics make good planners.

"Fad" image.   Who would want to take a new position that management might do away with any day? Who would want to take a new position that no one really understands is a key role that should endure?

Lack of supervisor support.   Who would want to take a position requiring peer communication with supervisors that the supervisors do not support?

Lack of support up and down.

Planner training—have trained planners

This entire handbook sets out in one of its objectives to train the maintenance planner in the vision and principles of effective planning. This handbook sets forth that body of knowledge. On a more specific level, this appendix, App. M, contains a section on planner training in an earlier section, and is shown in Fig. M.15. Finally, Chap. 11 cautions that the wrong person cannot be trained to become an effective planner just as "just anybody could not become a good mechanic, good lawyer, or good doctor if they just had enough of the right training."

Aids

OJT.   On the job training. Planners learn best from planning with someone who understands planning.

Having an already experienced planner on board (App. M).   Overwhelmingly the best way to bring new planners up to speed. Unusual to have a top planner already for a new planning organization.

Vision and general planning principles (Chap. 1, Chap. 2, Chap. 3, Chap. 4).

Classes.   The driver of the program can lead classes. Consultants also offer some training workshops and OJT follow-up.

Close supervision early in the program.   Close attention to the program is necessary to keep planners from drifting into bad habits.

Refresher training.

Weekly meetings to discuss issues and maintain consistency.   This would help keep everyone on the same page and allow a regular forum for concerns.

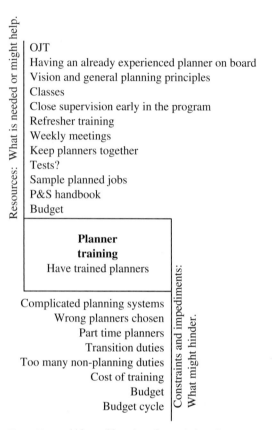

Figure M.15   Aids and barriers for training planners.

Keep planners together.   This allows reinforcing each other with good practices.

Tests?   (App. O). Appendix O offers a sample of what every planner should know. Do they know this material?

Sample planned jobs (App. D).   Appendix D has sample planned jobs so planners can get a feel for the level of detail and other plan criteria desired.

P&S Handbook.

Budget.   Outside training might need budgeting.

Barriers

Complicated planning systems.   Most planning organizations have overly complicated systems of estimating hours, setting forth job steps, and defining advance schedules.

Wrong planners chosen (Chap. 11, App. M).   The wrong planners cannot be trained to become good planners.

Part time planners.   Planners don't have enough time to properly plan.

Transition duties. Planners don't have enough time to properly plan.

Too many nonplanning duties. Planners don't have enough time to properly plan.

Cost of training.

Budget.

Budget cycle.

Urgent breakdowns—utilizing P&S in a reactive environment

Urgent breakdowns are the source of most planning system frustrations, which is shown in Fig. M.16. Chapter 4 resolves this issue.

Aids

Management support (Chap. 4). Management must support planning by insisting on scheduling and protecting planners from helping on jobs-in-progress to a great degree.

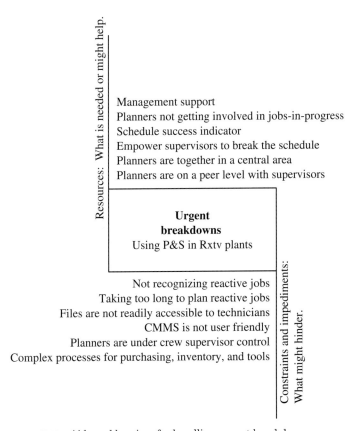

Resources: What is needed or might help.

Management support
Planners not getting involved in jobs-in-progress
Schedule success indicator
Empower supervisors to break the schedule
Planners are together in a central area
Planners are on a peer level with supervisors

Urgent breakdowns
Using P&S in Rxtv plants

Not recognizing reactive jobs
Taking too long to plan reactive jobs
Files are not readily accessible to technicians
CMMS is not user friendly
Planners are under crew supervisor control
Complex processes for purchasing, inventory, and tools

Constraints and impediments: What might hinder.

Figure M.16 Aids and barriers for handling urgent breakdowns.

Planners not getting involved in jobs-in-progress (Chap. 2-Principle 2).

Schedule success indicator pushed as a gauge of plant reactiveness. Schedule success is not a measure of supervisor's blind obedience. Don't tie supervisor pay to schedule success. Concentrate on starting each week with a schedule and each day with a schedule.

Empower supervisors to break the schedule. It's okay to break the daily schedule for emergencies and the weekly schedule for urgent work.

Planners are together in a central area. Keeping the planners together helps keep them on task (future work).

Planners are on a peer level with supervisors. This keeps them from being redirected to help jobs-in-progress.

Barriers

Not recognizing reactive jobs. Not planning them first.

Taking too long to plan reactive jobs. Making crews wait. Putting too much detail in plans.

Files are not readily accessible to technicians for on-the-job research (Chap. 8). If technicians are not to interrupt planners for help on jobs-in-progress, they must be able to access and use plant processes without planner help.

CMMS is not user friendly (Chap. 9, App. L). If technicians are not to interrupt planners for help on jobs-in-progress, they must be able to access and use plant processes without planner help.

Planners are under crew supervisor control. (Chap. 2-Planning Principle 1). Planners become involved in jobs-in-progress and do not have time to plan.

Complex processes for purchasing, inventory, and tools. Only planners can adequately use processes for purchasing, inventory, tools, etc. They are too difficult for technicians or supervisors to use for jobs-in-progress. If technicians are not to interrupt planners for help on jobs-in-progress, they must be able to access and use plant processes without planner help.

Technician interruptions—deal with planner distractions

Technician interruptions go hand-in-hand with urgent breakdowns as a significant barrier to effective planning. Basically, management needs to set a clear policy protecting the planners as much as possible from interruptions concerning jobs-in-progress, as shown in Fig. M.17. The policy should explain the feedback loop nature of planning and that it's "okay" if the current job plan is not perfect; the technician should give effective feedback to help "the next job." Chapter 2-Principle 2 explains that planners should work on future work and Chap. 5 gives guidelines for effective technician feedback.

Aids

Communication (Chap. 4). Planning that allows imperfect jobs to be improved through a feedback system is not "natural" and must be thoroughly explained to technicians.

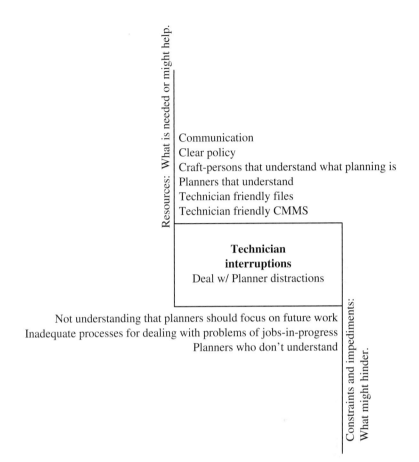

Figure M.17   Aids and barriers for handling technician interruptions.

Clear policy.

Craft-persons that understand what planning is.

Planners that understand.

Technician friendly files (Chap. 8).   If technicians are not to interrupt planners for help on jobs-in-progress, they must be able to access and use plant processes without planner help.

Technician friendly CMMS.   If technicians are not to interrupt planners for help on jobs-in-progress, they must be able to access and use plant processes without planner help.

Barriers

Not understanding that planners should focus on future work.   Planners should not be in the business of helping jobs already in-progress or they cannot plan future work. Then a vicious cycle starts where all jobs-in-progress need extensive help because planners never had time to plan them.

Inadequate processes for dealing with problems of jobs-in-progress.   Craft-persons must have ready access to peers, O&M manuals, supervisors, purchasing assistance, and other resources for resolving problems without having to interrupt planners focusing on future work.

Planners who don't understand.

Equipment tags—have tags on equipment

Having tags on equipment is an almost mandatory prerequisite for effective planning because it defines the filing system, and is shown in Fig. M.18. Chapter 2-Principle 3 addresses component level files using the tag numbers, Chap. 8 addresses filing systems, and App. K covers equipment tagging exclusively.

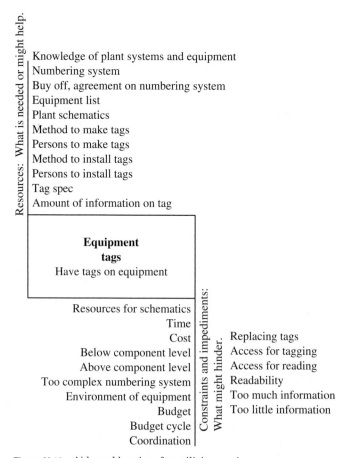

Figure M.18   Aids and barriers for utilizing equipment tags.

Aids

Knowledge of plant systems and equipment.

Numbering system.

Buy off, agreement on numbering system.

Equipment list.

Plant schematics.

Method to make tags.

Persons to make tags.

Method to install tags.

Persons to install tags.

Tag spec. Sizes, materials.

Amount of information on tag. In addition to the number, it is helpful to operations and maintenance to see the equipment name or function.

Barriers

Resources for schematics. This could be a large undertaking.

Time.

Cost.

Below component level. Each bearing on a pump is too low a level to be helpful for filing.

Above component level. An entire gas system might be too high a level to be helpful for filing.

Too complex numbering system (App. K). Ten to fifteen characters might be the limit of tag numbering sensibility.

Environment of equipment. This affects how tags might be hung or if they might be readable later.

Budget.

Budget cycle.

Coordination of different department responsibilities.

Replacing tags after equipment changed in maintenance. Experience shows that wiring the tags onto equipment rather that gluing them has advantages in craft-persons being willing to make sure they get onto the correct equipment.

Access for tagging.

Access for reading.

Readability.

Too much information on tags. The tag number and equipment name is usually sufficient. Much more information might slow or complicate the tagging process.

Too little information on tags. The tag number and equipment name is usually sufficient.

Files—have effective files

Having a filing system is a mandatory prerequisite for effective planning, which is shown in Fig. M.19. Chapter 2-Principle 3 addresses component level files and Chap. 8 addresses filing systems

Aids

Work order system (App. J).   Mandatory.

Work order forms (App. J).   Mandatory (unless CMMS is paperless).

Job feedback (Chap. 5).   Obtaining feedback from the field improves future work.

Component level files (Chap. 2-Principle 3).   Mandatory. Only files for individual pieces of equipment allow efficient retrieval of information for job planners.

Equipment numbers.   Mandatory. These must match tags hung in the field on equipment and what is in any CMMS.

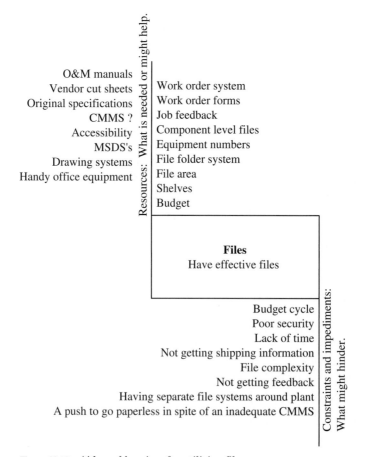

Figure M.19   Aids and barriers for utilizing files.

File folder system. Folders, forms. Needed and helpful.

File area. Needed.

Shelves. Needed.

Budget.

O&M manuals (Chap. 8). Items that would be kept in the filing system, not necessarily in the component level files.

Vendor cut sheets (Chap. 8). Items that would be kept in the filing system, not necessarily in the component level files.

Original specifications (Chap. 8). Items that would be kept in the filing system, not necessarily in the component level files.

CMMS? (Chap. 9, App. L) This could supplement or replace the need for a file system.

Accessibility. Close proximity for planners and for technicians.

MSDS's (Chap. 8). Items that would be kept in the filing system, not necessarily in the component level files.

Drawing systems. Helpful, although planners typically deal with individual pieces of equipment. Drawings are more typically managed or kept by an engineering group.

Handy office equipment to use with file information. Necessary items such as copiers, staplers, fax machines, labels, tables.

Barriers

Budget cycle.

Poor security. Technicians should be able to access files, but not feel free to take them without authorization.

Lack of time to file information. Planners do not realize importance of filing feedback or have too many other nonplanning duties. The plant may also not have enough planners. (Chap. 2-Principle 1).

Not getting shipping information. Often the data included with the shipped equipment is the most specific and important.

File complexity. Simple open files are the best for easily viewing files. The numbering system might also be a problem.

Not getting feedback (Chap. 5). Technicians might not provide very good feedback to aid future jobs.

Having separate file systems around plant. Different groups not using the planning file system could cause data inconsistency or planner information being out of date.

CMMS? (Chap. 9, App. L). If the CMMS cannot satisfy all of the planning needs, it may be confusing to decide which items to file and which items to store electronically.

A push to go paperless in spite of an inadequate CMMS.

Purchasing—buy timely nonstock parts

Chapter 5 addresses the need of planning to purchase nonstock parts and App. A addresses parts in general, and this is shown in Fig. M.20.

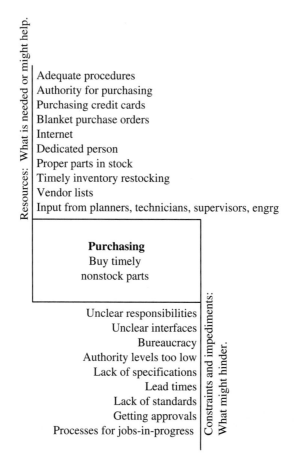

Figure M.20   Aids and barriers for purchasing activities.

Aids

Adequate procedures for nonstock purchases.

Authority for purchasing.   Arrange for planners to have adequate authority or easy access to approvals to buy most parts.

Blanket purchase orders.   Vendors that have standing purchase orders can deliver supplies rapidly.

Purchasing credit cards.   These can speed purchasing.

Internet.   Planners should have ready internet access. More and more items can be found easily through this medium.

Dedicated person for parts expediting?   This would be helpful for larger organizations.

Proper parts in stock.   This makes nonstock purchasing focus on the fewer exception items.

Timely inventory restocking procedures.   Fewer stock-outs limit extra purchasing activities.

Vendor lists.   Planners need a reference of available purchasing outlets.

Input from planners, technicians, supervisors and engineering. Planners need avenues to ask for specification criteria.

Barriers

Unclear responsibilities.   This can delay purchasing.

Unclear interfaces.   Such as who makes the particular decisions about what types of motors are standard.

Bureaucracy.

Authority levels too low.

Lack of specifications on what to buy.

Lead times.   Lead times are the time it takes a vendor to deliver a product after it receives a purchase order. The internal purchasing process itself can extend the time in which the plant receives an item.

Lack of standards.

Getting approvals.

Processes for jobs-in-progress.   Technicians and supervisors might not be able to purchase or not have access to purchasing help without unduly interrupting planners.

Work order system—have an effective foundation

A work order system serves as the foundation of the planning system as explained in App. A and is shown in Fig. M.21. Chapter 5 addresses how planners use work orders and App. J presents an entire example work order manual.

Aids

Management support.   Management has to insist on its use. Sometimes a "no work order means no work will be done" policy is necessary.

Prescribed process.   The work order routing processes need to be defined for normal and emergency work.

Forms.   Principally a work order form or format itself.

Tagged equipment (App. K).   Having tagged equipment greatly helps define which equipment is involved. This is much preferred over requiring originators to drill down in a CMMS hierarchy.

Deficiency tags (Chap. 5).   This is a helpful device for marking equipment and providing the operator with a field data collection paper device.

CMMS? (Chap. 9, App. L).   This has great potential for helping all parts of the work order process. It is most helpful as a work order origination point, but is less helpful for technicians in the field where paper forms may be more appropriate.

Barriers

Loss of ability to call maintenance without a work order.   Adherence to a work order system limits the ability to call in work as a special request.

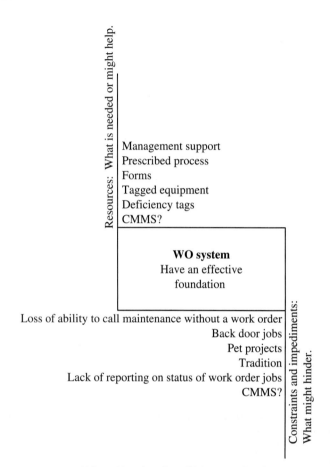

Figure M.21   Aids and barriers for utilizing a work order system.

Back door jobs.   Casual verbal work may reduce use of the formal work order system.

Pet projects.   Supervisors may ignore the work order system to do what they think is important.

Tradition.   It may be hard to convert the plant culture to using formal work orders.

Lack of reporting on status of work order jobs.   Persons become frustrated when they do use the work order system, but they see no progress on these jobs.

CMMS?   The CMMS may be inadequate or set up inadequately to provide the technicians, supervisors, or planners what they need.

CMMS—have a helpful computer system

A computerized maintenance management system (CMMS) is not a mandatory prerequisite for effective planning. However, it can provide a significant aid to planning, as shown in Fig. M.22. It can also be barrier

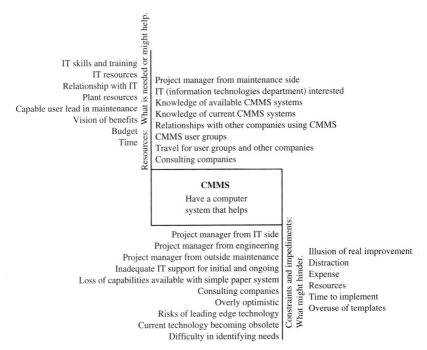

Figure M.22 Aids and barriers for utilizing a CMMs.

that can cripple a planning effort. Chapter 2-Principle 3 mentions cautions about computerization. The entirety of Chap. 9 is devoted to possible benefits and potential problems with a CMMS. Appendix L thoroughly examines the use and implementation of a CMMS.

Aids

Project manager from maintenance side. Most of these projects have a chance. It is more important to deliver what maintenance needs rather than just deliver something.

IT (Information Technologies Department) interested. The IT group has a lot going on and their participation in any CMMS system is vital.

IT skills and training. IT persons need overall computer expertise and specific training on the CMMS.

IT resources. CMMS projects and maintenance requires significant IT labor and hardware beyond the software itself.

Relationship with IT. The maintenance department is hard pressed to define all of its requirements for the specific CMMS (such as "what field length should this be?") and requires a willingness of the IT group to participate in continuous improvement at maintenance's direction. T the same time, the IT group needs to be able to explain limitations of the software.

Plant resources. Significant user time must be invested.

Capable user lead in maintenance.   A maintenance person must be familiar enough with the intricacies of the software and with planning practices to advise and lead the IT effort.

Vision of benefits.   Better ability to find and manage inventory. Better reporting ability. Better ability to find work orders. Better ability to link information to equipment. Common database. Ability to generate and manipulate schedules. Better ability to manage PM generation.

Budget.   CMMS systems are very expensive.

Time.   CMMS projects take a long time to implement.

Knowledge of available CMMS systems.   What systems are available? What features do they have?

Knowledge of current CMMS systems.   What systems does the company currently use? What are its features?

Relationships with other companies using CMMS.   Sharing information with other users accelerates the learning curve.

CMMS user groups.   In addition to learning, these forums offer opportunities to influence the CMMS development and future features.

Travel for user groups and other companies.   Takes time and money, but well worth it.

Consulting companies.   May have specific expertise with specific CMMS system.

Barriers

Project manager from IT side.   Industry wisdom holds that the majority of the resultant systems are not satisfactory to maintenance. IT soon loses patience with maintenance "tinkering."

Project manager from engineering.   Could be undesirable if engineering is not within the maintenance organization.

Project manager from outside maintenance.   Too much emphasis with getting it done.

Illusion of real improvement.   The CMMS itself even when fully implemented and functioning may not satisfy the maintenance requirement.

Distraction.   Too much emphasis on getting the computer working instead of planning and scheduling properly.

Expense.   CMMS systems are very expensive.

Resources.   Time taken away from other duties. Expecting user leads to severely curtail or even abandon their other job responsibilities.

Time to implement.   CMMS systems take a long time to implement.

Overuse of templates.   CMMS systems may discourage development of customized plans for individual pieces of equipment.

Inadequate IT support for initial and ongoing.   IT may simply walk away after a half-hearted installation effort.

Loss of capabilities available with simple paper system.   CMMS may not adequately allow attaching documents or other functions.

Consulting companies.   Consultants may not understand maintenance processes as well as computers (or vice-versa).

Overly optimistic about what is possible.   The CMMS may not be capable even when fully functional.

Risks of leading edge technology.   The technology may fail to become fully functional. It may not be compatible with current hardware or software. It may be unstable and not reliable.

Current technology becoming obsolete.   Once a CMMS is in place, the underlying technology may change to the point where the software or hardware providers may no longer support their systems and their new versions may radically change in functioning and related benefit.

Difficulty in identifying needs.   This factor affects the project not only in the design phase ("What would you like it to do?"), but in the testing phase when the budget is exhausted ("What is good enough?").

Special Circumstances

Figure M.23 selects several of the special circumstances discussed earlier in this appendix, App. M, on which to develop an aids and barriers analysis.

Improve existing planning—turn around an existing group

Figure M.24 shows that this is an area similar to the first aids and barriers analysis done for first organizing a planning group. However, it is much more difficult to turn around an existing group because so many poor practices (bad habits) must be changed. It is very hard to back out of these practices. Not only do the past practices not deliver results, but in most cases, planning has a bad name and may not be given a second chance. Management must change a negative perception of planning in addition to the practices.

All of the subject matter and principles of planning and scheduling in this handbook address critical issues for proper behavior.

Aids

Someone in charge of the transition.

Two white knights.   Someone in management with the authority and someone in a position to direct the changes.

Special circumstances

- ♦ **Improve existing planning**
- ♦ **New plants or units**
- ♦ **Self-directed teams**

Figure M.23   Special situations to analyze.

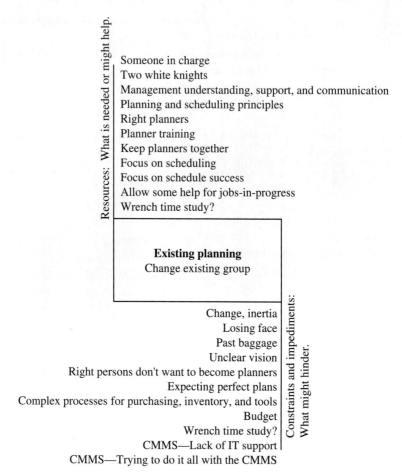

Figure M.24 Aids and barriers for redirecting an existing planning group.

Management understanding, support, and communication. Management needs to explain the new planning strategy and principles to entire workforce.

Planning and scheduling principles (Chap. 1, Chap. 2, Chap. 3, Chap. 4). Understanding is mandatory.

Right planners (Chap. 11, App. M). Mandatory.

Planner training. Very helpful.

Keep planners together. Helps ensure consistency.

Focus on scheduling (Chap. 3, Chap. 6). This is the area that directly improves productivity.

Focus on Schedule Success (Chap. 3). This measure drives good behavior in scheduling and planning.

Allow some help for jobs-in-progress. Allowing some planner interruption by technicians softens the turn-around.

Wrench time study? (Chap. 2-Principle 6). The results of a study may convince management, supervisors, and technicians to support a planning effort.

Barriers

Change, Inertia. It is hard to change existing practices.

Losing face. Planners and others may have considerable emotional investment in protesting a planning system that was set up incorrectly.

Past baggage. Previous planning system was sold as perfect jobs with all parts rather than feedback loop. Unclear vision (Chap. 1).

Right persons don't want to become planners (App. M).

Expecting perfect plans. Technicians were sold on the original planning program in that they would never have to hunt for parts or information.

Complex processes for purchasing, inventory, and tools. Only planners can adequately use processes for purchasing, inventory, tools, etc. They are too difficult for technicians or supervisors to use for jobs-in-progress.

Budget.

Wrench time study? This can be very unwelcome and intrusive.

CMMS – lack of IT support. IT is unwilling to invest the time to make any necessary changes.

CMMS – Trying to do it all with the CMMS. If the CMMS will not support a needed aspect of planning and scheduling, do that aspect manually or with other systems such as spreadsheets or email.

New plants or units—establish effective planning

Figure M.25 shows that new construction provides a good opportunity to begin operation with effective planning.

Aids

Establish planning group early. Having the planners in place before maintenance work is begun allows them to concentrate on collecting manuals and other basic file information and on equipment tagging.

Develop contractor relationships to enable collection of proper items.

Processes to receive information. These need to be established.

Contract details need to address planning needs. Coordination with the contracts group is necessary.

Tag equipment early. This takes resources.

Transfer knowledge from startup group. It would be helpful to include the startup group in initializing databases and files because of their knowledge.

Set up files (Chap. 8). Have these ready to receive information and as a guide on what to collect.

Resources: What is needed or might help.

Establish planning group early
Processes to receive information
Contract details
Tag equipment early
Transfer knowledge from startup group
Set up files
Set up some PM's
Resources to initialize CMMS

New plants
Establish
effective planning

Contract details not adequate
Contract details adequate, but not enforced
Contractors resist involvement
Timely receipt of O&M manuals
Start up group not given time or resources to participate
Generic as-builts
CMMS not in place

Constraints and impediments:
What might hinder.

Figure M.25   Aids and barriers for planning in new plants.

Set up some PM's.   This would also be a good time for formal RCM analysis (not normally a planning function).

Resources to initialize CMMS.

Barriers

Contract details not adequate.

Contract details adequate, but not enforced.

Contractors resist involvement.

Timely receipt of O&M manuals.   Many times these manuals are not received until well after startup and initial operation and troubleshooting.

Start up group not given time or resources to participate.   Management may be planning to disband this group without realizing their ability to help planning.

Generic as-builts.   The as-builts may not include adjustments made during startup.

CMMS not in place.

Self-directed teams—use planning and scheduling

A discussed earlier in this appendix, self-directed teams offer some unusual challenges to effective planning, but are worth considering because of the advent of TPM and other arrangements leading to more maintenance by operations personnel. Figure M.26 shows that the objective is not so much to use planning and scheduling as it is to achieve a higher productivity and maintenance effectiveness through the use of planning and scheduling.

Aids

Focus on schedule (Chap. 3).   Teams should still develop goals of work to accomplish.

Track Schedule Success (Chap. 3).   This measure drives good behavior in scheduling and planning.

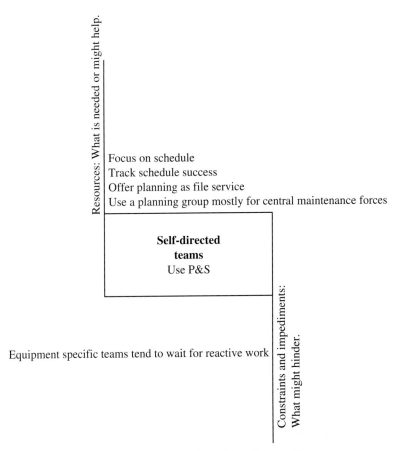

Figure M.26   Aids and barriers for planning with self-directed teams.

Offer planning as a file service (Chap. 2).

Use a planning group mostly for central maintenance forces that handle tasks that are beyond scope of teams.

Barriers
Equipment specific teams tend to wait for reactive work.

Summary

In summary, establishing a planning program along the lines of the purpose and principles of this book is not difficult to conceptualize. Management must spend time sharing the vision of planning and selecting the right persons for the planning positions. Attention must be given to the execution of the initial planning effort with careful communication and working together. Above all, management must want planning to work and give it support.

An existing planning program with minor misunderstandings might be helped by reviewing the issues addressed by the planning and scheduling principles. Organizations with more fundamental planning problems may be more difficult to reprogram. Certainly, active management involvement is key.

Planning may also help plants with other than traditional maintenance organizations within existing or new plants. Persons must consider their unique situations to adapt the planning principles as appropriate.

Formally analyzing a planning program helps one to recognize aids and barriers to allow taking appropriate action to insure success.

Finally, above all, someone must sponsor planning and remember why it is successful. Management must remember why the maintenance planning program works to guarantee its future success.

Example Formal Job Description for Planners

This appendix contains an example of a formal job description for maintenance planners. A company that develops a job description might find this example a useful starting point.

Maintenance Planner

The maintenance planner supports the function of the plant maintenance group by developing job plans and advance work schedules. Work requires good technical, analytical, and communication skills.

The planner reports to the Plant Maintenance Superintendent or higher level manager.

Duties

Assign codes to work orders.

Make field inspections and determine appropriate job work scopes.

Develop work plans for maintenance jobs with necessary information to allow efficient scheduling, assignment, and execution of maintenance work.

Create, use, and maintain plant history and other technical files to include information in current job plans to anticipate and avoid job delays.

Perform purchasing duties as required.

Stage materials and special tools as required.

Evaluate job feedback to improve future work.

Keep familiar with current safety rules and regulations.

Utilize CMMS computer software and other computer applications to support the maintenance function.

Provide technical assistance as required for maintenance personnel.

Prepare reports as required.

Perform related duties as required.

Minimum qualifications

Journeyman level or higher experience in designated craft.

Knowledge of methods, materials, tools, and equipment in the maintenance of a modern power station.

Knowledge of large industrial equipment such as fans, pumps, and boilers and their associated components such as bearings and actuators.

Knowledge of safety hazards and appropriate precautions applicable to work assignments.

Ability to apply technical knowledge to determine equipment problems and appropriate solutions in swiftly developing job plans.

Ability to perform basic mathematical computations involved in estimating and purchasing.

Ability to communicate clearly and concisely in written and verbal form.

Ability to maintain effective working relationships with other personnel including operators, vendors, contractors, subordinates, peers, and superiors.

Ability to utilize common office equipment such as copier, fax machine, telephone, and personal computer.

Ability to read, interpret, and apply information from files, drawings, catalogs, reports, and manuals.

Ability to maneuver continuously throughout plant to observe equipment. Ability to walk, sit, stoop, squat, climb ladders and stairs, and lift and carry objects up to 50 pounds in weight. Ability to perform simple grasping and fine finger manipulation of small objects. Ability to crawl on hands and knees to access and observe equipment.

Ability to work comfortably in confined spaces or exposed heights up to 250 feet with adequate protection.

Ability to wear respiratory protection devices for dust and solvents.

Must possess a valid state driver's license.

Example Training Tests

This appendix illustrates the type of knowledge a planner should be gaining when becoming familiar and adept with planning techniques. The review of answers to these tests helps clarify the planning principles. Planners should also review and practice the scheduling examples of Chap. 7, Advance Scheduling, for work hour forecasts, backlog work order sorting, and weekly allocation of work orders.

Maintenance Planning Test Number 1

Name:_____   Date:_____
(Closed Book, except for App. J: *Work Order System Manual.*)

1. 35% productivity means that for a 10-hour shift, how many hours are "nonproductive" on the average?
   a. 2¹/₂ hours
   b. 5¹/₂ hours
   c. 6¹/₂ hours
   d. 10 hours
2. Productive time includes which of the following?
   a. Walking to the job site.
   b. Carrying a gasket to the job site.
   c. Getting instructions from crew supervisor.
   d. Unbolting the casing on a pump.
3. The majority of a typical maintenance budget is:
   a. Wages and benefits
   b. Parts and supplies
   c. Special tools
4. Proactive maintenance is:
   a. Fixing equipment as soon as possible.
   b. Doing work to prevent equipment from developing serious problems.
   c. Being committed to quality work.

5. Maintenance planning improves _____ of the maintenance force work.
   a. Only efficiency
   b. Only quality
   c. Both productivity and effectiveness
6. Maintenance planning should be able to raise crew productivity from 35 up to 55%. This improvement is a _____ increase in productivity.
   a. 20%
   b. 35%
   c. 57%
   d. 100%
7. The purpose of planning is:

   _____

   _____

8. Which of the following is *not* a principle of planning?
   a. File system based on tag numbers.
   b. Quickly finding part numbers for jobs-in-progress.
   c. Planners use personal experience.
   d. Technicians are skilled in how to perform maintenance.
9. A planner should be able to plan for how many persons? _____
10. The outage code for a short duration outage is _____. (It helps file SNOW work.)
11. The work type code for trouble and breakdown is _____.
12. The department and crew code for operations crew A is _____.
13. The work type code for work to upgrade a piece of equipment is _____.
14. The group and system code for the condensate polisher is _____.
15. The group and system code for the demineralizer is _____.
16. The equipment type code for a control valve is _____.
17. The equipment type code for a pump is _____.
18. What should a planner do about making a minifile if there are not any tag numbers on the work order or the equipment?

    _____

    _____

19. Why is it important to file at the component level instead of the system level?

    _____

    _____

20. What is the primary supervisory effort over the planning department?
    a. Continual coordination meetings
    b. Direct supervision
    c. Procedures
    d. Selection and training of planners
    e. Indicators

Maintenance Planning Test Number 2

Name:_____   Date:_____
(Closed Book)

1. Name one distinguishing characteristic of a reactive-type work order.
2. Name one distinguishing characteristic of a proactive-type work order.
3. Name one distinguishing characteristic of a minimum maintenance work order.
4. Which of the following is *not* always part of the planned package?
   a. Who (which craft)
   b. What (what needs to be done)
   c. Where (where is equipment)
   d. How (how job needs to be done)
5. Which jobs would be planned first (both R-2 priority)?
   a. Reactive
   b. Proactive
6. What is the first thing a planner would do after selecting a reactive, extensive work order to plan?
   a. Go to the minifile.
   b. Make a field inspection.
   c. Plan the job.
   d. Check material availability.
7. What craftperson entry would the planner make for a job requiring an extra set of hands?
   a. Apprentice
   b. Trainee
   c. Helper
   d. Anyone
   e. Technician
8. What craftperson entry would the planner make for a job requiring pressure vessel welding?
   a. Technician
   b. Welder
   c. Helper
9. What craftperson entry would the planner make for a mechanical job merely requiring a responsible person?
   a. Mechanic
   b. Certified mechanic
   c. Technician
   d. Helper
10. The primary source of time estimates for jobs is:
   a. Planner judgment and experience
   b. Past work orders

    c. CMMS
    d. Time standards
    e. Supervisor estimate

11. What parts should the planner list on the work order form itself for a proactive work order?
    a. Only parts anticipated to be used
    b. All parts in the warehouse for that equipment
    c. All parts used on past jobs for that equipment
    d. All parts mentioned in the minifile

12. Must a planner find part information (such as the inventory number) even for an anticipated part if it is not already in the minifile for a reactive, extensive work order?
    a. Yes
    b. No

13. For a proactive, minimum maintenance work order, should a planner find the inventory number for an anticipated part if the information is not already in the minifile?
    a. Yes
    b. No

14. Using $25 for each work hour, what is the estimated labor cost for a job requiring two persons for 10 hours each?

15. What is the total job cost estimate for a job requiring two persons for 4 hours each, two gaskets, and insulation work? (Use $25 per work hour, $100 per gasket, and $900 for insulation cost.)

16. How does a planner get job assistance from engineering?
    a. Call one of the plant engineers.
    b. Send request to morning meeting.
    c. Through the Job Planning Coordinator.
    d. Through the Planning Supervisor.

17. How does a planner develop an estimate for insulation work?
    a. Through the planning supervisor.
    b. By rule of thumb.
    c. From planner experience.
    d. Through the purchaser.

18. Which is most correct for a job requiring special safety equipment?
    a. Include any special safety requirements with the job plan.
    b. Presume the technician is skilled and knows what to do for safety.

19. During the course of a job, what information would a planner *not* necessarily put in a new minifile?
    a. All past work orders kept in other files before planning was started.
    b. Worthwhile information that comes up during the job.
    c. The original of the work order form, if historically significant.
    d. All that concerns delays for that job.

20. Who fills out the actual cost sections of the work order form?
    a. Lead technician
    b. Crew supervisor
    c. Planner

Maintenance Planning Test Number 3

Name:_____   Date:_____
(Open Book)

1. Where would a planner look for pump impeller material information first?
    a. Technical files
    b. Minifile
    c. Vendor files
2. Use the sample Standard Plans in Chap. 8. For the "B" Fuel Oil Service Pump, what is the manufacturer's part number for the Casing Rings? _____
3. Use the sample Standard Plans in Chap. 8. What size chainfall is required for the Fuel Oil Pressure Temperature Control Valve Replacement? _____
4. What is the complete component tag number for the "A" Fuel Oil Service Pump on Fig. 8.9?
    a. 02-FC-028
    b. N02-FC-034
    c. 028
    d. 034
5. Which of the following files are in a proper shelf order?
    a. N02-FC-028, N02-FC-030, N02-FC-003
    b. N02-FC-002, N02-JX-021, S05-FC-002
6. What is the complete component tag number for the equipment designated by bubble 016 on Fig. 8.9? _____
7. Which pump is associated with relief valve N02-FC-029 on Fig. 8.9?
    a. "A" Pump
    b. "B" Pump
    c. "C" Pump
8. The use of Critical or Rotating Spares reduces service equipment downtime.
    a. True
    b. False
9. The planner uses which of the following to code whether the work is a failure or breakdown.
    a. Outage code
    b. Work type code

    c. Plan type code

    d. Crew code

10. When coding a work order, where does a planner determine which blanks are required to be filled out by the originator?

    a. Work Order System Manual

    b. Minifile

    c. Planning supervisor

11. Who creates the daily schedule?

    a. Crew supervisor

    b. Planning supervisor

    c. Planner

    d. Maintenance scheduler

12. What is the most important indicator of planning and scheduling performance?

    a. Planner productivity

    b. Planned coverage

    c. Type of work

    d. Minifiles made

    e. Wrench time

    f. Schedule compliance

13. A weekly schedule contained 1000 planned work hours for a crew to complete. The crew completed 750 of the planned work hours that were scheduled. The crew also started (but did not finish) jobs with 50 more of the scheduled work hours that will carry over into the next week. In addition, the crew completed 150 hours of high priority work orders not on the weekly schedule. The schedule compliance is____% for this crew.

14. Complying with the weekly schedule is always within the control of the crew supervisor.

    a. True

    b. False

(*Note:* Questions 15 through 20 use the general coding structure of Appendices J and K and use * as a wild cord. However, exact answers may depend on the specific CMMS the company uses.)

15. To find all the work orders in the computer for the Unit 2 condensate polishers, put____in the "Eq number:" field.

    a. N02-C

    b. N2-CP

    c. N02-CP*

    d. *CP*

16. When just approved by the morning meeting, ____ is in the work order "status" field.

    a. APPR

    b. WSCH

    c. INPRG

    d. CLOSE

    e. HOLD-MATL

17. To determine exactly what computer fields for which to enter data when finding a work order, consult the _____.

    a. Planning supervisor

    b. Work Order System Manual computer section

18. Persons can use the computer system to find all the work orders they have written.

    a. True

    b. False

19. Fill in the following fields to find all work orders that are planned and ready to be started on the Unit 2 B boiler feed pump (Equipment tag is N02-QB-002):

    Eq number: _____

    Status: _____

20. Fill in the following fields to find all Unit 2, waiting-to-be-scheduled work orders that can be done on a short notice outage in less than 24 hours:

    Eq number: _____

    Status: _____

    Outage code: _____

    Estimated duration: _____

Appendix

P

Questions for Managers to Ask to Improve Maintenance Planning

Someone once asked, "What is the plant manager responsible for?" The answer is "Everything."

To conclude the *Maintenance Planning and Scheduling Handbook,* this appendix illustrates the type of knowledge and information a maintenance manager uses when evaluating or implementing a planned maintenance process and culture. These are some of the key questions that managers must ask themselves and others and the concerns behind the questions.

1. Do we consider planning as essential?
2. Do we have a feel for what our wrench time probably is? (Do we know why maintenance planning helps instead just doing it because we have always done it or just wanting to do it because others do it? Do we think our wrench time without effective planning and scheduling is probably at or below 35%? Do we want at least 50% wrench time? Do the craftpersons have an idea of the concept of wrench time? Do we have a desire to keep technicians on job sites? Do we want planning because we want to free up the work force to do more work? Do we think that we need to schedule more work? Do we want planning and scheduling to help take us there? Do we want weekly schedules? Do we realize how much planning and scheduling will help?)
3. Is the planning system accountable to someone? (Do we accept indicators helping to drive us, such as planned coverage, wrench time, schedule compliance, days of planned backlog, and percentage of reactive work? Is 80% of our labor hours spent on planned jobs? Is schedule compliance being measured?)

4. How do we select a planner? (Do we spend enough effort to obtain the best person for a planner? Is the planner a highly respected and sought after position? Do we understand why a single planner is worth 17 persons? Do we think a person planning is as important as a person supervising? Does upper management understand why we need a planner position? Have the planners or the key persons directing planning been sufficiently empowered to make it happen?)

5. Where are the planners? (Do we avoid using planners as fill-in help for short-handed crews? Does the crew supervisor understand that for each person given over to planning, the crew receives 17 technicians in return? Are maintenance planners spending most of their time planning future work? Do planners avoid spending much time helping technicians with jobs that have already started? How do we know?)

6. How are the files arranged and used? (Are files established at the component level or are we trying to file by systems or some other means? Are we creating many minifiles? Do we insist on job feedback? Can planners easily show examples of some technician feedback that was put into the files? Would a technician in the middle of the night be able to find equipment information? Do we assure technicians that planning is not intending to use their feedback to develop plans so that contractors can replace the technicians?)

7. Do the pieces of equipment in the field have labels in place showing their equipment numbers? (Do the files use the same number? Does the equipment numbering system make sense? Is the equipment tagging group supporting the planners needs? Are equipment tags being placed in order of higher maintenance systems first? Do equipment tags look neat? Is the tagging group responsive to the planning group's suggestions?)

8. Are the planners planning all the work? (Are we trying to avoid overly detailed job procedures on initial plans? Do we realize that a scope from an accomplished technician that gives craft skill and hour estimates is an advantage over the original work request? Do we know we do not need O&M manuals on every job? Do the planners ultimately trust their instincts for determining time estimates? Do planners avoid estimating most jobs by the shift arrangement such as estimating 4 or 8 hours just because the work shift is 8 hours? Do we know that planning should not slow down the start of work on reactive jobs?)

9. Do we schedule for an entire week for each crew? (Are we willing to have the crews follow an advance allocation of work for a week? Does the 1-week schedule account for every available crew hour? Do we insist on proper use of the priority system? Do we support weekly and daily scheduling by not holding special meetings or events with less than a week's notice? Do supervisors assign work orders using the hours estimated by the planners? Is the percentage of labor hours spent on reactive work becoming less? Are we completing more proactive work orders than previously?)

10. Are we doing any staging of parts or tools? (Should we consider staging parts? Are there no parts currently staged for jobs that have already been completed? Does the right person stage the parts?)

11. How do we utilize the computer system? (Do we know that a computer system is not necessarily planning? Do we understand the proper role of a CMMS and how much it should help? Do we know a computer will not compensate for a poor maintenance process? Does the computer system do a daily backup of its records?)

12. Are we avoiding trying to obtain more out of planning than is there by itself? (Do we have a good work order system? Are technicians doing all work through work orders? Are we avoiding blanket work orders? Do we have clear guidelines on how to write work orders? Is there a standard work order form at each location? Does the work order form collect cost? Do the supervisors spend more time in the field coaching than in the office doing desk work? Are shops and tools useful for completing work? Is the storeroom service satisfactory for both having parts and allowing quick retrieval? Are unused items easily returned to the storeroom? Is the percentage of our work on PM measured? Is a majority of all work orders initiated during PMs? Are PMs oriented toward inspections? Do technicians write up work orders to correct other deficiencies when performing preventive maintenance? Are preventive maintenance jobs planned and scheduled? Is the feedback from preventive maintenance jobs reviewed? Does the computer cancel PMs if the crew does not complete them before the next ones are issued? Does PdM use the same equipment numbers as the planners? Is any PdM information going into the planning files? Is the percentage of our work on project work to improve equipment measured? Do we look for root causes after any failures? Has the maintenance group set up any equipment standards for the corporate project group to follow?)

13. Is planning "working"?

Contracting Out Work

This appendix provides a comprehensive analysis of contracting out work issues. This analysis is necessary because of the major trend of outsourcing, the fear technicians have that planning facilitates contractors, and the controversial nature of contracting disputes.

Bruce Hutchins of Lean SCM LLC (2005) says "For most companies, today's business model is simple: Core processes are retained, and noncore processes are outsourced." Whether or not maintenance is a core process, the fact that many companies do not consider it one is evident from the number of maintenance services they outsource. Just as companies compete to stay in business, maintenance organizations compete as well. They compete for the right to maintain company assets. They compete on the basis of being the best choice for company maintenance services. Contractors who would take over some or even all of the services offer considerable competition. Huge contractors continually approach senior management in large companies offering to perform their entire maintenance functions "better and at a lower cost." Furthermore, if the maintenance organization is competing, its individual employees are competing as well. Therefore, the issues involved in contracting out work are vital to maintenance personnel. Consequently, because maintenance planning plays an integral part in the strategy to make maintenance more effective and efficient ("better and at a lower cost"), this appendix serves an important role to address the interrelationships among maintenance, planning, and contracting out work. The message is clear; a company can make its own maintenance force better or have someone else do its maintenance.

In addition, as expressed by App. B Rule 1, technicians have a healthy fear of planning facilitating contractors. They resist management setting up a planning department and hesitate to provide good feedback on jobs they think planners would use to help future contractors. However, proper planning by its design does *not* facilitate extensive contracting. Planning strives to give technicians a head start, not give contractors a complete job plan. There are too many jobs for the planner to give sufficient details on every plan for contractors less familiar with plant equipment and processes. Planning hopes to be a technician library service and must have the feedback. Because of the genuine fear and need for technician support, it is appropriate to examine contracting thoroughly.

Finally, contracting out work is not a simple matter of the company deciding to hire another company to perform certain work that their own employees might have been able to accomplish. Arbitrator M. Fox, Jr., (1985) claims that "Subcontracting is one of the most fundamental and controversial issues in labor relations." Management and labor have interests that conflict and generate controversy. On one hand, a company wants to run its business as efficiently and economically as possible. On the other hand, employees desire job security. These interests sometimes conflict when management wants to contract out work. Labor groups routinely bring disputes over contracting out work to arbitration. Many companies *cannot* simply do as they please when it comes to contracting just as employees cannot simply perform work any way they want. Labor contracts direct company actions when it comes to contracting out work. Because of contractual requirements, it is helpful to examine arbitration cases to determine what work a company may or may not properly contract out to others.

This appendix addresses several aspects of contracting out work. Why would a company contract out work? What are some problems with contracting out work? What are some alternatives on how a company might contract out work? And finally, what are guidelines for contracting out work from the viewpoint of modern arbitration decisions?

Why Contract Out Work?

To answer the question "Why contract out maintenance?" one must first ask, "Why does a company do maintenance?" A company does maintenance to achieve a company objective. In very broad terms, that company objective is to make money. The company makes money with assets. Those assets must operate. Maintenance keeps assets operating. Thus, as Fig. Q.1 illustrates, a company does maintenance to keep assets operating so that it can make money with them, its objective. Furthermore, the general business wisdom that it is better to be effective than to be efficient, or that effectiveness comes before efficiency, applies in maintenance. Companies make much more money by keeping equipment operating at a high availability and efficiency than they make from skimping on maintenance expenditures. Ron Moore of The RM Group, Inc. (March 1998) who identifies many of the listed items below, notes that DuPont claims "Maintenance's contribution to uptime is worth 10 times the

Achieve a company objective:

Make a profit

With assets

Assets must run

Need maintenance

Figure Q.1   Where maintenance itself fits into the big picture.

potential for cost reduction." In other words, a company must foremost be concerned with effective maintenance and only secondarily concerned with efficient maintenance. Another way to put this would be to say, if a company is not keeping its assets operating, then it does not matter how much it saves on maintenance costs. Therefore, the reason a company would subcontract maintenance would be to procure a more effective maintenance effort and better asset performance.

A company may wish to contract out maintenance a number of reasons. The foremost reason is to achieve a company objective, primarily the best performance of assets. The following list shows this and other reasons.

Why Contract out Maintenance?

- Achieve Better Asset Performance
- Outsource Noncore Process
- Lower Cost
- Obtain Better Overall Expertise
- Reduce Risk and Variability of Maintenance Cost
- Obtain Unusual Skills for Unusual Work
- Obtain Distinct Expertise or Services
- Obtain Unusual Tools or Equipment
- Allow Easier Adding or Reducing of Staffing
- Lower Bureaucracy for Staffing
- Reduce Personnel Problems
- Outsource Lower Skilled Tasks
- Supplement Staffing for Special Events
- Obtain Help for Emergencies

The first reason above, to achieve better asset performance, may or may not be possible through contracting out maintenance work. However, the reasoning is proper. A company must carefully consider if contracting out all of maintenance or a portion of it would lead to better asset performance.

Management may feel that maintenance is simply not a core process and wish to outsource it. Many companies have successfully outsourced integral processes such as accounting, payroll, and even human resources.

Unfortunately, many companies look primarily at the third reason, cost savings, thinking that maintenance is not only a noncore process, but that it is a commodity as well. They think that anyone can provide the same effectiveness in maintenance. They think that their current maintenance effectiveness will continue if simply contracted to another agency. These companies do not even invest in their own maintenance organizations in terms of training, management attention, tooling, and processes and are surprised that asset performance is not in the top quartile. In reality, industry wisdom suggests that if a company focuses on lowering costs, reliability goes down. Yet, if a company focuses on increasing reliability, costs go down as stated in Fig. Q.2.

The company may realize that maintenance is not a commodity and the next reason, obtain better overall expertise, may be valid. Perhaps a company simply does not consider maintenance an area sufficient to justify management attention and professional staffing. Hiring a company that has devoted itself to modern maintenance practices and expertise to perform maintenance may raise the overall effectiveness of maintenance. In addition, hiring the original equipment manufacturer (OEM), may provide better expertise overall for maintaining a large system obtained from that OEM.

Brian Schimmoller, Associate Editor of *Power Engineering* (1998), points out the dramatic variability in maintenance yearly cost. Several major breakdowns may significantly increase the cost of 1 year's maintenance above another. A plant might lower its risk by outsourcing maintenance to a large maintenance contractor for a yearly fixed fee in which the contractor absorbs the risk by having contracts at many different plants.

Companies most commonly contract out maintenance work to obtain unusual skills for unusual work. This circumstance usually involves

Industry wisdom

- If focus on cost, then reliability goes down.
- If focus on reliability, then cost goes down.

Figure Q.2   The reliability-cost relationship.

only contracting out a portion of the plant's maintenance work, not all of it. A plant finds many situations in which an unusual maintenance challenge presents itself. The plant might normally only need welding services once or twice each year. Rather than keep a welder on staff, the company hires the services of another company to make a few welds each year. The concept of "pay as you go" definitely saves money over paying for largely unnecessary skills on staff continually. Chapter 5 includes a section on *Contracting out Work* describing how the maintenance and planning departments would interface with such needs for special skills.

Similarly, the plant may need other distinct expertise or services where it may not make sense for it to hire full time positions. This distinct expertise or other service may be professional or semiprofessional in nature such as predictive maintenance or cathodic protection servicing. A company may not have enough predictive maintenance work to justify full-time positions and might also wish to avail itself of professional companies that do the predictive maintenance work full-time working with many companies. Such cost sharing among companies might lower overall cost and increase the expertise of what might be possible to have in-house.

Along these same lines, a company might best contract for special tools or equipment used in maintenance only when needed. These might involve large cranes including their operators. While the occasional rental cost of a large crane would be expensive, it might be lower than the yearly cost of owning a crane.

Next, using a contractor might allow easier ramping up or down of staffing. The company may find it easier to ramp up for maintenance intensive times with contractors than to keep persons on staff. While hiring may not be a problem, companies may find it difficult to terminate satisfactory employees. On the other hand, a contractor with several contracts at several plants can keep good employees and shift resources among different plants as needed. George Connaughton of Augusta Service Company, Inc. (1993) notes that his experience shows that the occasional adding or reducing staff actually increases the overall competency level of the craft persons utilized. The less competent craft persons are not asked back.

A company's own bureaucracy may prevent it from responding adequately to hiring the necessary quantities and skills of maintenance personnel when needed.

Some companies may prefer to have a third party deal directly with maintenance persons to reduce its own personnel problems. A smaller plant may have limited human resource personnel while a larger contractor may be more capable of handling such issues.

The company may outsource less skilled maintenance work. While skilled positions and overall maintenance strategy might be core processes, certain tasks are not. Ron Moore suggests these lower skill tasks might include janitorial work and landscaping.

A company may benefit from having a contractor supplement its staff for special events and circumstances. Many plants use contractors to supplement forces for major outages. The plant needs more persons than it could normally keep on staff. In addition, a plant might suddenly find itself needing extra labor or expertise for an unexpected emergency. Plants may hire contractor help or expertise when the scope of an emergency goes beyond the capability of the plant. (Incidentally, this illustrates the true trade-off between reliability and cost. Plants readily pay handsomely for restoring critical assets back to service. Obviously, attention to reliability comes before cost. One wonders why some plants will not pay enough for *ongoing* maintenance, adequate to keep the assets from failing in the first place.)

All of these real and perceived benefits encourage plants to consider contracting. Additional circumstances exist as blends of these aspects. For example, some plants contract out scaffolding work because the work is not everyday, the work is labor intensive, the work can be done with lesser skilled labor under the supervision of other persons with special expertise in scaffolding safety regulations, and the work requires special equipment.

Problems with Contracting Out Work

Nonetheless, a company that makes its profits with assets, especially heavy or sophisticated equipment, may find that the expertise needed to keep that equipment operating, namely maintenance, is difficult to give away. Ron Moore (February 1998) has also identified many of the following potential problems with contracting.

Problems with Contracting out Maintenance

- May Lessen Asset Performance and Raise Cost
- May Outsource a Core Process
- Disclose Trade Secrets
- Cost to Coordinate and Manage Contractor
- More Interfaces
- Morale
- Not Sharing Knowledge
- Contractor Less Familiarity
- May Lose Best In-House Persons
- May Lose Best Contractor Persons
- Liability of Contractor On-Site
- In-House Contract May Prohibit

The greatest problem with contracting out work applies to anything purchased. A company may not get what it expected. Maintenance is an intricate combination of art and science requiring proper professional and management attention for best results. Many companies go from mediocre to terrible when trying to improve through relinquishing control of maintaining their vital assets.

The company may find that maintaining these assets was a core process after all. A core process gives a company its competitive edge in the market and certainly, the health of the assets plays a role. The more capital intensive is an industry, the more important it is for its assets to deliver.

Along the lines of a core process, Jim Humphries of Flour Corp. (2004) notes that close alliance with a contractor to maintain company assets may necessitate the disclosure of intellectual property or proprietary knowledge. The contractor should understand a company's process in order to maintain it. Moreover, if a company hires a large contractor for its effect, should the company turn over operations to the contractor as well as maintenance? The old adage that an operator can wreck a piece of equipment faster than maintenance can repair it is true. In addition, modern maintenance is moving toward more operator involvement in maintenance with programs such as, Total Productive Maintenance (TPM). Can a company and contractor have an agreement to create a maintenance effect without considering operations? Including operations as contracted work makes consideration of trade secrets and core processes even more relevant.

Adding a contractor requires coordination cost and effort. This includes management and follow-up work. As discussed in Chap. 5, planners and supervisors can coordinate single contracts for small specific tasks. However, massive contractor involvement requires more management attention. Turning over all maintenance responsibility requires extensive time setting up difficult contracts. Ron Moore (February 1998) stresses that "...the contractor must deliver an effect, not simply supply a service for a fee." Contracts agreeing on performance measures are much more difficult to establish than simple time and material contracts. Time and material based agreements have natural conflicts. The company wants the best maintenance for a reasonable cost. The contractor wants as much work as possible. In the case of time and material, the company must carefully monitor contractor performance, as if the company was the maintenance expert. In the case of performance agreements, the company must spend considerable time designing the agreements. In addition, many plants find that contractors do not deliver the quality as expected as discussed in the first problem listed above. Thus, in-house maintenance persons follow behind contractors checking quality and fixing problems.

Interfaces become more of a problem as more contractors enter the workforce. Will contractors enter their time on work orders specific to equipment so the plant can track asset maintenance? Will contractors keep this information themselves on their own systems? Will that move the knowledge base of the plant away from the owners? Will contractors deal directly with operations to request equipment clearances?

George Connaughton (1993) notes that morale of in-house persons becomes one of the main concerns for plants that mix in-house and contract labor. A difference in pay and benefits between in-house and contractors doing essentially the same work causes significant morale problems. In addition, in-house labor cannot help but be concerned over "outsiders doing their work," the subject of the arbitration section at the end of this appendix.

If in-house craft persons resent having contractors on site, they will certainly resist helping them. Sharing of knowledge makes maintenance much more effective. The experience of the workforce overall helps individuals on various tasks. Frequent communication among technicians increases productivity by overcoming trivial obstacles that stop jobs. Mixed forces of in-house and contract labor lose this benefit.

Outside contractors, by definition, would normally be less familiar with the plant. This is devastating to planning which depends on technicians being familiar with basic plant processes such as storerooms, purchasing, locations, and operation resources. While contractors can become familiar with the plant after a time, many contracting companies do not keep employees as long as a company might keep someone in-house, so a contract person might generally always be more on a learning curve.

Because of the morale issues and in-house persons seeing "their work" go away, technicians who are more competent may choose to leave. These "best" technicians can more easily obtain other jobs leaving the company with an overall less competent workforce. Similarly, company programs to downsize a portion of their workforce with rewards or compensation packages tend to see their best technicians leave first. A more aggressive program to terminate technicians who are less competent in favor of hiring more contractors further aggravates the morale issues of in-house persons remaining. Similarly, the contractor may have a harder time keeping competent technicians who can find other jobs with more security.

Jim Humphries (2004) also asks, "Can the service provider cause an environmental or safety incident for which you retain accountability or that will risk your company's reputation?" As a company gives over control to a contractor, it must consider accountability and liability issues. Does the contractor actually have a better safety record and environmental expertise than that of in-house? On the other hand, does the company lose some ability to train and control persons working at its site?

Finally, the company's labor contract may prohibit contracting out certain work. A contract is an agreement with the labor force that specifies each party's conduct. Contracts should reduce problems by setting forth agreements in advance. Unfortunately, many times companies and unions misunderstand these agreements. Companies might inadvertently contract out prohibited work and unions might raise concerns over perceived violations causing labor problems.

Alternative Forms of Contracting Out Work

This section considers several modes of contracting out maintenance making application of the above discussions. These modes include contracting out all of maintenance and operations, all of maintenance, all of the labor within maintenance, only lower skills, unusual tasks or other tasks requiring special expertise, and supplemental labor. It concludes with an appeal to upgrade or supplement in-house maintenance management expertise.

Contracting out all of maintenance and operations

For a company considering contracting out all of its operations and maintenance responsibilities, the great issue is not how to structure a performance based contact. The great issue is whether it truly would be better for a plant to give over control of its assets to someone else. The company takes a huge risk turning over control of its assets. Nevertheless, the above sections discuss the potential advantages and problems of contracting. The following paragraphs apply these considerations for such extreme contracting.

Companies that contract all of their maintenance and operations place the responsibility for asset performance on the contractors. The contracts should be based on achieving certain plant performance goals, not on time and material type thinking. Furthermore, as Jim Humphries (2004) suggests, the measure of achieving these goals should have a direct effect on company financial performance and the contractor should have significant control over the plant function to control or influence the performance. He also recommends the contractor be included in the design of the plan. Just as important, Humphries points out that the company must *want* the contractor to make the goals and recommends adding incentive awards that the company also wants the contractor to earn. The plant wants the high performance and is willing to pay for it. The plant does not want to save any money on maintenance if it means ending proper plant maintenance. In addition, if the contractor is responsible for total maintenance, certain areas (again such as janitorial and landscaping) should be subject to measures such as customer satisfaction.

The company must be truly willing to partner with the contractor sharing information, goals, and risks.

Doug Jones of Doug Jones and Associates (2005) presents an excellent consideration of partnering. Among other things, he advocates both parties must have trust, not being afraid to share information. They must be in the partnership for the long term, not getting caught up in short term prices. He also recommends short agreements with believable commitments.

Short-term agreements with overall maintenance organization have the problem of the contractor not being able to invest in equipment upgrades that would not pay back within the contract timeframe. If the contractor does not have the control over buying new equipment, the contractor must then convince plant management to spend company funds to reduce contractor cost, a conflict.

Planning fits into this scenario of contracted maintenance in the same place it does for company maintenance. The contractor should implement a planning function to increase contractor maintenance efficiency. The planning would function as in any maintenance environment, planning and scheduling work orders for craft technicians.

Contracting out all of maintenance

The scenario of contracting all of maintenance (but not operations) is more common for certain large maintenance contractors and OEM's. All of the preceding section recommendations and admonitions apply with the understanding that the contractor has less control over the plant achieving its goals without operations ownership. The separate operations group requires more interfaces and less control on the part of the contractor. The company has somewhat less risk because it maintains control over operations, but this extra interface may interfere with the contractor's efforts and success.

Contracting out all the labor within maintenance

Augusta Service Company (1993) contracts only the craft positions and keeps the supervisors and managers under company employment. This arrangement keeps the overall maintenance strategy under the control of the company, but allows the easier adding or reducing of craft positions to meet the workload. Not having any technicians work for the company avoids previous morale problems it had experienced in the past when it mixed in-house and contractor technicians. The company and contractor are careful to establish some job security and set competitive wages to keep competent technicians from leaving. This reduces the potential of not having skilled technicians familiar with the plant. Being significantly connected to a large contractor should also provide a company with easy access to extra forces to supplement the workforce during special circumstances.

Contracts are set on a time and material basis with the expectations that the contractor maintain certain numbers of persons and skill levels on-site. The time and material basis are appropriate because the company managers and supervisors direct the maintenance and are responsible for resulting asset performance.

Again, planning should fit directly into the maintenance organization. The company would employee the planners at the supervisor level, working for the company, not the contractor. The planners would plan work for competent technicians, schedule their work for the supervisors, and expect good feedback to improve future plans. This is basic planning without any modifications.

Contracting out lower skills

Management may decide to subcontract out lesser skilled work such as landscaping or plant cleaning. This might avoid the problems of mixing contractor and in-house persons of the same skill level. The company should take careful note of even this lesser skilled work currently performed by in-house personnel leading to labor problems. Will the in-house persons be dismissed? Will they be reclassified? Is it truly economical to contract it out or would it make appropriate "fill-in" work for existing persons? Perhaps the in-house persons would welcome management arranging for an outside contractor to do this work.

Planning's role does not change much with the addition of lesser "maintenance" work. Planners do not plan custodial work or landscaping whether in-house or not. Planning might be involved to obtain a quote and execute a purchase order for some of this type work, but it does not materially affect a planning department.

Contracting out unusual tasks or other tasks requiring special expertise

Maintenance organizations commonly conduct outsourcing for unusual tasks or other tasks requiring special expertise. Some labor conflicts occur when in-house persons disagree with the unavailability of properly skilled in-house resources. Nevertheless, the company saves money by contracting on an as-needed basis when there is not enough work to justify keeping such skills employed full-time. The plant should have contacts and processes for obtaining anticipated skilled expertise. This area is the most connected to a planning department. Chapter 5 discusses specifics on arranging, coordinating, and supervising such contractors.

Contracting out to supplement labor

Contracting out for supplemental labor may be the most controversial area for subcontracting. Companies do need access to more labor at times and cannot always wait for the existing workforce to "catch up." Outages of limited duration cannot be economically extended. Emergencies cannot wait. The company may be understaffed for the current workload and management may be unwilling to hire enough persons if it perceives not needing them for more than a few years. Management may be more willing to contract out work than to overwork its own workforce with extensive overtime. Management may desire to implement projects and equipment upgrades more quickly than the existing workforce can manage. Management may have access to contractors that can perform the desired work at a better price than its own employees can manage. The contractor might interest also the company in contracting work even if it is not necessarily less expensive than doing it in-house.

All of these circumstances cause labor great concern, but perhaps none more so than the last two. Contractors can sometime perform work more economically than can in-house labor. Isn't the in-house labor a "sunk cost" a company pays whether it uses it or not? Wouldn't it always be less expensive to use such "free" labor? Not necessarily. If the company does not have idle persons, it has economic choices. One choice is to delay current work to do the new work. Supposedly, the current work has an economic impact to the company if not done. Another choice is delaying the new work until the in-house force can get to it. However, the new work also has an economic impact to the company if not done. The company compares the value of delaying the new work versus delaying other work. The company may decide it wants all the work done because it wants to keep assets in-service so it can make a profit. An extension to this concern on the part of labor is that the company may realize it is paying more than the market rate for in-house labor and downsize through layoffs, terminations, or attrition in favor of using contractors.

In these situations, planning has difficulty. It is very difficult to plan for a mixed maintenance group. Planning typically cannot plan enough jobs in the level of detail necessary for craft-persons unfamiliar with the plant processes and equipment. Furthermore, technicians resist giving needed feedback to help planners under such circumstances. Planners have two directives in such situations. One, they should plan for the existing workforce to make it as productive as possible so that the existing workforce can complete all of management's projects and limit the lure of contracting out routine work. Second, they should continue to put the same level of detail in plans even going to contractors. The plans that do go to contractors should not be bid documents, but general scopes of work where the field supervisors over the respective areas may have to help the contractor personnel with plant familiarity. The supervisor might fill out the work orders giving the planners proper feedback for contractors and in-house technician work.

Increasing in-house maintenance management expertise

A management perception that it either does not have the expertise to manage effective maintenance or that a contractor could reduce cost leads to the wholesale contracting of maintenance. The possible reduction of cost has already been shown as a distraction, but companies seeking more expertise could easily consider upgrading their own in-house maintenance management. In light of the above, a company should strongly consider using an in-house workforce to avoid losing control of a possible core process. A logical alternative would be to upgrade the in-house knowledge and understanding of modern maintenance. Companies can do this by developing or hiring maintenance professionals to obtain optimum asset performance. These professionals can in turn contract with consultants to supplement their knowledge and expertise.

Arbitration Considerations for Contracting Out Work

The last section in this appendix explores arbitration rulings for contracting work disputes. Arbitration is a unique method for companies and unions to settle issues. Both parties agree to abide by the decisions of an impartial arbitrator saving time and expense over legal action within courts. One of arbitration's most unique attributes is the arbitrators do not develop compromises or alternate positions. Arbitrators do not seek to reach a reasonable middle ground between two extreme positions. They simply choose one or the other position presented. This encourages both parties make as reasonable as possible cases because the arbitrator is liable to reject an outrageous position. The arbitrator merely selects the position as offered that makes the most sense.

This examination is valuable not only for companies with contracts and organized labor, but also for those companies with neither one. Companies without contracts or unions should be interested because the rulings in these arbitration cases set forth principles of correct and honorable dealings with employees. Even companies without unions value their employees and try to avoid "patently unfair" practices. Furthermore, companies consist of employees and they should highly regard principles of proper conduct. Engaging the hearts and minds of

"Subcontracting is one of the most fundamental and controversial issues in labor relations."

M. Fox, Jr.

Figure Q.3 Maintenance management must understand this critical issue.

every employee in addition to their hands and bodies makes more productive and effective maintenance departments.

Shown in Fig. Q.3 and as mentioned before, Arbitrator M. Fox, Jr., claims that "Subcontracting is one of the most fundamental and controversial issues in labor relations." On one hand, a company wants to run its business as efficiently and economically as possible. On the other hand, employees desire job security. These interests conflict occasionally when management wants to contract out work.

This section conducts an extensive examination of 32 arbitration cases to illustrate cautions and guidelines for contracting out work. This investigation suggests certain guidelines that may be helpful in conducting company business. Endnotes to this chapter identify the individual cases.

The examination pursues what arbitrators consider the four primary aspects of contracting out work disputes. Each aspect is evaluated for cases with and without express contract provisions regarding contracting out work. As illustrated by Fig. Q.4, these interrelated aspects are impact on the employees, type of work or equipment involved, reasonableness and extent of the company's justification for the business decision, and good faith or honorable intention on the part of the employer.

Generally, if the company has violated any single provision of the contract, the arbitrator awards the case to the union, as stated in Fig. Q.5.

Arbitrators consider

♦ **Employee impact**

♦ **Type of work**

♦ **Business justification**

♦ **Good faith**

Figure Q.4 General areas of subcontracting disputes.

Words like
"more reasonable" and
"must discuss with union"
are critical.

Figure Q.5 Silent contracts do not address subcontracting directly.

On the other hand, if the contract contains provisions regarding contracting out and the company does not violate any of them, the arbitrator
usually rules for the company.

Many contracts are silent regarding subcontracting. These contracts
have no express contract provisions on contracting. As stated in Fig. Q.6
arbitrators consider that a contract is silent if it only contains a "recognition clause." A recognition clause merely states that the employees
work for the company. It does not give them exclusive right to all company work. Unions sometimes win cases involving silent contracts, but
arbitrators usually award to the company as long as there is some company economic justification.

Impact on employees

One of the important considerations that an arbitrator makes is what
impact the contracting out of work has on the employees. Impacts are usually detrimental to the employees. There are several types of adverse
impacts on employees from employers contracting out work to outside
companies. One of the most common impacts is the loss of overtime hours
that employees might have been able to work. Another common complaint is that employees already on layoff or only working part time
might have been able to come back to work full time. A more serious
impact is that of employees being laid off or actually losing their jobs
due to contracting out. Sometimes, companies reclassify employees into
other positions or eliminate entire classifications. In one case, the company denied employees matching hours and in another case, it denied
employees the opportunity to learn a new skill. In several cases studied, however, there are no adverse impacts at all on the employees.

A contract is considered
"silent" on subcontracting
if there is only a
recognition clause.

Figure Q.6 Words mean things.

Overtime.   Loss of overtime hours was a typical complaint of the union. In these cases, the employees claimed that had the company not contracted out the work, they could have performed the work themselves on overtime. Therefore, contracting the work made the employees lose money. Eight of the 32 cases studied involved loss of overtime. Of these eight cases, the union won five. Only three were decided in favor of management.

Four of the cases won by the union had actual contract language addressing the contracting of work. One of these cases had contract language that very specifically required while management could contract out work, management must allow at least one employee to work overtime whenever the contractor worked overtime. The company claimed that it was illegal for them to require someone to work when the company had no work for them to do. The arbitrator ruled that the extra employee on overtime was valuable in helping the contractor be conscious of specific plant safety considerations and the company could find work to fill the employee's spare time. The arbitrator was not very sympathetic toward the company trying to avoid specific contract language.[1] Another case the union won had specific contract language that prohibited contracting out unless it was "more" reasonable not to do so. The arbitrator felt the case could have gone either way except for the word "more." He felt this wording made the company obligated to go above and beyond the call of duty in evaluating why work should not go to the union, which the company did not do.[2] Another case with specific contract references won by the union was actually decided by a management letter. The contract did not allow contracting unless there were extenuating circumstances. There were extenuating circumstances, but the arbitrator ruled in favor of the union because the union produced a letter in which management had promised them all work on that system.[3] The other case's contract allowed contracting, but only if management held advance discussions with the union. The arbitrator ruled that a simple letter of notice was not "discussion" and awarded to the union.[4]

Two of the cases management won had specific contract language. In one of these cases, contracting out was not allowed if qualified employees were available. The arbitrator agreed with management the work involved a higher level of skill.[5] In the other case, the contract specifically did not allow contracting out to avoid paying overtime or if the employees could perform less expensively for maintenance work. The company won because it was able to show the work was not maintenance work, but related to optimizing recent construction work.[6]

Each of the preceding five cases involving loss of overtime had specific contract terms regarding contracting out. The arbitrator in each case did not evaluate whether the loss of overtime work was particularly severe on the employees, but rather addressed the specific requirements of the contract and other amplifying written documents. Where management violated any portion of the contract, the arbitrator tended to favor the union.

In the remaining two cases involving loss of overtime, however, the contract had no requirements regarding contracting. The company and the union each won one of these cases. In the case management won, the employee job description specifically included the work involved and management did not show any specific benefits of contracting out the work. Yet, the arbitrator awarded to management because of a letter management had sent to the union 2 years previously. The union had not responded to the letter. This letter specifically stated that management reserved the right to contract out this work.[7] Labor won the other case. In that case, the union successfully disputed management's claim that using its own employees would not have supported a required schedule.[8] Thus, in one case, the arbitrator allowed a past stated policy of being able to subcontract even though no benefit was shown, while in the other case the arbitrator awarded to the union because a perceived benefit was not factual. A reasonable explanation of this apparent discrepancy could be that in the former case, management had been upfront and let the union know how it planned to run its business. It would seem fair to require the union to voice timely concerns and not allow it to thwart long-range company plans later. In the latter case, management made no claim of involving the union in the planning stages.

Employees already not working.   More serious from an employee point of view is contracting work when there are employees already on layoff or only working part-time. Of the 32 cases reviewed, eight cases involved this type of situation. Management won six of these cases including the single case in which there was no contract language regarding contracting.

In the five cases management won where there was contract language, the arbitrators appeared to follow the same pattern as in the loss of overtime situation. The arbitrator considered the exact wording of the contract and any unusual circumstances with little regard to the seriousness of the employee impact. In two cases, the contract prohibited contracting for "customary" work, but past practice showed similar work was not always given to the employees. Therefore, the arbitrators decided the work was not customary.[9, 10] In another case, the contract did not allow contracting for maintenance work, but management showed how the work was part of previous construction work. Therefore, the arbitrator decided the work was not maintenance.[11] In the fourth case, management was able to show that circumstances "beyond their control" would have made using their own employees "doubly expensive." The contract only prohibited contracting without "consideration" of their own employees.[12] In the last case decided by Arbitrator Feldman, management was allowed to claim that they did not have sufficient supervisory personnel to use their own employees. That contract did not allow contracting unless there was a "lack of personnel."[13]

The union won two cases involving employees on layoff with specific contract language. One case was the same case in the loss of overtime claim where management's letter of notice was ruled not the equivalent of "discussion." The contract required upfront discussion.[14] In the other case, the contract only allowed contracting if there was a "lack of related services." Management claimed that it had too few supervisors to do the work in house. The union claimed that

> "The phrase 'related services,' ... was intended to apply to necessary bar-gaining unit employees, not to supervisors. Any other interpretation would totally negate the limitations on subcontracting because the Union has no control over employment of supervisors. The Company's view would allow it to subcontract any work just by dismissing supervisors and then plead-ing a lack of 'related services.'"

The arbitrator agreed because there was no indication of any other interpretation.[15] This ruling made in 1986 appears to contradict Arbitrator Feldman's 1982 decision in the preceding paragraph regard-ing lack of available supervision. Companies and unions should be wary of relying on either of these cases for guidance.

There remains one case involving employees on layoff. Management won the case. Although there was no reference to contracting in the contract, there was a "Memorandum of Understanding" that allowed contracting for new construction and large capital work. The arbitrator considered this memorandum and the large size of the project when he ruled for management.[16]

In all these cases involving employees on layoff, the arbitrator based his judgment on written documentation rather than much consideration of actual employee damage.

Loss of job.   Much more serious than not recalling already laid-off employees is when management terminates employees or lays off more employees. In three cases, employees lost their jobs. Two cases had con-tractual provisions regarding contracting and one did not. The only case the union won was one of the cases with contractual provisions. There were three cases where management did not discharge employees, but only laid them off. Management won all three.

Arbitration upheld management in only one of the two cases where management *terminated* employees and the contract had pertinent lan-guage. In the case decided in favor of management, the contract disal-lowed contracting if a job were lost. However, the work did not belong to the company. The company had stopped pursuing nonprofitable work and another company began to perform the work. The union really pur-sued a strange course by calling this work contracting.[17]

The union won the other case in which contractual language applied and an employee lost his job. In this case, the contract expressly did not allow contracting of plant protection work previously performed by employees. The company installed a remote camera system and contracted with off-site personnel for monitoring. The arbitrator did not allow contracting with upgraded new technology because he deemed the work to be the same work as previously performed.[18] Although it is difficult to generalize from a single case, it is interesting to note that the arbitrator made a judgment about the type of work in favor of the union when the case involved job loss. Previously when the dispute involved only overtime or employees already on layoff, the arbitrator readily agreed with management about the skill involved or construction nature of the work.

The remaining case involving job loss did not have a contract that addressed contracting. The company won this case. The case was very serious because a company decided to eliminate an entire trucking department in favor of using common carriers. The company proved it was more economical to use a common carrier. The arbitrator explained "... the Union here fears for its job security without fully appreciating that the best job security is with a successful, money making, firm/company."[19]

There were three cases involving *laying off* employees due to contracting. All three had no relevant contract provisions and management won each. In the first case, management reclassified a janitor into a higher paying parts handler job based on the economics of contracting out janitorial work. A few months after this move, management laid off a parts handler with less seniority due to lack of work. The arbitrator ruled that the reclassification and the later layoff were entirely separate moves, each made properly. The arbitrator did mention that the company was in serious financial straights, but did not link this with his decision.[20] In one of the other two cases, the lay off of a janitor was allowed in favor of contracting on strictly economic reasons. The arbitrator noted that the company had solicited the union for alternatives, but the union had offered none.[21] In the last case, the layoffs were determined to be caused by usual ups and downs of business and they were not related to the opening of a new plant (which the union claimed as contracting).[22]

Considering the information on job loss in addition to the previous cases on overtime provides an argument for extra caution for management in the area of job loss. In two of the six cases the union grieved on job loss, the work did not even belong to the company or was due to normal swings in business. There may be a trend for the union to present a grievance even in dubious circumstances involving layoffs or job loss. From the experience in the overtime situations, an arbitrator is very liable to use management violation of any contract provision as sufficient to decide in favor of the union. In addition, there may be some evidence to suggest an arbitrator would decide a borderline case based on the severity of the impact on the employee or the economic condition of the company. Management should exercise sensitivity to these considerations when contracting out work that may involve job loss or layoffs.

Employee classification.   Five cases had employees reclassified, that is, put into other jobs, because of contracting as the impact on the employees. Management lost one of the two cases that had contractual provisions against contracting, but won all three where no contract provisions were present.

In the case management won which had contractual provisions, the contract prohibited only layoffs for contracting. The contract also expressly allowed management to determine the number of employees in any classification. There was little question that management could eliminate a classification, especially as the reclassified employees had no loss of wages.[23]

In the case the union won which had contractual provisions, the contract clearly prohibited contracting for "goods and services within the normal scope of plant operations." Arbitration ruled that the associated trucking was a part of plant operations. Management could not reclassify all of its truckers into other jobs even without wage loss.[24]

Arbitration vindicated management in all three reclassification cases that had no contractual provisions about contracting out work. In each of these cases, management gave clear business reasons such as economics or customer appeal and there was no loss of wages.[25, 26, 27]

In these reclassification cases, the arbitrator tends to rely on strict interpretations of written documentation where possible. Management also appears to be free to reclassify employees with no loss of wages and no contract provisions. There is less guidance for reclassifications where loss of wages is involved.

Matching hours.   One case involved the contractual promise of matching hours. The arbitrator ruled for management that employees should not get matching hours because the annual inspection work was not routine and because it had been contracted out previously.[28]

Opportunity to learn.   An interesting case previously mentioned for loss of overtime also impacted the employees for loss of an opportunity to learn something new. The union won this case. The contract specifically prohibited contracting repair work unless it was "more" reasonable to do so. The contract also specifically permitted contracting of major reconstruction. The arbitrator ruled that management should have given this unusual and major repair welding assignment to its own employees. The employees did not normally weld nor did the company possess the special knowledge of how to accomplish the task. Arbitration decided this borderline case by the word "more" which meant that management should bend over backwards before contracting.[29]

No impact. Seven cases identified no adverse impact at all from the contracting out of work. Management won six of these cases and the union won only one. However, none of these cases were won or lost or the basis of having no adverse impact on the employees. Management won the two cases with no contractual provisions.[30,31] Management won two cases with contractual provisions, but no violations and no adverse employee impact.[32,33] The remaining two cases management won had contract provisions and not only showed no adverse impact, but actually showed growth in the bargaining units (a benefit, though not due to the subcontracting).[34,35] Labor did win one case that made no mention of employee impact. Labor won based on contractual provisions.[36]

Arbitrators seem to decide cases involving no adverse employee impact either by contract interpretation of the type of work involved or by virtue of the benefit to the company. The following sections discuss these issues.

Work type and equipment

Another important consideration an arbitrator makes is the type of work and equipment involved. This consideration usually falls into two broad categories. These categories are whether or not the employees customarily or normally do the work.

There are many different reasons for arbitrators to consider work not to be customary or normal for the employees. Typically, this work is unusual in some respect or infrequently performed. Construction work is one example. In addition, work may be a new type of method for doing old or new tasks. Sometimes, the extensiveness of the work with regard to time duration or amount of work is a factor. The level of skill required of the employees may contribute. Occasionally, the location or the type of equipment used helps to classify the work. The union was able to win cases even with work that employees did not customarily or normally perform and without specific contract provisions.

The following sections examine cases involving customary or normal work and unusual or infrequent work.

Normal, customary work. There were seven cases involving normal or customary work that had specific contract sections addressing contracting.

Management won five of these cases. In three of them, the contact specifically prohibited contracting if employees either were on layoff or would lose their jobs. Arbitration ruled because no layoffs or job losses occurred, management was free to subcontract. These cases involved ditch digging by ditch diggers,[37] cemetery security by guards,[38] and cable splicing by cable splicers.[39] Clearly, the employees normally did the work. In each case, however, management did offer other justification. In another case, the contract expressly forbade the contracting out of any "product" normally handled by the meat department of a grocery store. Management began having chickens prepackaged by an outside contractor. The butchers still handled the product albeit by a new "procedure." The arbitrator upheld that the company did not violate the contract. The arbitrator also noted in this case that previous attempts by the union to adjust that particular wording had failed in negotiations.[40] Arbitrators use such evidence to aid their interpretations of contracts and they are sensitive not to grant an interpretation that past negotiation has specifically avoided. The contract in this last case also merely required the company to "give consideration" to the employees before contracting. The company convinced the arbitrator that they had done so in contracting based on economics.[41]

The union won two cases that had pertinent contract provisions. In the first case, the contract did not allow contracting of work that unit employees presently performed. The company had attempted to classify the contracting as new type work since the contractor was doing the old work in a new manner with new technology.[42] In the second case, the arbitrator considered an incidental service, trucking, as customary work. The contract did not allow contracting of work within the normal scope of the plant. The arbitrator ruled for the union.[43]

There were seven other cases that involved customary or normal work and did not have any contractual provisions regarding contracting. The union won only one of these cases.

Management won six cases. In three of them, management claimed cost savings[44, 45] or increased customer appeal.[46] In one case, management claimed a lack of workspace[47] and in another, better performance by contracting.[48] The last case is the most unusual. In this case, there was no doubt the employees customarily performed the work. In fact, the company had even scheduled the employees to begin the work. However, management then contracted out the work with no apparent company benefit and loss of overtime to the employees. This case was the one previously discussed wherein the arbitrator leaned heavily on a prior management letter stating a right to contract out that particular type work.[49]

In the one case the union won with regular type work and no contract provisions, the nature of the work was the key. The arbitrator ruled that the work required no special knowledge or tools so management should not have contracted out the work and avoided employee overtime.[50]

Unusual, infrequent work. For noncustomary work in work type and equipment cases, arbitrators considered new construction, new type work, the extensiveness of the work, the level of the skill, the location of the work, and the skill required for the work.

Arbitrators typically do not consider *new construction work* to belong to the union for an operating plant unless the contract so specifies. Four of the cases investigated involved construction. Only one had no contract provisions regarding contracting. Management won this case, but only two of the three cases with contractual provisions.

There were three cases with contractual provisions. In two of these cases, it was not clear that the work was indeed construction. The first case involved warrantee work that management could obtain for free. The arbitrator said warrantee repairs were an extenuating circumstance contractually allowable to be contracted. Nevertheless, the arbitrator ruled in favor of the union because of a previous management letter stating that management would give all work on the system of the employees.[51] Management won the second case because the work was to modify a recent construction project to achieve planned performance. The arbitrator ruled that the task was not normal work protected by the contract.[52] The last case's contract allowed the contracting of new construction only if qualified personnel were not available. The work in question was clearly construction, but arbitration allowed management to claim a lack of qualified supervisory personnel.[53]

Only one construction work case involved no contractual terms. Arbitration decided in favor of management because of the benefit of company flexibility to pursue its other work. It is interesting that the arbitrator noted it was not relevant the contracting began during an employee strike.[54]

Besides construction, two cases involved *new type work*. Management won both of these cases and both had relevant contract provisions regarding contracting. In the first case, bus company employees did not get the opportunity to operate or maintain vans for handicapped riders. The arbitrator ruled that the smaller vehicles and gasoline engines (versus diesel) made the work sufficiently different from normal work that the contract prohibited contracting.[55] The other case presented a combination of new type work with customary work. The element of new type work played a role in persuading the arbitrator in spite of very clear restrictions against contracting customary work.[56]

The *extensiveness of the work* was a factor in three cases of noncustomary type work. In the one case labor won, management disregarded very specific contract regulations. Management claimed that the required miniscule amount of chaperoning of contractors by employees simply was unnecessary work. The arbitrator declined to overlook specific contract provisions.[57] Neither of the two cases management won had any contract language. In one case, a routine type maintenance task was much larger than usual, so management was able to contract the work. The arbitrator upheld the company even though there were employees on layoff and employees had started the task.[58] In the last case, management was able to contract peak summer work or grass cutting around the plant.[59]

The *level of skill involved* in the contracted task was critical in two cases. Both cases had contractual terms and the company and the union each won one case. The case ruled in favor of management had a contract which only allowed contracting if qualified personnel were not available. Management claimed a higher level of welding skill was necessary for safety reasons on a tank weld. The union did not prove otherwise.[60] The union won the other case. Management was not allowed to hold that a clause allowing contracting only if no "capable employees" were available included supervisors of which none was available.[61]

Arbitration included the actual *location of the work* in defining the type of work and subsequent ruling in two cases involving noncustomary work. Each case had contractual language on contracting. The one the company won had a contract very specific as to naming a particular truck route. The arbitrator classified a new truck route being used as different work, but he also considered that no jobs were lost by the former truck drivers.[62] The case won by the union involved an unusual situation and a strike. The contract entitled the employees to all "regular work" which included the handling of deliveries after receipt at the plant. Since the plant had been on strike earlier, another location had received deliveries. The arbitrator ruled that handling the deliveries at the new location belonged to the union.[63]

Equipment type played a role in two interesting cases. In the earlier truck driver route case, the arbitrator ruled in favor of management partially because the trucks used on the alternate contracted route were different tanker type trucks instead of smaller delivery trucks.[64] In contrast, the union won other case that also had contract provisions. The contractor used company owned welding equipment. This fact was not relevant to the question of skill that decided the case.[65]

Several points summarize the type of work and equipment considerations. As expected, the arbitrators rely heavily on the written word as much as possible. The most difficult part of this aspect involves relating the specific work to the specific contract terms. Surprisingly, the role of past practice does not seem to play an important role in this determination. One would think that past practice would be an excellent rule of thumb for deciding if work was customarily the work of the employees.

Three explanations may account for this discrepancy. First, the cases brought to arbitration are usually first of a kind. Parties might not bring cases easily decided by past practice as far as arbitration. Second, the sample of cases reviewed may have missed the pertinent cases. Third, and most likely, perhaps the doctrine of past practice is so self-evident that it is usually not expressly mentioned. Concerning contracts with no contracting out provisions, the arbitrator is liable to judge the case from a scrutiny of the company's business basis for the decision. Finally, in regarding type of work considerations, it would seem likely that customary or normal work would be more favorable to a union ruling, but there is insufficient evidence in these cases to verify this stance.

Reasonableness and extent justified by employer

The following sections examine the 32 cases with respect to company justification of the decision to contract out work. The aspect is also an important consideration from an arbitrator's point of view.

The justifications offered by the companies fell into several groups. Companies claimed cost savings in almost half the cases. One company claimed a belief in increased customer appeal. A number of companies offered lack of available personnel as a reason. One company made a successful point that the work was experimental. Only two companies raised past practice as critical points. A few companies did not even bring up the benefit to the company and a couple of companies explained that the work did not actually belong to the company in the first place. The following sections examine each of these categories including whether or not the contracts addressed contracting out work.

Cost savings.   Cost savings cases fell into two subcategories. Often the savings were straightforward claims of money saved, but several cases brought forth claims of better performance without any elaboration on the money involved. There were eleven of the former type cases and six of the latter.

Of the 11 *lower cost* cases, there were six cases where the contracts specifically dealt with contracting out work and five where the contracts were silent. The arbitrators upheld the union in three of the six specific contract cases, but in none of the five silent contract cases.

Management won three cases involving lower cost with contract specifics. In one case, arbitration deemed management's claim of "economy" not relevant because the contract forbade contracting out only if there were employees on layoff. Thus, arbitration decided the case solely on employee status.[66] In the other two cases, the cost savings were relevant. In one case, the contract required only "consideration" of the employees before contracting. A reading of the case revealed only a statement that "costs would have doubled" which was acceptable for a favorable ruling.[67] In the other case, although the contract only addressed customary work and the work was a new type, the arbitrator closely examined a dispute over the actual savings. Because there were "some savings," management won.[68]

In two of the cases with contract terms that the union won, the deciding factor was strictly the type of work involved. The work involved was not restricted to in-house work according to the contract or in one case other written agreements. The arbitrator did not consider relevant management claims of having free warrantee work[69] or avoiding having nonproductive employees[70] by contracting. In the third case with contract terms the union won, the contract expressly restricted the type of work from contracting. Nevertheless, the arbitrator did go so far as to mention that the economic data was "inconclusive."[71]

In most of these cases where the contract addressed contracting, the arbitrator usually allowed any contracting not specifically excluded by the contract. However, exceptions did exist and arbitration resolved them by considering company savings. The magnitude of the savings was not particularly important.

On the other hand, management won all five lower cost type cases where there were no pertinent contract terms. There were three cases where the company contracted regular work of the bargaining unit. In the first, the company saved $1000 per year versus no employee impact.[72] In the second, the company saved $7700 per year versus an employee reclassification. The arbitrator also mentioned the poor economic condition of the company.[73] In the third, the company estimated savings of $100,000 per year versus a whole group of employees losing their jobs.[74] In the other two cases, arbitration upheld management in disputes involving different type work. In the first, management saved $56,000 with employees already on layoff.[75] In the second, management's assertion of "economics" prevailed over zero impact to the employees.[76] In both cases, the arbitrator appeared to give equal weight to the type of work and the savings. From the first three cases with customary work, however, it seemed that if the economics favored the company, the type of work was not very important if there was no contract language on contracting.

Instead of outright lower costs, six cases addressed the economics in the form of *better performance, efficiency, or technology.* Arbitration upheld the union position in two of these cases. Management won two of these cases that had contractual provisions regarding contracting out work. In one case, the contract forbade contracting only if employees were laid off. No employees were laid off, but management bolstered their case with a lengthy list of how the current guards had been negligent or incompetent in their duties. The arbitrator ruled entirely on the employee impact basis of no layoffs.[77] In contrast, the arbitrator lauded a company's attempt to increase "efficiency" in the second case. Nevertheless, the arbitrator relied heavily on the interpretation of the contract terms as derived from previous negotiation records. This interpretation led him to judge the case as a type of work case.[78]

The union also won two cases that involved performance, efficiency, or technology with specific contract language. These cases were both decided as type of work cases wherein the arbitrator compared the type of work to exactly what the contract restricted from being contracted. In one case, the arbitrator disregarded vague claims of better performance and efficiency as "inconclusive economic data."[79] In the other case, the arbitrator did not allow company use of new technology because the contract specifically addressed work already being done by unit employees as sacred.[80]

Management won both of the cases that did not involve contract language. In the first case, management presented a detailed case of how much theft had occurred under the existing employee guards. The arbitrator allowed management to reclassify the existing guards to less desirable jobs and contract the security work. It appeared that arbitration would have upheld the company even for employee discharge.[81] In the final case, the arbitrator allowed the company to contract irregular work to permit a streamlined organization better able to pursue its main work. Nevertheless, in this case, there was no employee impact.[82]

For performance, efficiency, or new technology cases, there seemed to be several trends. First, when management violated express contract provisions in any manner, the union won. Second, when management did not violate express contract provisions, management won, even with little or no justification. Third, when contract provisions were absent, the company appeared to present a better case as to why it was reasonable to contract out the work.

Customer appeal.   Closely allied with the cases on cost and performance was a case that had no contractual requirements and the arbitrator upheld the company because of a reasonable belief that customer appeal would be enhanced. The company felt that contracting grain inspections to a third party would make inspections appear less biased. Nonetheless, there was not a significant employee impact.[83]

Availability.   The availability of qualified personnel was a not-so-reliable justification proposed by management in seven arbitration cases. Management won only three of these cases. The union won four cases including the only one that involved no contractual terms. These seven cases included schedule concerns, skill concerns, and supervision concerns.

The union won both cases where the company claimed there were *insufficient personnel to meet a schedule.* In the first case, the contract allowed contracting only if management first discussed the plan with the union. The arbitrator ruled for the union on two connected issues. First, he claimed that a simple letter of notice was not "discussion," and second, the arbitrator carefully evaluated the project and its necessary labor hours. He felt that there were indeed few enough labor hours necessary that the company could have used the employees. The net result is that the "discussion" provision of the contract placed a higher burden on the company to present a "reasonable" case for contracting.[84] In the other schedule type case, the arbitrator ruled in favor of the union even though contract terms were absent regarding contracting. In this case involving normal work, the arbitrator also evaluated the project and felt that the employees had the necessary overtime hours available.[85]

On the other hand, management won two out of three cases involving *employee skill*. All three cases had contracting provisions present in the contract. In one case management won, the contract specifically allowed contracting only if qualified personnel were not available. The arbitrator ruled for the company's plausible claim of higher welding skill needed for safety, which the union did not dispute.[86] Management also claimed the contracted work needed a higher skill for safety in the second case. The contract addressed skill involved as well as type of work. The arbitrator ruled for management mostly on the type of work basis.[87] The union won the last case even though the arbitrator agreed that the employees did not have the necessary skill. Nevertheless, he ruled that there was a chance for them to learn something. The arbitrator stated that this situation was a borderline case and he felt persuaded by the words "more reasonable" in disallowing management's claim. No safety aspect was present, but there existed the potential for damaging an expensive piece of equipment.[88]

This appendix already discussed the conflicting cases where management claimed a *lack of supervisors* being a valid reason for subcontracting. The contracts expressly allowed contracting in case of a lack of personnel[89] or related services.[90] Management won the former case and the union won the latter.

Companies usually win arbitration with silent contracts if they have ANY justification

Figure Q.7 Unions are not insulated from subcontracting merely by having a contract.

Overall, management is on less sure justification grounds claiming availability concerns rather than economic reasons. Scheduling is most likely to invite scrutiny with skill not far behind. It is uncertain how the arbitrator might rule on a supervisory availability case.

Experimental. In one case with contractual provisions regarding contracting, management based a portion of its justification on the new work being experimental. The company did not want to involve their employees in a program with an uncertain future. The arbitrator took note of this fact, but ruled for management largely on the grounds of type of work and no adverse employee impact.[91]

Past practice. Past practice only came into significant play twice. In both cases, the company and the arbitrator relied heavily on past practice to show that work was not customary for the bargaining units. However, even after such determination, the arbitrator ruled the case involving contractual terms as a type of work case[92] and the case with a silent contract as an economics case.[93]

There are two implications here. If the contract restricts contracting out customary work, the company may show past contracting practice as proof of the work not being customary. However, for a silent contract, the arbitrator may want to inspect the economics rather than just the past practice.

No benefit. The company claimed no specific benefit at all in only four of the 32 cases. The lack of company benefit did not come into play in any of these cases. Arbitration decided the three cases with contractual provisions strictly as type of work cases, two in favor of management[94, 95] and one in favor of labor.[96] Arbitration also upheld management in the case without contractual provisions. The only employee impact was a loss of overtime opportunity. A close reading of this case found the arbitrator giving attention to a past letter whereby management claimed this right.[97] Thus, for all of the cases of no company benefit, it appeared the arbitrators ruled on the basis of written documentation.

Not company work. Companies also successfully claimed the contracted work not company work.[98, 99] The key to these cases was that while the companies may have benefited somewhat from the work, they were not paying the contractor directly or indirectly.

Good faith

The final area of consideration in the 32 cases is that of good faith. Good faith involves the motives of the employer in being fair with the employees. The governing principles or good faith according to Arbitrator Hunter are the legitimacy of the employer's business reason, the employer's intent with regard to the employees and bargaining unit, and the impact on the employees.[100] The legitimacy and the impact issues have been addressed. The intent of the employee could use some further discussion.

In only one case did the arbitrator address the value of the moral issue as being more important than the contract. Arbitrator Feldman thoroughly researched the contract and kept management from procedurally dismissing a late filing. He stated that the "issue" should be more important than the "procedure." However, he was still able to rule by the contract terms.[101]

Three cases show that the burden of proof is on the union for proving the company had a dishonorable motive. For example,

"To the extent that the governing standards require proof, no violation can be found."[102]

"...the arbitrator finds no basis or cause to discredit Company honesty, sincerity, integrity, and possible accuracy."[103]

And in the third case, the arbitrator considered as honorable management's simple belief of customer appeal even though the subsequent facts showed no increase at all.[104]

The union claimed a lack of company integrity in three cases to no avail. In one, it claimed the company was avoiding contract terms by helping supervise a contractor.[105] In another, employees had even been scheduled to do the work before the contracting.[106] In the last, the union simply made an unsupported claim of bad faith which the arbitrator did not even address.[107]

One arbitrator saw previous discussions with the union where management involved the union in evaluating alternatives as an explicit act of good faith. Regardless, the arbitrator decided the case on an economics basis for management.[108] Past discussions did not help management win another case in which the work was contractually prohibited from being contracted. The union won this case.[109]

Overall, there were not any cases examined where management appeared as a malicious group attempting to beat down the union or the employees. Of course, the company would be sure to put on its best face for arbitration. Arbitrators resolved the cases mostly on tangible facts without much attempt to embrace the subjective factor of good faith, even in borderline cases.

Summary of arbitration.   In summary, most of the arbitrators consider four primary areas when judging cases of contracting out work. These areas are employee impact, type of work, the reasonableness of a company's business decision, and good faith. Above all, however, the arbitrator makes every effort to adhere to any contractual provisions or other written documentation.

An arbitrator's instinct is to avoid ANY subjective factors and rely on contract and other written word.

Figure Q.8 Arbitrators rely heavily on written agreements in contracts and other documents.

For cases in which the contract has express contract provisions regarding contracting out, a clear trend exists. As a rule, if the company has violated any single provision of the contract, the arbitrator awards the case to the union. The union wins regardless of the employee impact, type of work, or decision reasonableness. On the other hand, if the contract contains provisions regarding contracting out and the company does not violate any of them, the arbitrator usually rules for the company. He makes this ruling also without regard for employee impact, type of work, or reasonableness.

A different situation arises when there are no express contract provisions on contracting. There are exceptions, but the arbitrators usually award to the company in spite of employee impacts such as job loss as long as there is some company economic justification. Arbitrators consider that a contract is silent on the issue of contracting if it only contains a "recognition clause." As stated in Fig. Q.9 a recognition clause merely states that the employees work for the company. It does not give them exclusive right to all company work. In addition, the fact that the contract is silent about contracting does not bar contracting out work.

There also appears a higher probability of a union bringing a case to arbitration if job loss is involved whether or not there is any merit in the grievance.

A "recognition clause" merely states that the employees work for the company. It does not give them exclusive right to all company work.

Figure Q.9 Even with silent contracts, companies should have a reason to subcontract work.

Finally, good faith itself as it relates to employer intention does not seem to be too significant. This factor is a subjective consideration that the arbitrators usually discuss as an added reason to uphold the company. The cases make it very clear that the burden of proof is on the grievant to prove bad faith on the part of the company.

Summary

A company should consider contracting out any of its maintenance work only if it can achieve better asset performance and not on the basis of illusory lower cost alone. A company might consider a number of strategies for contracting out maintenance work. The most common form of contracting is when companies hire special expertise or skills not routinely kept on staff or hire extra labor needed for limited special times of extreme maintenance. Contracting out entire maintenance programs carries larger risks for companies with significant assets because maintenance might well be a core process or the contracted maintenance might end up both less effective and more expensive. A better solution for overall maintenance might be upgrading professional skills to manage effective in-house maintenance. Contracting for routine services that mixes same skill in-house and contract labor leads to significant morale problems. These latter situations lead to most labor arbitration cases.

Arbitration settles most contracting labor disputes based on specific contract or written terms without regard to business economics when the documents address contracting. When contracts are silent regarding contracting, arbitrators generally rule on the basis of company economic justification. Companies should heed arbitration guidelines to do what is right by their employees and avoid labor problems. Companies should follow the written word of their labor documents and, where contracts are silent regarding subcontracting, they should have proper business reasons for contracting out work.

Planning can play an essential role in any form of company contracting out of maintenance. An entire maintenance program run by a contractor should install planning to increase its maintenance efficiency. When contracting out only special skills, planning can arrange purchase orders to have the special expertise on hand for in-house supervisors to manage. Mixed labor contracting situations cause planning the most problems. They make technicians fear losing work and increase their resistance to planning. In these situations, planning should continue to plan all the jobs keeping to a standard of not spending too much time including extensive details. Planners would expect the supervisors to guide less familiar contractors in the field and to provide proper feedback on contractor work. The planners also expect the supervisors to help gain good feedback on work from in-house technicians. The best service a planning program can provide is to help in-house maintenance personnel become more efficient and reduce management's tendency to look outside for supplemental maintenance labor.

References

(All references are cases as recorded in the Commerce Clearing House. Inc. Labor Arbitration Awards.)

1. U.S. Industrial Chemical Co., 85-2ARB8482 (J. Caraway, 1985).
2. Latrobe Steel Co., 85-1ARB8308 (R. Creo, 1985).
3. United Aircraft Products, Inc., 82-2ARB8414 (H. Dworking, 1982).
4. Rockwell International Corp., 85-1ARB8217 (H. Wren, 1985).
5. Williamette Industries, Inc., 85-2ARB8396 (M. Fox. Jr., 1985).
6. GAF Corp., 81-2ARB8560 (R. Penfield. 1981).
7. General Dynamics, 78-1ARB8136 (W. Gray, 1978).
8. Mead Paper Corp., 85-2ARB8458 (C. Ipavec. 1985).
9. Laurel Run Mining Co., 85-1ARB8247 (M. Feldman, 1985).
10. United Telephone Company of Missouri, 84-lARB8278 (M. Berger, 1984).
11. GAF Corp.
12. Lever Brothers Co., 79-lARB8128 (Collins, 1979).
13. Reitz Coal Co., 82-2ARB8384 (H. Felman, 1982).
14. Rockwell International Corp.
15. Manville Forest Products Corp., 85-2ARB8559 (D. Nolan, 1985)
16. Reynolds Metal Co., 84-2ARB8450 (Blinn, 1984).
17. Orange and Rockland Utilities, Inc., 79-2ARB8455 (Robins, 1979).
18. Lobdell Emery Manufacturing Co., 86-1ARB8054 (Daniel, 1986).
19. Continental Fibre Drum, Inc., 86-1ARB8218 (M. Hart, 1986).
20. American Motors Sales Corp., 84-lARB8215 (J. Thornell, 1984).
21. Fruehauf Corp., 83-1ARB8309 (D. Crane, 1983).
22. West Virginia Armature Co., Inc., 77-1ARB8114 (J. Hunter, 1977).
23. Greenwood Cemetery, 77-2ARB8538 (M. Glushien, 1977).
24. American Paper Co., 78-1ARB8063 (R. Ables, 1978).
25. Singer Co., 78-2ARB8280 (Kossoff, 1978).
26. Central Soya Co., Inc., 84-2ARB8524 (Craver, 1984),
27. American Motors Sales Corp.
28. International Paper, 84-2ARB8585 (Dunn, 1984).
29. Latrobe Steel Co.
30. Philips Petroleum Co., 80-1ARB8125 (H. Finston, 1980).
31. Binswanger Glass Co., 83-1ARB8246 (L. Holman, 1983).
32. Mid-America Dairyman, Inc., 84-2ARB8546 (Yarowsky, 1984).
33. Safeway Stores, Inc., 83-2ARB8432 (D. Allen, Jr., 1983).
34. Pennzoil Co., 81-2ARB8617 (Amis, 1981).
35. Central Ohio Transit Authority 78-2ARB8336 (M. Handsaker, 1978).
36. Jeffboat, Inc., 85-2ARB8400 (W. Mulhall, 1985).
37. Orange and Rockland Industries, Inc.
38. Greenwood Cemetery
39. United Telephone Company of Missouri
40. Safeway Stores, Inc.
41. Lever Brothers Co.
42. Lobdell Emery Manufacturing Co.
43. American Paper Co.
44. Fruehauf Corp.
45. Continental Fibre Drum, Inc.
46. Central Soya Co., Inc.
47. West Virginia Armature Co,
48. Singer Co.
49. General Dynamics
50. Mead Paper Corp.

51. United Aircraft Products, Inc.
52. GAF Corp.
53. Reitz Coal Co.
54. Binswanger Glass Co.
55. Central Ohio Transit Authority
56. Laurel Run Mining Company
57. U.S. Industrial Chemical Co.
58. Reynolds Metal Co.
59. Phillips Petroleum Co.
60. Williamette Industries, Inc.
61. Manville Forest Products, Inc.
62. Pennzoil Co.
63. Jeffboat, Inc.
64. Pennzoil Co.
65. Latrobe Steel Co.
66. United Telephone Company of Missouri
67. Lever Brothers Co.
68. Central Ohio Transit Authority
69. United Aircraft Products, Inc.
70. U.S. Industrial Chemical Co.
71. American Paper Co.
72. Fruehauf Corp.
73. American Motors Sales Corp.
74. Continental Fibre Drum, Inc.
75. Reynolds Metal Co.
76. Phillips Petroleum Co.
77. Greenwood Cemetery
78. Safeway Stores, Inc.
79. American Paper Co.
80. Lobdell Emery Manufacturing Co.
81. Singer Co.
82. Binswanger Glass Co.
83. Central Soya Co., Inc.
84. Rockwell International Corp.
85. Mead Paper Corp.
86. Central Soya Co., Inc.
87. GAF Corporation
88. Latrobe Steel Co.
89. Reitz Coal Co.
90. Manville Forest Products Corp.
91. Central Ohio Transit Authority
92. International Paper
93. Phillips Petroleum Co.
94. International Paper
95. Laurel Run Mining Co.
96. Jeffboat, Inc.
97. General Dynamics
98. Orange and Rockland Utilities, Inc.
99. Mid-America Dairyman, Inc.
100. West Virginia Armature Co., Inc.
101. Reitz Coal Co.
102. United Telephone Company of Missouri
103. Continental Fibre Drum, Inc.
104. Central Soya Co., Inc.
105. Philips Petroleum Co.
106. General Dynamics
107. Singer Co.
108. Fruehauf Corp.
109. Lobdell Emery Manufacturing Co.

Concise Text of Missions, Principles, and Guidelines

This appendix concisely recaps the text of the many principles and guidelines pulled together from throughout the book and helpfully puts them into a single place for reference.

Maintenance Planning Mission Statement

The Planning Department increases the Maintenance Department's ability to complete work orders. Work plans avoid anticipated delays, improve on past jobs, and allow scheduling. Advance scheduling allows supervisors to assign and control the proper amount of work. A work crew is ready to go immediately to work upon receiving a planned and scheduled assignment because all instructions, parts, tools, clearances, and other arrangements are ready. The right jobs are ready to go.

Maintenance Planning Principles

1. The planners are organized into a separate department from the craft maintenance crews to facilitate specializing in planning techniques as well as focusing on future work.

2. The Planning Department concentrates on future work—work that has not been started—in order to provide the Maintenance Department at least 1 week of work backlog that is planned, approved, and ready to execute. This backlog allows crews to work primarily on planned work. Crew supervisors handle the current day's work and problems. Any problems that arise after the commencement of any job are resolved by the craft technicians or supervisors. After every job completion, feedback is given by the lead technician or supervisor to the Planning Department. The feedback consists of any problems, plan changes, or

other helpful information so that future work plans and schedules might be improved. The planners ensure that feedback information gets properly filed to aid future work.

3. The Planning Department maintains a simple, secure file system based on equipment tag numbers. The file system enables planners to utilize equipment data and information learned on previous work to prepare and improve work plans, especially on repetitive maintenance tasks. The majority of maintenance tasks are repetitive over a sufficient period of time. File cost information assists making repair or replace decisions. Supervisors and plant engineers are trained to access these files to gather information they need with minimal planner assistance.

4. Planners use personal experience and file information to develop work plans to avoid anticipated work delays and quality or safety problems. As a minimum, planners are experienced, top level technicians that are trained in planning techniques.

5. The Planning Department recognizes the skill of the crafts. In general, the planner's responsibility is "what" before "how." The planner determines the scope of the work request including clarification of the originator's intent where necessary. The planner then plans the general strategy of the work (such as repair or replace) and includes a preliminary procedure if there is not one already in the file. The craft technicians use their expertise to make the specified repair or replacement. The planners and technicians work together over repeated jobs to develop better procedures and checklists.

6. Wrench time is the primary measure of work force efficiency and of planning and scheduling effectiveness. Wrench time is the proportion of available-to-work time during which craft technicians are not being kept from productively working on a job site by delays such as waiting for assignment, clearance, parts, tools, instructions, travel, coordination with other crafts, or equipment information. Work that is planned before assignment reduces unnecessary delays during jobs and work that is scheduled reduces delays between jobs.

Maintenance Scheduling Principles

1. Job plans providing number of persons required, lowest required craft skill level, craft work hours per skill level, and job duration information are necessary for advance scheduling.

2. Weekly and daily schedules must be adhered to as closely as possible. Proper priorities must be placed on new work orders to prevent undue interruption of these schedules.

3. A scheduler develops a 1-week schedule for each crew based on a craft hours available forecast that shows highest skill levels available, job priorities, and information from job plans. Consideration is also

made of multiple jobs on the same equipment or system and of proactive versus reactive work available.

4. The 1-week schedule assigns work for every available work hour. The schedule allows for emergencies and high priority, reactive jobs by scheduling a sufficient amount of work hours on easily interrupted tasks. Preference is given to completing higher priority work by underutilizing available skill levels over completing lower priority work.

5. The crew supervisor develops a daily schedule one day in advance using current job progress, the 1-week schedule and new high priority, reactive jobs as a guide. The crew supervisor matches personnel skills and tasks. The crew supervisor handles the current day's work and problems even to rescheduling the entire crew for emergencies.

6. Wrench time is the primary measure of workforce efficiency and of planning and scheduling effectiveness. Work that is planned before assignment reduces unnecessary delays during jobs, and work that is scheduled reduces delays between jobs. Schedule compliance is the measure of adherence to the 1-week schedule and its effectiveness.

Guidelines for Deciding if Work is Proactive or Reactive

Proactive work heads off more serious work later. If the damage is already done, the work is reactive.

Reactive:

1. Where equipment has actually broken down or failed to operate properly.

2. Priority-1 jobs are defined as urgent and so they are reactive.

Proactive:

1. Work done to prevent equipment from failing.

2. Any PM job.

3. Generally predictive maintenance initiated work that if done in time will prevent equipment problems.

4. Project work.

Guidelines for Deciding if Work is Extensive or Minimum Maintenance

It is not cost effective to spend much time planning certain work. In these cases, simple scoping and abbreviated information is all that is necessary. This work is considered *minimum maintenance*.

Minimum maintenance:
Minimum maintenance work must meet all of the following conditions:

1. Work has no historical value.

2. Work estimate is not more than 4 total work hours (e.g., two persons for 2 hours each or one person for 4 hours).

3. While parts may be required, no ordering or even reserving of parts is necessary.

Extensive maintenance:
All other work is considered "extensive."

Guidelines for Deciding Whether to Stage Parts or Tools

Always stage anticipated items that are:

1. Nonstock and purchased especially for the job.

2. Certainly needed for the job and there is little likelihood any other items from the same place will be needed. Example: Job to replace air filter.

Favor staging anticipated items where:

1. There is high likelihood that the item will be needed.

2. There is a low likelihood that other items from the same place will be needed.

3. Technician time is valuable.

4. Technician time is limited.

5. Persons to stage items are readily available.

6. Equipment downtime is valuable.

7. Equipment downtime is limited.

8. Distance to the storeroom or tool room is excessive.

9. Availability or accessibility of storeroom or tool room is limited.

10. It is relatively difficult later to transport items to the site if they are not staged.

11. Item is easily returnable to storeroom or tool room.

12. The item is disposable if unused or lower in value than would be worth technicians' time to return.

13. There is some experience with planning and scheduling.

14. There is high maturity and sophistication of the planners to anticipate items correctly.

15. There is high confidence that the job will start the week or day scheduled.

16. The storeroom frequently has the wrong parts.

17. The storeroom is unstaffed or not very secure.

18. The storeroom carries very low stock levels.

19. The storeroom has a lot of stock-outs.

Do not stage items that are:

1. For unscheduled jobs unless a nonstock item was exclusively obtained for a job.

2. Difficult or impractical to move repeatedly due to size or storage requirements.

3. Difficult or impractical to move repeatedly due to legal tracking requirements.

Guidelines for Craft Technicians to Provide Adequate Job Feedback

1. Identify quantity of persons and specific craft and grade of each person. Identify the names of the persons.

2. Identify labor hours of each person. Give start and finish times of job. Explain any variance from the plan estimates if greater or less than 20%.

3. Thoroughly describe the problem if not accurately specified by the plan.

4. Thoroughly describe the action taken if the job did not proceed according to the plan. Report any special problems and solutions.

5. Identify actual quantities of parts used and report stock numbers if not given by the plan.

6. Identify actual special tools used or made if not given by the plan.

7. Return the original work order and all attachments provided by planning. Include any field notes and return any datasheets that the technician filled out whether or not planning provided them.

8. Return updated drawings.

9. Note any changes to equipment technical information such as new serial numbers and model numbers and names. Return any

manufacturer's information or literature that was received with any new parts being installed. This information is especially vital and often cannot otherwise be determined to help future maintenance.

10. Include any other information such as bearing clearances (radial and thrust), wear ring clearance, shaft runout clearance, bearing to cap clearances, coupling condition and gap clearance.

11. Make any recommendations to help future plans.

Glossary

aging Automatically increasing the priority or attention given to an incomplete work order the older it becomes. The concept of aging is that an older work order should get higher attention than a similar work order only recently written. This practice may help a crew with low productivity, but hinder a crew with already high productivity. The former crew might increase its productivity to include the aged work order. The latter crew might have to drop a more serious work order to accommodate the aged, though less important, work order.

backlog Amount of identified work on work orders either by number of work orders or work hours.

blankets or blanket work orders Standing work orders not specifying specific work but allowing the collection of work hours for time accounting.

book hours Standard hours by which maintenance personnel might be paid regardless of actual hours spent.

clearing equipment or systems Tagging or locking out the necessary valves or devices surrounding equipment to be worked on so that the equipment is safe. Also may involve draining or otherwise making equipment ready to be maintained.

CMMS Computerized maintenance management system.

component level file A file made for a specific piece of equipment rather than for a system or group of equipment. Also called a *minifile* to emphasize its size relative to a system file.

confined space An area with limited access and potential respiratory hazard requiring a special permit to enter.

COPQ (Cost of Poor Quality) This is the calculated cost of the poor performance of the company based on its current performance versus perfection.

critical factors Inputs necessary for different parts of a process that do significantly affect the overall results of the process. The significance of identifying critical factors is that companies can only control process results by manipulating critical factors of the process.

corrective maintenance Work to restore an equipment to proper operating condition before failure or breakdown occurs.

CPM (Critical Path Method) A schedule technique allowing determination of the overall time of a large maintenance by arranging and sequencing tasks with preceding and succeeding events.

deficiency tag An information tag hung by the requester of maintenance work to identify the equipment.

equivalent availability Equivalent availability factor (EAF). EAF is a common utility performance measure of how much generating capacity is actually available over a given period for producing power. When a unit is only available (whether it is running or not) for less than full load, the equivalent amount of its full load availability is counted. For example, a unit having boiler feedwater pump problems and only available for half-load for the entire month would have an equivalent availability of 50%. Likewise, a unit available for full load the entire month of 30 days except for a full outage lasting three of those days would have an equivalent availability of 90%. The equivalent availability of an entire plant site or utility would be calculated by weighting the units by megawatt capacity. Equivalent availability is normally expressed as a percent such as 90%, although technically the factor itself would be shown as 0.90.

extensive maintenance   Maintenance that takes more than a few hours and may have historically significant data.

equipment file, equipment history file   See *component level file.*

front end loading   Spending time and resources ensuring that equipment will be suitable for maintenance before the equipment is purchased or installed.

GPM   Gallons per minute.

I&C   Instrument and controls craft.

infant mortality   The failure of a component or equipment soon after initial installation or after a maintenance operation. There is a higher percentage chance that equipment will need repair at these times than during the remainder of its time in service.

insource   Using in-house resources to obtain a service or material. Making equipment or components with in-house labor.

job tool card   A record kept by a tool room to keep track of tools issued to jobs rather than individual technicians.

KPI (Key Process Indicator)   This is the same as the term metrics except that the company has has determined this metric to be particularly important.

metrics   Indicators or measures of maintenance performance.

minifile   See *component level file.*

minimum maintenance   Maintenance work requiring less than a few hours and not historically significant.

MSDS   Material Safety Data Sheet describing dangers and safety procedures for a specific hazardous substance.

MTBF (mean time between failure)   A calculation showing the trend of average time periods between failures of the same equipment over time or showing the time periods between failure of different equipment. For example, "The MTBF of the circuit has improved from three weeks to over 2 months."

O&M manuals   Operation and maintenance manuals provided by the equipment manufacturers or suppliers giving basic details on operation and maintenance of the equipment. They usually included suggested preventive maintenance tasks, troubleshooting guides, and identification of parts and special tools.

OEM   Original Equipment Manufacturer.

OJT (on-the-job training)   Many times either the best or the only training an employee receives is while actually doing the job either alone or under the eye of an experienced co-worker.

operations, production   The plant personnel responsible for operating the plant equipment and systems.

operations coordinator   A specific person in the operations group specifically tasked to help maintenance planners.

originator   A person that writes a work order or other request for maintenance work.

outage   A condition of being out of service and unavailable for operation. Refers to the condition of the entire plant unit, not the individual piece of equipment.

overhaul   Normally a major outage requiring a significant amount of time and planning.

piece workers, piece work   An arrangement where technicians receive pay based on how many units are produced.

pigeon-holing   Estimating a job's time requirements by referring to a table or index of similar jobs and making adjustments for particular job differences.

planner, maintenance planner   The person responsible for preparing work orders for execution through applying the principles of maintenance planning.

planning, maintenance planning   The preparatory work to make work orders ready to execute. This term may involve scheduling depending on how it is used.

planned job, planned work, job plan   The product of the planner including all the preparatory work done.

planner active file   A planner may keep the original work order in an active file and take a copy into the field for scoping purposes.

PM optimization   A similar process to RCM to develop preventive maintenance tasks and frequencies to reduce likely failure modes. Makes significant use of existing PMs as a starting point.

point of diminishing returns   The point where an additional action or task provides a benefit, but the benefit would not outweigh the cost of the additional action or task.

potential factors   Inputs necessary for different parts of a process that might significantly affect the overall result of the process.

predictive maintenance   The use of technologically sophisticated devices as vibration trend analysis to predict future impending equipment problems.

preventive maintenance   Time- or interval-based maintenance designed to head off or detect equipment problems.

proactive maintenance   Maintenance performed to head off failure and breakdown.

product life cycle   The economic curves showing profitability of any given product for sale. Initially profit is low as the product is developed. Next, profit increases as sales increase. Then profit begins to drop as market substitute products and other competition become common. The company may attempt to modify the product and willingly take some drop in profit before the effect of market forces. This modified product eventually achieves greater profitability than possible from the initial product. An analogy can be made to playing tennis. One can become a fairly good player without using correct techniques. But when the player decides to adopt correct grips, stances, and other techniques, the player's ability will first drop as the new skills are learned. However, the player should be able eventually to rise above the old level of ability.

project work, project maintenance   Modifying or improving a piece of equipment or system.

RCM (reliability centered maintenance)   A process or system to evaluate equipment and develop preventive maintenance tasks and frequencies to reduce likely failure modes.

reactive maintenance   Maintenance work performed as a response to a failure, breakdown, or other urgent equipment situation.

scoping, scope   Determining the job scope, studying and defining what work a job requires, deciding the magnitude of the work involved. For an individual job, this requires determining the objective of the job. For an outage, this involves determining all the jobs involved in the outage.

shakedown   Returning units or plant processes back to service after maintenance activities. The shakedown process identifies discrepancies with the maintenance work that require attention.

SNOW   An acronym for short notice outage work.

SPC (Statistical Process Control)   SPC uses mathematics to calculate whether a process is "in control" or functioning without a trend or special problems by measuring variance or the sigma level of its results over time.

stretch goals   Goals that are hard to reach, but not seen as impossible.

stockout   A measure of how many times the storeroom is out of stock for an item when that item is requested. Stockouts do not necessarily measure that a storeroom is either out of a material or has a less than desirable quantity on hand.

sunk cost   A sum already spent by the company that cannot be retrieved whether it receives a return or not. The company cannot retrieve the cost regardless of future decisions. Many companies errantly decide to "throw good money after bad" because they have already made an investment.

vicious cycle   A situation where the solution to a circumstance creates another problem in a chain that makes the original problem worse. An example is when a plant attempts to save labor by reducing preventive maintenance. The incidence of equipment failure rises requiring more labor and leaving no time for ever doing preventive maintenance. The plant finally saves no labor and has worse equipment reliability.

Bibliography

Baldridge, Rick (2005). Cargill, personal conversation with the author.

Baldwin, Robert (1995). Planning for the future. *Maintenance Technology,* 8(4): 8.

Beck, Larry (1996). Editorial: Your CEO needs your help. *Engineer's Digest,* July 1996: 6.

Berger, David (1997). CMMS software review: A detailed guide to help you find the software that best meets your needs. *Plant Services Special Supplement,* January 1997: 5.

Berger, David (2003). Defining "user friendly." *Plant Services,* August 2003: 27–29.

Berger, David (2004). It's all about the people. *Plant Services,* October 2004: 32.

Berger, David (2005). Keys to success with CMMS. *Plant Services,* January 2005: 33.

Bradley, Larry (2005). Technical Superintendent of St. John's River Power Park, personal conversation with the author.

Brown, Gifford (1993). Maintenance prevention. Keynote address presented at Society for Maintenance and Reliability Professionals Annual Conference, 2–3 October, Nashville, Tenn.

Commerce Clearing House, Inc. Labor Arbitration Awards.

Connaughton, George E. (1993). One plant's experience with contract maintenance. Paper presented at Society for Maintenance and Reliability Professionals Annual Conference, 2–3 October, Nashville, Tenn.

Crossan, John (1997). Experiences in a corporate maintenance improvement initiative. Discussion during paper presented at Society for Maintenance and Reliability Professionals Annual Conference, 5–8 October, Pittsburgh, Pa.

Crew, Wayne A. PE (1999). Using incentive-based contract maintenance. *Maintenance Technology,* February 1999: 26–27.

Day, John E. Jr. PE (1993). Maintenance vision: Total Proactive Maintenance. Paper presented at Society for Maintenance and Reliability Professionals Annual Conference, 2–3 October, Nashville, Tenn.

Fox, M. Jr. (1985). Williamette Industries, Inc. 85-2ARB8396. Commerce Clearing House, Inc. Labor Arbitration Awards.

Humphries, Jim PE (2004). Get on the same page. *Plant Services,* October 2004: 51–56.

Hutchins, Greg (2005). Armor yourself against outsourcing. *Quality Digest,* March 2005: 42–47.

Idhammar, Christer (1998). Results oriented maintenance: What the best performers do different from others. *Iron and Steel Engineer,* June 1998: 58.

Jones, Doug (2005). Making partnerships real. Keynote speech presented at Association of Iron and Steel Technology Specialty Conference, March 10, Mobile, Ala.

Kapp, Karl (1996). The USA Principle. *Manufacturing Systems,* August 1996.

Kelley, Eleanor (1993). Case studies. Paper presented at Society for Maintenance and Reliability Professionals Annual Conference, 2–3 October, Nashville, Tenn.

Mintzberg, Henry (1983). *Structure in Fives: Designing Effective Organizations.* Englewood Cliffs, N.J.: Prentice-Hall.

Mobley, Keith (1997). Barriers to plant performance: real or imagined? *Plant Services,* 18(7): 37–39.

Moore, Ron (1998). Switch in contract maintenance proves costly. *Maintenance Technology,* February 1998: 26–30.

Moore, Ron (1998). Using maintenance contractors effectively. *Maintenance Technology,* March 1998: 13–15.

Naisbitt, John (1984). *Megatrends, ten new directions transforming our lives.* New York, NY. Warner Books, Inc.

Nicholas, Jack and Drew Troyer (2005). Procedure-based maintenance. Paper presented at Reliability World Conference, April 27, San Antonio, Tx.

Oas, Arne (2005). How to really achieve ROI. *MRO Today,* December 2004/January 2005: 35.

Paulson, Stephen (1988). Lecture on organizational theory in University of North Florida's MBA program, January 1988. Jacksonville, Fla.

Peterson, Brad (1998). The central issue: to centralize or decentralize maintenance. *Maintenance Technology,* December 1998: 29–31.

Phillippi, Nicholas (1997). Evaluating maintenance information management systems. *Maintenance Technology,* April 1997: 24.

Reimer, Ron (1997). The use of metrics to manage proactive maintenance. Paper presented at Society for Maintenance and Reliability Professionals Annual Conference, 5–8 October, Pittsburgh, Pa.

Riggs, Warren and Edwin Harrington (1995). Mechanical integrity at Tennessee Eastman Division. Paper presented at Society for Maintenance and Reliability Professionals Annual Conference, 2–4 October, Chicago, Ill.

Schimmoller, Brian K. (1998). Outsourcing plant maintenance. *Power Engineering,* February 1998: 16–22.

Senge, Peter M. (1990). *The Fifth Discipline: The Art and Practice of the Learning Organization,* 1st edition. New York: Currency Doubleday.

Smith, Adam (1776). The Wealth of Nations. Edited volume by Edwin Cannan 1994 New York, NY: The Modern Library, Random House, Inc. pp 4–5.

Sowell, Thomas (1993). That's the way it is with bureaucrats, politicos and kids. *Florida Times Union,* June 6, 1993, A-15.

Spielman, Joe (1997). Maintenance and reliability as a competitive advantage. Keynote address presented at Society for Maintenance and Reliability Professionals Annual Conference, 5–8 October, Pittsburgh, Pa.

Stewart, Steve (1997). Strategies to optimize the impact of a PM program. Paper presented at Society for Maintenance and Reliability Professionals Annual Conference, 5–8 October, Pittsburgh, Pa.

Sutherland, Sandy and Ian Gordon (1997). A manufacturing strategy based on maintenance. Paper presented at Society for Maintenance and Reliability Professionals Annual Conference, 5–8 October, Pittsburgh, Pa.

Tognazzini, Bruce (2005). First principles of interaction design. World wide web http://www.asktog.com/basics/first principles.html. Accessed April 7, 2005.

Wireman, Terry (1996). Maintenance basics: the first step toward zero breakdowns. *Engineer's Digest,* August 1996: 36–41.

Yourdon, Edward (2004). *Death march, second edition.* Upper Saddle River, NJ. Prentice Hall Professional Technical Reference.

Young, Keith (2001). Getting it right and keeping it right: the use of procedures and configuration management in maintenance. Paper included in JEA workshop, July 10, Jacksonville, Fl.

Young, Phillip (1997). Roundtable discussion at Society for Maintenance and Reliability Professionals Annual Conference, 5–8 October, Pittsburgh, Pa.

Index

About the Author

Richard D. Palmer, PE, MBA, CMRP, better known as Doc Palmer, with this handbook bridges the significant gap between the well-publicized benefits of maintenance planning and the achievement of the benefits in actual practice. Doc Palmer has more than 23 years of industrial experience with the past 17 directly in a maintenance role. He also has a master's degree in business from the University of North Florida and a degree in engineering from the Georgia Institute of Technology. In addition, he holds a professional engineering license in the State of Florida. This maintenance background with a technical and business education gives him a unique perspective on the maintenance environment and challenges. As an actual practitioner within a company's maintenance organization, he recognized and developed the necessary principles, and then led the grassroots' establishment of a successful planning program. Doc Palmer is the author of numerous well-received articles and presentations and is a recognized authority on the establishment of successful maintenance planning. He is also a Certified Maintenance and Reliability Professional. Doc Palmer works in the maintenance department of an electric company and lives in Neptune Beach, Florida with his wife, Nancy, and two daughters, Abby and Brittany.